Innovative Extraction Techniques and Hyphenated Instrument Configuration for Complex Matrices Analysis

Innovative Extraction Techniques and Hyphenated Instrument Configuration for Complex Matrices Analysis

Special Issue Editors

Marcello Locatelli
Simone Carradori
Andrei Mocan

MDPI • Basel • Beijing • Wuhan • Barcelona • Belgrade

MDPI

Special Issue Editors
Marcello Locatelli
University "G. d'Annunzio" of Chieti-Pescara
Italy

Simone Carradori
University "G. d'Annunzio" of Chieti-Pescara
Italy

Andrei Mocan
Iuliu Haţieganu University of Medicine and Pharmacy
Romania

Editorial Office
MDPI
St. Alban-Anlage 66
4052 Basel, Switzerland

This is a reprint of articles from the Special Issue published online in the open access journal *Molecules* (ISSN 1420-3049) in 2018 (available at: https://www.mdpi.com/journal/molecules/special_issues/complex_matrices_analysis)

For citation purposes, cite each article independently as indicated on the article page online and as indicated below:

LastName, A.A.; LastName, B.B.; LastName, C.C. Article Title. *Journal Name* **Year**, *Article Number*, Page Range.

ISBN 978-3-03921-084-8 (Pbk)
ISBN 978-3-03921-085-5 (PDF)

Contents

About the Special Issue Editors

Marcello Locatelli Number of articles: 121; h-index: 28; Total number of citations: 1953. From 1 November 2008 Confirmed Researcher (full-time) at the University "G. d'Annunzio" Chieti-Pescara, Department of Pharmacy.

The research activity is aimed at the development and validation of chromatographic methods (according to international guidelines, ICH) for the qualitative and quantitative determination of biologically active molecules in complex matrices both of human and animal origin (eg serum, plasma, bile , liver tissues, hypothalamus, kidneys, faeces, urine), both in cosmetic formulations and in food and environmental samples. In addition to the study of all the processes related to the preanalytic stages such as sampling, extraction and purification, separation, enrichment, the application of conventional analytical and hyphenated methods (HPLC-MS and HPLC-MS / MS) for accurate determination , sensitive and selective of biologically active molecules.

These procedures have been applied to different analytes: glucosamine, 5-amino-salicylic acid, natural or synthetic bioactive compounds, anti-inflammatory, drugs and their associations, fluoroquinolones, secondary metabolites of vegetable origin and food supplements, heavy metals, finding also application in clinical and pre-clinical studies aimed at the evaluation of quantitative, pharmacokinetic, bioequivalence and absorption, distribution, metabolism and excretion (ADME) profiles of the analyzed analytes, in order also to characterize new systems for conveying the active principle to improve their pharmacological properties. In the development of the method, predictive and chemometric models were applied both for the optimization of extraction protocols and for the final processing of the data. Particular attention is paid to innovative (micro) extraction techniques and new instrumental configurations for quantitative analysis in a complex matrix.

Scientific activity is proven by more than 129 publications in international peer-review journals, 12 book chapters, more than 13 oral communications to congresses and more than 110 poster communications. He has been and is a reviewer of more than 96 International peer-reviewed Journals.

Member of the Italian Chemical Society (SCI, card number 13779), of the American Chemical Society (ACS, card number 30617260), and of the Italian Society of Phytochemistry (SIF). It is included in the list of external experts for the evaluation of e-Cost research projects (European Cooperation in Science & Technology). He is a reviewer for the MIUR for National Projects (SIR) and is included in the REPRISE Register (Register of Expert Peer Reviewers for Italian Scientific Evaluation) in the "Basic Research" section. Referee for the VQR 2011-2014. He is referee for other universities for proposals, through competitive procedures, for the allocation of University funds for the activation of research grants (University of Insubria 2016, University of Florence 2017, and University of Insubria 2018).

He is a member of the Editorial Board for the following Journals:

1. "Molecules" section "Analytical Chemistry" (MDPI, ISSN 1420-3049)

2. "Current Bioactive Compounds" (Bentham Science Publisher, ISSN: 1875-6646 Online, ISSN: 1573-4072 Print)

3. "American Journal of Modern Chromatography" (Columbia International Publishing, ISSN: 2374-5479 Online)

4. "Journal of Selcuk University Science Faculty" (ISSN: 2458-9411)

5. "Review in Separation Sciences" (eISSN: 2589-1677)

6. "Cumhuriyet Science Journal" (ISSN 2587-2680; e-ISSN 2587-246X).

He is Associate Editor of the magazine "Frontiers in Pharmacology" section "Ethnopharmacology" (ISSN: 1663-9812) and Reviewer Editor of the journal "Frontiers in Oncology" section "Pharmacology for anti-cancer drugs" (ISSN: 2234-943X). He is a member of the Scientific Committee of the journal "Scienze e Ricerche" (ISSN 2283-5873), published by the Italian Book Association.

Guest Editor for more than 9 Special Issues:ì in International peer-reviewed Journals.

Simone Carradori, after completing graduation in "Drug Chemistry and Technology", obtained his PhD in "Pharmaceutical Sciences" at the University of Rome "La Sapienza" (Italy). He collaborates with several departments abroad. Currently he is assistant professor at the Department of Pharmacy of "G. d'Annunzio" University of Chieti-Pescara (Italy). The scientific activity is mainly focused on the synthesis and characterization of heterocyclic derivatives as well as extraction of natural compounds with potential biological activity, and is documented from several papers in international peer-reviewed journals, one European patent and participations in numerous conferences.

Andrei Mocan, after graduating in "Pharmacy" obtained his PhD in "Pharmaceutical Sciences" with the distinction "Summa cum laude" at the "Iuliu Hațieganu" University of Medicine and Pharmacy from Cluj-Napoca. He is currently collaborating with several departments abroad and made several research stays in Germany, Portugal or Italy. Currently he is a senior lecturer in the department of Pharmaceutical Botany from "Iuliu Hațieganu" University of Medicine and Pharmacy from Cluj-Napoca, as well as working into a chromatography lab in a research institute. His scientific activity is mainly focused on pharmaceutical biology, valorization of traditional medicinal and edible plants and fungi, extraction optimization of bioactive compounds from plant materials, experimental design applied to extraction and process optimization, bioactivity and chemical characterization of natural products, development of new nutraceuticals based on medicinal plants and fungi, natural products as enzyme inhibitors. He was distinguished with several national awards in science and his scientific activity is documented by several papers in international peer-reviewed journals, and participations in numerous conferences.

Preface to "Innovative Extraction Techniques and Hyphenated Instrument Configuration for Complex Matrices Analysis"

The interest in complex analytical techniques has been growing in the last period due to the renewed necessity for analyzing complex biological matrices like herbal medicinal products and biological fluids. This particular necessity has developed in the area of life/medical sciences for quality control and standardization as well as to reveal potential molecules that serve as biological markers.

Natural product research has increased considerably since the '90s as a tool for providing new chemical entities, as a consequence of several outstanding developments in the areas of separation methods, spectroscopic techniques, and a broad range of sensitive bioassays. From a historical point of view, natural product-based drug discovery has been dominated by medicinal plants as original matrices for the discovery of new compounds [1]. Traditionally used medicinal plants were the first source of medicines and have maintained a crucial role in drug discovery and development [2]. Furthermore, extraction is considered a fundamental process used for the separation and recovery of active molecules from different matrices and converts the real matrix into a sample suitable for subsequent analytical procedures [3]. Extraction is crucial when it comes to targeting a specific class of natural molecules, as physicochemical properties of natural products are extremely variable. Furthermore, extraction techniques have to be adapted to plant parts and types of tissue matrices, as in many cases some classes of natural biomolecules can be mostly found in specific plant parts, i.e, rhizomes, flowers, leaves, stigmas, buds, etc. As such, various extraction methods have been developed and tested to meet all the issues raised above [3].

The Special Issue gathered in this printed book was proposed by three Guest Editors, all of them professors of pharmaceutical sciences, one analytical chemist and two pharmacists, doing research and teaching in the fields of analytical chemistry, medicinal chemistry, and pharmaceutical botany. This along with the description of the Special Issue allowed us to consider a broad range of submissions and concluded with the selection of 18 manuscripts after being peer-reviewed. The published papers (one review and 17 original research articles) were submitted by research groups from different countries that fit the aims and scopes of our Special Issue [4]. We would like to thank all contributors and colleagues who chose to publish their works here as well as the reviewers who dedicated their time, effort, and expertise to evaluating the submissions and assuring the high quality of the published work. We would also like to thank the publisher MDPI and the editorial staff of the journal for their constant and professional support as well as for their invitation to edit this Special Issue [4].

Finally, we would like to thank all authors and readers and we hope that the content of this book will offer new perspectives and ideas to initiate and continue research further.

References

1. Ahn, K. The worldwide trend of using botanical drugs and strategies for developing global drugs. *BMB Rep.* **2017**, *50*, 111–116.

2. Newman, D.J.; Cragg, G.M. Natural Products as Sources of New Drugs from 1981 to 2014. *J. Nat.*

Prod. **2016**, *79*, 629–661, doi:10.1021/acs.jnatprod.5b01055.

3. Belwal, T.; Ezzat, S.M.; Rastrelli, L.; Bhatt, I.D.; Daglia, M.; Baldi, A.; Prasad, H.; Erdogan, I.; Kumar, J. A critical analysis of extraction techniques used for botanicals: Trends, priorities, industrial uses and optimization strategies. *Trends Anal. Chem.* **2018**, *100*, 82–102, doi:10.1016/j.trac.2017.12.018.

4. Locatelli, M.; Carradori, S.; Mocan, A. Innovative Extraction Techniques and Hyphenated Instrument Configuration for Complex Matrices Analysis. *Molecules* **2018**, *23*, 2391, doi:10.3390/molecules23092391.

Marcello Locatelli, Simone Carradori, Andrei Mocan
Special Issue Editors

molecules

MDPI

Editorial

Innovative Extraction Techniques and Hyphenated Instrument Configuration for Complex Matrices Analysis

Marcello Locatelli [1],*, Simone Carradori [1],* and Andrei Mocan [2],*

1   Department of Pharmacy, University "G. d'Annunzio" of Chieti-Pescara, Via dei Vestini 31,
    66100 Chieti, Italy
2   Department of Pharmaceutical Botany, Faculty of Pharmacy, "Iuliu Haţieganu" University of Medicine and
    Pharmacy, 400337 Cluj-Napoca, Romania
*   Correspondence: m.locatelli@unich.it (M.L.); simone.carradori@unich.it (S.C.); amocanm@gmail.com (A.M.)

Received: 11 September 2018; Accepted: 17 September 2018; Published: 18 September 2018

This special issue was proposed by three Co-Guest-Editors with complementary expertise in the fields of Analytical Chemistry, Medicinal Chemistry, and Pharmaceutical Botany to better understand the most recent techniques to extract, isolate, characterize, and biologically evaluate natural occurring compounds from complex matrices (plant extracts, biological fluids). The interest in this research field is demonstrated by relevant literature in high impact factor journals such as Molecules (http://www.mdpi.com/journal/molecules/special_issues), which promoted this special issue with an emphasis on the most innovative approaches to the matter.

The complexity of the topic requires knowledge of analytical chemistry, extraction procedures, validation of statistical approaches, botany, and chemical/enzymatic stability of natural compounds. We selected 18 manuscripts (one review and 17 research articles) submitted by researchers from different countries that fit the aims and scope of our mission. We are also grateful to all the contributors and colleagues/reviewers who devoted their precious time and expertise to finalize this special issue. Lastly, we want to thank MDPI publisher and the Editorial staff of the journal for their constant and professional support.

Samanidou's research group, who are strongly involved in the development of innovative extraction analyses under a rigorous validation method, described exhaustively the "state of the art" of Ionic Liquids (ILs) in the extraction procedures [1]. Pros and cons were considered and justified the role of ILs in miniaturized microextraction techniques, such as solid-phase microextraction (SPME), dispersive liquid-liquid microextraction (DLLME), single-drop microextraction (SDME), stir bar sorptive extraction (SBSE), and stir cake sorptive extraction (SCSE). The versatility of ILs, beyond their use as extraction solvents, is characterized by the evidence that they could provide alternative advantages as intermediate solvents, mediators, and desorption solvents [2].

The other 17 research articles can be divided into three main groups.

The first one is related to the application of validated methods for the detection and quantification of drugs or metabolites in real samples/complex matrices. Panderi et al. [3] studied an accurate and precise determination of metformin and rosuvastatin in human plasma by HILIC-ESI/MS (Hydrophilic Interaction Liquid Chromatography-Electrospray Ionization Mass Spectrometry), limiting the sample preparation process and the chromatographic run time. These procedures were also applied for their suitability in the routine analysis of plasma samples from eight patients under this therapeutic treatment. He et al. [4] reported the determination by High Performance Liquid Chromatography-Quadrupole Time-of-Flight Mass Spectrometry (HPLC-Q-TOF-MS) using three important branched-chain ketoacids (α-ketoisocaproate, α-keto-β-methylvalerate and α-ketoisovalerate) in serum and muscle samples.

The second group of articles dealt with the application of innovative analytical techniques for environmental purposes. Huang et al. [5] used GC-MS and GC-O (Gas Chromatography-Mass Spectrometry/Olfactometry) for the identification of volatile compounds as an attempt to monitor indoor air quality. Zhenh et al. [6] proposed a daily monitoring of yttrium and rare earth elements (YREEs) in seawater by ICP-MS (Inductively Coupled Plasma-Mass Spectrometry) coupled to a cheap flow injection system online and to a specific pre-concentration step.

The third group of research articles analyzed plant and food matrices, characterized by a high economic, ethnopharmacological, and health-promoting value. The first three articles [7–9] tried to better understand the parameters influencing the extraction of polyphenols, alkaloids, and gelatin from natural sources. The authors compared and implemented their procedures by adding enzymes (actinidin) or specific substances (magnetite). Other important papers explore exhaustively by means of innovative equipment such as UPLC-MS (Ultra Performance Liquid Chromatography-Mass Spectrometry) [10], HSCCC (High-Speed Counter-Current Chromatography) [11], NIR (Near Infrared spectroscopy) [12], and UPLC-qTOF MS/UPLC-QqQ MS [13,14] plants and their derived products. Lastly, some research articles are devoted not only to the recovery and full characterization of plant metabolites, but also to the assessment of their biological activity against a panel of pharmacologically relevant targets (acetylcholinesterase, tyrosinase, α-amylase, sirtuin 1, hematopoiesis and hemostasis, skin-whitening ability) [15–19].

References

1. Kissoudi, M.; Samanidou, V. Recent advances in applications of ionic liquids in miniaturized microextraction techniques. *Molecules* **2018**, *23*, 1437. [CrossRef] [PubMed]
2. Diuzheva, A.; Carradori, S.; Andruch, V.; Locatelli, M.; De Luca, E.; Tiecco, M.; Germani, R.; Menghini, L.; Nocentini, A.; Gratteri, P.; et al. Use of innovative (micro)extraction techniques to characterize Harpagophytum procumbens root and its commercial food supplements. *Phytochem. Anal.* **2018**, *29*, 233–241. [CrossRef] [PubMed]
3. Antonopoulos, N.; Machairas, G.; Migias, G.; Vonaparti, A.; Brakoulia, V.; Pistos, C.; Gennimata, D.; Panderi, I. Hydrophilic Interaction Liquid Chromatography-Electrospray Ionization Mass Spectrometry for therapeutic drug monitoring of Metformin and Rosuvastatin in human plasma. *Molecules* **2018**, *23*, 1548. [CrossRef] [PubMed]
4. Zhang, Y.; Yin, B.; Li, R.; He, P. Determination of branched-chain keto acids in serum and muscles using High Performance Liquid Chromatography-Quadrupole Time-of-Flight Mass Spectrometry. *Molecules* **2018**, *23*, 147. [CrossRef] [PubMed]
5. Liu, R.; Wang, C.; Huang, A.; Lv, B. Characterization of odors of wood by Gas Chromatography-Olfactometry with removal of extractives as attempt to control indoor air quality. *Molecules* **2018**, *23*, 203. [CrossRef] [PubMed]
6. Zhu, Z.; Zheng, A. Fast determination of Yttrium and Rare Earth Elements in seawater by Inductively Coupled Plasma-Mass Spectrometry after online flow injection pretreatment. *Molecules* **2018**, *23*, 489. [CrossRef] [PubMed]
7. Yang, L.; Tian, J.; Meng, J.; Zhao, R.; Li, C.; Ma, J.; Jin, T. Modification and characterization of Fe_3O_4 nanoparticles for use in adsorption of alkaloids. *Molecules* **2018**, *23*, 562. [CrossRef] [PubMed]
8. Ahmad, T.; Ismail, A.; Ahmad, S.A.; Khalil, K.A.; Leo, T.K.; Awad, E.A.; Imlan, J.C.; Sazili, A.Q. Effects of ultrasound assisted extraction in conjugation with aid of actinidin on the molecular and physicochemical properties of bovine hide gelatin. *Molecules* **2018**, *23*, 730. [CrossRef] [PubMed]
9. Boutaoui, N.; Zaiter, L.; Benayache, F.; Benayache, S.; Carradori, S.; Cesa, S.; Giusti, A.M.; Campestre, C.; Menghini, L.; Innosa, D.; et al. Qualitative and quantitative phytochemical analysis of different extracts from Thymus algeriensis aerial parts. *Molecules* **2018**, *23*, 463. [CrossRef] [PubMed]
10. Zhang, Y.; Xiong, H.; Xu, X.; Xue, X.; Liu, M.; Xu, S.; Liu, H.; Gao, Y.; Zhang, H.; Li, X. Compounds identification in Semen Cuscutae by Ultra-High-Performance Liquid Chromatography (UPLCs) coupled to Electrospray Ionization Mass Spectrometry. *Molecules* **2018**, *23*, 1199. [CrossRef] [PubMed]

11. He, J.; Fan, P.; Feng, S.; Shao, P.; Sun, P. Isolation and purification of two isoflavones from Hericium erinaceum mycelium by High-Speed Counter-Current Chromatography. *Molecules* **2018**, *23*, 560. [CrossRef] [PubMed]

12. Gavan, A.; Colobatiu, L.; Mocan, A.; Toiu, A.; Tomuta, I. Development of a NIR method for the in-line quantification of the Total Polyphenolic Content: A study applied on Ajuga genevensis L. dry extract obtained in a fluid bed process. *Molecules* **2018**, *23*, 2152. [CrossRef] [PubMed]

13. Chen, S.; Lin, J.; Liu, H.; Gong, Z.; Wang, X.; Li, M.; Aharoni, A.; Yang, Z.; Yu, X. Insights into tissue-specific specialized metabolism in Tieguanyin tea cultivar by untargeted metabolomics. *Molecules* **2018**, *23*, 1817. [CrossRef] [PubMed]

14. Chen, S.; Li, M.; Zheng, G.; Wang, T.; Lin, J.; Wang, S.; Wang, X.; Chao, Q.; Cao, S.; Yang, Z.; et al. Metabolite profiling of 14 Wuyi Rock tea cultivars using UPLC-QTOF MS and UPLC-QqQ MS combined with chemometrics. *Molecules* **2018**, *23*, 104. [CrossRef] [PubMed]

15. Melucci, D.; Locatelli, M.; Locatelli, C.; Zappi, A.; De Laurentiis, F.; Carradori, S.; Campestre, C.; Leporini, L.; Zengin, G.; Picot, C.M.N.; et al. A comparative assessment of biological effects and chemical profile of Italian Asphodeline lutea extracts. *Molecules* **2018**, *23*, 461. [CrossRef] [PubMed]

16. Qi, J.-J.; Yan, Y.-M.; Cheng, L.-Z.; Liu, B.-H.; Qin, F.-Y.; Cheng, Y.-X. A novel flavonoid glucoside from the fruits of Lycium ruthenicun. *Molecules* **2018**, *23*, 325. [CrossRef] [PubMed]

17. Hu, Y.; Cui, X.; Zhang, Z.; Chen, L.; Zhang, Y.; Wang, C.; Yang, X.; Qu, Y.; Xiong, Y. Optimisation of ethanol-reflux extraction of saponins from steamed Panax notoginseng by Response Surface Methodology and evaluation of hematopoiesis effect. *Molecules* **2018**, *23*, 1206. [CrossRef] [PubMed]

18. Liu, H.; Pan, J.; Yang, Y.; Cui, X.; Qu, Y. Production of minor ginenosides from Panax notoginseng by microwave processing method and evaluation of their blood-enriching and hemostatic activity. *Molecules* **2018**, *23*, 1243. [CrossRef] [PubMed]

19. Dai, C.-Y.; Liu, P.-F.; Liao, P.-R.; Qu, Y.; Wang, C.-X.; Yang, Y.; Cui, X.-M. Optimization of flavonoids extraction process in Panax notoginseng stem leaf and a study of antioxidant activity and its effects on mouse melanoma B16 cells. *Molecules* **2018**, *23*, 2219. [CrossRef] [PubMed]

molecules

MDPI

Article

Process Optimization for Improved Phenolic Compounds Recovery from Walnut (*Juglans regia* L.) Septum: Phytochemical Profile and Biological Activities

Marius Emil Rusu [1,†], **Ana-Maria Gheldiu** [2,†], **Andrei Mocan** [2,*], **Cadmiel Moldovan** [2], **Daniela-Saveta Popa** [3,*], **Ioan Tomuta** [1] and **Laurian Vlase** [1]

[1] Department of Pharmaceutical Technology and Biopharmaceutics, Faculty of Pharmacy, "Iuliu Hatieganu" University of Medicine and Pharmacy, 8 Victor Babes, 400012 Cluj-Napoca, Romania; marius.e.rusu@gmail.com (M.E.R.); tomutaioan@umfcluj.ro (I.T.); laurian.vlase@umfcluj.ro (L.V.)

[2] Department of Pharmaceutical Botany, Faculty of Pharmacy, "Iuliu Hatieganu" University of Medicine and Pharmacy, 8 Victor Babes, 400012 Cluj-Napoca, Romania; Gheldiu.Ana@umfcluj.ro (A.-M.G.); moldovan.cadmiel@yahoo.com (C.M.)

[3] Department of Toxicology, Faculty of Pharmacy, "Iuliu Hatieganu" University of Medicine and Pharmacy, 8 Victor Babes, 400012 Cluj-Napoca, Romania

[*] Correspondence: mocan.andrei@umfcluj.ro (A.M.); dpopa@umfcluj.ro (D.-S.P.); Tel.: +40-374-834005 (A.M.); +40-721-563469 (D.-S.P.)

[†] These authors have contributed equally.

Received: 10 October 2018; Accepted: 23 October 2018; Published: 30 October 2018

Abstract: Plant by-products can be valuable sources of polyphenol bioactive compounds. Walnut (*Juglans regia* L.) is a very important tree nut rich in biologically active molecules, but its septum was scarcely researched. Experimental data indicated a hypoglycemic effect of septum extracts, with almost no details about its phytochemical composition. The main objectives of this study were: (1) to obtain walnut septum (WS) extracts with high content in bioactive compounds and antioxidant activity based on an original experimental design; (2) characterization of the phytochemical profile of the WS extracts using HPLC-MS/MS; (3) evaluation of the biological potential of the richest polyphenolic WS extract. The variables of the experimental design were: extraction method (maceration and Ultra-Turrax extraction), temperature, solvent (acetone and ethanol), and percentage of water in the solvent. The first quantifiable responses were: total phenolic content, total flavonoid content, condensed tannins, and ABTS antioxidant capacity. The phytochemical profile of lyophilized extracts obtained by Ultra-Turrax extraction (UTE), the most efficient method, was further determined by HPLC-MS/MS analysis of individual polyphenolic and phytosterols compounds. It is the first study to assay the detailed composition of WS in hydrophilic and lipophilic compounds. The biological potential of the richest polyphenolic WS extract was also evaluated by FRAP and DPPH antioxidant capacity and the inhibition of tyrosinase, an enzyme involved in the browning in fruits and vegetables, skin wrinkles and aging. Conclusion: The phytochemical profile of the analyzed extracts proves that WS can be a valuable source of biologically active compounds (polyphenols) for food and/or pharmaceutical industry and warrant the continuation of current research in further evaluating its bioactive potential.

Keywords: walnut septum; polyphenols; phytosterols; HPLC-MS/MS; Ultra-Turrax extraction; biological activity; antioxidant activity; experimental design; optimization; phytochemicals

1. Introduction

Each year the food industry creates a substantial amount of waste and serious issues are associated with its disposal. Coupled with the tendency of the consumers to avoid foods prepared with chemical origin preservatives, many studies have been recently conducted, intended to find natural alternatives, such as plant by-products, rich in bioactive compounds with high potential for health and pharmaceutical industry [1,2].

In the last decades, the number of people with body mass problems increased in the world obesogenic culture. Overweight and obesity are increasingly seen as major concerns for human health [3]. Processed food, the so called "junk food", with high content of carbohydrates, fats, and salt, is linked to overweight and obesity via several mechanisms [4]. Excessive body weight, associated with several pro-inflammatory cytokines (e.g., leptin, interleukin 6, interleukin 8, tumor necrosis factor-alpha), and a chronic, low-grade inflammation [5], is seen as a major risk factor for obesity-associated diseases, such as metabolic dysfunction [6], diabetes [7], cardiovascular diseases [8], and cancers [9], including endometrial [10], breast [11], gastrointestinal [12], pancreatic [13], prostate [14], hepatic [15], renal [16], colorectal [17].

Epidemiological studies and clinical trials demonstrated that diets with high intake of plant origin foods (vegetables, fruits, nuts) can safeguard against excessive weight-related diseases and offer powerful protection for the cardiovascular, gastrointestinal, and immune systems [18,19]. Phytochemicals, including carotenoids, glucosinolates, and polyphenols, work synergistically to reduce inflammation and oxidation, providing defense against initiation and evolution of ailments [20]. Phenolic acids and flavonoids, the major contributors of the polyphenols group, act as natural antioxidants decreasing the risk of degenerative diseases [21]. Polyphenols are compounds which donate electrons or hydrogen atoms to reactive radicals preventing the degradation of vital molecules or cellular damage [22]. Besides their role as antioxidants in the detoxifying system with a scavenging role against reactive oxygen or nitrogen species, plant polyphenols can take part in the enzymatic pathways involved in the energetic balance or act as signaling molecules in the cell [23]. In addition to the antioxidant activity, several studies [24,25] confirmed the antimicrobial activity of the polyphenols, making them a good substitute to antibiotics and chemical preservatives.

Walnut (*Juglans regia* L.), a valued crop of high economic importance, represents a good source of nutritional and nutraceutical compounds [26]. Besides the well-known antioxidant, antibacterial, and anti-inflammatory bioactivity of the walnut kernel [27], several studies proved that walnut leaves [28] and green husk [29] could induce the same great health benefits. Walnut membrane septum, another by-product of this valuable plant, was traditionally used as a cold remedy or cough suppressant, presented a hypoglycemic activity in an experimental animal model [30], and improved blood profile in murine experiments [31]. Walnut septum extracts had no acute or subchronic toxicity in rat [32]. However, to the best of our knowledge the phytochemical profile of walnut membrane septum has not been reported in the literature.

The aim of the study was the determination of phenolic and phytosterol compounds from the walnut septum based on an experimental design. Extraction method, solvent, temperature, and water percentage, the variables of the study, were combined with statistical tools and analysis using LC-MS/MS in order to determine the optimal extraction conditions, identification, and quantification of main phenolic and phytosterol molecules from septum. Several methods were employed to determine the antioxidant capacity (ABTS, DPPH, and FRAP) and the enzymatic inhibitory activity.

2. Materials and Methods

2.1. Chemicals

The reagents used in this study were: vanillin (99%), sodium carbonate, ferric chloride, 6-hydroxy-2,5,7,8-tetramethylchromane-2-carboxylic acid (Trolox) (97%), diammonium 2,2′-azino-bis(3-ethylbenzothiazoline-6-sulfonate) (ABTS) (>98%), 2,2-diphenyl-1-(2,4,6-trinitro-phenyl) hydrazine

(DPPH), 2,4,6-Tris(2-pyridyl)-*S*-triazine (TPTZ) ($\geq$99%), dimethyl sulfoxide (DMSO) ($\geq$99%), phosphate buffer, mushroom tyrosinase, 3,4-Dihydroxy-L-phenylalanine (L-DOPA) ($\geq$98%), and kojic acid were purchased from Sigma (Sigma Aldrich Chemie GmbH, Schnelldorf, Germany). Folin–Ciocâlteu reagent, hydrochloric acid (37%), acetone, ethanol, methanol were purchased from Merck (Darmstadt, Germany). Aluminum chloride ($\geq$98%) was purchased from Carl Roth (Karlsruhe, Germany). All reagents were of analytical grade and all solvents were of LC grade. Water was of Milli-Q-quality.

The standards used for both spectrophotometric and LC-MS/MS analysis were: quercetin ($\geq$95%), hyperoside (quercetin 3-D-galactoside) ($\geq$97%), isoquercitrin (quercetin 3-β-D-glucoside) ($\geq$98%), quercitrin (quercetin 3-rhamonoside) ($\geq$78%), (+)-catechin ($\geq$96%), (−)-epicatechin ($\geq$90%), vanillic acid ($\geq$97%), syringic acid ($\geq$95%), protocatechuic acid (3,4-dihydroxybenzoic acid) ($\geq$97%), campesterol (~65%), ergosterol ($\geq$95%), and stigmasterol (~95%) purchased from Sigma-Aldrich, gallic acid ($\geq$98%) purchased from Merck (Darmstadt, Germany), and beta-sitosterol ($\geq$80%) purchased from Carl Roth (Karlsruhe, Germany).

2.2. Plant Samples

Walnuts (*Juglans regia* L.) of high quality were provided by an organic orchard in Bucium, Maramureş County, Romania. In the autumn of 2016, walnuts were harvested and kept in a dark, airy shelter, at temperatures ~0 °C. At the beginning of March 2017, the unshelled walnuts were delivered to the Faculty of Pharmacy, "Iuliu Hatieganu" University of Medicine and Pharmacy Cluj-Napoca, Romania, and identified by Dr. Andrei Mocan from the Department of Pharmaceutical Botany. A voucher specimen was deposited in the Herbarium of this Department. The unshelled walnuts were cracked and the walnut septum (WS) removed from the hard shells just prior to the extractions.

2.3. Samples

WS was ground in a coffee grinder (Argis, RC-21, Electroarges SA, Curtea de Argeş, Romania) for 5 min. Then, the ground septum powder was screened through a 200 μm Retsch sieve.

2.4. Preparation of Extracts

The extraction process was carried out based on a D-optimal experimental design developed by Modde software, version 11.0 (Sartorius Stedim Data Analytics AB, Umeå, Sweden) using four variable factors: preparation method, temperature, solvent, and percentage of water in solvent (Table 1).

Table 1. Independent and dependent variables of experimental design evaluated for walnut septum extracts.

Variables	Level		
	−1	0	1
Independent variables (factors)			
Extraction method (X_1)	Ultra-turrax		Maceration
Temperature (°C) (X_2)	20	30	40
Solvent (X_3)	Acetone		Ethanol
Water in solvent (%, *v/v*) (X_4)	5	25	50
Dependent variables (responses)			
Total phenolic content (TPC, mg GAE/g dw [1]) (Y_1)			
Total flavonoid content (TFC, mg QE/g dw [2]) (Y_2)			
Condensed tannin content (CTC, mg CE/g dw [3]) (Y_3)			
Total antioxidant activity (TAA, mg TE/g dw [4]) (Y_4)			

[1]—mg GAE/g dw = gallic acid equivalents per dry weight of walnut septum; [2]—mg QE/g dw = quercetin equivalents per dry weight of walnut septum; [3]—mg CE/g dw = catechin equivalents per dry weight of walnut septum; [4]—mg TE/g dw = trolox equivalents per dry weight of walnut septum.

WS was weighed (2 g) and mixed with the extraction solvent (20 mL) in Falcon tubes. The Ultra-Turrax extraction (UTE) was performed in two steps: using an Ultra-Turrax homogenizer (T 18; IKA Labortechnik, Staufen, Germany) for 2 min (1 min at 9500 rpm and 1 min at 13,500 rpm) [33] and again 2 min using a Vortex RX-3 (Velp Scientifica, Usmate, Italy). The homogenate was centrifuged (Hettich, Micro 22R, Andreas Hettich GmbH & Co., Tuttlingen, Germany) 15 min at 3000 rpm, maintaining the extraction temperature. The supernatant was carefully separated, and the solvent removed under vacuum at 40 °C using a rotary evaporator (Hei-VAP, Heidolph Instruments GmbH & Co., Schwabach, Germany). The dry residue was taken up in water, placed in amber glass vials, and lyophilized (Advantage 2.0, SP Scientific, Warminster, PA, USA).

For the maceration method, WS (2 g) was added to Erlenmayer flasks with the extraction solvent and kept for 10 days at 20, 30, and 40 °C (Conterm Oven, JP Selecta S.A., Barcelona, Spain) and stirred twice daily. After 10 days, the samples were centrifuged (Hettlich Micro 22R, Andreas Hettlich GmbH & Co. KG, Tuttlingen, Germany) 10 min at 5300 rpm, maintaining the extraction temperature. Then, the supernatant was separated, the solvent evaporated and the remaining water removed as seen before. After lyophilization, the samples (for both extraction methods) were stored at room temperature.

For further determinations, lyophilized extract was dissolved in EtOH 70% (10 mg/mL). All assays were executed in triplicate.

2.5. Quantitative Determinations of Total Bioactive Compounds

2.5.1. Total Phenolic Content

The total phenolic content (TPC) of the WS extracts was determined by Folin-Ciocâlteu spectrophotometric method according to a method described previously [34]. In brief, in a 96 well plate, 20 µL of each sample (WS extracts diluted 5 times) were mixed with 100 µL of FC reagent (diluted 1:10). After 3 min, 80 µL of sodium carbonate solution (7.5% *w/v*) was added to the wells. The plate was incubated for 30 min in the dark at room temperature. A Synergy HT Multi-Detection Microplate Reader with 96 well plates (BioTek Instruments, Inc., Winooski, VT, USA) was used to measure the absorbance at 760 nm against a solvent blank. Gallic acid was used as a reference standard, and the content of phenolics was expressed as gallic acid equivalents (GAE) per dry weight of septum (mg GAE/g dw).

2.5.2. Total Flavonoid Content

The total flavonoid content (TFC) of the WS extracts was determined according to a method described previously [35]. In a 96 well plate, 100 µL of sample extracts were added to 100 µL of 2% $AlCl_3$ aqueous solution. The plate was incubated for 15 min in the dark at room temperature. The absorbance at 420 nm was measured against a solvent blank. The TFC was expressed as quercetin equivalents (QE) per dry weight (dw) of vegetal material (mg QE/g dw).

2.5.3. Condensed Tannin Content

The condensed tannin content (CTC) in WS extracts was determined according to a modified version of the vanillin assay described before [36,37]. Briefly, in a 96 well plate, 50 µL of sample WS extracts were added to 250 µL 0.5% vanillin in 4% concentrated HCl in methanol. The plate was incubated for 20 min in the dark at 30 °C. The absorbance at 500 nm was measured against a solvent blank. The condensed tannins were expressed as catechin equivalents (CE) per dry weight (dw) of vegetal material (mg CE/g dw).

2.6. Phytochemical Analysis by LC-MS/MS

The phytochemical profile of lyophilized WS extracts obtained by UTE method was assessed by liquid chromatography coupled with mass spectrometry in tandem (LC-MS/MS). The experiment was carried out using an Agilent 1100 HPLC Series system (Agilent, Santa Clara, CA, USA) equipped with degasser, binary gradient pump, column thermostat, auto sampler, and UV detector. The HPLC system was coupled with an Agilent Ion Trap 1100 SL mass spectrometer (LC/MSD Ion Trap VL).

2.6.1. Identification and Quantification of Polyphenolic Compounds

A previously LC-MS/MS method [38–41] was slightly modified (replacing of sodium phosphate with acetic acid in the mobile phase) and applied for the identification of 18 polyphenols in the sample WS extracts: caftaric acid, gentisic acid, caffeic acid, chlorogenic acid, *p*-coumaric acid, ferulic acid, sinapic acid, hyperoside, isoquercitrin, rutozid, myricetol, fisetin, quercitrin, quercetin, patuletin, luteolin, kaempferol, and apigenin. In brief, chromatographic separation was performed on a reverse-phase analytical column (Zorbax SB-C18, 100 mm × 3.0 mm i.d., 3.5 µm) with a mixture of methanol: 0.1% acetic acid (*v*/*v*) as mobile phase and a binary gradient. The elution started with a linear gradient, beginning with 5% methanol and ending at 42% methanol at 35 min; isocratic elution followed for the next 3 min with 42% methanol; rebalancing in the next 7 min with 5% methanol. The flow rate was 1 mL/min, the column temperature 48 °C and the injection volume was 5 µL.

The detection of the compounds was performed on both UV and MS mode. The UV detector was set at 330 nm until 17 min (for the detection of polyphenolic acids, then at 370 nm until 38 min to detect flavonoids and their aglycones. The MS system operated using an electrospray ion source in negative mode (capillary +3000 V, nebulizer 60 psi (nitrogen), dry gas nitrogen at 12 L/min, dry gas temperature 360 °C). The chromatographic data were processed using ChemStation and DataAnalysis software from Agilent, USA.

Another LC-MS method was used to identify other six polyphenols in WS extracts: epicatechin, catechin, syringic acid, gallic acid, protocatechuic acid, and vanillic acid. The chromatographic separation was performed on the same analytical column as mentioned before (Zorbax SB-C18, 100 mm × 3.0 mm i.d., 3.5 µm) with a mixture of methanol: 0.1% acetic acid (*v*/*v*) as mobile phase and a binary gradient (start: 3% methanol; at 3 min: 8% methanol; at 8.5 min: 20% methanol; keep 20% methanol until 10 min then rebalance column with 3% methanol). The flow rate was 1 mL/min and the injection volume was 5 µL. The detection of the compounds was performed on MS mode (Table 2). The MS system operated using an electrospray ion source in negative mode (capillary +3000 V, nebulizer 60 psi (nitrogen), dry gas nitrogen at 12 L/min, dry gas temperature 360 °C). All identified polyphenols were quantified both in the WS extracts and hydrolyzed WS extracts (equal quantities of extract and 4 M HCl kept 30 min on 100 °C water bath) on the basis of their peak areas and comparison with a calibration curve of their corresponding standards (epicatechin, catechin, syringic acid, gallic acid, protocatechuic acid, vanillic acid, hyperoside, isoquercitrin, quercitrin). The results were expressed as milligrams of phenolic per gram of dry weight of septum extract.

Table 2. Detection and quantification of certain polyphenols by the new LC-MS method developed in view of their analysis in walnut septum extracts.

Polyphenol	Monitored Ion (*m/z*)	Retention Time (min)	Calibration Range (*n* = 8) (µg/mL)	Coefficient of Linearity (R^2)	Accuracy (Bias, %)
Epicatechin	289	9.0	0.3–21.5	0.9922	90.7–112.1
Catechin	289	6.0	0.3–21.5	0.9974	94.3–108.9
Gallic acid	169	1.5	0.3–22.2	0.9987	96.4–108.6
Syringic acid	197	8.4	0.3–21.0	0.9997	90.5–105.5
Protocatechuic acid	153	2.8	0.3–23.9	0.9977	87.0–112.2
Vanillic acid	167	6.7	0.3–21.1	0.9993	95.6–105.6

2.6.2. Identification and Quantification of Phytosterols

The pytosterols in the septum extracts were determined according to a method described previously [42,43]. In brief, chromatographic separation was performed on a Zorbax SB-C18 (100 mm × 3.0 mm i.d., 5 µm) column (Agilent Technologies) with a mixture of methanol:acetonitrile

(10:90, v/v) and isocratic elution, at 45 °C with a flow rate of 1 mL/min. The detection of analytes was performed in the multiple reaction monitoring (MRM) mode for the quantification of phytosterols, positive ion detection, using an ion trap mass spectrometer equipped with an atmospheric pressure chemical ionization (APCI) source (capillary −4000 V, nebulizer 60 psi (nitrogen), vaporizer 400 °C, dry gas nitrogen at 7 L/min, dry gas temperature 325 °C).

Four external standards were used for quantification: beta-sitosterol, stigmasterol, campesterol, and ergosterol. The identified phytosterols (beta-sitosterol and campesterol) were quantified on the basis of their peak areas and comparison with a calibration curve of their corresponding standards. The results were expressed as milligrams phytosterols per gram of dry weight of septum extract.

2.7. Antioxidant Activity Assays

2.7.1. ABTS Radical Cation Scavenging Activity

The antiradical activity of WS extracts was determined according to the trolox equivalent antioxidant capacity (TEAC) assay described previously [35,44]. The scavenging activity against ABTS radical cation (2,2′-azino-bis(3-ethylbenzothiazoline)-6-sulphonic acid) was assessed and used to plot the trolox calibration curve. The total antioxidant activity (TAA) according to TEAC assay was expressed as trolox equivalents (TE) per gram of dry lyophilized extract (mg TE/g dw extract). This assay was used during the screening phase of the study for the evaluation of total antioxidant activity of the 23 samples obtained by either maceration or UTE method.

2.7.2. DPPH Radical Scavenging Activity

The antiradical activity of WS extracts was assessed using a method previously described [45]. The capacity to scavenge the free radical DPPH was determined in a 96 well plate mixing 30 µL of sample solution with a 0.004% methanolic solution of DPPH for 30 min in the dark. The absorbance at 517 nm was measured against a solvent blank. Trolox was used as a reference standard and the results were expressed as trolox equivalents per gram of dry lyophilized extract (mg TE/g dw extract). This assay was performed on the richest polyphenolic WS extract.

2.7.3. FRAP Assay

The reduction capacity of the WS extract was evaluated by FRAP (ferric reducing antioxidant power) assay that analyzes the blue-colored Fe^{2+}-TPTZ formed by the reduction of Fe^{3+}-TPTZ. A method previously described [46] was used with slight modifications. In brief, 25 µL of sample were incubated with 175 µL FRAP reagent (300 mM acetate buffer, pH 3.6: 10 mM TPTZ in 40 mM HCl: 20 mM $FeCl_3 \cdot 6H_2O$ in 40 mM HCl, 10:1:1, $v/v/v$) in a 96 well plate for 30 min in the dark. Trolox was used as an external standard (calibration curve obtained for 0.01–0.10 mg/mL) and the absorbance was measured at 593 nm. The results were expressed as trolox equivalents per gram of dry lyophilized extract (mg TE/g dw extract). This assay was done on the richest polyphenolic WS extract.

2.8. Tyrosinase Inhibitory Activity

The tyrosinase inhibitory activity of WS extract was evaluated by a 96-well microplate method previously described [47] with slight changes. Briefly, four wells were designated (WS lyophilized extract dissolved in water containing 5% DMSO) as follows: (A) 66 mM phosphate buffer, pH 6.6 (PB) (120 µL) and mushroom tyrosinase in the same buffer, 46 U/mL (MT) (40 µL); (B) only PB (160 µL); (C) PB (80 µL), MT (40 µL) and the sample (40 µL); (D) PB (120 µL) and the sample (40 µL). After 10 min incubation at room temperature, 2.5 mM L-DOPA prepared in PB (40 µL) was added in all wells. The microplate was kept again at room temperature for 20 min and the absorbance was measured at 475 nm. The tyrosinase inhibitory activity was assessed using kojic acid as an external standard (0.01–0.10 mg/mL). The inhibition percentage of enzymatic activity was calculated by the following equation: $[(A - B) - (C - D)] \times 100/(A - B)$. The results were expressed as milligram kojic acid equivalents per gram of dry lyophilized extract (mg KAE/g dw extract). This evaluation was carried out for the richest polyphenolic WS extract.

2.9. Identification of the Experimental Conditions to Obtain WS Extracts Rich in Phytochemicals

During the screening step, the quantifiable responses TPC, TFC, CTC, TAA according to TEAC assay, were analyzed by the Modde software, version 11.0, to identify the optimal extraction conditions. For the optimization step, individual phenolic and phytosterol levels were evaluated and the independent factors investigated were working temperature, organic solvent, and percentage of water in solvent mixture. The responses were identification and quantification of each quantified phytochemical compound: epicatechin, catechin, syringic acid, gallic acid, protocatechuic acid, vanillic acid, hyperoside, isoquercitrin, quercitrin, campesterol, beta-sitosterol.

2.10. Statistical Analysis

All samples were analyzed in triplicate (n = 3) and the results were expressed as the mean $\pm$ Standard Deviation (SD).

3. Results and Discussion

3.1. Fitting of the Experimental Data with the Models

The independent and dependent variables of experimental design evaluated for WS extraction yield during the screening step are shown in Table 1. The independent variables (factors) were the extraction method, working temperature, organic solvent, and percent of water in solvent mixture. The dependent variables (responses) were TPC, TFC, CTC, and TAA. The matrix of the experimental design generated by the Modde software, version 11.0, along with the responses obtained after performing all the experimental runs are given in Table 3.

Table 3. Matrix of experimental design and experimental results for total phenolic content (TPC), total flavonoid content (TFC), condensed tannin content (CTC), and total antioxidant activity (TAA) of walnut septum extracts based on a factorial design.

Sample Code	Run Order	Factorial Design with Coded Values				Determination (Experimental Results)			
		X_1	X_2	X_3	X_4	Y_1 (TPC)	Y_2 (TFC)	Y_3 (CTC)	Y_4 (TAA)
N1	9	Ultra-turrax	40	Acetone	5	32.60 ± 1.24	3.91 ± 0.18	126.70 ± 0.74	89.69 ± 0.48
N2	13	Ultra-turrax	20	Acetone	5	14.01 ± 1.53	1.85 ± 0.07	63.97 ± 0.63	39.55 ± 1.45
N3	17	Ultra-turrax	30	Acetone	25	50.51 ± 3.55	7.61 ± 0.64	181.74 ± 1.11	146.51 ± 2.40
N4	14	Ultra-turrax	30	Acetone	25	59.52 ± 10.99	9.76 ± 0.23	237.20 ± 3.22	174.28 ± 8.22
N5	10	Ultra-turrax	30	Acetone	25	61.75 ± 5.30	9.12 ± 1.11	227.71 ± 0.71	163.46 ± 4.42
N6	20	Ultra-turrax	30	Acetone	25	28.62 ± 1.20	4.04 ± 0.13	56.60 ± 0.56	55.51 ± 11.84
N7	18	Ultra-turrax	20	Acetone	50	34.80 ± 5.32	5.81 ± 0.07	74.04 ± 0.81	101.28 ± 2.58
N8	5	Ultra-turrax	40	Acetone	50	67.03 ± 9.76	8.99 ± 0.09	235.77 ± 7.47	168.62 ± 9.68
N9	6	Ultra-turrax	20	Ethanol	5	18.10 ± 1.46	4.08 ± 1.71	85.81 ± 0.16	61.14 ± 2.74
N10	19	Ultra-turrax	27	Ethanol	5	34.65 ± 0.96	4.79 ± 0.48	156.77 ± 0.14	102.77 ± 4.31
N11	15	Ultra-turrax	40	Ethanol	5	48.37 ± 3.90	7.05 ± 1.96	184.07 ± 1.36	122.35 ± 2.18
N12	4	Ultra-turrax	40	Ethanol	50	22.80 ± 1.89	2.40 ± 0.13	37.70 ± 0.03	41.53 ± 5.27
N13	16	Ultra-turrax	20	Ethanol	50	45.03 ± 2.64	6.51 ± 0.39	131.92 ± 0.22	120.18 ± 3.01
N14	22	Maceration	20	Acetone	5	13.29 ± 0.48	1.20 ± 0.04	2.98 ± 0.13	11.21 ± 0.61
N15	23	Maceration	40	Acetone	5	6.64 ± 4.26	0.82 ± 0.05	17.89 ± 0.11	17.67 ± 0.38
N16	8	Maceration	30	Acetone	25	24.37 ± 1.64	5.53 ± 0.06	16.60 ± 0.06	68.47 ± 1.66
N17	11	Maceration	40	Acetone	50	31.27 ± 5.24	7.11 ± 0.19	15.05 ± 0.52	77.68 ± 7.89
N18	3	Maceration	20	Acetone	50	13.97 ± 2.53	1.84 ± 0.04	55.14 ± 0.16	41.23 ± 0.14
N19	21	Maceration	40	Ethanol	5	17.27 ± 2.43	2.04 ± 0.42	1.14 ± 0.04	20.66 ± 3.47
N20	7	Maceration	20	Ethanol	5	25.04 ± 2.50	2.86 ± 0.24	86.90 ± 0.50	58.66 ± 1.52
N21	1	Maceration	40	Ethanol	33	14.30 ± 2.89	2.10 ± 0.06	28.09 ± 0.06	36.22 ± 0.59
N22	12	Maceration	20	Ethanol	50	29.08 ± 5.01	6.13 ± 0.15	18.18 ± 0.03	82.36 ± 1.49
N23	2	Maceration	40	Ethanol	50	16.63 ± 5.59	4.78 ± 2.75	2.26 ± 0.09	23.02 ± 3.38

X_1, extraction method; X_2, temperature (°C); X_3, solvent; X_4, water in solvent (%, v/v). Y_1, TPC—total phenolic content expressed as mg GAE/g dw = gallic acid equivalents per dry weight of walnut septum; Y_2, TFC—total flavonoid content expressed as mg QE/g dw = quercetin equivalents per dry weight of walnut septum; Y_3, CTC—condensed tannin content expressed as mg CE/g dw = catechin equivalents per dry weight of walnut septum; Y_4, TAA—total antioxidant activity expressed as mg TE/g dw = trolox equivalents per dry weight of walnut septum. Data are shown as mean ± SD (standard deviation).

As it can be observed from the results, the extraction yields of TPC, TFC, CTC, as well as the TAA, were influenced by the extraction method and factors evaluated in the experimental design.

For evaluation of the partial least squares regression (PLS) for fitting of the experimental data with the experimental design, R^2 and Q^2 were used as statistical parameters. The goodness of fit is overestimated by the value of R^2, describing the percent of the variation of the response explained by the model, and underestimated by the value of Q^2, representing the percent of the variation of the response predicted by the model according to cross validation. The two aforementioned statistical parameters are the most reliable for describing the model validity; high values and a difference of no more than 0.2–0.3 between these two indicate a high predictive power of a good model. Furthermore, the reproducibility of the model was evaluated considering the variation of the response under the same experimental conditions (pure error) in comparison with the total variation of the response.

The summary of fit for the responses evaluated in the screening step is presented in Table 4 and the regression coefficients are given in Table 5.

Table 4. Optimization of extraction parameter for fitted factorial model by analysis of variance (ANOVA).

Quantifiable Responses	Reproducibility	Source	Degrees of Freedom	Sum of Squares	Mean Square	F Value	p Value
Total phenolic		Regression	8	4838.0	604.7	4.90	0.006
content (Y_1)	0.86	Lack of fit	10	1481.1	148.1	3.60	0.159
($R^2 = 0.75$, $Q^2 = 0.52$)		Pure error	3	123.2	41.05		
Total flavonoid		Regression	7	97.52	13.93	3.26	0.028
content (Y_2)	0.88	Lack of fit	11	57.09	5.19	5.79	0.087
($R^2 = 0.61$, $Q^2 = 0.37$)		Pure error	3	2.68	0.89		
Condensed tannin		Regression	7	112,775	16,110.8	8.29	0.001
content (Y_3)	0.82	Lack of fit	11	23,769.4	2160.8	1.88	0.329
($R^2 = 0.80$, $Q^2 = 0.63$)		Pure error	3	3435.2	1145.8		
Total antioxidant		Regression	7	41,969.5	5995.6	6.21	0.002
activity (Y_4)	0.92	Lack of fit	11	12,939.3	1176.3	6.18	0.080
($R^2 = 0.75$, $Q^2 = 0.59$)		Pure error	3	570.4	190.1		

R^2, coefficient of determination; *F*-value, Fischer's ratio; *p*-value, probability; Q^2, goodness of prediction.

Table 5. Regression equation coefficients.

Effect	Responses			
	Y_1 (Total Phenolic Content)	Y_2 (Total Flavonoid Content)	Y_3 (Condensed Tannin Content)	Y_4 (Total Antioxidant Activity)
Constant	38.859	4.844	92.917	80.521
X_1 (M)	−10.595	−1.281	−63.611	−33.412
X_1 (UTE)	10.595	1.281	63.611	33.412
X_2 (Temperature)	1.551	0.2914	1.763	1.892
X_3 (Acetone)	1.737	0.4991	14.528	11.771
X_3 (Ethanol)	−1.737	−0.4991	−14.528	−11.771
X_4 (Water %)	5.534	1.196	5.550	14.822
$X_4 \times X_4$	−8.261	-	-	-
X_1 (M) $\times X_2$	-	-	−12.390	-
X_1 (UTE) $\times X_2$	-	-	12.390	-
$X_2 \times X_3$ (Acetone)	4.818	0.7329	-	13.769
$X_2 \times X_3$ (Ethanol)	−4.819	−0.7329	-	13.769
X_1 (M) $\times X_3$ (Acetone)	−3.291	−0.6606	−20.223	−13.364
X_1 (M) $\times X_3$ (Ethanol)	3.291	0.6606	20.223	13.364
X_1 (UTE) $\times X_3$ (Acetone)	3.291	0.6606	20.223	13.364
X_1 (UTE) $\times X_3$ (Ethanol)	−3.291	−0.6606	−20.223	−13.364
X_3 (Acetone) $\times X_4$	5.023	0.8522	26.089	15.728
X_3 (Ethanol) $\times X_4$	−5.023	−0.8522	−26.089	−15.728

M—maceration; UTE—ultra-turrax extraction. For data in bold, *p*-value was <0.005, therefore statistically significant.

The experimental setup is appropriate for the purpose of the study and, by working under the same experimental conditions, the replicates generated similar responses, this statement being supported by the reproducibility values > 0.82. The response variation is considered by the developed models ($R^2 > 0.61$) and the predictive capacity was found to be adequate ($Q^2 > 0.37$). The analysis of variance (ANOVA), shown in Table 4, supports the statistical significance of the model, with p-value in the range of 0.001 to 0.028 and F-values between 3.26 and 8.29. According to the results given in Table 4, the fitting models were found to be adequate to describe the experimental data, taking into account that the values for the lack of fit were not significant in extent to the pure error (3.60 for TPC, 5.79 for TFC, 1.88 for CTC, and 6.18 for TAA).

The independent and dependent variables of experimental design evaluated for WS extracts during the optimization step are shown in Table 6. The independent factors were temperature, organic solvent, and percentage of water in solvent mixture. The responses were identification and quantification of the following bioactive compounds: epicatechin, catechin, syringic acid, gallic acid, protocatechuic acid, vanillic acid, hyperoside, isoquercitrin, quercitrin, campesterol, beta-sitosterol.

Table 6. Independent and dependent variable of experimental design evaluated for bioactive compounds from walnut septum extracts.

Variables	Level		
	−1	0	1
Independent variables (factors)			
Temperature (°C) (X_1)	20	30	40
Solvent (%, v/v) (X_2)	Acetone		Ethanol
Water in solvent (%, v/v) (X_3)	5	25	50
Dependent variables (responses)			
Epicatechin (µg/g dw) (Y_1)			
Catechin (µg/g dw) (Y_2)			
Syringic acid (µg/g dw) (Y_3)			
Syringic acid (µg/g dw) (Y_3)			
Gallic acid (µg/g dw) (Y_4)			
Protocatechuic acid (µg/g dw) (Y_5)			
Vanillic acid (µg/g dw) (Y_6)			
Hyperoside (µg/g dw) (Y_7)			
Isoquercitrin (µg/g dw) (Y_8)			
Quercitrin (µg/g dw) (Y_9)			
Campesterol (µg/g dw) (Y_{10})			
Beta-sitosterol (µg/g dw) (Y_{11})			

All units are expressed as µg identified compound per gram of dry weight walnut extract.

For the matrix of the experimental design the same Modde software and version was used as in the screening step, and the responses obtained after performing all the experimental runs are given in Table 7. For fitting the experimental data with the experimental design, the same statistical parameters were determined as mentioned previously (R^2, Q^2, regression, lack of fit, and pure error) and they are presented in Table 8. By analyzing the results shown in Table 8, the fitting models were adequate to describe the experimental data, considering the values of the reproducibility, lack of fit, and pure error. The regression coefficients for bioactive compounds determined in WS extracts are shown in Table 9.

Table 7. Matrix of experimental design for bioactive compounds recovery from walnut septum extracts.

Sample Code	Run Order	Factorial Design with Coded Values			Determination (Experimental Results)										
		X_1	X_2	X_3	Y_1	Y_2	Y_3	Y_4	Y_5	Y_6	Y_7	Y_8	Y_9	Y_{10}	Y_{11}
N1	9	40	Acetone	5	6.091	288.29	2.392	29.074	3.441	2.229	32.726	71.290	583.86	106.22	9932.57
N2	13	20	Acetone	5	2.703	138.59	1.021	13.655	2.117	1.680	13.110	24.039	216.02	42.702	9080.36
N3	17	30	Acetone	25	10.493	447.67	4.215	55.240	13.388	6.219	46.858	99.334	980.69	131.36	26,461.16
N4	14	30	Acetone	25	11.463	468.62	5.010	61.002	11.575	8.881	43.083	109.42	894.89	140.26	19,546.42
N5	10	30	Acetone	25	8.540	408.01	5.154	48.422	12.845	7.173	40.641	94.494	852.79	162.52	22,146.69
N6	20	30	Acetone	25	10.093	396.95	7.111	148.16	28.108	8.905	35.702	68.449	694.32	36.067	5929.59
N7	18	20	Acetone	50	5.136	250.65	3.221	31.031	8.427	3.537	24.933	54.283	495.81	ND	1338.98
N8	5	40	Acetone	50	12.540	597.65	5.202	79.584	9.943	5.577	67.329	103.60	1073.04	292.07	31,018.16
N9	6	20	Ethanol	5	3.533	152.11	1.560	16.679	2.498	1.723	19.014	37.562	326.99	114.36	15,243.10
N10	19	26	Ethanol	5	6.861	273.05	2.381	27.903	4.400	3.022	36.926	75.113	629.22	162.84	22,277.34
N11	15	40	Ethanol	5	8.556	329.04	2.853	31.983	9.645	5.728	33.095	70.960	695.23	8.988	1175.42
N12	4	40	Ethanol	50	9.800	596.98	14.711	130.95	138.58	40.277	32.288	77.528	867.83	104.04	21,736.13
N13	16	20	Ethanol	50	3.238	79.930	6.860	63.446	86.115	11.476	28.768	49.137	449.96	171.06	28,934.75

X_1, temperature (°C); X_2, solvent; X_3, water in solvent (%, v/v). Y_1—Epicatechin; Y_2—Catechin; Y_3—Syringic acid; Y_4—Gallic acid; Y_5—Protocatechuic acid; Y_6—Vanillic acid; Y_7—Hyperoside; Y_8—Isoquercitrin; Y_9—Quercitrin; Y_{10}—Campesterol; Y_{11}—Beta-sitosterol. All responses are expressed as μg bioactive compound per gram of dry weight walnut septum. ND—not determined.

13

Table 8. Optimization of extraction parameter for fitted factorial model by analysis of variance for bioactive compounds in walnut septum extracts (ANOVA).

Quantifiable Responses	Reproducibility	Source	Degrees of Freedom	Sum of Squares	Mean Square	F Value	p Value
Epicatechin (Y_1) ($R^2 = 0.91$, $Q^2 = 0.55$)	0.86	Regression	5	1.18×10^{-4}	2.36×10^{-5}	16.040	0.001
		Lack of fit	4	5.88×10^{-6}	1.47×10^{-6}	0.9956	0.523
		Pure error	3	4.43×10^{-6}	1.48×10^{-6}		
Catechin (Y_2) ($R^2 = 0.94$, $Q^2 = 0.65$)	0.95	Regression	5	3.09×10^{-1}	6.19×10^{-2}	24.345	0.001
		Lack of fit	4	1.44×10^{-2}	3.60×10^{-3}	3.1984	0.183
		Pure error	3	3.38×10^{-3}	1.13×10^{-3}		
Syringic acid (Y_3) ($R^2 = 0.79$, $Q^2 = 0.44$)	0.87	Regression	4	1.19×10^{-4}	2.99×10^{-5}	7.5251	0.008
		Lack of fit	5	2.72×10^{-5}	5.44×10^{-6}	3.5928	0.161
		Pure error	3	4.54×10^{-6}	1.51×10^{-6}		
Gallic acid (Y_4) ($R^2 = 0.97$, $Q^2 = 0.67$)	0.96	Regression	6	1.15×10^{-2}	1.92×10^{-3}	37.712	0.001
		Lack of fit	3	1.76×10^{-4}	5.85×10^{-5}	1.4762	0.428
		Pure error	2	7.93×10^{-5}	3.97×10^{-5}		
Protocatechuic acid (Y_5) ($R^2 = 0.93$, $Q^2 = 0.68$)	0.96	Regression	4	1.84×10^{-2}	4.60×10^{-3}	27.368	0.001
		Lack of fit	5	1.16×10^{-3}	2.31×10^{-4}	3.6946	0.156
		Pure error	3	1.88×10^{-4}	6.26×10^{-5}		
Vanillic acid (Y_6) ($R^2 = 0.84$, $Q^2 = 0.19$)	0.81	Regression	5	8.37×10^{-5}	1.67×10^{-5}	5.4394	0.043
		Lack of fit	3	1.18×10^{-5}	3.92×10^{-6}	2.1522	0.333
		Pure error	2	3.64×10^{-6}	1.82×10^{-6}		
Hyperoside (Y_7) ($R^2 = 0.88$, $Q^2 = 0.34$)	0.87	Regression	6	1.93×10^{-3}	3.21×10^{-4}	7.6743	0.013
		Lack of fit	3	1.86×10^{-4}	6.19×10^{-5}	2.8308	0.208
		Pure error	3	6.55×10^{-5}	2.18×10^{-5}		
Isoquercitrin (Y_8) ($R^2 = 0.98$, $Q^2 = 0.76$)	0.92	Regression	6	7.86×10^{-3}	1.31×10^{-3}	41.990	0.001
		Lack of fit	3	4.00×10^{-5}	1.33×10^{-5}	0.2295	0.870
		Pure error	2	1.16×10^{-4}	5.80×10^{-5}		
Quercitrin (Y_9) ($R^2 = 0.92$, $Q^2 = 0.63$)	0.78	Regression	5	7.32×10^{-1}	1.46×10^{-1}	16.425	0.001
		Lack of fit	4	1.92×10^{-2}	4.80×10^{-3}	0.3334	0.842
		Pure error	3	4.32×10^{-2}	1.44×10^{-2}		
Campesterol (Y_{10}) ($R^2 = 0.91$, $Q^2 = 0.34$)	0.95	Regression	6	6.34×10^{-2}	1.06×10^{-2}	8.6364	0.016
		Lack of fit	3	5.61×10^{-3}	1.87×10^{-3}	7.2591	0.123
		Pure error	2	5.15×10^{-4}	2.57×10^{-4}		
Beta-sitosterol (Y_{11}) ($R^2 = 0.87$, $Q^2 = 0.25$)	0.88	Regression	6	9.76×10^2	1.63×10^2	5.6232	0.039
		Lack of fit	3	1.20×10^2	40.1	3.2875	0.242
		Pure error	2	24.4	12.2		

R^2, coefficient of determination; F-value, Fischer's ratio; p-value, probability. Q^2, goodness of prediction.

Table 9. Regression equation coefficients for bioactive compounds determined in walnut septum extracts.

Effect						Response					
	Y_1	Y_2	Y_3	Y_4	Y_5	Y_6	Y_7	Y_8	Y_9	Y_{10}	Y_{11}
	Epicatechin	Catechin	Syringic Acid	Gallic Acid	Protocatechuic Acid	Vanillic Acid	Hyper oside	Iso quercitrin	Quercitrin	Campesterol	Beta-sitosterol
Constant	**0.00936**	**0.0244**	**0.00489**	**0.0589**	**0.0279**	**0.007207**	**0.03906**	**0.0947**	**0.815**	**0.141**	**22.6**
X_1 (Temperature)	**0.00257**	**0.125**	**0.00127**	**0.0157**	**0.00618**	**0.00154**	**0.00968**	**0.02044**	**0.198**	**0.0277**	2.13
X_2 (Acetone)	0.000325	0.0153	−0.00115	−0.008201	−0.0223	**−0.00173**	0.00214	0.00286	0.01066	−0.000295	−1.91
X_2 (Ethanol)	−0.000325	−0.0153	0.00115	0.008201	0.0223	**0.00173**	−0.00214	−0.00286	−0.01066	0.000295	1.91
X_3 (Water %)	**0.00117**	**0.0612**	**0.00197**	**0.0223**	**0.01902**	**0.00245**	**0.00623**	**0.00977**	**0.116**	**0.0295**	**5.31**
$X_1 \times X_1$	**−0.00199**	**−0.0679**	-	**−0.01015**	-	−0.00175	−0.00522	−0.0253	**−0.159**	−0.0275	**−6.206**
$X_1 \times X_3$	-	**0.05056**	-	0.00972	-	-	-	-	-	**0.0285**	3.75
$X_1 \times X_2$ (Acetone)	-	-	-	-	-	-	**0.004901**	0.00397	-	**0.0593**	**5.77**
$X_1 \times X_2$ (Ethanol)	-	-	-	-	-	-	**−0.004901**	−0.00397	-	**−0.0593**	**−5.77**
X_2 (Acetone) $\times X_3$	0.000853	-	−0.00133	−0.00896	−0.0219	−0.00153	0.00461	**0.00541**	0.0576	-	-
X_2 (Ethanol) $\times X_3$	−0.000853	-	0.00133	0.00896	0.0219	0.00153	−0.00461	**−0.00541**	−0.576	-	-

For data in bold, *p*-value was <0.005, therefore statistically significant.

3.2. The Influence of Studied Variables on TPC, TFC, CTC, TAA, and Individual Bioactive Compounds

The different working conditions for walnut septum extracts are shown in Table 3. A number of 13 samples were obtained by UTE method, while 10 samples were obtained by maceration. The working temperature was in the range of 20 to 40 °C, the two solvents used were acetone and ethanol mixed with water in various proportions.

The results for TPC, TFC, and CTC for the 23 walnut septum extracts are depicted in Figure 1, and the results for TAA of the same 23 WS extracts are shown in Figure 2. As it can be observed, acetone presents higher extraction power for the bioactive compounds, while the method with higher extraction efficiency was UTE.

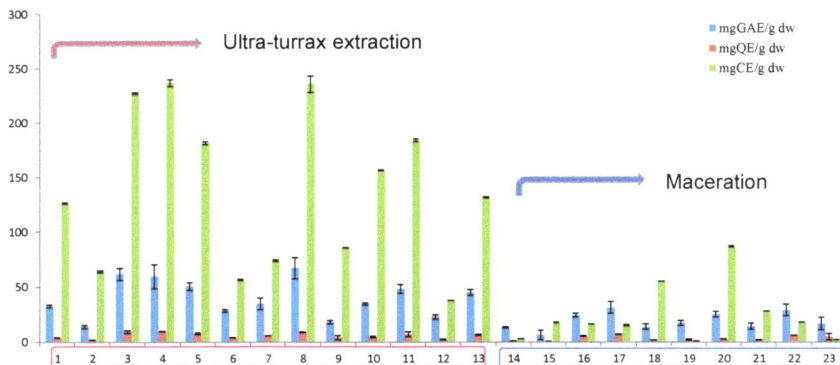

Figure 1. Total phenolic content (gallic acid equivalents, GAE), total flavonoid content (quercetin equivalents, QE), and condensed tannin content (catechin equivalents, CE) of analyzed walnut septum extracts.

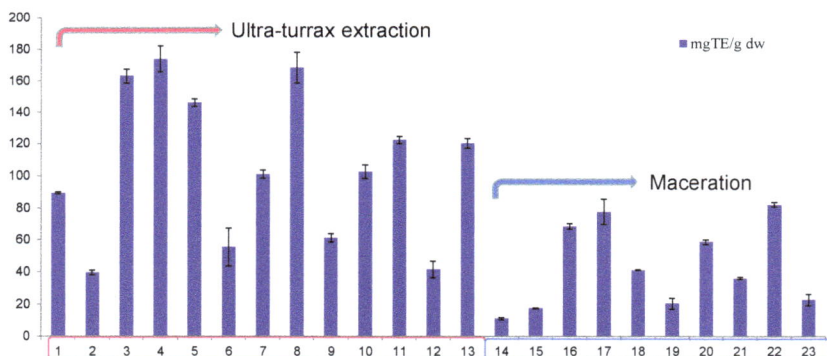

Figure 2. The total antioxidant activity evaluated through ABTS radical cation scavenging activity assay (expressed as Trolox equivalents, TE) of analyzed walnut septum extracts.

Longer extraction time period and high amount of solvent are involved in maceration. In this case, phenolic compounds may also suffer oxidation, hydrolysis, and ionization of the molecules. These could be reasons for the observed results regarding the two extraction methods that were used. Moreover, the ultrasound energy can leach the bioactive compounds of interest, thus increasing the yield of extraction [48].

There was a good correlation between the content of bioactive compounds from WS extracts and their antioxidant activity. The extracts presenting the highest content of these compounds exhibited the highest TAA and these extracts were obtained by UTE method. Therefore, the 13 extracts obtained by this method were further analyzed in terms of the factors influencing their extraction efficiency.

The polyphenols in WS extracts, which were determined and quantified by HPLC/MS, are summarized in Table 8. The compounds found in the highest amount were catechin and quercitrin. The two phytosterols determined in the analyzed samples were campesterol and beta-sitosterol. Forwards, the manner in which the extraction yield of the main bioactive compounds is influenced by the working conditions is briefly presented. Moreover, the influence of working conditions on the bioactive compounds extraction yield from WS samples are presented as scaled and centered coefficient plots in Figure 3. In addition, the response surfaces for predicting the extraction yield of bioactive compounds from WS extracts with respect to the evaluated working conditions are shown in Figure 4.

For epicatechin (Y_1), the highest extraction yield is obtained when working at high temperature. The high percentage of water in acetone influences its extraction to a lesser extent. For this bioactive compound, the use of ethanol in the extraction mixture is not favorable. Catechin (Y_2) extraction is influenced by temperature and solvent. More precisely, for this compound the best working conditions would be high temperature and high percentage of water in solvent, according to the response surface generated by Modde software. For syringic acid (Y_3) extraction, the optimum extraction conditions are high temperature and high amount of water in ethanol as solvent. The extraction of gallic acid (Y_4) can be optimized if working at high temperature and high amount of water in the solvent mixture. In this case, a higher extraction yield can be obtained if ethanol is used instead of acetone. For protocatechuic acid (Y_5), all the evaluated factors seemed to have a statistically significant influence upon the extraction power, the most important working parameter being the amount of water in ethanol. Vanillic acid (Y_6) extraction power increases with the increase of temperature and water percentage in solvent. If acetone is used in mixture with water, then the working temperature does not influence the extraction yield. On the contrary, if a mixture of ethanol with water is used, then an increase in temperature will lead to an increase in the extraction power. Hyperoside (Y_7) is best extracted in the following working conditions: high temperature, high percentage of water in solvent, and acetone as solvent. For isoquercitrin (Y_8), the extraction yield is statistically significant influenced by solvent mixture temperature, percentage of water in the solvent mixture, and the organic solvent. The use of acetone has a positive influence on the recovery of isoquercitrin, while ethanol has a negative influence, both of them being statistically significant. With regard to quercitrin (Y_9), the most important working parameters were found to be the temperature and, to a lesser extent, the amount of water in the solvent.

Figure 3. *Cont.*

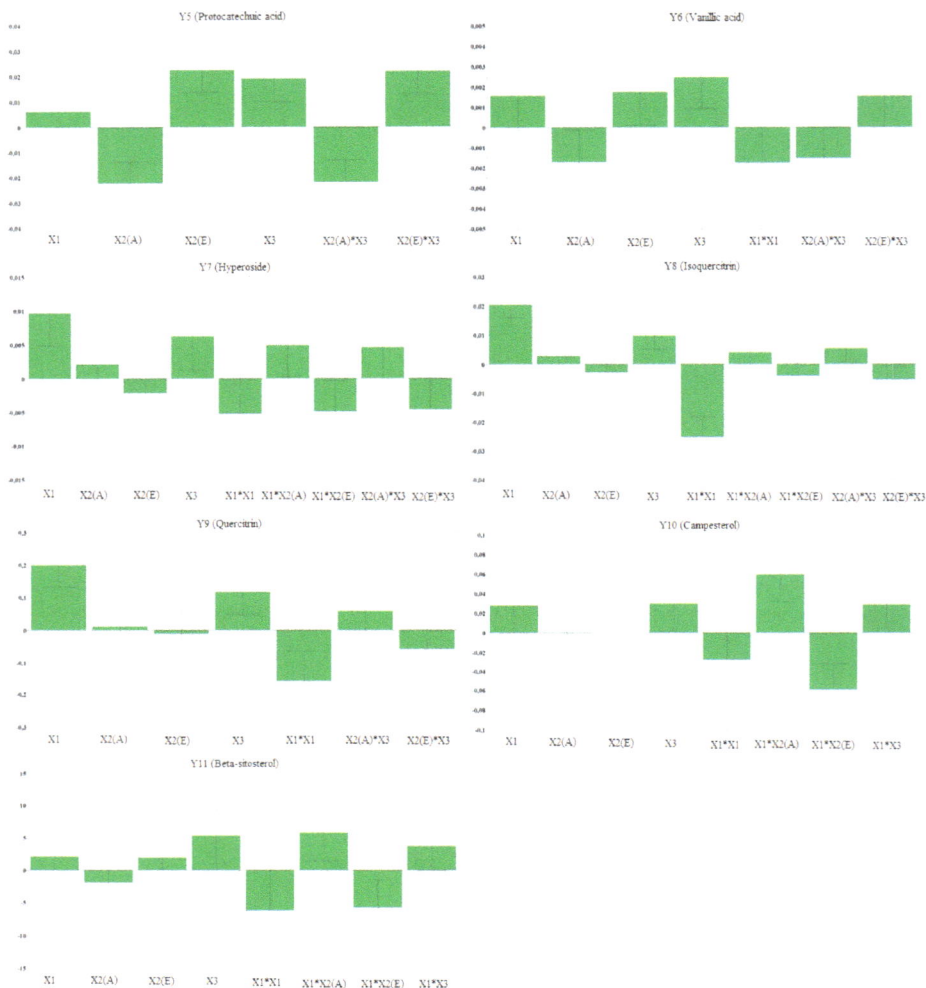

Figure 3. Influence of working conditions on the bioactive compounds recovery from walnut septum extracts, presented as scaled and centered coefficient plots. X_1—temperature (°C); X_2(A)—solvent type (acetone), X_2(E)—solvent type (ethanol); X_3—water % in mixture with solvent; Y_1, Y_2, Y_3, Y_4, Y_5, Y_6, Y_7, Y_8, Y_9, Y_{10}, Y_{11}—dependent variables (bioactive compounds) according to Table 5.

For the two phytosterols analyzed, the influence of the extraction conditions is different from those evaluated for the recovery of polyphenols. More precisely, for campesterol (Y_{10}) and beta-sitosterol (Y_{11}), the highest impact on the extraction yield is attributed to percentage of water in mixture with acetone and ethanol. For both sterols, acetone mixed with a high percentage of water had a positive influence, whereas ethanol displayed a negative influence on the recovery of these two bioactive compounds.

After the analysis of all the evaluated responses and the manner by which each factor influences the extraction yield for the evaluated bioactive compounds, the Modde software generated the optimal extraction conditions for each evaluated bioactive compound, which are given in Table 10. In general, the best working conditions with the highest extraction power for epicatechin, catechin, hyperoside, quercitrin, campesterol, beta-sitosterol are a temperature of 40 °C and a mixture of solvent, acetone and water in equal proportions.

The phytochemical profile of the analyzed extracts proved that WS can be a valuable source of biologically active compounds for food and/or pharmaceutical industry and warrant the continuation of current research in further evaluating its bioactive potential.

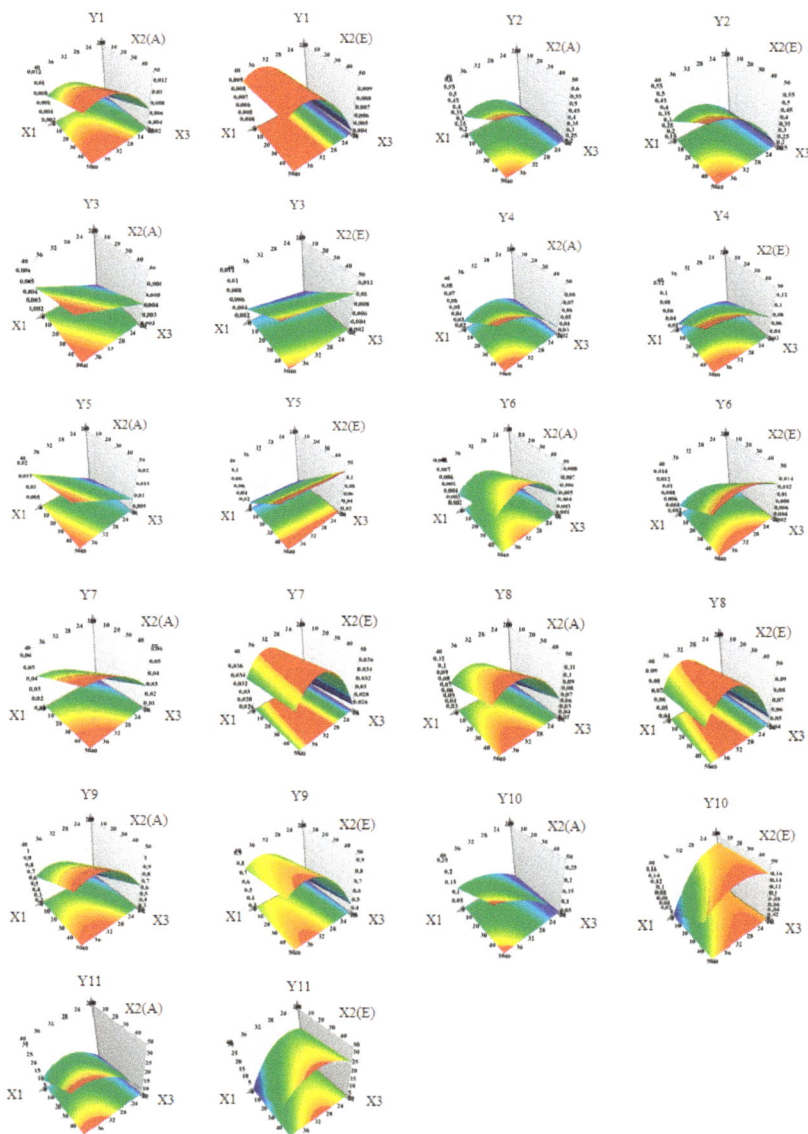

Figure 4. Response surface for predicting the bioactive compounds recovery from walnut septum extracts with respect to: X_1—temperature (°C); X_2(A)—solvent type (acetone), X_2(E)—solvent type (ethanol); X_3—water % in mixture with solvent (the regions in red represent the domains of working conditions that assure a maximum extraction yield for the evaluated bioactive compounds).

Table 10. Optimum experimental conditions for improved recovery of bioactive compounds from walnut septum extracts obtained by ultra-turrax extraction.

Evaluated	TPC [1]	TFC [2]	CTC [3]	TAA [4]	Epi-catechin	Catechin	Syringic Acid	Gallic Acid	Proto-catechuic Acid	Vanillic Acid	Hyper-oside	Iso-quercitrin	Quercitrin	Campesterol	Beta-sitosterol
Temperature	40 °C	30 °C	30 °C	30 °C	40 °C	40 °C	40 °C	30 °C	40 °C	40 °C	40 °C	30 °C	40 °C	40 °C	40 °C
Solvent	Acetone	Acetone	Acetone	Acetone	Acetone	Acetone	Ethanol	Acetone	Ethanol	Ethanol	Acetone	Acetone	Acetone	Acetone	Acetone
Water %	50%	25%	25%	25%	50%	50%	50%	25%	50%	50%	50%	25%	50%	50%	50%
Determined	67.03 ± 9.76	9.76 ± 0.23	237.20 ± 3.22	174.28 ± 8.22	12.450	597.647	14.711	148.164	138.58	40.277	67.329	109.42	1073.04	292.07	3108.16

[1] TPC—total phenolic content expressed as mg GAE/g dw = gallic acid equivalents per dry weight of walnut septum; [2] TFC—total flavonoid content expressed as mg QE/g dw = quercetin equivalents per dry weight of walnut septum; [3] CTC—condensed tannin content expressed as mg CE/g dw = catechin equivalents per dry weight of walnut septum; [4] TAA—total antioxidant activity expressed as mg TE/g dw = trolox equivalents per dry weight of walnut septum. Data are shown as mean ± SD (standard deviation). All determined amount of bioactive compounds are expressed as μg bioactive compound per gram of dry weight walnut septum.

3.3. Quantitative Determinations of Total Bioactive Compounds

Numerous studies revealed that phenolic compounds can be found in tree nut species and their health benefits might be attributed to the phenolic profiles and antioxidant activity [49,50]. Most of the phenolic content is found in the tree nut by-products [51], therefore the interest for this research domain.

Polar solvents are considered the best ones for phenolics extraction, while non-polar solvents (e.g., pentane, hexane, chloroform, diethyl ether) are frequently used for the extraction of less polar constituents, such as tocotrienols and tocopherols, carotenoids and chlorophylls. As expected from previous studies [29,52], the binary-solvent systems extracted more phenolic compounds than the mono-solvent systems. This fact correlates with the differences in the polarity of the extraction mixtures used and solubility of phenolic compounds in them. The mixture of two polar protic solvents, water and ethanol, is less effective than the mixture of a polar protic solvent (water) and a polar, relatively acidic, aprotic solvent (acetone).

As mentioned before, the UTE method exhibited higher extraction yields than maceration in terms of evaluated bioactive compounds and TAA.

3.3.1. Total Phenolics

A clear difference can be seen between the two richest phenolic compounds extracts based on extraction method: 67.03 ± 9.76 mg GAE/g dw for Ultra-Turrax (run order 5) and 31.27 ± 5.24 mg GAE/g dw for maceration method (run order 11) (Table 3).

We could not find any results for WS in the literature, therefore values for nuts and other by-products were used as comparison. In one study, the TPC in nuts varies from 1.03 to 16.50 mg GAE/g, with pecans, walnuts, and pistachios presenting the highest values [53], while Alasalvar and Bolling (2015) found values at 15.50 to 16.25 mg GAE/g walnut [54]. Another study, performed on walnut seed and by-products, presented TPC mean values of 116.22 ± 3.76 mg GAE/g seed extract (21.43 mg/g walnut seed after taking into account the extraction yields), 94.39 ± 5.63 mg GAE/g leaf extract, and 50.18 ± 2.69 mg GAE/g green husk extract [55]. Akbari et al. (2012) obtained TPC mean values of 52.05 ± 1.27 mg GAE/g, 24.68 ± 0.43 mg GAE/g, and 18.04 ± 0.42 mg GAE/g, for walnut pellicle, hull, and shell, respectively [56], while Shah et al. (2018) acquired quantities from 37.61 to 46.47 mg GAE/g dw walnut leaf extract [57].

Comparing the data from our study with those found in the literature, TPC in WS has equivalent values with those determined for other walnut by-products, such as green husks or leaves. It is evident that WS can be an important source of polyphenolic compounds.

3.3.2. Total Flavonoids

Flavonoids, important polyphenolic compounds in tree nuts, have been associated with several health promoting properties, such as anti-inflammatory, antioxidant, anticancer, antiviral, antibacterial, and hepatoprotective [58]. The highest TFC value was 9.76 ± 0.23 mg QE/g dw walnut septum (run order 14, Table 3), approximately 10 times lower than total extractable phenolics. In one recent study the total flavonoids content of the walnut leaf extract ranged from 5.52 to 28.48 mg QE/g [57], while Mocan et al. (2018), researching *Prunus domestica* leaves, found TFC values between 36.60 ± 2.90 and 60.32 ± 4.12 mg QE/g leaf extract [59]. However, an objective comparison between the results is quite difficult, due to different matrixes and extraction protocols.

3.3.3. Condensed Tannins

Condensed tannins or proanthocyanidins (oligomeric and polymeric forms of flavan-3-ols) are usually quantified using the vanillin assay. The highest CTC value was 237.20 ± 3.22 mg CE/g dw septum (run order 14, Table 3), comparable to those found in pecan nut shell [25,60], but much higher than results obtained in almond kernels [61] or hazelnut and its skin [62]. Knowing that the reactivity of vanillin with catechin is different from that of vanillin with tannins [36], and because the use of catechin as standard in matrices with high content of tannins may under- or overestimate their concentration [60], these values should be viewed with some cautiousness. Despite a potential overestimation, clearly walnut septum is a valuable natural source of condensed tannins.

Proanthocyanidins are extensively metabolized by gut microbiota to valerolactone intermediates and hydroxybenzoic acids, an important aspect of their bioavailability [63]. A major challenge that influences the bioavailability of these health promoting compounds is their bioaccessibility, the amount which is released from the food matrix in the lumen of the GI tract, and as a result available for absorption [63].

3.4. Identification and Quantification of Individual Polyphenols

From the 18 phenolic compounds analyzed by the validated LC-MS/MS method, gentisic acid, *p*-coumaric acid, ferulic acid, hyperosid, isoquercitrin, quercitrin, and quercetin were identified in the WS extracts. Only hyperoside, isoquercitrin, and quercitrin, found in high amounts, were quantified.

All the six polyphenols (epicatechin, catechin, syringic acid, gallic acid, protocatechuic acid, and vanillic acid) analyzed by the other LC-MS method were identified and quantified in WS extracts. This method showed a good linearity ($R^2 > 0.9922$) and accuracy (<15%) over the calibration range (Table 2).

After samples hydrolysis, the amount of epicatechin, syringic acid, gallic acid, protocatechuic acid, and vanillic acid increased, while catechin decreased (see Table 11). This registered increase is most probably due to their release from the septum matrix. The particular case of epicatechin increase in parallel with catechin decrease could be attributed to heat-related epimerization of one compound into its optic isomer during the hydrolysis process [26,64].

This is the first study that identifies and quantifies the polyphenolic compounds present in WS.

3.5. Identification and Quantification of Phytosterols

The content of phytosterols in the analyzed WS extracts is presented in Table 8. The richest extracts in beta-sitosterol (31.02 mg/g dw septum) and campesterol (0.292 mg/g dw septum) were obtained using 50% aqueous of acetone (run order 7, Table 3). Martinez et al. (2010) identified β-sitosterol at 0.772 to 2.52 mg/g walnut oil and campesterol at 0.044 to 0.121 mg/g walnut oil [65].

3.6. Antioxidant Activity

The antioxidant activity of tree nuts or some of their by-products has been previously reported [66–68], but there are no references in the literature about WS antioxidant activity.

Knowing that there are limits in the bioavailability of polyphenols, caused by the extensive catabolism and phase 2 metabolism, it is questionable that polyphenols are responsible for a direct in vivo antioxidant function. However, polyphenols might function through upregulation of antioxidant activity [49]. Ellagitannins, important polyphenols in walnuts with known antioxidant and anti-inflammatory bioactivity, are hydrolyzed to ellagic acid and then converted to urolithins by gut microflora, having a potential role against initiation and progression of several illnesses, including cancer, neurodegenerative, and cardiovascular diseases [27].

Table 11. Quantitative evaluation of the recovery of main bioactive compounds in non-hydrolyzed and hydrolyzed samples of walnut septum extracts.

Sample Code/Bioactive Compound	Non-Hydrolyzed Sample						Hydrolyzed Samples					
	Epicatechin	Catechin	Syringic Acid	Gallic Acid	Protocatechuic Acid	Vanillic Acid	Epicatechin	Catechin	Syringic Acid	Gallic Acid	Protocatechuic Acid	Vanillic Acid
N1	0.006	0.288	0.002	0.029	0.003	0.002	0.249	0.282	0.047	1.918	0.065	0.036
N2	0.003	0.139	0.001	0.014	0.002	0.002	0.097	0.089	0.031	1.084	0.037	0.027
N3	0.010	0.448	0.004	0.055	0.013	0.006	0.356	0.377	0.088	3.537	0.132	0.097
N4	0.011	0.469	0.005	0.061	0.012	0.009	0.134	0.083	0.096	4.194	0.100	0.092
N5	0.009	0.408	0.005	0.048	0.013	0.007	0.316	0.286	0.091	3.543	0.145	0.103
N6	0.010	0.397	0.007	0.148	0.028	0.009	0.247	0.244	0.099	3.814	0.168	0.128
N7	0.005	0.251	0.003	0.031	0.008	0.004	0.266	0.311	0.044	1.943	0.074	0.053
N8	0.013	0.598	0.005	0.080	0.010	0.006	0.544	0.555	0.108	4.436	0.140	0.071
N9	0.004	0.152	0.002	0.017	0.002	0.002	0.149	0.146	0.031	1.447	0.040	0.027
N10	0.007	0.273	0.002	0.028	0.004	0.003	0.227	0.215	0.063	2.484	0.090	0.056
N11	0.009	0.329	0.003	0.032	0.010	0.006	0.254	0.273	0.064	2.487	0.101	0.069
N12	0.010	0.597	0.015	0.131	0.139	0.040	0.141	0.156	0.110	2.738	0.277	0.130
N13	0.003	0.080	0.007	0.063	0.086	0.011	0.000	0.037	0.046	1.622	0.223	0.076

All determined amount of bioactive compounds are expressed as mg bioactive compound per gram of dry weight walnut septum.

23

3.6.1. ABTS Radical Cation Scavenging Activity Assay

The antioxidant activity against the stable synthetic ABTS radical cation of different WS extracts is summarized in Table 3 and depicted in Figure 2. This assay is based on electron transfer reactions to evaluate radical scavenging activity of various compounds. The highest antioxidant activity was found for the 75% acetone extract (run order 14) at 174.28 $\pm$ 8.22 mg TE/g dw septum, followed by the 50% acetone extract (run order 5) at 168.62 $\pm$ 9.68 mg TE/g dw septum, both samples obtained by UTE method (Table 3). As mentioned, the antioxidant activity is positively influenced by the fact that these two extracts had the highest content of phenols, flavonoids, and condensed tannins. In other related sources, the ABTS reported scavenging activity was 83.46–93.08% in raw walnuts, 78.3 mg TE/g dw raw pecans, 84.9–93.6% in raw hazelnuts, 309–1375 µmol TE/g hazelnut skin [69], 3.36 mmol TE/g pecan kernel and 8.24 mmol TE/g pecan shell crude extracts [70], 3063–3573 µmol TE/100 g dw natural hazelnut [71]. Nevertheless, because of different ways of expression and/or different preparation method, it is not possible to compare the present results with those from the literature.

3.6.2. DPPH Radical Scavenging Assay

The DPPH (2,2-diphenyl-1-picrylhydrazyl) radical scavenging assay was used to evaluate the ability of septum extracts to scavenge this stable free radical. The change in absorbance at 517 nm is employed as a measure of the scavenging effect of a particular extract for DPPH radicals. The absorbance will decrease faster if the antioxidant activity of the extract (in terms of hydrogen atom-donating capacity) is more potent. Antioxidant molecules can reduce DPPH free radicals and change them to a colorless product resulting in a decreased absorbance [72].

In our research, the in vitro DPPH radical scavenging activity was 255.89 mg TE/g of septum extract obtained by Ultra-Turrax with an equal acetone/water volume solution. In other studies on related matrices, the values for the DPPH radical scavenging activity were 14.2 mmol TE/100 g fw (fresh weight) natural walnuts [71], 2.11 mmol TE/g pecan kernel and 4.80 mmol TE/g pecan shell crude extracts [70]. The percentage of DPPH discoloration found by Slatnar et al. (2014) ranged from 60.0 to 96.4% (782.5 to 1682.5 µM trolox/kg) for kernels, 63.0 to 73.2% (312.1 to 810.6 µM trolox/kg) for pellets and 17.7 to 29.9% (870.0 to 1430.2 µM trolox/g) for oil [73]. There are no available data regarding the DPPH assay for WS and a comparison with the results of other researchers, because of different type of samples and measurement units, is not possible.

3.6.3. FRAP (Ferric-Reducing Antioxidant Power) Assay

Reducing power can function as a significant sign of the antioxidant activity and is usually evaluated based on measurement of the conversion of Fe^{3+} to Fe^{2+} in the presence of antioxidants. In our study, the reducing power for the richest polyphenolic septum extract achieved using UTE method and water/acetone (1:1) at 30 °C, was 400.97 mg TE/g septum extract.

As determined by the FRAP assay, the total antioxidant activity of walnut (*Juglans regia*), attributed primarily to their high phenolic content [74], ranked second only to rose hips (*Rosa canina*) among various fruits and foodstuffs [63]. FRAP activity data for related matrices were 418.92 µM Fe^{2+}/g to 1067.94 µM Fe^{2+}/g fw walnut leaves [57], 95.4 µmol TAE/g to 181.2 µmol TAE/g dw walnut kernel [75], or 454 µmol Fe^{2+}/g walnut [76]. As for the other assays, no information was found regarding septum FRAP activity, therefore an objective comparison between the results is unlikely.

Our results on the topic of the antioxidant action of walnut septum are in agreement with previous reports showing a direct relationship between TPC and antioxidant activity [57,77]. Phenolic compounds can act as free radicals scavengers, hydrogen donors, reducing agents, singlet oxygen quenchers, and metal chelators [55] and are mainly responsible for the walnut antioxidant activity [78]. There are differences in the antioxidant activity between the phenolic compounds most likely due to the number of hydroxyls present in the aromatic ring. Zhang et al. (2009) found that compounds with five hydroxyl groups, such as catechin and epicatechin, were the most active free radical scavengers [79].

These two flavonoids may also present cardiovascular benefits [80], improve blood pressure [81], and positively affect total and low-density lipoprotein cholesterol [82]. Gallic acid, with three hydroxyls showed higher antioxidant activity than protocatechuic acid, with two hydroxyls, or syringic and vanillic acids, with only one hydroxyl [79]. Hyperoside, quercitrin, and isoquercitrin are glycosides formed from the flavonoid quercetin and different types of carbohydrates. Like quercetin, they exhibit antioxidant activity, acting as scavengers of free radicals, but because of the sugar portion of the molecule they are more soluble in water than quercetin, provide superior absorption, and are more bioavailable to the body [83]. In our study, the highest bioactive compounds values recovered from WS extracts via the experimental design correspond to the samples with the highest antioxidant activity, a positive correlation in line with the aforementioned literature findings.

Considering that walnuts possess the highest antioxidant activity in all assays (ABTS, DPPH, FRAP) among nuts [54] and assessing the biological assays of the present study we conclude that WS could be a useful functional ingredient in food technology and pharmaceutical industry.

3.7. Tyrosinase Inhibitory Activity

Tyrosinase, a copper-containing enzyme, responsible for the oxidation of tyrosine to L-DOPA and the hydroxylation of L-tyrosine, is involved in several cellular processes, such as biosynthesis of melanin, insect molting, or browning of damaged fruits and vegetables. In humans, melanin regulates skin color and plays a protective role by absorbing ultraviolet sunlight and removing reactive oxygen species from the skin [84]. However, overproduction of melanin in the skin may results in hyperpigmentation or hypermelanosis, characterized by melasma and age spots. Also, over accumulation of melanin in the brain, via oxidation of dopamine, is implicated in the pathogenesis of Parkinson's disease and related neurodegenerative disorders [85]. Thus, tyrosinase inhibition may not only alleviate skin hyperpigmentation and browning progression in food, but also inhibit wrinkle formation, improve neurodegeneration associated with Parkinson's disease, and slow down aging [86].

Several recent studies [87,88] aimed to find natural sources of tyrosinase inhibitors in order to replace the synthetic ones, but to the best of our knowledge there is no previous study of tyrosinase inhibitory activity of WS. In the present study, the tyrosinase inhibitory activity of 50% aqueous acetone lyophilized septum extract was 129.98 ± 3.03 mg KAE/g. In other plant matrices, the tyrosinase inhibitory activity was 30.5 ± 1.7 mg KAE/g extract of *Pseudosempervivum* plant [89], 16.81 ± 0.58 mg KAE/g extract of *Lycium* leaves [34], or 31.46 ± 0.19 mg KAE/g extract of *Lycium* berries [90]. Based on these results, we conclude that walnut septum may be a powerful alternative source for natural tyrosinase inhibitors very convenient for the food, pharmaceutical, or cosmetic industry. It can be used to obtain different formulations for preventing the aforementioned disorders.

4. Conclusions

This study aimed to characterize septum extracts of walnut (*Juglans regia*) and to describe the optimum experimental conditions for maximizing the extraction efficiency of the bioactive compounds found in this less-studied by-product of this species, in the light of its traditional uses as a remedy for colds and coughs. Specifically, we focused to obtain walnut septum extracts with high content in bioactive compounds (phenols, flavonoids, condensed tannins), having antioxidant and enzyme inhibitory activity, based on an experimental design, and to characterize the phytochemical profile of the extracts using HPLC-MS/MS.

In order to determine the optimal extraction conditions of the main phenolics and phytosterols, several parameters, such as extraction method, solvent, temperature, water percentage, were combined, and they were coupled with statistical tools and chemical analysis (LC-MS/MS). The content in phenolic compounds, tannins, and phytosterols was correlated with the evaluated antioxidant and tyrosinase inhibitory activities. The antioxidant activity of the extracts was assessed using several methods (ABTS, DPPH, and FRAP), and the results showed good antiradical effects. Regarding the tyrosinase inhibitory activity, walnut septum extract showed very good results, therefore, this by-product can be further used in cosmetic products to treat skin hyperpigmentation as well as to inhibit wrinkle formation.

The aforementioned results confirm the use of walnut septum as a health-promoting agent. Further research is needed in order to deeply comprehend the bioavailability of the bioactive molecules and the involved metabolic pathways.

Author Contributions: Conceptualization, M.E.R., L.V., A.M. and D.-S.P.; Methodology, M.E.R., A.M., I.T. and D.-S.P.; Software, M.E.R., A.-M.G. and I.T.; Formal Analysis, M.E.R., A.-M.G., A.M., C.M., L.V., D.-S.P.; Investigation, M.E.R., A.M. and L.V.; Supervision, D.-S.P., L.V.; Writing-Original Draft Preparation, M.E.R., D.-S.P. and A.-M.G.; Writing-review and editing, M.E.R., A.-M.G., A.M., C.M., D.-S.P., I.T. and L.V.

Funding: This work was supported by "Iuliu Hatieganu" University of Medicine and Pharmacy in Cluj-Napoca, Romania through a PhD grant (PCD No. 3066/48/01.02.2018).

Conflicts of Interest: The authors declare no conflict of interest.

References

1. Soccol, C.R.; da Costa, E.S.F.; Letti, L.A.J.; Karp, S.G.; Woiciechowski, A.L.; de Souza Vandenberghe, L.P. Recent developments and innovations in solid state fermentation. *Biotechnol. Res. Innov.* **2017**, *1*, 52–71. [CrossRef]

2. Zhang, S.; Wang, S.; Huang, J.; Lai, X.; Du, Y.; Liu, X.; Li, B.; Feng, R.; Yang, G. High-specificity quantification method for almond-by-products, based on differential proteomic analysis. *Food Chem.* **2016**, *194*, 522–528. [CrossRef] [PubMed]

3. Kempf, K.; Martin, S.; Döhring, C.; Dugi, K.; von Wolmar, C.W.; Haastert, B.; Schneider, M. The epidemiological Boehringer Ingelheim Employee study-part I: Impact of overweight and obesity on cardiometabolic risk. *J. Obes.* **2013**. [CrossRef] [PubMed]

4. Murakami, K.; Livingstone, M.B.E. Eating Frequency Is Positively Associated with Overweight and Central Obesity in US Adults. *J. Nutr.* **2015**, *145*, 2715–2724. [CrossRef] [PubMed]

5. Franceschi, C.; Campisi, J. Chronic inflammation (Inflammaging) and its potential contribution to age-associated diseases. *J. Gerontol. A Biol. Sci. Med. Sci.* **2014**, *69*, S4–S9. [CrossRef] [PubMed]

6. Leon-Cabrera, S.; Solís-Lozano, L.; Suárez-Álvarez, K.; González-Chávez, A.; Béjar, Y.L.; Robles-Díaz, G.; Escobedo, G. Hyperleptinemia is associated with parameters of low-grade systemic inflammation and metabolic dysfunction in obese human beings. *Front. Integr. Neurosci.* **2013**, *7*. [CrossRef] [PubMed]

7. Garg, S.K.; Maurer, H.; Reed, K.; Selagamsetty, R. Diabetes and cancer: Two diseases with obesity as a common risk factor. *Diabetes Obes. Metab.* **2014**, *16*, 97–110. [CrossRef] [PubMed]

8. Bastien, M.; Poirier, P.; Lemieux, I.; Després, J.P. Overview of epidemiology and contribution of obesity to cardiovascular disease. *Prog. Cardiovasc. Dis.* **2014**, *56*, 369–381. [CrossRef] [PubMed]

9. Divella, R.; De Luca, R.; Abbate, I.; Naglieri, E.; Daniele, A. Obesity and cancer: The role of adipose tissue and adipo-cytokines-induced chronic inflammation. *J. Cancer* **2016**, *7*, 2346–2359. [CrossRef] [PubMed]

10. Mauland, K.K.; Eng, Ø.; Ytre-Hauge, S.; Tangen, I.L.; Berg, A.; Salvesen, H.B.; Salvesen, Ø.O.; Krakstad, C.; Trovik, J.; Hoivik, E.A.; et al. High visceral fat percentage is associated with poor outcome in endometrial cancer. *Oncotarget* **2017**, *8*, 105184–105195. [CrossRef] [PubMed]

11. Gui, Y.; Pan, Q.; Chen, X.; Xu, S.; Luo, X.; Chen, L. The association between obesity related adipokines and risk of breast cancer: A meta-analysis. *Oncotarget* **2017**, *8*, 75389–75399. [CrossRef] [PubMed]

12. Cani, P.D.; Jordan, B.F. Gut microbiota-mediated inflammation in obesity: A link with gastrointestinal cancer. *Nat. Rev. Gastroenterol. Hepatol.* **2018**. [CrossRef] [PubMed]

13. Eibl, G.; Cruz-Monserrate, Z.; Korc, M.; Petrov, M.S.; Goodarzi, M.O.; Fisher, W.E.; Habtezion, A.; Lugea, A.; Pandol, S.J.; Hart, P.A.; et al. Diabetes Mellitus and Obesity as Risk Factors for Pancreatic Cancer. *J. Acad. Nutr. Diet.* **2018**, *118*, 555–567. [CrossRef] [PubMed]

14. Ferro, M.; Terracciano, D.; Buonerba, C.; Lucarelli, G.; Bottero, D.; Perdonà, S.; Autorino, R.; Serino, A.; Cantiello, F.; Damiano, R.; et al. The emerging role of obesity, diet and lipid metabolism in prostate cancer. *Future Oncol.* **2017**, *13*, 285–293. [CrossRef] [PubMed]

15. Petrick, J.; Freedman, N.; Demuth, J.; Yang, B.; Van Den Eeden, S.; Engel, L.; McGlynn, K. Obesity, diabetes, serum glucose, and risk of primary liver cancer by birth cohort, race/ethnicity, and sex: Multiphasic health checkup study. *Cancer Epidemiol.* **2016**, *42*, 140–146. [CrossRef] [PubMed]

16. Zhang, J.; Chen, Q.; Li, Z.-M.; Xu, X.-D.; Song, A.-F.; Wang, L.-S. Association of body mass index with mortality and postoperative survival in renal cell cancer patients, a meta-analysis. *Oncotarget* **2018**, *9*, 13959–13970. [CrossRef] [PubMed]

17. Aravani, A.; Downing, A.; Thomas, J.D.; Lagergren, J.; Morris, E.J.A.; Hull, M.A. Obesity surgery and risk of colorectal and other obesity-related cancers: An English population-based cohort study. *Cancer Epidemiol.* **2018**, *53*, 99–104. [CrossRef] [PubMed]

18. Grosso, G.; Yang, J.; Marventano, S.; Micek, A.; Galvano, F.; Kales, S. Nut consumption on all-cause, cardiovascular, and cancer mortality risk: A systematic review and meta-analysis of epidemiologic studies. *Am. J. Clin. Nutr.* **2015**, *101*, 783–793. [CrossRef] [PubMed]

19. Aune, D.; Keum, N.; Giovannucci, E.; Fadnes, L.; Boffetta, P.; Greenwood, D.; Tonstad, S.; Vatten, L.; Riboli, E.; Norat, T. Nut consumption and risk of cardiovascular disease, total cancer, all-cause and cause-specific mortality: A systematic review and dose-response meta-analysis of prospective studies. *BMC Med.* **2016**, *14*. [CrossRef] [PubMed]

20. Hever, J.; Cronise, R.J. Plant-based nutrition for healthcare professionals: Implementing diet as a primary modality in the prevention and treatment of chronic disease. *J. Geriatr. Cardiol.* **2017**, *14*, 355–368. [CrossRef] [PubMed]

21. Shahidi, F.; Ambigaipalan, P. Phenolics and polyphenolics in foods, beverages and spices: Antioxidant activity and health effects-A review. *J. Funct. Foods* **2015**, *18*, 820–897. [CrossRef]

22. Oroian, M.; Escriche, I. Antioxidants: Characterization, natural sources, extraction and analysis. *Food Res. Int.* **2015**, *74*, 10–36. [CrossRef] [PubMed]

23. Bjørklund, G.; Chirumbolo, S. Role of oxidative stress and antioxidants in daily nutrition and human health. *Nutrition* **2017**, *33*, 311–321. [CrossRef] [PubMed]

24. Smeriglio, A.; Denaro, M.; Barreca, D.; Calderaro, A.; Bisignano, C.; Ginestra, G.; Bellocco, E.; Trombetta, D. In vitro evaluation of the antioxidant, cytoprotective, and antimicrobial properties of essential oil from *Pistacia vera* L. Variety Bronte Hull. *Int. J. Mol. Sci.* **2017**, *18*, 1212. [CrossRef] [PubMed]

25. do Prado, A.; da Silva, H.; da Silveira, S.; Barreto, P.; Vieira, C.; Maraschin, M.; Ferreira, S.; Block, J. Effect of the extraction process on the phenolic compounds profile and the antioxidant and antimicrobial activity of extracts of pecan nut [*Carya illinoinensis* (Wangenh) C. Koch] shell. *Ind. Crop. Prod.* **2014**, *52*, 552–561. [CrossRef]

26. Rusu, M.E.; Gheldiu, A.-M.; Mocan, A.; Vlase, L.; Popa, D.-S. Anti-aging potential of tree nuts with a focus on phytochemical composition, molecular mechanisms and thermal stability of major bioactive compounds. *Food Funct.* **2018**, *9*, 2554–2575. [CrossRef] [PubMed]

27. Sánchez-González, C.; Ciudad, C.J.; Noé, V.; Izquierdo-Pulido, M. Health benefits of walnut polyphenols: An exploration beyond their lipid profile. *Crit. Rev. Food Sci. Nutr.* **2017**, *57*, 3373–3383. [CrossRef] [PubMed]

28. Vieira, V.; Prieto, M.A.; Barros, L.; Coutinho, J.A.P.; Ferreira, O.; Ferreira, I.C.F.R. Optimization and comparison of maceration and microwave extraction systems for the production of phenolic compounds from *Juglans regia* L. for the valorization of walnut leaves. *Ind. Crops Prod.* **2017**, *107*, 341–352. [CrossRef]

29. Fernández-Agulló, A.; Pereira, E.; Freire, M.S.; Valentão, P.; Andrade, P.B.; González-Álvarez, J.; Pereira, J.A. Influence of solvent on the antioxidant and antimicrobial properties of walnut (*Juglans regia* L.) green husk extracts. *Ind. Crop. Prod.* **2013**, *42*, 126–132. [CrossRef]

30. Dehghani, F.; Mashhoody, T.; Panjehshahin, M. Effect of aqueous extract of walnut septum on blood glucose and pancreatic structure in streptozotocin-induced diabetic mouse. *Iran. J. Pharmacol. Ther.* **2012**, *11*, 10–14.

31. Ramishvili, L.; Gordeziani, M.; Tavdishvili, E.; Bedineishvili, N.; Dzidziguri, D.; Kotrikadze, N. The effect of extract of greek walnut (*Juglans regia* L.) septa on some functional characteristics of erythrocytes. *Georg. Med. News* **2016**, *261*, 51–57.

32. Ravanbakhsh, A.; Mahdavi, M.; Jalilzade-Amin, G.; Javadi, S.; Maham, M.; Mohammadnejad, D.; Rashidi, M.R. Acute and subchronic toxicity study of the median septum of *Juglans regia* in Wistar rats. *Adv. Pharm. Bull.* **2016**, *6*, 541–549. [CrossRef] [PubMed]

33. Christopoulos, M.; Tsantili, E. Storage of fresh walnuts (*Juglans regia* L.)-Low temperature and phenolic compounds. *Postharvest Biol. Technol.* **2012**, *73*, 80–88. [CrossRef]
34. Mocan, A.; Zengin, G.; Simirgiotis, M.; Schafberg, M.; Mollica, A.; Vodnar, D.C.; Crişan, G.; Rohn, S. Functional constituents of wild and cultivated Goji (*L. barbarum* L.) leaves: Phytochemical characterization, biological profile, and computational studies. *J. Enzym. Inhib. Med. Chem.* **2017**, *32*, 153–168. [CrossRef] [PubMed]
35. Mocan, A.; Schafberg, M.; Crisan, G.; Rohn, S. Determination of lignans and phenolic components of *Schisandra chinensis* (Turcz.) Baill. using HPLC-ESI-ToF-MS and HPLC-online TEAC: Contribution of individual components to overall antioxidant activity and comparison with traditional antioxidant assays. *J. Funct. Foods* **2016**, *24*, 579–594. [CrossRef]
36. Price, M.L.; Van Scoyoc, S.; Butler, L.G. A Critical Evaluation of the Vanillin Reaction as an Assay for Tannin in Sorghum Grain. *J. Agric. Food Chem.* **1978**, *26*, 1214–1218. [CrossRef]
37. Alasalvar, C.; Karamać, M.; Amarowicz, R.; Shahidi, F. Antioxidant and antiradical activities in extracts of hazelnut kernel (*Corylus avellana* L.) and hazelnut green leafy cover. *J. Agric. Food Chem.* **2006**, *54*, 4826–4832. [CrossRef] [PubMed]
38. Meda, R.N.; Vlase, L.; Lamien-Meda, A.; Lamien, C.E.; Muntean, D.; Tiperciuc, B.; Oniga, I.; Nacoulma, O.G. Identification and quantification of phenolic compounds from *Balanites aegyptiaca* (L.) Del (Balanitaceae) galls and leaves by HPLC-MS. *Nat. Prod. Res.* **2011**, *25*, 93–99. [CrossRef] [PubMed]
39. Mocan, A.; Vlase, L.; Raita, O.; Hanganu, D.; Paltinean, R.; Dezsi, S.; Gheldiu, A.M.; Oprean, R.; Crisan, G. Comparative studies on antioxidant activity and polyphenolic content of *Lycium barbarum* L. and *Lycium chinense* Mill. leaves. *Pak. J. Pharm. Sci.* **2015**, *28*, 1511–1515. [CrossRef] [PubMed]
40. Pop, C.E.; Pârvu, M.; Arsene, A.L.; Pârvu, A.E.; Vodnar, D.C.; Tarcea, M.; Toiu, A.M.; Vlase, L. Investigation of antioxidant and antimicrobial potential of some extracts from *Hedera helix* L. *Farmacia* **2017**, *65*, 624–629.
41. Babotă, M.; Mocan, A.; Vlase, L.; Crisan, O.; Ielciu, I.; Gheldiu, A.M.; Vodnar, D.C.; Crişan, G.; Pãltinean, R. Phytochemical analysis, antioxidant and antimicrobial activities of *Helichrysum arenarium* (L.) Moench. and *Antennaria dioica* (L.) Gaertn. flowers. *Molecules* **2018**, *23*, 409. [CrossRef] [PubMed]
42. Vlase, L.; Parvu, M.; Parvu, E.A.; Toiu, A. Phytochemical analysis of *Allium fistulosum* L. and *A. ursinum* L. *Dig. J. Nanomater. Biostruct.* **2012**, *8*, 457–467.
43. Toiu, A.; Mocan, A.; Vlase, L.; Pârvu, A.E.; Vodnar, D.C.; Gheldiu, A.M.; Moldovan, C.; Oniga, I. Phytochemical composition, antioxidant, antimicrobial and in vivo anti-inflammatory activity of traditionally used Romanian *Ajuga laxmannii* (Murray) Benth. ("Nobleman's beard"-barba împăratului). *Front. Pharmacol.* **2018**, *9*. [CrossRef] [PubMed]
44. Shahidi, F.; Alasalvar, C.; Liyana-Pathirana, C.M. Antioxidant phytochemicals in hazelnut kernel (*Corylus avellana* L.) and hazelnut byproducts. *J. Agric. Food Chem.* **2007**, *55*, 1212–1220. [CrossRef] [PubMed]
45. Mocan, A.; Fernandes, Â.; Barros, L.; Crişan, G.; Smiljković, M.; Soković, M.; Ferreira, I.C.F.R. Chemical composition and bioactive properties of the wild mushroom: *Polyporus squamosus* (Huds.) Fr: A study with samples from Romania. *Food Funct.* **2018**, *9*, 160–170. [CrossRef] [PubMed]
46. Damiano, S.; Forino, M.; De, A.; Vitali, L.A.; Lupidi, G.; Taglialatela-Scafati, O. Antioxidant and antibiofilm activities of secondary metabolites from *Ziziphus jujuba* leaves used for infusion preparation. *Food Chem.* **2017**, *230*, 24–29. [CrossRef] [PubMed]
47. Masuda, T.; Fujita, N.; Odaka, Y.; Takeda, Y.; Yonemori, S.; Nakamoto, K.; Kuninaga, H. Tyrosinase inhibitory activity of ethanol extracts from medicinal and edible plants cultivated in okinawa and identification of a water-soluble inhibitor from the leaves of *Nandina domestica*. *Biosci. Biotechnol. Biochem.* **2007**, *71*, 2316–2320. [CrossRef] [PubMed]
48. Hilbig, J.; Alves, V.R.; Müller, C.M.O.; Micke, G.A.; Vitali, L.; Pedrosa, R.C.; Block, J.M. Ultrasonic-assisted extraction combined with sample preparation and analysis using LC-ESI-MS/MS allowed the identification of 24 new phenolic compounds in pecan nut shell [*Carya illinoinensis* (Wangenh) C. Koch] extracts. *Food Res. Int.* **2018**, *106*, 549–557. [CrossRef] [PubMed]
49. Bolling, B.W. Almond Polyphenols: Methods of Analysis, Contribution to Food Quality, and Health Promotion. *Compr. Rev. Food Sci. Food Saf.* **2017**, *16*, 346–368. [CrossRef]

50. Esposito, T.; Sansone, F.; Franceschelli, S.; Del Gaudio, P.; Picerno, P.; Aquino, R.P.; Mencherini, T. Hazelnut (*Corylus avellana* L.) Shells Extract: Phenolic Composition, Antioxidant Effect and Cytotoxic Activity on Human Cancer Cell Lines. *Int J. Mol. Sci.* **2017**, *18*, 392. [CrossRef] [PubMed]

51. Prgomet, I.; Gonçalves, B.; Domínguez-Perles, R.; Pascual-Seva, N.; Barros, A.I.R.N. Valorization Challenges to Almond Residues: Phytochemical Composition and Functional Application. *Molecules* **2017**, *22*, 1774. [CrossRef] [PubMed]

52. Albuquerque, B.R.; Prieto, M.A.; Vazquez, J.A.; Barreiro, M.F.; Barros, L.; Ferreira, I.C.F.R. Recovery of bioactive compounds from *Arbutus unedo* L. fruits: Comparative optimization study of maceration/microwave/ultrasound extraction techniques. *Food Res. Int.* **2018**, *109*, 455–471. [CrossRef] [PubMed]

53. Bolling, B.W.; McKay, D.L.; Blumberg, J.B. The phytochemical composition and antioxidant actions of tree nuts. *Asia Pac. J. Clin. Nutr.* **2010**, *19*, 117–123. [PubMed]

54. Alasalvar, C.; Bolling, B. Review of nut phytochemicals, fat-soluble bioactives, antioxidant components and health effects. *Br. J. Nutr.* **2015**, *113*, S68–S78. [CrossRef] [PubMed]

55. Carvalho, M.; Ferreira, P.J.; Mendes, V.S.; Silva, R.; Pereira, J.A.; Jerónimo, C.; Silva, B.M. Human cancer cell antiproliferative and antioxidant activities of *Juglans regia* L. *Food Chem. Toxicol.* **2010**, *48*, 441–447. [CrossRef] [PubMed]

56. Akbari, V.; Jamei, R.; Heidari, R.; Esfahlan, J.A. Antiradical activity of different parts of Walnut (*Juglans regia* L.) fruit as a function of genotype. *Food Chem.* **2012**, *135*, 2404–2410. [CrossRef] [PubMed]

57. Shah, U.N.; Mir, J.I.; Ahmed, N.; Jan, S.; Fazili, K.M. Bioefficacy potential of different genotypes of walnut *Juglans regia* L. *J. Food Sci. Technol.* **2018**, *55*, 605–618. [CrossRef] [PubMed]

58. Santos, A.; Barros, L.; Calhelha, R.C.; Dueñas, M.; Carvalho, A.M.; Santos-Buelga, C.; Ferreira, I.C.F.R. Leaves and decoction of *Juglans regia* L.: Different performances regarding bioactive compounds and in vitro antioxidant and antitumor effects. *Ind. Crop. Prod.* **2013**, *51*, 430–436. [CrossRef]

59. Mocan, A.; Diuzheva, A.; Carradori, S.; Andruch, V.; Massafra, C.; Moldovan, C.; Sisea, C.; Petzer, J.P.; Petzer, A.; Zara, S.; et al. Development of novel techniques to extract phenolic compounds from Romanian cultivars of *Prunus domestica* L. and their biological properties. *Food Chem. Toxicol.* **2018**. [CrossRef] [PubMed]

60. de la Rosa, L.A.; Alvarez-Parrilla, E.; Shahidi, F. Phenolic compounds and antioxidant activity of kernels and shells of Mexican pecan (*Carya illinoinensis*). *J. Agric. Food Chem.* **2011**, *59*, 152–162. [CrossRef] [PubMed]

61. Lin, J.T.; Liu, S.C.; Hu, C.C.; Shyu, Y.S.; Hsu, C.Y.; Yang, D.J. Effects of roasting temperature and duration on fatty acid composition, phenolic composition, Maillard reaction degree and antioxidant attribute of almond (*Prunus dulcis*) kernel. *Food Chem.* **2016**, *190*, 520–528. [CrossRef] [PubMed]

62. Lainas, K.; Alasalvar, C.; Bolling, B.W. Effects of roasting on proanthocyanidin contents of Turkish Tombul hazelnut and its skin. *J. Funct. Foods* **2016**, *23*, 647–653. [CrossRef]

63. Chang, S.K.; Alasalvar, C.; Bolling, B.W.; Shahidi, F. Nuts and their co-products: The impact of processing (roasting) on phenolics, bioavailability, and health benefits-A comprehensive review. *J. Funct. Foods* **2016**, *26*, 88–122. [CrossRef]

64. Payne, M.J.; Hurst, W.J.; Miller, K.B.; Rank, C.; Stuart, D.A. Impact of fermentation, drying, roasting, and dutch processing on epicatechin and catechin content of cacao beans and cocoa ingredients. *J. Agric. Food Chem.* **2010**, *58*, 10518–10527. [CrossRef] [PubMed]

65. Martinez, M.L.; Labuckas, D.O.; Lamarque, A.L.; Maestri, D.M. Walnut (*Juglans regia* L.): Genetic resources, chemistry, by-products. *J. Sci. Food Agric.* **2010**, *90*, 1959–1967. [CrossRef] [PubMed]

66. Schlörmann, W.; Birringer, M.; Böhm, V.; Löber, K.; Jahreis, G.; Lorkowski, S.; Müller, A.K.; Schöne, F.; Glei, M. Influence of roasting conditions on health-related compounds in different nuts. *Food Chem.* **2015**, *180*, 77–85. [CrossRef] [PubMed]

67. Figueroa, F.; Marhuenda, J.; Zafrlla, P.; Martínez-Cachá, A.; Mulero, J.; Cerdá, B. Total phenolics content, bioavailability and antioxidant capacity of 10 different genotypes of walnut (*Juglans regia* L.). *J. Food Nutr. Res.* **2016**, *55*, 229–236.

68. Alasalvar, C.; Karamać, M.; Kosińska, A.; Rybarczyk, A.; Shahidi, F.; Amarowicz, R. Antioxidant activity of hazelnut skin phenolics. *J. Agric. Food Chem.* **2009**, *57*, 4645–4650. [CrossRef] [PubMed]

69. Taş, N.G.; Gökmen, V. Phenolic compounds in natural and roasted nuts and their skins: A brief review. *Curr. Opin. Food Sci.* **2017**, *14*, 103–109. [CrossRef]

70. de la Rosa, L.A.; Vazquez-Flores, A.A.; Alvarez-Parrilla, E.; Rodrigo-García, J.; Medina-Campos, O.N.; Ávila-Nava, A.; González-Reyes, S.; Pedraza-Chaverri, J. Content of major classes of polyphenolic compounds, antioxidant, antiproliferative, and cell protective activity of pecan crude extracts and their fractions. *J. Funct. Foods* **2014**, *7*, 219–228. [CrossRef]

71. Arcan, I.; Yemeniciog, A. Antioxidant activity and phenolic content of fresh and dry nuts with or without the seed coat. *J. Food Compost. Anal.* **2009**, *22*, 184–188. [CrossRef]

72. Delgado, T.; Malheiro, R.; Pereira, J.A.; Ramalhosa, E. Hazelnut (*Corylus avellana* L.) kernels as a source of antioxidants and their potential in relation to other nuts. *Ind. Crop. Prod.* **2010**, *32*, 621–626. [CrossRef]

73. Slatnar, A.; Mikulic-Petkovsek, M.; Stampar, F.; Veberic, R.; Solar, A. HPLC-MS^{n} identification and quantification of phenolic compounds in hazelnut kernels, oil and bagasse pellets. *Food Res. Int.* **2014**, *64*, 783–789. [CrossRef] [PubMed]

74. Panth, N.; Paudel, K.R.; Karki, R. Phytochemical profile and biological activity of *Juglans regia*. *J. Integr. Med.* **2016**, *14*, 359–373. [CrossRef]

75. Christopoulos, M.V.; Tsantili, E. Effects of temperature and packaging atmosphere on total antioxidants and colour of walnut (*Juglans regia* L.) kernels during storage. *Sci. Hort.* **2011**, *131*, 49–57. [CrossRef]

76. Chen, C.Y.; Blumberg, J.B. Phytochemical composition of nuts. *Asia Pac. J. Clin. Nutr.* **2008**, *17*, 329–332. [PubMed]

77. Wong, W.H.; Lee, W.X.; Ramanan, R.N.; Tee, L.H.; Kong, K.W.; Galanakis, C.M.; Sun, J.; Prasad, K.N. Two level half factorial design for the extraction of phenolics, flavonoids and antioxidants recovery from palm kernel by-product. *Ind. Crop. Prod.* **2015**, *63*, 238–248. [CrossRef]

78. Fu, M.; Qu, Q.; Yang, X.; Zhang, X. Effect of intermittent oven drying on lipid oxidation, fatty acids composition and antioxidant activities of walnut. *LWT-Food Sci. Technol.* **2016**, *65*, 1126–1132. [CrossRef]

79. Zhang, Z.; Liao, L.; Moore, J.; Wu, T.; Wang, Z. Antioxidant phenolic compounds from walnut kernels (*Juglans regia* L.). *Food Chem.* **2009**, *113*, 160–165. [CrossRef]

80. Hooper, L.; Kay, C.; Abdelhamid, A.; Kroon, P.A.; Cohn, J.S.; Rimm, E.B.; Cassidy, A. Effects of chocolate, cocoa, and flavan-3-ols on cardiovascular health: A systematic review and meta-analysis of randomized trials. *Am. J. Clin. Nutr.* **2012**, *95*, 740–751. [CrossRef] [PubMed]

81. Ellinger, S.; Reusch, A.; Stehle, P.; Helfrich, H.P. Epicatechin ingested via cocoa products reduces blood pressure in humans: A nonlinear regression model with a Bayesian approach. *Am. J. Clin. Nutr.* **2012**, *95*, 1365–1377. [CrossRef] [PubMed]

82. Khalesi, S.; Sun, J.; Buys, N.; Jamshidi, A.; Nikbakht-Nasrabadi, E.; Khosravi-Boroujeni, H. Green tea catechins and blood pressure: A systematic review and meta-analysis of randomised controlled trials. *Eur. J. Nutr.* **2014**, *53*, 1299–1311. [CrossRef] [PubMed]

83. Pei, J.; Chen, A.; Zhao, L.; Cao, F.; Ding, G.; Xiao, W. One-pot synthesis of hyperoside by a three-enzyme cascade using a UDP-galactose regeneration system. *J. Agric. Food Chem.* **2017**, *65*, 6042–6048. [CrossRef] [PubMed]

84. Biswas, R.; Mukherjee, P.K.; Chaudhary, S.K. Tyrosinase inhibition kinetic studies of standardized extract of *Berberis aristata*. *Nat. Prod. Res.* **2016**, *30*, 1451–1454. [CrossRef] [PubMed]

85. Tan, X.; Song, Y.H.; Park, C.; Lee, K.W.; Kim, J.Y.; Kim, D.W.; Kim, K.D.; Lee, K.W.; Curtis-Long, M.J.; Park, K.H. Highly potent tyrosinase inhibitor, neorauflavane from *Campylotropis hirtella* and inhibitory mechanism with molecular docking. *Bioorg. Med. Chem.* **2016**, *24*, 153–159. [CrossRef] [PubMed]

86. Malik, W.; Ahmed, D.; Izhar, S. Tyrosinase Inhibitory Activities of *Carissa opaca* Stapf ex Haines Roots Extracts and Their Phytochemical Analysis. *Pharmacogn. Mag.* **2017**, *13*, S544–S548. [CrossRef] [PubMed]

87. Uysal, S.; Zengin, G.; Aktumsek, A.; Karatas, S. Chemical and biological approaches on nine fruit tree leaves collected from the Mediterranean region of Turkey. *J. Funct. Foods* **2016**, *22*, 518–532. [CrossRef]

88. Quispe, Y.N.; Hwang, S.H.; Wang, Z.; Lim, S.S. Screening of Peruvian medicinal plants for tyrosinase inhibitory properties: Identification of tyrosinase inhibitors in *Hypericum laricifolium* Juss. *Molecules* **2017**, *22*, 402. [CrossRef] [PubMed]

89. Savran, A.; Zengin, G.; Aktumsek, A.; Mocan, A.; Glamoćlija, J.; Ćirić, A.; Soković, M. Phenolic compounds and biological effects of edible *Rumex scutatus* and *Pseudosempervivum sempervivum*: Potential sources of natural agents with health benefits. *Food Funct.* **2016**, *7*, 3252–3262. [CrossRef] [PubMed]

90. Mocan, A.; Moldovan, C.; Zengin, G.; Bender, O.; Locatelli, M.; Simirgiotis, M.; Atalay, A.; Vodnar, D.C.; Rohn, S.; Crişan, G. UHPLC-QTOF-MS analysis of bioactive constituents from two Romanian Goji (*Lycium barbarum* L.) berries cultivars and their antioxidant, enzyme inhibitory, and real-time cytotoxicological evaluation. *Food Chem. Toxicol.* **2018**, *115*, 414–424. [CrossRef] [PubMed]

Sample Availability: Samples of the extracts are not available from the authors.

molecules

MDPI

Communication

Ultra-High-Performance Liquid Chromatography Tandem Mass Spectrometry Assay for Determination of Endogenous GHB and GHB-Glucuronide in Nails

Francesco Paolo Busardò [1],*, Massimo Gottardi [2], Anastasio Tini [3], Claudia Mortali [4], Raffaele Giorgetti [1] and Simona Pichini [4]

[1] Dep. of Excellence-Biomedical Sciences and Public Health, University "Politecnica delle Marche" of Ancona, 60020 Ancona, Italy; r.giorgetti@univpm.it
[2] Comedical S.r.L., 38100 Trento, Italy; Massimo.gottardi@comedical.biz
[3] Unit of Forensic Toxicology (UoFT), Department of Anatomical, Histological, Forensic and Orthopedic Sciences, Sapienza University of Rome, 00167 Rome, Italy; Anastasio.tini78@gmail.com
[4] National Centre on Addiction and Doping, Istituto Superiore di Sanità, 00167 Rome, Italy; claudia.mortali@iss.it (C.M.); simona.pichini@iss.it (S.P.)
* Correspondence: fra.busardo@libero.it; Tel.: +39-071-220-6212

Received: 29 September 2018; Accepted: 15 October 2018; Published: 18 October 2018

Abstract: Background: The short chain fatty acid gamma-hydroxybutyric acid (GHB) is a precursor, and the metabolite of gamma-aminobutyric acid is commonly used as an illegal recreational drug of abuse. **Methods**: An ultra-high-performance liquid chromatography tandem mass spectrometry was developed and validated for endogenous GHB and its glucuronide in nails, to complement hair in forensic contexts for a retrospective detection of psychotropic drugs consumption. **Results**: GHB endogenous values for children and adolescents, adult females, and adult males in fingernails ranged from 0.3 to 3.0, 3.2, and 3.8 ng/mg, respectively, and toenails values ranged from 0.3 to 1.8, 2.0, and 2.4 ng/mg, respectively. In the three different groups, values of GHB in fingernails were statistically higher than those in toenails. GHB glucuronide could only be detected in finger nails with values ranging from 0.08 to 0.233, 0.252 and 0.243 in children and adolescents, adult females and adult males, respectively. **Conclusions**: The validated method was efficaciously applied to real finger and toe nails specimens from a population of males and females non GHB consumers. A preliminary cut-off of 5.0 ng/mg nail for endogenous GHB and 0.5 ng/mg for endogenous GHB-Gluc in the general population was proposed.

Keywords: ultra-high-performance liquid chromatography tandem mass spectrometry; GHB; GHB glucuronide; nails; endogenous values

---

1. Introduction

The short chain fatty acid gamma-hydroxybutyric acid (GHB) is a precursor and metabolite of gamma-aminobutyric acid (GABA) and behaves as an inhibitory neurotransmitter in the central nervous system.

Its sodium salt, sodium oxybate, is approved as an adjuvant medication for detoxification and withdrawal of alcohol dependence (Alcover®) in some countries and for the treatment of narcolepsy-associated cataplexy (Xyrem®) [1].

Nevertheless, the most common use of GHB is as an illegal recreational "club" drug marketed for its ability to produce euphoria and sexual arousal [2]. In particular, this drug is gaining importance in combination with other psychoactive and non-psychoactive drugs such as mephedrone, methamphetamine, erectile dysfunction agents and alkyl nitrites (or poppers) in the context of "chemsex": intentional or non-intentional intake of certain psychoactive and non-psychoactive drugs in the context of rave parties eventually followed by sexual encounters with the aim of aiding and/or enhancing the sexual relationship, mainly in homosexual settings [3,4].

The dual nature of endogenous neurotransmitter and exogenous pharmacologically active compound makes the proof of GHB intake a difficult assignment [5].

Cut-offs have been proposed for traditional biological matrices (e.g., blood and urine) to objectively discriminate exogenous drug consumption from endogenous values in antemortem and post-mortem samples [6]. However, in both biological fluids, GHB presents a short window of detection (around 5 h blood and less than 12 h urine) [7,8], so GHB and its glucuronide (GHB-Gluc) have been investigated in hair as potential biomarkers of GHB single and repeated intake [5]. For this purpose, baseline hair GHB values in the general population have been established to distinguish even a single intake, e.g., in GHB-facilitated sexual assaults [5]. However, the currently accepted approach to documenting a single GHB exposure in hair suggests "to use each subject as its own control" [5].

Nail testing can be accomplished to replace and/or complement hair testing in forensic contexts for the retrospective detection of psychotropic drug consumption. As the nail grows, ingested xenobiotics are incorporated into the keratin matrix, where they can be gathered for protracted periods of time (3–5 months in fingernails, and 8–14 months in toenails), allowing retrospective detection of drug consumption [9,10].

Considering that GHB has never been investigated in nails, we sought to develop and validate an easily applicable and fast ultra-high-performance liquid chromatography tandem mass spectrometry (UHPLC–MS/MS) method with rapid sample preparation for the determination of endogenous GHB and GHB-Gluc in nails. Secondly, finger and toe nails were compared to look for any eventual difference in GHB endogenous concentrations.

2. Results and Discussion

2.1. UHPLC–MS/MS and Validation Parameters

Representative chromatograms obtained following the extraction of GHB and GHB-gluc from child finger nails, female finger nails and male finger nails are shown in Figure 1A–C. Retention times of the two analytes were: 1.81 min for GHB-gluc and 1.87 min for GHB. A chromatographic run was completed in 10 min.

Determination coefficients (r^2) of calibration curves were equal to or higher than 0.99 in all cases. Calculated LOD and LOQ values were suitable for the aim of the present study (Table 1). Precision and accuracy of LOQs always showed coefficients of variation lower than 20%. For all the other QC samples, the intra- and inter-assay precision and accuracy values met the internationally established acceptance criteria (Table 2) [11,12].

Table 1. Calibration method for GHB and GHB-Glucuronide in nails.

Analyte	Calibration Line		Determination Coefficient (r^2)	LOD [a]	LOQ [a]
	Slope [a]	Intercept [a]		ng/mg	ng/mg
GHB	0.112 ± 0.0015	0.007 ± 0.023	0.997 ± 0.002	0.10	0.30
GHB-Glucuronide	0.171 ± 0.01	0.031 ± 0.04	0.996 ± 0.003	0.02	0.07

[a] Mean and S.D. of five replicates on three subsequent working days.

No significant analyte degradation was noted after one and three months storage of nails at room temperature, with differences being less than 10% from the initial GHB and GHB-Gluc concentration.

There was no carryover detected when injecting drug-free samples after the calibration curve's highest point. No additional peaks from eventual endogenous substances from the keratin matrix were observed following the injection of drug-free nails. Similarly, none of the most common psychoactive drugs (cannabinoid, cocaine, opiates, amphetamines type-stimulants) or common psychoactive medications (e.g., benzodiazepines and antidepressants) interfered with the assay. No significant ion suppression (less than 10% analytical signal suppression) due to matrix effect occurred during chromatographic runs.

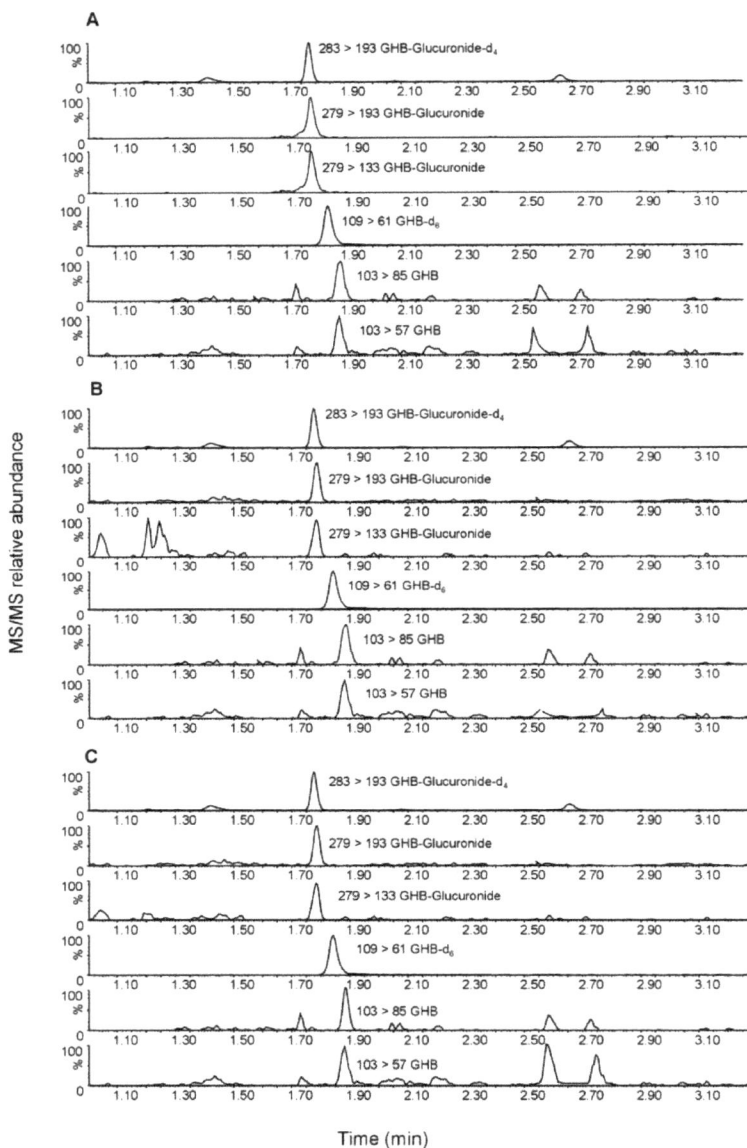

Figure 1. UHPLC–MS/MS chromatogram of an extract of (**A**) child finger nails containing 0.91 ng/mg GHB, 0.15 ng/mg GHB-gluc and 0.5 ng/mg ISs; (**B**) female finger nails containing 1.42 ng/mg GHB and 0.21 ng/mg GHB-gluc; (**C**) male fingernail containing 1.63 ng/mg GHB, 0.92 ng/mg GHB-gluc and 0.5 ng/mg ISs.

Table 2. Intra-assay (n = 5) and inter-assay (n = 15) precision, accuracy (n = 15) and recovery of GHB and GHB-Glucuronide in nails.

Analyte	Intra-Assay Precision (% CV)			Inter-Assay Precision (% CV)			Accuracy (% Error)			Recovery (%)
	Low QC *	Medium QC *	High QC *	Low QC	Medium QC	High QC	Low QC	Medium QC	High QC	
GHB	2.4	2.3	2.0	3.6	4.1	2.0	5.0	3.8	2.5	96.5
GHB-Glucuronide	2.1	0.5	1.2	2.4	1.0	1.1	8.4	3.4	2.0	87.8

* Low QC: 1 ng/mg GHB and 0.1 ng/mg GHB-Glucuronide. Medium QC: 5 ng/mg GHB and 0.5 ng/mg GHB-Glucuronide. High QC: 8 ng/mg GHB and 0.8 ng/mg GHB-Glucuronide.

2.2. Analysis of Biological Samples

As reported above, the validated UHPLC–MS/MS method was used to measure endogenous values of GHB and GHB-Gluc in finger and toe nails collected from 30 children and adolescents, 30 adult females and 30 adult males (Table 3). From the obtained results, it can be said that in the case of GHB, similar values were measured for children and adolescents, adult females and adult males, with fingernail values ranging from 0.3 to 3.0, 3.2 and 3.8 ng/mg, respectively, and toenails values ranging from 0.3 to 1.8, 2.0 and 2.4 ng/mg, respectively. In the three different groups, values of GHB in fingernails were statistically higher than those in toenails. GHB glucuronide could only be detected in finger nails with values ranging from 0.08 to 0.233, 0.252 and 0.243 in children and adolescents, adult females and adult males, respectively. With these values available, we can propose a preliminary cut-off of 5.0 ng/mg nail for endogenous GHB and 0.5 ng/mg nail for GHB-gluc in the general population. This value necessarily has to be substantiated in a higher number of individuals with different age, gender and ethnicity.

Table 3. GHB and GHB-gluc mean values in finger and toe nails collected from 30 children and adolescents, 30 adult females and 30 adult males.

	Children and Adolescents		Adult Females		Adult Males	
NAILS GHB (ng/mg nail)	Fingernails	Toenails	Fingernails	Toenails	Fingernails	Toenails
MIN	0.30	0.30	0.30	0.30	0.30	0.30
MAX	3.00	1.80	3.20	2.00	3.80	2.40
MEAN	1.51	0.91	1.62	0.96	1.78	1.13
SD	0.87	0.42	0.80	0.44	0.99	0.58
p value Fingernails vs. toenails	<0.01		<0.001		<0.01	
NAILS GHB-gluc (ng/mg nail)						
MIN	0.082	ND	0.080	ND	0.082	ND
MAX	0.233	ND	0.252	ND	0.243	ND
MEAN	0.160	ND	0.153	ND	0.166	ND
SD	0.038	ND	0.048	ND	0.038	ND

MIN: minimum value. MAX: maximum value. SD: standard deviation. ND: not detected (under LOD value).

3. Material and Methods

3.1. Chemicals and Materials

GHB, pure standard (>99%) was purchased from Sigma-Aldrich (Milan, Italy). GHB-d_6, used as internal standard (IS), was supplied as a methanolic solution by Sigma-Aldrich (Milan, Italy). Standards of the O-glucuronide derivative of GHB (GHB-Gluc) and its deuterium-labeled IS (GHB-Gluc-d_4) were synthesized by Pedersen et al. [13] and provided by the Department of Drug Design and Pharmacology, Faculty of Health and Medical Sciences, University of Copenhagen (Prof. D.S. Pedersen). Deuterium-labeled GBH sodium salt and its glucuronide are reported in Figure 2.

deuterium labelled GBH sodium salt

deuterium labelled GBH glucuronide

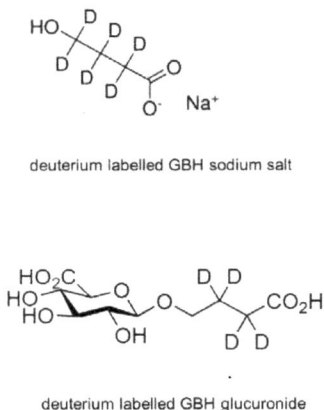

Figure 2. Deuterium labelled GBH sodium salt and deuterium-labeled GHB glucuronide.

VMA-TM3 (acidic aqueous buffer) reagent for nails digestion, diluent, washing solution and multimatrix eluent were obtained from Comedical s.r.l. (Trento, Italy). Oasis PRiME HLB solid phase extraction columns were from Waters, Milano, Italy. Ultrapure water and all other reagents of UHPLC–MS grade were acquired from Sigma-Aldrich (Milan, Italy).

3.2. Calibration Standards and Quality Control Samples

Methanolic standard solutions (10 and 1 mg/mL) and working solutions (10 and 1 µg/mL) of GHB and GHB-Gluc were stored at −20 °C.

Calibration standards of GHB and GHB-Gluc from limit of quantification (LOQ) to 10 ng GHB and 1 ng/mg GHB-Gluc per mg nails were prepared by daily spiking blank neonates nails to test linearity for each analytical batch.

Quality Control (QC) samples at three concentrations (low, medium, high) spanning the linear dynamic ranges of the calibration curves were also prepared by daily spiking blank nails with volumes of GHB and GHB-Gluc standard solutions appropriate to each analytical batch in order to check validation parameters (e.g., accuracy, precision, recovery, etc.) [11,12].

3.3. Biological Sample Collection and Preparation

GHB-free nails were generously donated by personnel of the research institutions participating in the study and by their relatives. Nail donors signed an informed consent for themselves and their children, when the latter also provided their nails. Since nails are waste material, spontaneously donated by participants, no ethical approval was required.

In detail, finger and toe nails were collected from 30 children and adolescents (range 5–16 years), 30 adult females (range 22–56 years) and 30 adult males (21–54 years). Nails were stored in paper envelopes at ambient temperatures until analysis. Nails were clipped as close to the nail bed as comfortable over a clean sheet of paper to collect the clippings, in order to obtain 2–3 mm of clippings from each of the 10 digits. Neonate nails, used as blank-free matrix, were donated by the Clinic Hospital of Barcelona as discharged material.

Nails were firstly washed with 2 mL dichloromethane and dried under nitrogen. A 25 mg sample was combined with 5 µL mixture of ISs (0.5 ng/mg), 0.5 mL Comedical VMA-TM3® Reagent and kept at 100 °C for 1 h. Nails were then discarded, and the liquid mixture underwent solid phase extraction with Oasis PRiME HLB. Specifically, 0.5 mL mixture added to 0.5 mL Comedical Diluent® was loaded in an HLB column, washed with 0.5 mL Comedical Washing Solution®, dried under nitrogen, and eluted with 0.5 mL Comedical Multimatrix Eluent®. The eluted mixture was diluted 1:10 with water and 1 µL injected into UHPLC-MS/MS system.

3.4. UHPLC–MS/MS

A Waters Acquity UPLC I-Class chromatograph (Waters, Milano, Italy) was used interfaced to a TQ-S micro triple quadrupole mass spectrometer (Waters). Analytes separation was achieved using an HSS T3 reversed phase C18 column (1.8 µm, 2.1 × 150 mm) from Waters.

A mixture of 95% solvent A (0.1% formic acid in water) and 5% solvent B (methanol) was used as mobile phase in isocratic conditions at a flow rate of 0.35 mL/min for 2 min. Then, the column was washed for 5 min, bringing solvent A to 5%, and then the mobile phase was re-equilibrated to initial conditions in 3 min.

A triple quadrupole mass spectrometer was used to detect analytes operating in multiple reaction monitoring (MRM) mode via negative electrospray ionization (ESI) The applied ESI conditions were as reported: capillary voltage -2.0 kV, desolvation temperature 650 °C, source temperature 150 °C, cone gas flow rate 20 L/h, desolvation gas flow rate 1200 L/h and collision gas flow rate 0.13 mL min^{-1}. Cone voltage and collision energy were 20 V to 45 V. MRM transitions were the following: for GHB m/z 103 > 57 and m/z 103 > 85; for GHB-glucuronide m/z 279 > 193 and m/z 279 > 133; for GHB-d$_6$ 109 > 61 and for GHB-Gluc-d$_4$ 283 > 193. Transitions in bold were used for quantification. MS/MS spectra of GHB and GHB-Gluc are reported in Figure 3.

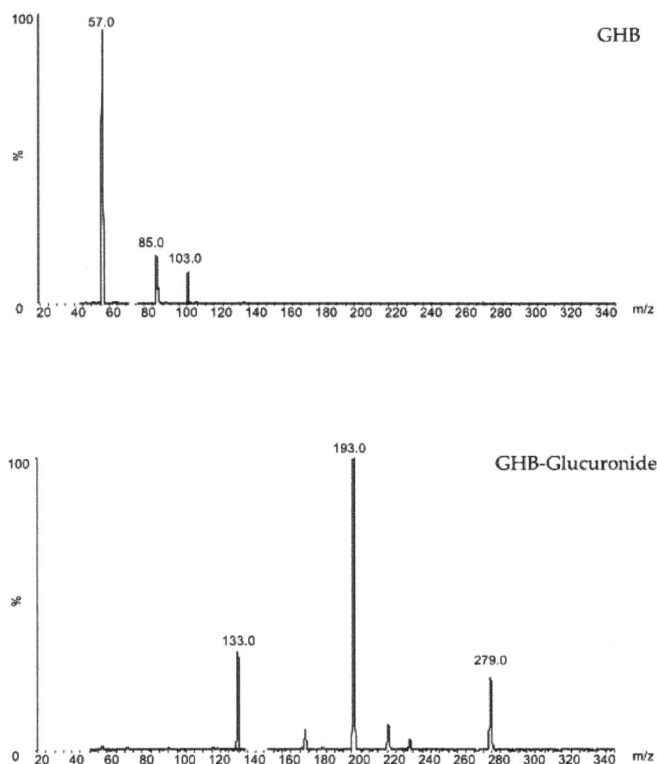

Figure 3. MS/MS spectra of GHB and GHB-Gluc.

3.5. Validation of Analytical Method

The method developed following international criteria [11] was tested in a validation protocol following the most recent standard practices [12]. Linearity, limits of detection and quantification, accuracy, precision, recovery selectivity, carryover, stability and matrix effect were determined as previously described [14]. Five different daily replicates of calibration points and QC samples (low, medium, and high QCs) were used to calculate validation parameters along three successive working days. The above reported samples were obtained by spiking opportune concentrations of analytes under investigation and ISs in neonates nails, which were pre-analyzed and did not show measurable concentrations of GHB and GHB-Gluc.

4. Conclusions

We described a UHPLC–MS/MS assay which for the first time allowed the simultaneous identification and quantification of GHB and GHB-Glu in nails. The method was validated and applied to the determination of endogenous GHB and GHB-Gluc values in children and adolescent, adult women and adult males. For the first time, baseline values of GHB in nails have been preliminarily suggested, prompting investigation on a larger number of individuals from the general population of non GHB consumers.

Author Contributions: Conceptualization, F.P.B., S.P. and M.G.; Methodology, M.G., A.T. and C.M.; Software, M.G., A.T. and C.M.; Validation, M.G., A.T. and C.M.; Formal Analysis, M.G., A.T. and C.M.; Investigation, F.P.B., S.P. and M.G.; Resources, F.P.B., S.P. and M.G. Data Curation, M.G., A.T. and C.M.; Writing-Original Draft Preparation F.P.B., S.P. and M.G.; Writing-Review & Editing, F.P.B., S.P. A.T., C.M. and M.G.; Visualization, F.P.B., S.P. A.T., C.M. and M.G.; Supervision, C.M.; Project Administration, F.P.B., S.P. A.T., C.M. and M.G.; Funding Acquisition, F.P.B., S.P. and M.G.

Funding: The investigation was carried out with intramural funding of National Centre on Addiction and doping and Section of Legal Medicine, Università Politecnica delle Marche.

Financial and Competing Interest Disclosure: No authors have competing interests in this study.

Acknowledgments: The authors would like to thank Comedical S.r.L. and Emilia Marchei for technical assistance.

Conflicts of Interest: The authors declare no conflict of interest.

References

1. Busardo, F.P.; Jones, A.W. GHB Pharmacology and Toxicology: Acute Intoxication, Concentrations in Blood and Urine in Forensic Cases and Treatment of the Withdrawal Syndrome. *Curr. Neuropharmacol.* **2015**, *13*, 47–70. [CrossRef] [PubMed]

2. Bellis, M.A.; Hughes, K.; Bennett, A.; Thomson, R. The role of an international nightlife resort in the proliferation of recreational drugs. *Addiction* **2003**, *98*, 1713–1721. [CrossRef] [PubMed]

3. Giorgetti, R.; Tagliabracci, A.; Schifano, F.; Zaami, S.; Marinelli, E.; Busardò, F.P. When "Chems" Meet Sex: A Rising Phenomenon Called "ChemSex". *Curr. Neuropharmacol.* **2017**, *15*, 762–770. [CrossRef] [PubMed]

4. Pichini, S.; Marchei, E.; Pacifici, R.; Marinelli, E.; Busardò, F.P. Chemsex intoxication involving sildenafil as an adulterant of GHB. *Drug Test. Anal.* **2017**, *9*, 956–959. [CrossRef] [PubMed]

5. Busardò, F.P.; Pichini, S.; Zaami, S.; Pacifici, R.; Kintz, P. Hair testing of GHB: An everlasting issue in forensic toxicology. *Clin. Chem. Lab. Med.* **2018**, *56*, 198–208. [CrossRef] [PubMed]

6. Busardò, F.P.; Kyriakou, C. GHB in Biological Specimens: Which Cut-off Levels Should Be Taken into Consideration in Forensic Toxicological Investigation? *Recent Pat. Biotechnol.* **2014**, *8*, 206–214. [CrossRef] [PubMed]

7. Kintz, P.; Goulle, J.P.; Cirimele, V.; Ludes, B. Window of detection of gamma-hydroxybutyrate in blood and saliva. *Clin. Chem.* **2001**, *47*, 2033–2034. [PubMed]

8. Verstraete, AG. Detection times of drugs of abuse in blood, urine, and oral fluid. *Ther. Drug Monit.* **2004**, *26*, 200–205. [CrossRef] [PubMed]

9. Solimini, R.; Minutillo, A.; Kyriakou, C.; Pichini, S.; Pacifici, R.; Busardò, F.P. Nails in Forensic Toxicology: An Update. *Curr. Pharm. Des.* **2017**, *23*, 5468–5479. [CrossRef] [PubMed]

10. Palmeri, A.; Pichini, S.; Pacifici, R.; Zuccaro, P.; Lopez, A. Drugs in nails: Physiology, pharmacokinetics and forensic toxicology. *Clin. Pharmacokinet.* **2000**, *38*, 95–110. [CrossRef] [PubMed]

11. Wille, S.M.R.; Coucke, W.; De Baere, T.; Peters, F.T. Update of Standard Practices for New Method Validation in Forensic Toxicology. *Curr. Pharm. Des.* **2017**, *23*, 5442–5454. [CrossRef] [PubMed]

12. Peters, F.T.; Wissenbach, D.K.; Busardò, F.P.; Marchei, E.; Pichini, S. Method Development in Forensic Toxicology. *Curr. Pharm. Des.* **2017**, *23*, 5455–5467. [CrossRef] [PubMed]

13. Nymann Petersen, I.; Langgaard Kristensen, J.; Tortzen, C.; Breindahl, T.; Sejer Pedersen, D. Synthesis and stability study of a new major metabolite of gamma-hydroxybutyric acid. *Beilstein J. Org. Chem.* **2013**, *9*, 641–646. [CrossRef] [PubMed]

14. Busardò, F.P.; Kyriakou, C.; Marchei, E.; Pacifici, R.; Pedersen, D.S.; Pichini, S. Ultra-high performance liquid chromatography tandem mass spectrometry (UHPLC-MS/MS) for determination of GHB, precursors and metabolites in different specimens: Application to clinical and forensic cases. *J. Pharm. Biomed. Anal.* **2017**, *137*, 123–131. [CrossRef] [PubMed]

Sample Availability: Samples of the compounds are no more available from the authors. They have been used for analysis.

molecules

MDPI

Article

Optimization of Flavonoids Extraction Process in *Panax notoginseng* Stem Leaf and a Study of Antioxidant Activity and Its Effects on Mouse Melanoma B16 Cells

Chun-Yan Dai [1,2,3,4,†], **Peng-Fei Liu** [1,2,3,4,†], **Pei-Ran Liao** [1,2,3,4], **Yuan Qu** [1,2,3,4], **Cheng-Xiao Wang** [1,2,3,4], **Ye Yang** [1,2,3,4,*] and **Xiu-Ming Cui** [1,2,3,4,*]

[1] College of Life Science and Technology, Kunming University of Science and Technology, Kunming 650500, China; daichunyankm@126.com (C.-Y.D.); ragnarok928@sina.com (P.-F.L.); westpp@126.com (P.-R.L.); quyuan2001@126.com (Y.Q.); wcx1192002@126.com (C.-X.W.)
[2] Yunnan Key Laboratory of Sustainable Utilization of Panax Notoginseng, Kunming 650500, China
[3] Laboratory of Sustainable Utilization of Panax Notoginseng Resources, State Administration of Traditional Chinese Medicine, Kunming 650500, China
[4] University Based Provincial Key Laboratory of Screening and Utilization of Targeted Drugs, Kunming 650500, China
[*] Correspondence: yangyekm@163.com or yangye@kmust.edu.cn (Y.Y.); sanqi37@vip.sina.com (X.-M.C.); Tel.: +86-183-8715-6001 (Y.Y.); +86-183-871-86037 (X.-M.C.)
[†] These authors contribute equally to this work.

Academic Editors: Marcello Locatelli, Simone Carradori and Andrei Mocan
Received: 7 June 2018; Accepted: 22 June 2018; Published: 01 September 2018

Abstract: The *Panax notoginseng* (*P. notoginseng*) stem leaf is rich in flavonoids. However, because of a lack of research on the flavonoid extraction process and functional development of *P. notoginseng* stem leaf, these parts are discarded as agricultural wastes. Therefore, in this study, we intend to optimize the extraction process and develop the skin-whitening functions of *P. notoginseng* stem leaf extracts. The extraction process of the stem and leaf of *P. notoginseng* flavonoid (SLPF) is optimized based on the Box–Behnken design (BBD) and the response surface methodology (RSM). The optimum extraction conditions of the SLPF are as follows: the extraction time, the ethanol concentration, the sodium dodecyl sulfate (SDS) content and the liquid material ratio (v/w, which are 52 min, 48.7%, 1.9%, and 20:1, respectively. Under the optimal extraction conditions, the average total SLPF content is 2.10%. The antioxidant activity and anti-deposition of melanin of mouse B16 cells of *P. notoginseng* stem leaf extracts are studied. The results indicate that the EC_{50} values of reducing activity, 2,2-diphenyl-1-picrylhydrazyl (DPPH) free radical scavenging activities, the superoxide anion removal ability, and the 2,2-azino-bis-3-ethylbenzthiazoline-6-sulphonic acid (ABTS) free radical removal ability are 7.212, 2.893, 2.949, and 0.855 mg/mL, respectively. The extracts IC_{50} values of the tyrosinase and melanin synthesis are 0.045 and 0.046 mg/mL, respectively. Therefore, the optimal processing technology for the SLPF obtained in this study not only increases its utilization rate, but also decreases material costs. The extracts from the *P. notoginseng* stem leaf may be developed as food or beauty products.

Keywords: antioxidant; flavonoids; mouse melanoma B16 cells; *Panax notoginseng*; surfactant

---

1. Introduction

Panax notoginseng (*P. notoginseng*) (Burk.) F. H. Chen is a perennial plant belonging to the Araliaceae. It has underground roots that are dried for use as medicine, and was included in the

Chinese pharmacopoeia. The *P. notoginseng* stem leaf has a similar medicinal value as the root. It is reported in the Pents'ao Kang Mu that the stem leaf is suitable for the treatment of fractures, bruises, and bleeding. Modern studies have also confirmed that the *P. notoginseng* stem leaf has a medicinal value similar to that of the root for the blood system, the cardiovascular system, the nervous system, and the metabolic system [1]. Common products made from the *P. notoginseng* stem leaf include Qiye Shen'an Pian, Jiang Zhi Ling capsules, and Qiye Shen'an capsules.

Although the stem leaf of *P. notoginseng* has been officially certified as food due to their high nutritional and high medicinal values (DBS 53/024-2017) [2], little attention has focused on promoting these plant parts for food. Preliminary statistics have shown that the annual output of the stem leaf of *P. notoginseng* is 1500 tons and only 5% of this amount is utilized [3]. Therefore, encouraging the development of the stem leaf resources of *P. notoginseng* can greatly increase the income of *P. notoginseng* producers and reduce environmental pollution, thus increasing the socio-economic benefits.

The most effective active ingredients extracted from the stem leaf of *P. notoginseng* are the total saponins and flavonoid glycosides, which account for 4–6% and 0.54–2.49%, respectively [4]. Flavonoids are important because of their well-defined pharmacological activities, which include antioxidant, liver-protective, and anti-tumor activities [5]. The most commonly used organic solvents for extracting flavonoids are ethanol and methanol [6]. For example, Uysal et al. [7] successfully extracted flavonoid-containing components from *Cotoneaster integerrimus* using methanol. Wang et al. [8] used 70% ethanol to extract flavonoids from the leaves of acer truncatum. The extraction rate was higher, and the extract was easier to concentrate and dry. It is considered that ethanol extraction is a better extraction method. At present, only Zhang et al. [4,9] extracted SLPF by a microwave alkali water method and a hot dipping method. However, the ultrasonic method can make the extract continue to shock, contribute to the solute diffusion, shorten the extraction time greatly, and improve the extraction rate of total flavonoids and the use of raw materials, which is a relatively new method for flavonoid extraction [6]. Therefore, the efficient extraction of the SLPF and research on the use of the product for cosmetics and health care are of great significance for the further development of this industry.

Surfactants are widely used as an auxiliary in the extraction of active components; they increase the extraction efficiency, shorten the extraction time, and increase the solubility of the active water-insoluble ingredients in water. In addition, the application of surfactants reduces the use of organic solvents and their cost, optimizes the target components during the extraction process, and improves the purity of the active ingredients. Most of the surfactants used for extraction are non-toxic. Sodium dodecyl sulfate (SDS), Triton X-100, Tween-20, Tween-80, and Span-20 are commonly used surfactants that can significantly increase the extraction rate of flavonoids [10–13].

The response surface methodology (RSM) has been widely used for process optimization, and the Box–Behnken design (BBD) is one statistical model of the response surface design methods. The BBD represents an independent quadratic design that does not contain an embedded factorial or fractional factorial design. Compared with other design methods, the BBD is easy to design and to analyze statistically; therefore, it is widely used in the extraction process optimization of flavonoids [14]. For instance, the microwave-assisted extraction variables were optimized using the RSM for optimal recoveries of total flavonoid content [15]. In addition, some researchers obtained a better technological condition to extract total flavones from Flos Populi [16] and Coriandrum sativum seeds [17] by using a BBD of the RSM.

Previously, the flavonoid extraction methods have been reported [18], but so far, the extraction process of SLPF has not yet been formed. In this study, the ultrasound-assisted extraction process of SLPF was optimized by using the BBD to determine the content of total flavonoids as an index for the first time. The independent variables, i.e., a surfactant type, dosage, liquid material ratio (*v*/*w*, extraction time, and an ethanol concentration, are set as single factors. Therefore, the objectives of the present work were (i) to optimize the ultrasound-assisted extraction process of total flavonoids; (ii) to evaluate the antioxidant activity of the extracts; (iii) to evaluate the effect of the extracts on B16 cell activity; (iv) to explore the feasibility of the application of the SLPF in cosmetics, healthcare products, and the pharmaceutical industry; (v) to provide a reference for the development of *P. notoginseng* stem leaf extract products and reduce the discharge of agricultural wastes into the environment.

2. Results and Discussion

2.1. Effect of Surfactant Types on Extraction Efficiency of SLPF

In addition to the control group (CK) (water instead of surfactants), different types of surfactants (1.5%) were used in other groups by maintaining liquid material ratio (*v*/*w* of 15:1 and using a 40% ethanol concentration. The total flavonoid content was determined after ultrasonic extraction at 80 °C for 40 min; the results are shown in Figure 1. It can be seen that the surfactants improved the extraction efficiency of the SLPF, and the highest content was extracted by the SDS. The SLPF content using SDS was increased by 12.8% compared to the CK. Therefore, SDS was used in the subsequent experiments.

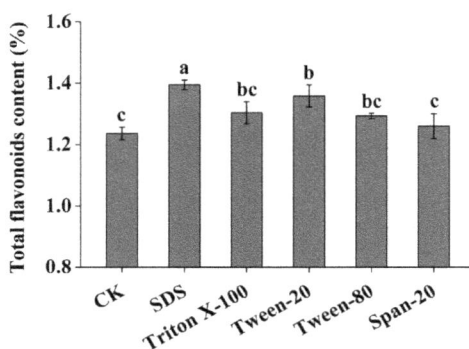

Figure 1. Effects of different types of surfactants on contents of SLPF. Different letters imply significant differences at $p < 0.05$.

2.2. Single Factor Experiments

The result showed that the extraction efficiency of SLPF increased with the increase in the extraction time; at liquid material ratio (*v*/*w*) of 15:1, the ethanol concentration was 40% and the SDS content was 1.5% (Figure 2A). The yield of the SLPF content decreased when the extraction time was longer than 40 min. The SLPF content increased when the ethanol concentration increased from 30% to 50%, but decreased gradually when the ethanol concentration was higher than 50%. The highest yield of the SLPF content was 23.1% for extraction time of 40 min, liquid material ratio (*v*/*w*) of 15:1, and an SDS content of 1.5% (Figure 2B). When the extraction time was 40 min, the liquid material ratio (*v*/*w*) was 15:1, and the ethanol concentration was 40%, the SLPF content continued to increase with the SDS content from 0.5% to 2%. When the SDS content was increased from 0.5% to 2%, and the SLPF content was increased by 10% (Figure 2C). The total flavonoid content was highest when the liquid material ratio (*v*/*w*) was 20:1, the extraction time was 40 min, the ethanol concentration was 40%, and the SDS content was 1.5% (Figure 2D). The SLPF content was 1.5 times higher than that when liquid material ratio (*v*/*w*) of 5:1 was used, and decreased when the liquid material ratio was greater than 20:1. Therefore, for the subsequent BBD experiments, extraction time ranging from 40 to 60 min, ethanol concentrations ranging from 40% to 60%, SDS contents ranging from 1.5% to 2.5%, and the liquid material ratio (*v*/*w*) ranging from 15:1 to 25:1 were used.

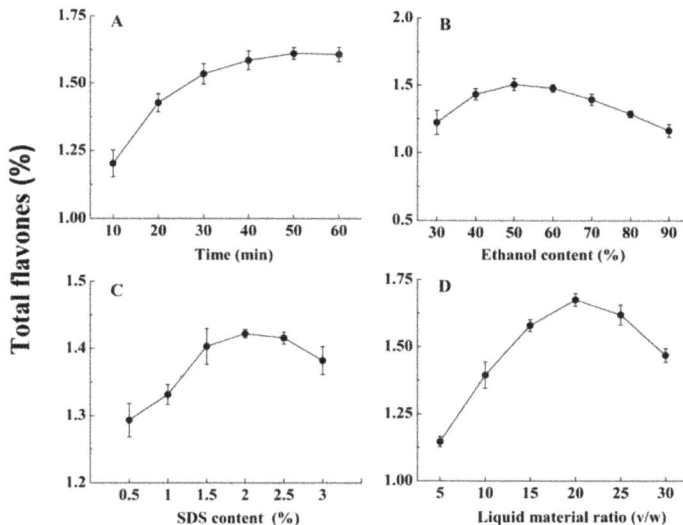

Figure 2. The single factor experiments. Total flavones (%) as a function of (**A**) ultrasound time (min); (**B**) ethanol concentrations (%); (**C**) SDS contents (%) and (**D**) liquid material ratios (*v/w*).

2.3. Model Fitting and Optimization of SLPF

2.3.1. Model Fitting

The designed matrix, the results of the analysis of variance, and the adequacy and fitness of the models are shown in Table 1. A multiple regression analysis was conducted using the Design–Expert software. The relationship between the response variable (the content of SLPF) and the independent variables was expressed by the following second-order polynomial equation:

$$Y = -30.678 + 0.487A + 0.304B + 4.596C + 0.853D - 0.001AB - 0.020AC - 0.005AD + 0.010BC - 0.001BD - 0.076CD - 0.003A^2 - 0.002B^2 - 0.683C^2 - 0.010D^2 \tag{1}$$

where Y is the SLPF content; A, B, C, and D represent the extraction time, the ethanol concentration, the SDS content, and the liquid material ratio, respectively.

Table 1. Experimental design and results for response surface analysis.

Std. Order	Run Order	Variable				Total Flavonoids Y (%)	
		A	B	C	D	Observed	Predicted
25	1	50 (0)	50 (0)	2 (0)	20:1 (0)	2.127	2.117
8	2	50 (0)	50 (0)	2.5 (1)	25:1 (1)	1.384	1.422
11	3	40 (−1)	50 (0)	2 (0)	25:1 (1)	1.781	1.776
22	4	50 (0)	60 (1)	2 (0)	15:1 (−1)	1.615	1.662
12	5	60 (1)	50 (0)	2 (0)	25:1 (1)	1.369	1.364
20	6	60 (1)	50 (0)	2.5 (1)	20:1 (0)	1.579	1.562
1	7	40 (−1)	40 (−1)	2 (0)	20:1 (0)	1.441	1.448
6	8	50 (0)	50 (0)	2.5 (1)	15:1 (−1)	1.794	1.844
10	9	60 (1)	50 (0)	2 (0)	15:1 (−1)	1.949	1.922
17	10	40 (−1)	50 (0)	1.5 (−1)	20:1 (0)	1.535	1.584
9	11	40 (−1)	50 (0)	2 (0)	15:1 (−1)	1.335	1.306
21	12	50 (0)	40 (−1)	2 (0)	15:1 (−1)	1.630	1.626

Table 1. *Cont.*

Std. Order	Run Order	Variable				Total Flavonoids Y (%)	
		A	B	C	D	Observed	Predicted
27	13	50 (0)	50 (0)	2 (0)	20:1 (0)	2.127	2.117
2	14	60 (1)	40 (−1)	2 (0)	20:1 (0)	1.762	1.811
16	15	50 (0)	60 (1)	2.5 (1)	20:1 (0)	1.711	1.663
7	16	50 (0)	50 (0)	1.5 (−1)	25:1 (1)	1.973	1.924
23	17	50 (0)	40 (−1)	2 (0)	25:1 (1)	1.694	1.678
28	18	50 (0)	50 (0)	2 (0)	20:1 (0)	2.145	2.117
14	19	50 (0)	60 (1)	1.5 (−1)	20:1 (0)	1.666	1.683
29	20	50 (0)	50 (0)	2 (0)	20:1 (0)	2.074	2.117
15	21	50 (0)	40 (−1)	2.5 (1)	20:1 (0)	1.669	1.618
4	22	60 (1)	60 (1)	2 (0)	20:1 (0)	1.496	1.490
13	23	50 (0)	40 (−1)	1.5 (−1)	20:1 (0)	1.832	1.846
18	24	60 (1)	50 (0)	1.5 (−1)	20:1 (0)	1.879	1.885
5	25	50 (0)	50 (0)	1.5 (−1)	15:1 (−1)	1.627	1.590
19	26	40 (−1)	50 (0)	2.5 (1)	20:1 (0)	1.634	1.659
26	27	50 (0)	50 (0)	2 (0)	20:1 (0)	2.114	2.117
24	28	50 (0)	60 (1)	2 (0)	25:1 (1)	1.486	1.522
3	29	40 (−1)	60 (1)	2 (0)	20:1 (0)	1.697	1.649

2.3.2. Analysis of Response Surface

The residual analysis of the response surface optimization model was conducted using a graphical analysis tool. It is important to test the uniformity of the error variance of the model. If the results exhibit a good fit with the predicted values, the fitted curve of the experimental and predicted values is linear. Moreover, it is important to evaluate if the distribution of the residuals is normal when determining the model accuracy. When the residuals have a normal distribution, the fitted curve of the residuals is linear [19]. In addition, if the predictive values of the residuals have a random distribution, the homogeneity of the residual variance is consistent with the optimization requirements [20].

In this study, the quadratic model had the best fit for the data (R^2 value of 0.9818) (Figure 3A) [21]. In addition, the normal probability plot of the residuals (Figure 3B) and the plot of the residuals versus the predicted response (Figure 3C) were used to validate the adequacy of the model. The data showed that the points in the normal plot formed a straight line (Figure 3B). This indicated that the data exhibited no deviation from normality. In addition, the plot of the residuals versus the predicted response showed a random distribution, which indicated that the model was adequate and did not violate the assumption of independence or constant variance (Figure 3C) [22].

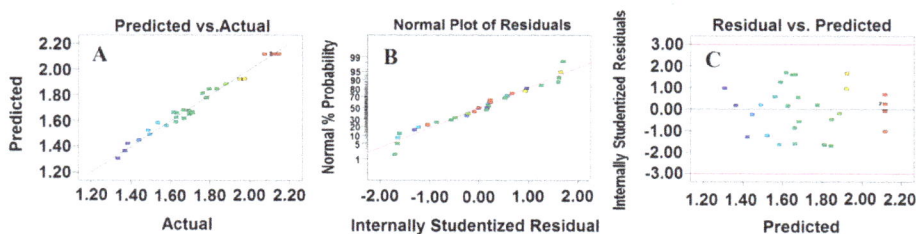

Figure 3. Predicted total flavonoid content versus experimental total flavonoids content (**A**); normal probability plots as a function of residuals (**B**) and plots of the residuals versus the predicted response (**C**).

The results showed that a high proportion of the total variance was explained by the quadratic regression model (R^2 of 0.982), indicating that the model is suitable for the optimization of the extraction parameters [23]. The R^2_{Adj} of 0.964 was relatively close to the R^2_{Pred} of 0.902, corroborating that the model is significant [14]. The value of the signal-to-noise ratio (R_{SN}) was 24.553, which was well above 4. The result indicated that this model could be used for the design space. Moreover, the coefficient of variation (CV) was low (2.657%), indicating that the model fitted the experimental data satisfactorily and can be used to predict the outcome within the range of the data. The regression coefficients are listed in Table 2. In the SLPF model, A, B, C, AB, AC, AD, BC, CD, A^2, B^2, C^2, and D^2 were significantly different ($p < 0.05$), but D and BD were not significantly different ($p > 0.05$).

Table 2. The results of analysis on variance (ANOVA) for the effects of variables.

Source	Sum of Squares	DF	Mean Square	F Value	p Value Prob > F	Significance
Model	1.589	14	0.114	53.850	<0.0001	***
A	0.031	1	0.031	14.757	0.0018	**
B	0.011	1	0.011	5.026	0.0417	*
C	0.046	1	0.046	21.826	0.0004	***
D	0.006	1	0.006	2.750	0.1195	
AB	0.068	1	0.068	32.313	<0.0001	***
AC	0.040	1	0.040	18.891	0.0007	***
AD	0.263	1	0.263	124.984	<0.0001	***
BC	0.011	1	0.011	5.165	0.0393	*
BD	0.009	1	0.009	4.360	0.0555	
CD	0.143	1	0.143	67.750	<0.0001	***
A^2	0.487	1	0.487	230.873	<0.0001	***
B^2	0.386	1	0.386	182.935	<0.0001	***
C^2	0.189	1	0.189	89.704	<0.0001	***
D^2	0.409	1	0.409	194.231	<0.0001	***
Residual	0.030	14	0.002			
Lack of Fit	0.027	10	0.003	3.858	0.1025	
Pure Error	0.003	4	0.001			
Cor Total	1.619	28				
Std. Dev.	0.046		R^2	0.982		
Mean	1.728		R^2 Adj	0.964		
C.V. (%)	2.657		R^2 Pred	0.902		
PRESS	0.158		R$_{SN}$	24.553		

DF = degree of freedom. PRESS = predicted residual sum of squares. Cor Total = correlation total. R^2 Adj = adjusted R^2. Std. Dev. = standard deviation. R^2 Pred = predicted R^2. C.V. (%) = coefficient of variation percent. R$_{SN}$ = signal-to-noise ratio.

A 3D surface graph (Figure 4A$_1$–D$_1$) of the total flavonoid content and the contour curve (Figure 4A$_2$–D$_2$) of the two tested variables were generated from the final model to describe the interactions between the independent variables and the optimal process parameters. Each graph was completed on the condition that the other factors were kept at their respective zero levels each time [24]. If the contour plot has a circular shape, the interactions between the corresponding factors are negligible. The elliptical shape of the contour plot indicates that the interactions between the variables contribute to the content of total flavonoids at a significant level [25].

Based on the data shown in Table 2 and Figure 4, in the SLPF model, the ranking of the interaction effect between the independent variables from high to low was CD–AD–AB–AC. The effects of C (the SDS content) and D (the liquid material ratio (v/w)) on the SLPF content were shown as a 3D plot and an associated contour plot. The elliptical shape of the contour plot illustrated a significant ($p < 0.0001$) correlation between C and D, which contributed to the different SLPF contents (Table 2; Figure 4D$_1$,D$_2$). The SLPF contents increased (2.13%), when the SDS content was increased from 1.5% to 2.0% and the liquid material ratio (v/w) was increased from 15:1 to 20:1 (Figure 4D$_1$,D$_2$). It is important that the extraction process is economical and feasible and the modeling approach facilitates the reduction in the cost and time during future mass production.

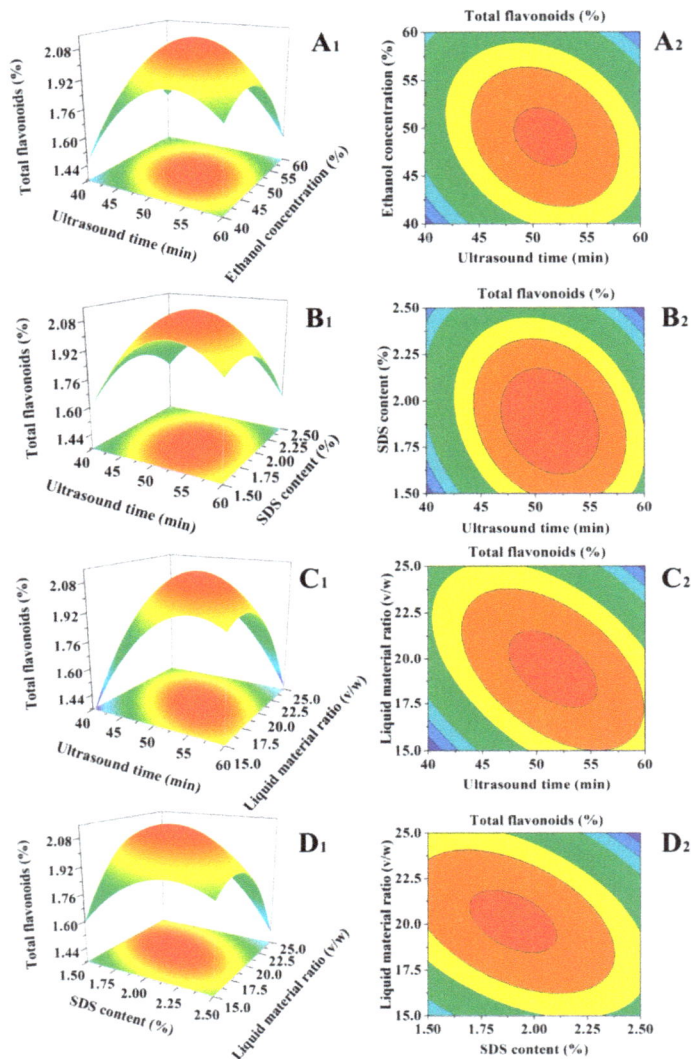

Figure 4. Response surface for the effect of operating parameters on total flavonoid contents; (**A₁**) ultrasound time vs. ethanol concentration; (**B₁**) ultrasound time vs. an SDS content; (**C₁**) ultrasound time vs. liquid material ratio (v/w); (**D₁**) an SDS content vs. liquid material ratio (v/w). (**A₂–D₂**) are their contour plots, respectively.

2.3.3. Validation of the Model

To validate the proposed model, various reaction conditions were selected within the range of the variables (Table 3). The optimal process parameters for obtaining a high total flavonoid content obtained by Equation (1) were extraction time of 51.84 min, an ethanol concentration of 48.68%, an SDS content of 1.88%, and liquid material ratio (v/w) of 19.81:1; this resulted in a total flavonoid content of 2.13%. The accuracy of the model was confirmed based on three assays under optimum conditions. The average total flavonoid content was 2.10%, which showed no significant difference from the predicted value. This indicated that the experimental value was very close to the predicted value, which proved that the results were reasonable and reliable; the model of Equation (1) was considered to be satisfactory and accurate for predicting the SLPF content. This approach does not only improve the extraction efficiency, but also increases the utilization rate of the *P. notoginseng* stem leaf.

Table 3. The optimization criteria of the maximum total flavonoid content and the results of model validation.

Name	Goal	Lower Limit	Upper Limit	Lower Weight	Upper Weight
A: Ultrasound time (min)	is in range	40	60	1	1
B: Ethanol concentration (%)	is in range	40	60	1	1
C: SDS content (%)	is in range	1.5	2.5	1	1
D: Liquid material ratio (v/w)	is in range	15	25	1	1
Total flavonoids content (%)	maximize	1.33502	100	1	1
Category	Ultrasound time (min)	Ethanol concentration (%)	SDS content (%)	Liquid material ratio (v/w)	Total flavonoids content (%)
Predicted value	51.84	48.68	1.88	19.81:1	2.13 ± 0.008a
Experimental value	52	48.7	1.9	20:1	2.10 ± 0.017a

Note: The little letters "a" means that there is no significant difference between the predicted value and the experimental value of the total flavonoid content at $p < 0.05$.

2.4. Antioxidant Activities of the P. notoginseng Stem Leaf Extracts

The reduction ability of the *P. notoginseng* stem leaf extracts is positively related to its antioxidant activity; therefore, it can be used as an important evaluation index of antioxidant activity [26]. The reduction abilities of the extracts and of the ascorbic acid increased with the concentration and showed a significant positive correlation with the concentration (Figure 5A). The EC_{50} of the extracts and ascorbic acid were 7.212 and 0.0287 mg/mL, respectively. These results suggested that *P. notoginseng* extracts are an effective electron donor capable of reacting with free radicals to convert them into more stable products.

The DPPH radical is a relatively stable free radical and can accept electrons or hydrogen atoms to form stable molecules. Hydroxyl radicals possess the highest oxidation ability and can damage biological molecules. Therefore, DPPH and hydroxyl radicals have been widely used as an important evaluation index of antioxidant activity [27]. The DPPH radical scavenging activities increased with the concentration of the extracts and ascorbic acid, and they all showed a significant positive correlation with the concentration. The EC_{50} values of the extracts and ascorbic acid were 2.893 and 0.0938 mg/mL, respectively (Figure 5B). The hydroxyl radical scavenging activities had a similar trend as the DPPH radical for the treatments of the extracts or the ascorbic acid, and the EC_{50} values of the extracts and ascorbic acid were 1.514 and 0.0902 mg/mL, respectively (Figure 5C). These results indicated that the *P. notoginseng* extracts have a high ability to remove DPPH or hydroxyl radicals.

Superoxide anions can lead to the occurrence of peroxides in the plasma membrane [28]. Therefore, the ability to remove superoxide anions is extremely important for an antioxidant. In this study, the superoxide anion scavenging rate increased with the concentration of the extracts and ascorbic acid, and the removal rate was also correlated with the concentration (Figure 5D). The EC_{50} values of the extracts and ascorbic acid were 2.949 and 0.140 mg/mL, respectively, and the highest removal rates were 94.90% and 93.73%, respectively.

In addition, the antioxidant activity of the extracts is generally reflected in the ability to remove ABTS+ free radicals [29]. The scavenging effect of the extracts and ascorbic acid on the ABTS+ free radicals increased with the concentration (Figure 5E), and there was a positive correlation between the scavenging rate and the concentration of the extracts or the ascorbic acid. The EC_{50} values of the *P. notoginseng* extracts and the ascorbic acid were 0.855 and 0.205 mg/mL, respectively.

Many studies have found that flavonoids possess good antioxidant properties. The results of Farombi et al. [30] showed that an extract of *Garcinia kola* seeds exhibited a 57% scavenging effect on superoxide at a concentration of 1 mg/mL, an 85% scavenging effect on hydrogen peroxide at a concentration of 1.5 mg/mL, and an 89% scavenging effect on DPPH at a concentration of 2 mg/mL. Srisawat et al. [31] demonstrated the superoxide anion radical scavenging activities of rice extracts EC_{50} values in the range of 0.6–5 mg/mL, which was similar to the results of this study. Overall, this research also indicated that the *P. notoginseng* stem leaf extracts possess a strong antioxidant activity.

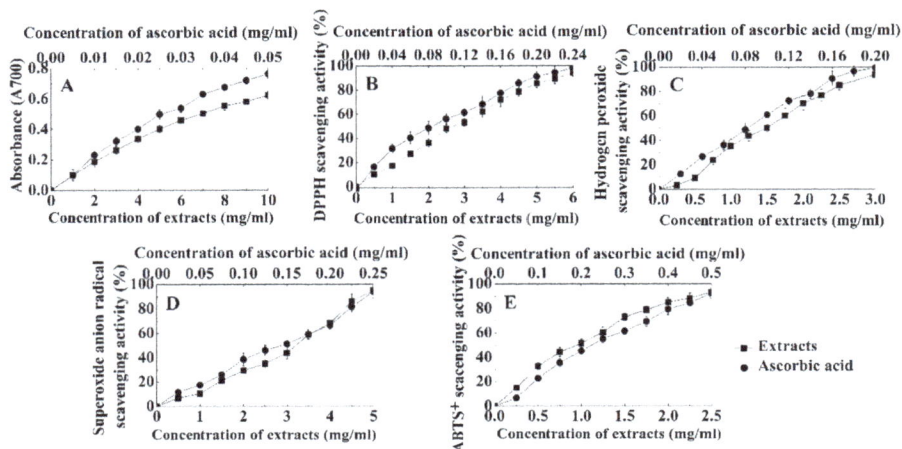

Figure 5. Antioxidant activity of the extracts from *P. notoginseng* stem leaf and ascorbic acid at different concentrations. (**A–E**) were the ferric ions reducing activity, the DPPH radical scavenging rate, the hydroxyl radical scavenging rate, the superoxide anion scavenging rate and the ABTS+ radical scavenging rate, respectively.

2.5. Effect of Extracts on B16 Cell Activity

The safety of the *P. notoginseng* stem leaf extracts is extremely important in the investigation of cell viability, cellular tyrosinase activity and melanin levels in B16 melanoma cells. The results showed that the cell viability was about 85% at a concentration of 0.032 mg/mL and that the IC_{50} value of the cell viability was 0.405 mg/mL (Figure 6A). The *P. notoginseng* stem leaf extracts were not especially cytotoxic to B16 cells. Therefore, extract concentrations lower than 0.032 mg/mL were used to perform the subsequent experiments.

Tyrosinase is a rate-limiting, regulatory melanogenic enzyme that is involved in the melanin synthesis pathway [32]. The results indicated that the extracts significantly reduced the tyrosinase activity and melanin production in the murine cells in a dose-dependent manner (Figure 6B). There were positive correlations between the extracted content and the inhibition ratio of the tyrosinase or the melanin synthesis (R^2 = 0.9162 and R^2 = 0.9118). The IC_{50} values of the inhibition ratio of the tyrosinase and melanin synthesis were 0.045 mg/mL and 0.046 mg/mL, respectively (Figure 6B).

Melanin plays a decisive role in skin color. The formation of melanin is regulated by tyrosinase, which is a major rate-limiting enzyme that is regulated by free radicals. Detailed studies have shown that flavonoids are rather potent inhibitors, and can directly inhibit tyrosinase and act on the distal part of the melanogenesis oxidative pathway [33]. Our results indicated that the *P. notoginseng* stem leaf extracts exhibited good performance in reducing the tyrosinase activity and melanin production. It is evident that *P. notoginseng* stem leaf extracts may have great prospects as health food and a cosmetic product because of its excellent performance in anti-pigment deposition.

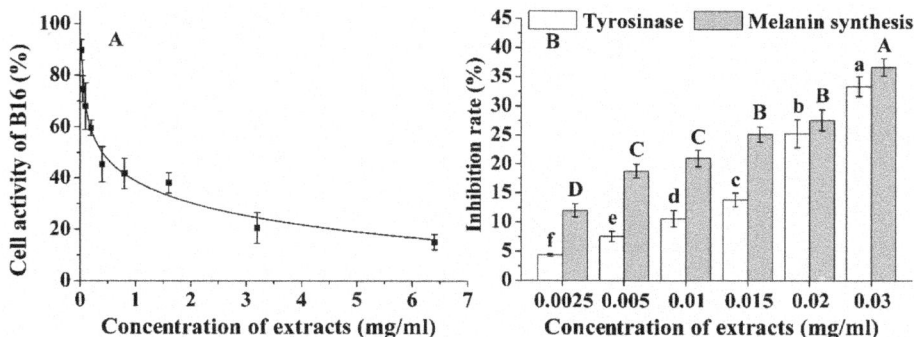

Figure 6. The cell activity (**A**) and the inhibition rate of tyrosinase and melanin synthesis (**B**) in B16 cells. Capital letters indicate differences in the inhibition rate of melanin synthesis under different treatment; lowercase letters indicate differences in the inhibition rate of tyrosinase under different treatment ($p < 0.05$).

3. Materials and Methods

3.1. Plant Materials

Fresh stem leaf of *P. notoginseng* was bought from the international trading center of *P. notoginseng*, Wenshang, Yunnan province, in October 2017, and then washed in distilled water, and surface water was removed. The fresh stem leaf of *P. notoginseng* was dried to a constant weight in 60 °C, and were then powdered for further use.

3.2. Chemicals and Reagents

Na_2HPO_4, NaH_2PO_4, TCA (trichloroacetic acid), $FeCl_3$, $FeSO_4$, ascorbic acid, pyrogallol, H_2O_2 were obtained from Sinopharm Chemical Reagent Co., Ltd. (Shanghai, China) DPPH (2,2-diphenyl-1-picrylhydrazyl), ABTS, α-MSH, L-DOPA and MTT were purchased from Sigma Chemical Co. (St. Louis, MO, USA), PBS (Biological Industries, Beit-Haemek, Israel), Dulbecco's Modifed Eagle Medium (DMEM) culture and fetal bovine serum (FBS; Gibco, NY, USA), and trypsin were obtained from Biological Industries, Israel. DMSO and Triton X-100 were obtained from MP Biomedicals and Amresco in USA, respectively. All the other reagents were of analytical grade. Ultrapure water was obtained from the Millipore system (Bedford, MA, USA).

3.3. Optimization of Extraction Condition Selection

Ultrasound-assisted extraction of total flavonoids from the *P. notoginseng* stem leaf was optimized by the RSM. The content of SLPF was set as the index to judge the extraction condition. According to the extracting technology, the surfactants, the extraction time, the ethanol concentration, the SDS content and the liquid material ratio are key factors. The highest values in the experiments were the optimal central values for the SLPF content. Then, they were used to study the best extraction process technology of SLPF by the method of RSM. Single factor treatment was as follows:

The *P. notoginseng* stem leaf was (1) extracted using different surfactants (Table 4), ultrasound extraction for 40 min by maintaining the liquid material ratio (15:1, v/w), and 40% ethanol concentration; (2) extracted by ultrasound for 10, 20, 30, 40, 50 and 60 min when the liquid material ratio was 15:1 (v/w), the concentration of ethanol was 40%, and the content of SDS was 1.5%; (3) extracted by 30, 40, 50, 60, 70, 80 and 90% ethanol for 40 min when the liquid material ratio was at 15:1 (v/w), and the SDS content was 1.5%; (4) extracted by 0.5, 1, 1.5, 2, 2.5 and 3% SDS for 40 min when the liquid material ratio (v/w) was 15:1 (v/w), and the content of ethanol was 40%; (5) extracted with different liquid material ratios (5:1, 10:1, 15:1, 20:1, 25:1 and 30:1; v/w) for 40 min by maintaining the concentration of ethanol was 40%, and the content of SDS was 1.5%. Then, the content of SLPF in extracts was determined.

Table 4. Surfactant types.

Surfactant	SDS	Triton X-100	Tween-20	Tween-80	Span-20
HLB	40	14.6	16.7	15	8.6
Types	Anionic	Nonionic	Nonionic	Nonionic	Nonionic

3.4. Experimental Design of RSM

Based on the results analysis of single factor experiments, the SLPF extraction condition from *P. notoginseng* stem leaf was developed and optimized using three-levels, four-factor BBD, combined with RSM (Design–Expert Software, trial version 8.1.0; Stat-Ease Inc., Minneapolis, MN, USA). Each independent variable was coded at three levels between −1 and +1, where the time (A); the ethanol concentration (B); the SDS content (C); and the liquid material ratio (D) were changed in the ranges shown in Table 5. The SLPF content was the response variable. For statistical calculations, a second-order quadratic equation was used as follows: [34].

$$Y = \beta_0 + \sum_{i=1}^{4} \beta_i X_i + \sum_{i=1}^{4} \beta_{ii} X_i^2 + \sum_{i=1}^{3} \cdot \sum_{j=i+1}^{4} \beta_{ij} X_i X_j + \sum_{i=1}^{4} \beta_{iii} X_i^3 \tag{2}$$

where Y is the predicted response, β_0 is an intercept coefficient, β_i is the linear terms, β_{ii} is the squared terms, β_{ij} is the interaction terms, and X_i and X_j represent the coded levels of independent variables.

Table 5. Levels and codes of independent variable used for response surface analysis.

Factors	Codes	Levels		
		−1	0	1
Ultrasound time (min)	A	40	50	60
Ethanol concentration (%)	B	40	50	60
SDS content (%)	C	1.5	2	2.5
Liquid material ratio (v/w)	D	15:1	20:1	25:1

3.5. Total Flavonoid Assay

The content of total flavonoids was determined based on an established method with some modifications, and rutin was selected as the reference standard [35]. The dried rutin was weighed 10 mg accurately, dissolved with 20 mL ethanol and transferred into a 100 mL volumetric flask by adding water to scale. A series of standard solutions of rutin in the range of 0.0–6.0 mL were precisely transferred to 25 mL volumetric flasks. Then, 1 mL of 5% $NaNO_2$ was added; after 6 min, 1.0 mL of 10% $Al(NO_3)_3$ was added and the solution was kept for 6 min at room temperature; 10 mL of 4% NaOH was added to the volumetric flasks. This solution was then diluted with water solution to 25 mL and kept for 15 min at room temperature. The absorbance against a blank sample without rutin (A) was measured at the wavelength of 510 nm using a UV-2600 spectrophotometer (Shanghai Yuanxi Instruments Co. Ltd., Shanghai, China) to obtain the linear response (A = 4.6971B − 0.0072, R^2 = 0.9992). B was the total flavonoids content (mg) at the wavelength of 510 nm. An aliquot of 0.1 mL of the samples were diluted with methanol to 50 mL and was then determine the concentration of total flavonoids at the spectrophotometer. The total flavonoid content was calculated according to the following equation:

$$y = \frac{m \times V}{M} \times 100\% \tag{3}$$

where m is total flavonoid content, V is the volume of extraction solution and M is the quality of extracts.

3.6. In Vitro Antioxidant Activity Assays

We researched the antioxidant activities using different concentrations of extracts. L-ascorbic acid was used to prepare a standard solution and compared with the samples. All the samples were conducted for 5 times.

3.6.1. Reducing Power Assay

The ferrous-ion-reducing ability of extracts assay was carried out according to Park's [36] method with slight modifications. Briefly, different concentrations of extracts were mixed with a sodium phosphate buffer (2.5 mL, 0.2 mg/mL, pH 6.6) and potassium ferricyanide (2.5 mL, 1%, w/v) in the test tube, followed by incubation in a water bath at 50 °C for 20 min. Then, trichloroacetic acid (2.5 mL, 10%, w/v) was added to the mixture to end the reaction. In addition, the solution was centrifuged at 3000 rpm for 10 min. Finally, the supernatant (2.5 mL) was mixed with the deionized water and the FeCl$_3$ (0.5 mL, 0.1%, w/v). The reaction mixture was left at 25 °C for 10 min in the dark and determined the absorbance at 700 nm.

3.6.2. DPPH Radical Scavenging Capacity

The DPPH radical scavenging capacity was determined by Qu's method with some slight modifications [37]. Different concentrations of extracts (1 mL) were mixed with 1 mL of a 0.002 mg/mL DPPH solution. After the mixtures were incubated at room temperature in the dark for 0.5 h, the discolorations were measured at 517 nm. The scavenging percentage was calculated by the following equation:

$$\text{DPPH radical scavenging activity } (\%) = (1 - \frac{A_s}{A_0}) \times 100\% \tag{4}$$

where A_s means the absorbance of a sample, and A_0 means the blank control solution without a sample.

3.6.3. Detection of ABTS$^+$ Scavenging Assay

ABTS scavenging activity of extracts was quantified by Wang's method with some modifications [38]. ABTS$^+$ (0.007 mg/mL) was dissolved into a potassium persulphate solution (0.00245 mg/mL). The reaction mixture was left at 25 °C for 12 h in the dark before use. The mixture was diluted with 75% of ethanol to adjust its absorbance to 0.70 $\pm$ 0.02 at 734 nm. Extracts with different concentrations (0.1 mL) were added into 3.9 mL of this ABTS$^+$ solution, and were then incubated at room temperature for 10 min. The absorbance was measured at 734 nm. Ascorbic acid was used as the reference compound for measuring the ABTS$^+$ scavenging activity. The scavenging effect of ABTS$^+$ was defined as:

$$\text{Scavenging ability } (\%) = (1 - \frac{A_s}{A_0}) \times 100\% \tag{5}$$

where A_s means the absorbance of a sample and A_0 means the blank control solution without a sample.

3.6.4. Assay of Hydroxyl Radical Scavenging Activity

The radical scavenging capability was measured according to a previously described method with minor modifications [39]. Two milliliter of FeSO$_4$ (0.006 mg/mL) and 2.0 mL different concentrations of the extracts were mixed. Then, 2.0 mL H$_2$O$_2$ (0.006 mg/mL) was added in the mixture to start the reaction. After shaken, the mixture was incubated at 37 °C for 30 min. Then, 2.0 mL of salicylic acid (0.006 mg/mL) was added, mixed and incubated at 37 °C for 10 min. The absorbance was measured at

510 nm. Ascorbic acid was used for comparison. The capability of hydroxyl radical scavenging activity was calculated by the following equation:

$$\text{Scavenging ability } (\%) = (1 - \frac{A_1 - A_2}{A_0}) \times 100\% \tag{6}$$

where A_1 and A_2 are the absorbance of a sample (with and without hydrogen peroxide, respectively), and A_0 was the absorbance of a background solution.

3.6.5. Detection of Superoxide Anion Scavenging Activity

The superoxide anion radical scavenging activity of extracts was assayed following the methods described by İlhami Gülçin [40]. The beginning of reaction was to put 2 mmol NADH (20 μL) into the mixture (180 μL), which contained different concentrations of extracts (10 μL), and 1 mmol NBT (20 μL), 0.1 mmol PMS (20 μL), a 250 mmol potassium phosphate buffer (pH 7.4, 40 μL) and water (90 μL), and was then incubated at 25 °C for 20 min. The absorbance of the resulting solution was measured at 570 nm, and the capability of superoxide anion scavenging activity was calculated by the following equation:

$$\text{Superoxide anion radical scavenging activity } (\%) = (1 - \frac{A_s}{A_c}) \times 100\% \tag{7}$$

where A_s and A_c are the absorbance of a sample and a control solution, respectively.

3.7. Cell Experiment

3.7.1. Cell Culture

Mouse melanoma cell lines, B16 (obtained from Kunming Institute of Zoology (CAS), Kunming, Yunnan, China) were cultured in DMEM supplemented with 10% fetal bovine serum and 1% penicillin/streptomycin. The cells were incubated in a humidified atmosphere of 5% CO_2 at 37 °C.

Effects of the extractive on viability of B16 cells were determined using the MTT assay [41]. Cells were seeded into a 6-well plate (10^5 cells/well) and incubated with samples at different concentrations. Cells were labelled with the MTT solution at 37 °C for 3 h after 48 h of incubation. The violet formazan precipitates were dissolved in DMSO, and then the absorbance of each well, using a microplate reader with a 630 nm reference, was determined at 590 nm.

3.7.2. α-MSH Treatment

B16 cells were cultured for 24 h at a density of 10^5 cells/mL in 6-well plates, which contained 10% fetal bovine serum and 1% penicillin/streptomycin at 37 °C in a humidified atmosphere with 5% CO_2. The medium was substituted by a fresh supplement (α-MSH) and different concentrations of extracts and incubated for 48 h. All experiments reported in next sections were performed with these stimulate cells by using the above procedure.

3.7.3. Intracellular Tyrosinase Activity

Tyrosinase enzyme activity was estimated by measuring the rate of L-DOPA oxidation [42]. α-MSH-stimulated cells were treated in the absence or the presence of samples (0.0025–0.03 mg/mL) for 48 h. The cells were then solubilized in a phosphate buffer (0.1 M; pH 6.8) containing Triton X-100 (0.1%). Cellular lysates were centrifuged at 12,000 rpm at 4 °C for 20 min. The cellular extract was incubated with L-DOPA (1.25 mM), and the absorbance was followed spectrophotometrically at 475 nm until the reaction finished.

3.7.4. Melanin Content Assay

The α-MSH-stimulated cells were cultured at a density of 10^5 cells/mL in 6-well plates for 48 h in the absence or the presence of samples (0.0025–0.03 mg/mL). The cells were washed twice in PBS and dissolved in NaOH (1 mL, 10%) at 100 °C for 1 h. The cell lysates were centrifuged at 1000 g for 10 min, and then the absorbance value of the supernatant was measured at 405 nm using a standard curve of synthetic melanin. The inhibition rate of melanin synthesis of the cells was estimated as a percentage of the control culture [43].

3.8. Statistical Analysis

All experiments data were processed with Microsoft Excel software 2010 and the curves were fitted with Origin 8.0. Statistical analysis was carried out by using SPSS 19.0.

4. Conclusions

Ultrasound-assisted extraction enhanced the extraction efficiency of total flavonoids. The optimum extraction conditions of SLPF were: ultrasound-assisted extraction time of 52 min, an ethanol concentration of 48.7%, a surfactant SDS content of 1.9%, and liquid material ratio (v/w of 20:1; this resulted in a 2.10% SLPF content. It was recommended that producers adjusted these data for the extraction of SLPF based on the specific conditions to achieve large-scale and efficient production. The *P. notoginseng* stem leaf extracts exhibited a strong antioxidant activity and induced a decrease in the melanin synthesis by inhibiting the tyrosinase activity. Our results promote the use of *P. notoginseng* stem leaf extracts for cosmetic preparations and food uias a skin whitening agent.

Author Contributions: Y.Y., X.-M.C. and Y.Q. conceived and designed the experiments; C.-Y.D., P.-F.L. and P.-R.L. performed the experiments and analyzed the data; C.-Y.D. and Y.Y. drafted the manuscript; C.-Y.D., P.-F.L. and C.-X.W. revised the manuscript; X.-M.C. was also responsible for raising the fund.

Funding: This research was funded by the Ministry of Science and Technology of Yunnan Province, China (No. 2017ZF014), Standardization of Traditional Chinese medicine project (ZYBZH-C-YN-58), China Postdoctoral Science Foundation (No. 167358), the Key Special Project of National Key Research and Development Program (2017YFC1702500).

Conflicts of Interest: The authors declare no conflicts of interest in this work.

Abbreviations

ABTS	2,2-azino-bis-3-ethylbenzthiazoline-6-sulphonic acid
BBD	Box-Behnken design
DPPH	2,2-diphenyl-1-picrylhydrazyl
RSM	Response surface methodology
SLPF	stem and leaf of *P. notoginseng* flavonoid

References

1. Lv, Q.Y. Chemical composition and pharmacological action of Panax notoginsegnoside in the leaves. *Lishizhen Med. Mater. Med. Res.* **2006**, *17*, 2065–2066. (In Chinese)
2. Yunnan provincial health and Family Planning Commission. *Dried Panax notoginseng Stem and Leaf*; DBS 53/024-2017; Yunnan Standard Press: Kunming, China, 2017.
3. Geng, J.L.; Shen, Y.; Song, X.W. Development Studies on Saponins in *Panax notoginseng* Leaves. *J. Yunnan Coll. Tradit. Chin. Med.* **2001**, *24*, 7–8.
4. Zhang, Z.X.; Song, G.B. Extraction technology of total flavonoids in the stalk and leaves of *Panax notoginseng*. *Biotechnology* **2006**, *16*, 65–67.
5. Wang, T.; Guo, R.; Zhou, G.H.; Zhou, X.D.; Kou, Z.Z.; Sui, F.; Li, C.; Tang, L.Y.; Wang, Z.J. Traditional uses, botany, phytochemistry, pharmacology and toxicology of *Panax notoginseng* (Burk.) F.H. Chen: A review. *J. Ethnopharmacol.* **2016**, *188*, 234–258. [CrossRef] [PubMed]

6. Zhang, Z.X. *Preliminary Study on Flavonoids in Stalk and Leaves of Panax notoginseng*; Chongqing University: Chongqing, China, 2005.

7. Uysal, A.; Zengin, G.; Mollica, A.; Gunes, E.; Locatelli, M.; Yilmaz, T.; Aktumsek, A. Chemical and biological insights on *Cotoneaster integerrimus*: A new (−)-epicatechin source for food and medicinal applications. *Phytomedicine* **2016**, *23*, 979–988. [CrossRef] [PubMed]

8. Wang, L.; Ma, X.; Wang, S.; Wang, X. A study on the extraction of flavonoids from the leaves of acer truncatum. *J. Northwest For. College* **1997**, *12*, 64–67.

9. Zhang, Z.X.; Zhang, S.X.; Song, G.B. Study of microwave-alkali water method for extraction of total flavone from Panax notoginseng stalks and leaves. *Pharm. Biotechnol.* **2005**, *12*, 389–392.

10. Gong, K.; Zhan, X.L.; Guo, Z.T.; Xue, J.P.; Jin-Bu, L.I.; Sheng, W. Study on extraction of total flavonoids from pomegranate leaf by surfactant. *Sci. Technol. Food Ind.* **2014**, *23*, 261–264, 322.

11. Dong, S.G. Process research of the surfactant-assisted extraction of total flavonoids extraction from *Mulbery* leaves. *J. Anhui Agric. Sci.* **2011**, *39*, 19778–19780.

12. Eng, A.T.; Heng, M.Y.; Ong, E.S. Evaluation of surfactant assisted pressurized liquid extraction for the determination of glycyrrhizin and ephedrine in medicinal plants. *Anal. Chim. Acta* **2007**, *583*, 289–295. [CrossRef] [PubMed]

13. Yang, H.; Yan, Z.; Xiaoli, L.U. Surfactant assisted ultrasonic extraction of flavonoids from peanut hull. *China Oils Fats* **2012**, *37*, 57–60.

14. Fan, T.; Hu, J.G.; Fu, L.D.; Zhang, L.J. Optimization of enzymolysis-ultrasonic assisted extraction of polysaccharides from *Momordica charabtia* L. by response surface methodology. *Carbohydr. Polym.* **2015**, *115*, 701–706. [CrossRef] [PubMed]

15. Alara, O.R.; Abdurahman, N.H.; Olalere, O.A. Optimization of microwave-assisted extraction of flavonoids and antioxidants from Vernonia amygdalina leaf using response surface methodology. *Food Bioprod. Process.* **2018**, *107*, 36–48. [CrossRef]

16. Sheng, Z.L.; Wan, P.F.; Dong, C.L.; Li, Y.H. Optimization of total flavonoids content extracted from flos populi using response surface methodology. *Ind. Crop. Prod.* **2013**, *43*, 778–786. [CrossRef]

17. Zeković, Z.; Vidović, S.; Vladić, J.; Radosavljević, R.; Cvejin, A.; Elgndi, M.A.; Pavlić, B. Optimization of subcritical water extraction of antioxidants from coriandrum sativum, seeds by response surface methodology. *J. Supercrit. Fluids* **2014**, *95*, 560–566. [CrossRef]

18. Zengin, G.; Uysal, A.; Aktumsek, A.; Mocan, A.; Mollica, A.; Locatelli, M.; Mahomoodally, M.F. Euphorbia denticulata lam.: A promising source of phyto-pharmaceuticals for the development of novel functional formulations. *Biomed. Pharmacother.* **2017**, *87*, 27–36. [CrossRef] [PubMed]

19. Preece, D.; Montgomery, D. *Design and Analysis of Experiments*; Wiley: New York, NY, USA, 2010.

20. Draper, N.R.; Smith, H. *Applied Regression Analysis*; Wiley: New York, NY, USA, 1966.

21. Babaki, M.; Yousefi, M.; Habibi, Z.; Mohammadi, M. Process optimization for biodiesel production from waste cooking oil using multi-enzyme systems through response surface methodology. *Renew. Energy* **2017**, *105*, 465–472. [CrossRef]

22. Noordin, M.Y.; Venkatesh, V.C.; Sharif, S.; Elting, S.; Abdullah, A. Application of response surface methodology in describing the performance of coated carbide tools when turning aisi 1045 steel. *J. Mater. Process. Technol.* **2004**, *145*, 46–58. [CrossRef]

23. Sivaraman, B.; Shakila, R.J.; Jeyasekaran, G.; Sukumar, D.; Manimaran, U.; Sumathi, G. Antioxidant activities of squid protein hydrolysates prepared with papain using response surface methodology. *Food Sci. Biotechnol.* **2016**, *25*, 665–672. [CrossRef]

24. Amari, A.; Gannouni, A.; Chlendi, M.; Bellagi, A. Optimization by response surface methodology (RSM) for toluene adsorption onto prepared acid activated clay. *Can. J. Chem. Eng.* **2008**, *86*, 1093–1102. [CrossRef]

25. Muralidhar, R.V.; Chirumamila, R.R. A response surface approach for the comparison of lipase production by Canida cylindracea using two different carbon sources. *Biochem. Eng. J.* **2001**, *9*, 17–23. [CrossRef]

26. Fan, L.; Li, J.; Deng, K.; Ai, L. Effects of drying methods on the antioxidant activities of polysaccharides extracted from *Ganoderma lucidum*. *Carbohydr. Polym.* **2012**, *87*, 1849–1854. [CrossRef]

27. Wu, S.; Huang, X. Preparation and antioxidant activities of oligosaccharides from *Crassostrea gigas*. *Food Chem.* **2017**, *216*, 243–246. [CrossRef] [PubMed]

28. Qiao, D.; Ke, C.; Hu, B.; Luo, J.G.; Ye, H.; Sun, Y.; Yan, X.Y.; Zeng, X.X. Antioxidant activities of polysaccharides from *Hyriopsis cumingii*. *Carbohydr. Polym.* **2009**, *78*, 199–204. [CrossRef]

29. Parsa, A.; Salout, S.A. Investigation of the antioxidant activity of electrosynthesized polyaniline/reduced graphene oxide nanocomposite in a binary electrolyte system on abts and dpph free radicals. *J. Electroanal. Chem.* **2016**, *760*, 113–118. [CrossRef]

30. Farombi, E.O.; Akanni, O.O.; Emerole, G.O. Antioxidant and scavenging activities of flavonoid extract (kolaviron) of *Garcinia kola* seeds. *Pharm. Biol.* **2002**, *40*, 107–116. [CrossRef]

31. Srisawat, U.; Panunto, W.; Kaendee, N.; Tanuchit, S.; Itharat, A.; Lerdvuthisopon, N.; Hansakul, P. Determination of phenolic compounds, flavonoids, and antioxidant activities in water extracts of thai red and white rice cultivars. *J. Med. Assoc. Thail.* **2010**, *7*, 83–91.

32. Hearing, V.J. Biochemical control of melanogenesis and melanosomal organization. *J. Investig. Dermatol. Symp. Proc.* **1999**, *4*, 24–28. [CrossRef] [PubMed]

33. Chang, T.S. An updated review of tyrosinase inhibitors. *Int. J. Mol. Sci.* **2009**, *10*, 2440–2475. [CrossRef] [PubMed]

34. Rajasimman, M.; Murugaiyan, K. Sorption of nickel by hypnea valentiae: Application of response surface methodology. *World Acad. Sci. Eng. Technol.* **2011**, *1*, 7.

35. Karaden'iz, F.; Burdurlu, H.S.; Koca, N.; Soyer, Y. Antioxidant activity of selected fruits and vegetables grown in turkey. *Turkish J. Agric. For.* **2005**, *29*, 297–303.

36. Park, S.Y.; Je, J.Y.; Hwang, J.Y.; Ahn, C.B. Abalone protein hydrolysates: Preparation, angiotensin i converting enzyme inhibition and cellular antioxidant activity. *Prev. Nutr. Food Sci.* **2015**, *20*, 176–182. [CrossRef] [PubMed]

37. Qu, Y.; Li, C.; Zhang, C.; Zeng, C.; Fu, C. Optimization of infrared-assisted extraction of *Bletilla striata* polysaccharides based on response surface methodology and their antioxidant activities. *Carbohydr. Polym.* **2016**, *148*, 345–353. [CrossRef] [PubMed]

38. Wang, J.L.; Yang, W.; Tang, Y.Y.; Xu, Q.; Huang, S.L.; Yao, J.; Zhang, J.; Lei, Z.Q. Regioselective sulfation of artemisia sphaerocephala polysaccharide: Solution conformation and antioxidant activities in vitro. *Carbohydr. Polym.* **2016**, *136*, 527–536. [CrossRef] [PubMed]

39. Zhong, X.K.; Jin, X.; Lai, F.Y.; Lin, Q.S.; Jiang, J.G. Chemical analysis and antioxidant activities in vitro, of polysaccharide extracted from *Opuntia ficus indica*, Mill. cultivated in China. *Carbohydr. Polym.* **2010**, *82*, 722–727. [CrossRef]

40. Gülçin, İ. Antioxidant activity of caffeic acid (3,4-dihydroxycinnamic acid). *Toxicology* **2006**, *217*, 213–220. [CrossRef] [PubMed]

41. Tada, H.; Shiho, O.; Kuroshima, K.; Koyama, M.; Tsukamoto, K. An improved colorimetric assay for interleukin 2. *J. Immunol. Methods* **1986**, *93*, 157–165. [CrossRef]

42. Skandrani, I.; Pinon, A.A.; Ghedira, K.; Chekir, G.L. Chloroform extract from *Moricandia arvensis*, inhibits growth of b16-f0 melanoma cells and promotes differentiation in vitro. *Cell Prolif.* **2010**, *43*, 471–479. [CrossRef] [PubMed]

43. Uchida, R.; Ishikawa, S.; Tomoda, H. Inhibition of tyrosinase activity and melanin pigmentation by 2-hydroxytyrosol. *Acta Pharm. Sin. B* **2014**, *4*, 141–145. [CrossRef] [PubMed]

Sample Availability: Samples of the extracts are available from the authors.

molecules

MDPI

Article

Development of a NIR Method for the In-Line Quantification of the Total Polyphenolic Content: A Study Applied on *Ajuga genevensis* L. Dry Extract Obtained in a Fluid Bed Process

Alexandru Gavan [1], Liora Colobatiu [1,*], Andrei Mocan [2], Anca Toiu [3,†] and Ioan Tomuta [4,†]

[1]   Department of Medical Devices, Iuliu Hatieganu University of Medicine and Pharmacy, 4 Louis Pasteur Street, Cluj-Napoca 400439, Romania; gavan.alexandru@umfcluj.ro

[2]   Department of Pharmaceutical Botany, Iuliu Hatieganu University of Medicine and Pharmacy, 23 Gheorghe Marinescu Street, Cluj-Napoca 400337, Romania; mocan.andrei@umfcluj.ro

[3]   Department of Pharmacognosy, Iuliu Hatieganu University of Medicine and Pharmacy, 12 Ion Creanga Street, Cluj-Napoca 400010, Romania; atoiu@umfcluj.ro

[4]   Department of Pharmaceutical Technology and Biopharmacy, Iuliu Hatieganu University of Medicine and Pharmacy, 41 Victor Babes Street, Cluj-Napoca 400012, Romania; tomutaioan@umfcluj.ro

*   Correspondence: mihaiu.mihaela@umfcluj.ro; Tel.: +40-745-412-447

†   Joint last co-authorship.

Received: 21 July 2018; Accepted: 24 August 2018; Published: 27 August 2018

Abstract: This study describes an innovative *in-line* near-infrared (NIR) process monitoring method for the quantification of the total polyphenolic content (TPC) of *Ajuga genevensis* dry extracts. The dry extract was obtained in a fluidized bed processor, by spraying and adsorbing a liquid extract onto an inert powder support. NIR spectra were recorded continuously during the extract's spraying process. For the calibration of the *in-line* TPC quantification method, samples were collected during the entire process. The TPC of each sample was assessed spectroscopically, by applying a UV-Vis reference method. The obtained values were further used in order to develop a quality OPLS prediction model by correlating them with the corresponding NIR spectra. The final dry extract registered good flowability and compressibility properties, a concentration in active principles three times higher than the one of the liquid extract and an overall process yield of 85%. The average TPC's recovery of the NIR in-line prediction method, compared with the reference UV-Vis one, was 98.7%, indicating a reliable monitoring method which provided accurate predictions of the TPC during the process, permitting a good process overview and enabling us to establish the process's end point at the exact moment when the product reaches the desired TPC concentration.

Keywords: *Ajuga genevensis*; near-infrared spectroscopy; dry extract; fluid bed process; microNIR; in-line monitoring; total polyphenolic content

1. Introduction

The genus *Ajuga* L., from the family Lamiaceae, includes over 300 species of annual and perennial herbs in global distribution [1,2]. A number of six *Ajuga* species are mentioned in the spontaneous Romanian flora, these being mainly used in our traditional medicine due to their anti-inflammatory, wound healing, hepatoprotective, and anti-diarrheal pharmacological properties [3,4].

Phytochemical studies revealed the presence of multiple bioactive compounds in *Ajuga* species, including: phytoecdysteroids, *neo*-clerodane-diterpenes and diterpenoids, triterpenes, sterols, anthocyanidin-glucosides and iridoid glycosides, flavonoides, triglycerides, tocopherol and essential oils [3,5–7]. However, the active compounds of the same plant species may be different from one region to another, in terms of chemistry, pharmacology and toxicology. There is little information on the chemical composition of Romanian *Ajuga* species, only *A. reptans, A. genevensis* and *A. laxmannii* being previously analyzed [3,5,7,8].

In general, phytomedicines require exceptional chemical and physical stability, as well as low microbial growth in order to be used in therapy. Therefore, dry powder extracts are among the most suitable pharmaceutical forms of phytomedicines, due to their characteristics in the solid state, in addition to the possibility of easy preparation of tablets and capsules [9–11].

The fluid bed process is mostly used for drying and granulation purposes, being frequently an indispensable step in solid oral dosage form manufacturing. It usually consists in obtaining granules by spraying a binder solution over a fluidized powder bed [12]. However, the fluidized bed process can be also used as a fast method in order to obtain dry extracts by spraying and adsorbing liquid herbal hidroalcoholic extracts over a solid support [13].

The increasing demand for quality in herbal medicines has prompted a dire need for the quantitative analysis of bioactive compounds (including polyphenols) in herbs or medicinal plant extracts or herbal products by the use of efficient analytical methods [14].

Polyphenolic compounds are widely known today as antioxidant, antimicrobial, antiviral and anti-inflammatory agents, becoming desirable plant metabolites and being frequently used in the treatment of specific pathological conditions [15–18].

Different analytical methods such as high performance liquid chromatography (HPLC), gas chromatography (GC) or combinations of these analyses with mass spectrometry (MS) are currently employed in the determination of phenolic compounds [3,19–22]. Even though efficient, for they provide rapid separation and quantification of phenolic compounds, these techniques also require a lot of pre-processing steps before and during the analysis [17,22].

In recent years, the near-infrared (NIR) spectroscopy, especially combined with chemometric algorithm, has become an extensively used tool due to its fast, non-destructive, and low-cost characteristics, being widely applied in many unit operations of pharmaceutical manufacturing, including blending, granulation, fluidized bed drying, tablet-compression and coating, etc., [23]. NIR can be used *at-line* (measurement where the sample is removed, isolated from, and analyzed in close proximity to the process stream), *on-line* (measurement where the sample is diverted from the manufacturing process and may be returned to the process stream), as well as *in-line* (measurement where the sample is not removed from the process stream) [24]. Techniques based on NIR spectrometry have been used in fluidized bed granulation processes in order to determine the end-point of the process, as well as to monitor particle size and more frequently, moisture content (as water molecules strongly absorb NIR emissions on the 1400 and 1900 nm region) [23,25–27]. It has also been applied for the *in-line* quantification of film thickness of pharmaceutical pellets during fluid bed coating processes [28].

Recently, NIR spectroscopy (coupled with multivariate analyses) has also found excellent potential for quick and inexpensive quantification of total polyphenolic content (TPC) [29]. The technique has been recently applied (either *at-line* or *on-line*) to determine the content of phenolic compounds in yerba mate [22], the total content of emodin, chrysophanol, rhein, aloeemodin, and physcion in *Rhei Radix et Rhizoma* [30], for the simultaneous testing of total polyphenols, caffeine and free amino acids in Chinese tea from different categories [31], for measuring the phenolic composition of propolis [32], as well as for the total polyphenols quantification in *Acridocarpus orientalis* and *Moringa peregrina* [17].

Although, until now, *in-line* NIR spectroscopy technology has been mainly applied in the fluidized bed processes in order to monitor parameters such as particle size or moisture content, to the best of authors' knowledge, no such method has been yet developed in order to assess the TPC from dry extracts [23,33,34].

The aim of this study was to develop a manufacturing method for obtaining dry *A. genevensis* extracts by implementing a fluidized bed process, as well as to develop an innovative *in-line* NIR method for the quantification of their TPC.

2. Results and Discussion

The concentration of total polyphenols in the initially obtained liquid *A. genevensis* herbal extract was 1.88%. The liquid extract was further used in the fluidized bed preparation of the dry extract, as described in the "Materials and methods" section.

2.1. Dry Extract Characterization

The flow and compressibility properties of the obtained dry extract were determined according to the specifications of the European Pharmacopeia 9th ed. monograph, by measuring the untapped and tapped density and calculating the Carr's index (CI) and Hausner ratio (HR).

The registered mean values of the untapped and tapped density were 0.48 g/mL and 0.56 g/mL respectively. The HR of 1.16 and CI of 16% described good flow characteristics of the final dry extract.

The final moisture content of the obtained dry extract was 2.72%. The TPC of the 8 samples withdrawn during the extract's spraying phase are presented in Table 1. Based on the obtained results, the dry extract proved to be three times more concentrated in polyphenols compared to the liquid one.

The fluid extract's concentration of 1.88% represented 1.88 mg gallic acid/100 g dry herbal material, but taking into consideration the fact that the dry herbal material/solvent ratio was 10%, the liquid extract's concentration of 1.88% could also be expressed as 1.88 mg gallic acid/1 mL fluid extract. The concentration of the final dry extract obtained was 469.26 mg of gallic acid/100 g of dry extract (according to Table 1). Given that the fluid extract registered a density of 0.9595 g/mL, the 360 g of sprayed fluid extract represented 375 mL, the obtained 469.26 mg of gallic acid/100 g dry extract corresponding to 5.756 mg/mL fluid extract. Based on the fact that the 375 mL of fluid extract contained 6.768 mg gallic acid/mL fluid extract (according to the 1.88% gallic acid extract's concentration mentioned above), the overall process yield could be calculated. A high yield of 85% was registered, value reflecting an efficient process, which therefore allowed us to obtain a concentrated dry extract, with a minimum loss in active principles.

Table 1. Measured and near-infrared (NIR) predicted dry extract properties.

Sample	Spectra	Process Time (Min)	Reference UV-Vis Off-Line Method (mg Gallic Acid/ 100 g Dry Extract)	NIR-Chemometric In-Line Method (mg Gallic Acid/100 g Dry Extract)	Recovery (%)
1	30	05′	50.94	54.67	107.33
2	60	10′	106.79	86.93	81.40
3	90	15′	168.58	184.30	109.33
4	120	20′	215.75	203.30	94.23
5	150	25′	240.44	237.16	98.64
6	180	30′	324.81	340.44	104.81
7	210	35′	392.49	370.93	94.51
8	248	41′	469.26	468.05	99.74

2.2. NIR Spectra Analysis and Method Development

A NIR method has been developed for the *in-line* assessment of the TPC of the dry extract.

Firstly, all the acquired spectral data was imported into the SIMCA Multivariate Data Analysis (MVDA) software, in order to develop the spectral prediction model. A 2nd Savitzky–Golay (SG) derivative, with a quadratic polynomial order, was further applied to remove baseline effects and linear trends. This pre-processing technique also allowed the smoothing of the spectra and the improvement of the overall resolution of the information extracted from the spectra [35].

The raw reflectance spectra, as well as the spectra pre-processed with the 2nd SG derivative algorithm are illustrated in Figure 1.

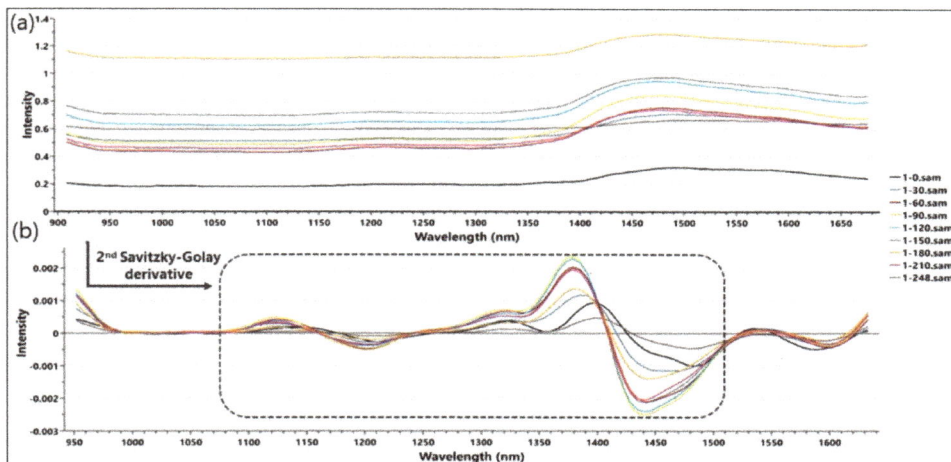

Figure 1. The raw (**a**) and 2nd derivative pre-processed (**b**) spectra registered during the fluid bed process and used for the development of the spectral prediction model.

The highest spectral intensity variations can be observed between 1070 nm and 1520 nm, a domain which was further selected for the development of the Principal Component Analysis (PCA) model, which registered a R2X of 0.981 for just one principal component (PC). In order to summarize the relationship among the variables, the loadings plot of the PC was plotted in Figure 2, overlapped with a 2nd derivative preprocessed spectrum. The loadings score values registered two significant peaks, inversely proportional with the two intensity peaks of the spectra. This observation suggests that the high intensity regions account for the fluctuation of the most substantial variation of the samples properties, represented by the TPC.

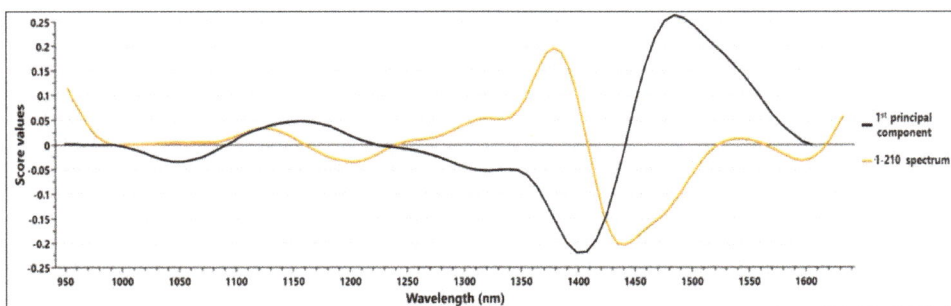

Figure 2. Loadings plot generated for the principal component analysis (PCA) model, overlapped with a preprocessed spectrum.

The same spectral region chosen for the PCA's model development has been further used for the development of the Orthogonal Partial Least Squares (OPLS) model. In order to do this, the 9 spectra (plotted in Figure 1), corresponding to the plain powder bed (t = 0) and the in-process withdrawn dry extract samples were correlated with the values representing their TPC assessed off-line by applying the UV-Vis reference method.

The calculated OPLS model included three factors: one predictive and two orthogonal ones. For the predictive fraction, high statistical parameters were registered, with R2X of 0.982 and Q2 of 0.951, showing good quality of the model. For the orthogonal (uncorrelated) fraction, just a low cumulated R2X of 0.077 was recorded, but which played an important role in the improvement of the overall quality of the model. The Root Mean Square Error of cross-validation (RMSEcv) corresponding to the three OPLS factors prediction model was 31.5 mg gallic acid/100 g dry extract, value which describes a good model, capable of delivering quality predictions.

Further, the scores scatter plot of the OPLS model was calculated and represented. This plot summarizes the relationship among the model's observations and allows the evaluation of their score value evolution. Figure 3 illustrates the predictive versus the first orthogonal component scores. It can be clearly noticed that the score values of the predictive component increase constantly starting with the first analyzed sample corresponding to the 30th spectra, to the last one corresponding to the 248th recorded spectra. In other words, the predictive component score values increase as the herbal extract adsorbing process goes on; i.e., the TPC increases.

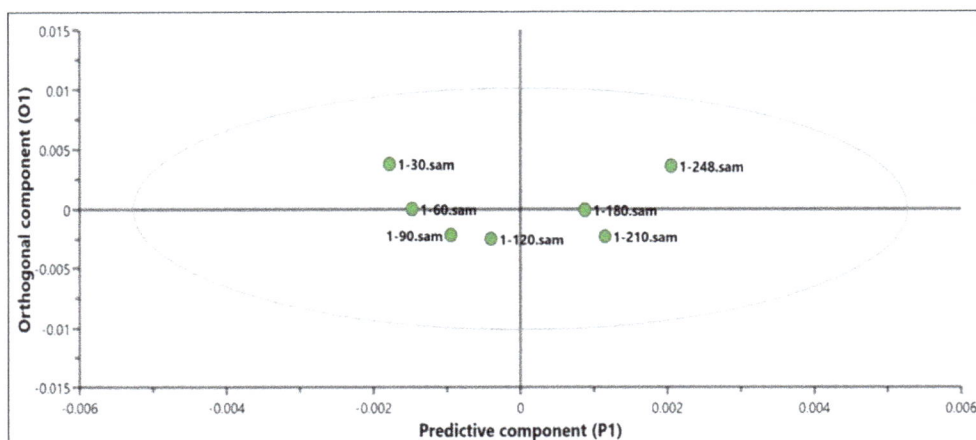

Figure 3. Scores scatter plot obtained from the orthogonal partial least squares (OPLS) analysis of the spectra corresponding to the 8 dry extract samples.

The orthogonal component scores evolution did not vary much along the monitored process, thus suggesting that this component did not significantly reflect the properties of the product.

In order to perform an internal cross-validation of the previously developed method, the OPLS model was further applied during the ongoing process, in order to predict the TPC throughout it.

The values obtained by applying the reference UV-Vis off-line method, as well as the NIR predicted results for the same samples are presented in Table 1. The NIR predictions were relatively accurate compared to the measured ones, registering an average recovery of 98.7%. The correlation coefficient (R2) between the measured and predicted values was 0.994, describing a reliable prediction model.

For a better visualization, the TPC predicted values and the 9 calibration sample values were plotted over time, as illustrated in Figure 4. At the beginning of the spraying process, for the first 5 min, the TPC in-line predictions were not very stable, causing high prediction fluctuations, a fact which can be explained by an inhomogeneous product at this specific point. Moreover, for the first 5 min, the predicted TPC values were on average way higher than the reference, errors caused by a higher quantity of residual solvent present at the beginning of the extract's spraying phase. As the process went on, the spraying and evaporation processes reached equilibrium, and their influence over the spectra became systematic. This systematic influence was assessed and eliminated with the use of MVDA during the development of the prediction model, resulting in the stabilization of the TPC predictions as the process went on and passed the first minutes.

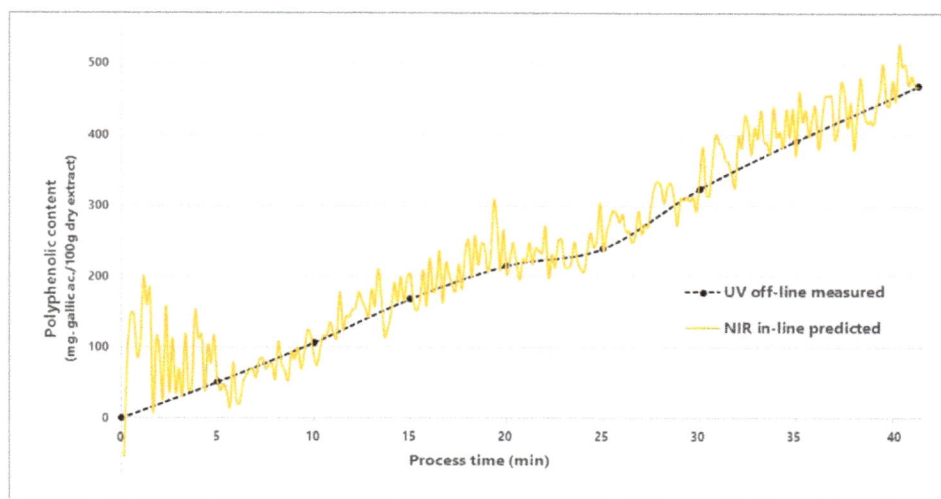

Figure 4. UV-Vis off-line measured vs. NIR in-line predicted total polyphenolic content.

Another observation made based on Figure 4 is that between minutes 20 and 25 the TPC increased at a slower pace compared with the rest of the process. This can be explained by the fact that, at some point, a small quantity of powder adhered to the NIR detector's surface, thus impeding the recording of the spectra of the powder bed. In order to remove the obstructive powder from the surface of the NIR detector, the spraying rate has been reduced from 9 g/min to approx. 5 g/min. Consequently, after a very short period of time (about 30–60 s), the adhered powder was set in motion and removed from the detector by the fluidized powder bed.

The previously described approach, used in order to predict the TPC of the obtained dry extract in real time, enhances the control of the overall process, also enabling the identification of any modifications of the liquid extract's spraying rate, which in this case represents a critical process parameter. Moreover, it allows the establishment of the process's end point exactly at the moment when the product reaches the desired TPC concentration.

3. Materials and Methods

3.1. Materials

The aerial parts of *A. genevensis* were harvested from wild populations from Cluj County, Romania, on July 2017, at full flowering stage. The voucher specimen of the studied plant was stored in the Herbarium of the Pharmacognosy Department of the Faculty of Pharmacy, Iuliu Hatieganu University of Medicine and Pharmacy Cluj-Napoca, Romania (accession number: AG-74).

The air-dried natural product was grounded to a fine powder and extracted with ethanol, by using 50 g of plant material and 500 mL 70% ethanol, at room temperature. Lactose monohydrate (Tablettose 80) was purchased from Meggle, (Wasserburg am Inn, Germany), while the microcrystalline cellulose (Avicel PH102) was obtained from FMC BioPolymer, (Philadelphia, PA, USA).

3.2. Dry Extract Preparation and Characterization

The dry *A. genevensis* extract was prepared in fluidized bed, using a laboratory scale Aeromatic Strea 1 (GEA, Kirchberg, Switzerland) fluid bed processor, by adsorbing 360 g of ethanolic extract on a solid support composed of 30 g lactose and 90 g microcrystalline cellulose (120 g lactose-cellulose solid support). The powder bed was fluidized with an air flow of 7.5–12 m^3/h, heated at an inlet temperature of just 30 °C, thus impeding the degradation of the product's active principles.

Firstly, the powder mixture was preheated and homogenized for 10 min. Afterwards, the alcoholic extract was sprayed from the top of the expansion vessel, with a spraying rate of 9 g/min, through a 0.8 mm nozzle, the formed product being further dried in the same apparatus, over a period of 8 min.

A total of 8 samples of approximately 3 g were withdrawn during the extract spraying phase, through a sampling port, which allowed the sample collection without interrupting the ongoing process.

The total content in polyphenols was determined for the fluid herbal extract sprayed during the process, as well as for the 8 in-process withdrawn samples. The TPC was registered spectrophotometrically, using the Folin–Ciocalteu method, with some modifications, as previously described by Gavan et al. [13]. The absorbance was measured at 760 nm, using a JASCO UV-VIS spectrophotometer (JASCO, Tokyo, Japan). The calibration curve was obtained by using different concentrations of gallic acid solutions, the TPC being expressed as mg gallic acid/100 g of *A. genevensis* dry extract.

The flow and compressibility properties of the obtained dry extract were determined by measuring the untapped and tapped density, by using a SVM tapped density tester (Erweka, Germany), followed by the calculation of the CI and HR [36].

The final moisture content of the dry extract was analyzed with the aid of an Ohaus MB45 (Ohaus, Parsippany, NJ, USA) humidity balance. Approximately 1 g of extract was placed onto the sample pan and subsequently dried at 80 °C, until the weight change was less than 0.2% over 10 min, the total percent of mass loss being automatically calculated.

In order to assess the performance of the developed process, the yield of the active principles adsorption was calculated using Equation (1).

$$\text{Yield (\%)} = \frac{\text{Adsorbed active principles content}}{\text{Introduced active principles content}} \times 100 \tag{1}$$

3.3. NIR Process Monitoring

The *in-line* monitoring of the fluid extract's spraying process was performed using a MicroNir Pat-U spectrometer (Viavi Solutions, San Jose, CA, USA). This apparatus incorporates a Linear Variable Filter enabling its size reduction and facilitating the device's direct attachment to the walls of the expansion vessel or to any other manufacturing apparatus, without using a fiber optic probe that could cause signal attenuation or dispersion. Therefore, the spectrometer is more stable and the measurements performed more reliable [37–41].

The apparatus was attached to the expansion vessel at the same height as the sampling port, so that the NIR detector could register the spectra by avoiding any interference with the process, as well as with the fluidized powder bed.

Spectra of the moving powder bed were registered in-line during the spraying of the liquid extract, without interrupting the ongoing process. The recordings have been performed continuously, at every 10 s, in reflectance mode, over the whole range of the spectrometer (950–1650 nm), with a resolution of 6 nm. Each spectra represented the average of 200 scans, recorded with an integration time of 7 ms per scan. All the above-mentioned parameters were set and controlled by the MicroNir spectrometer's own JDSU Pro software.

3.4. Spectral Data Analysis

In order to extract the desired information from the large amount of in-line registered spectral data, the SIMCA 14.0 (Sartorius Stedim, Umea, Sweden) software was used to perform the necessary multivariate data analysis.

Firstly, a PCA was performed in order to get an overview of the gathered spectral data, as well as to identify the spectral domains which are specific to the changes related to the polyphenolic content.

Secondly, an OPLS model was developed for the prediction of the TPC. This kind of model separates the X-specific spectral systematic variation into predictive and orthogonal (uncorrelated) fractions. An optimal number of OPLS factors was chosen based on the highest fraction of X variation modeled in the component (R2X), on the fraction of Y variation predicted according to cross-validation, by using the X model (Q2) and low RMSEcv and, in the same time, by avoiding the overfitting of the model [42].

The previously developed OPLS model allowed the *in-line* monitoring of the loading of the active principles on the solid support, thus enabling the control of the process and the establishment of the end point of the spraying process exactly at the moment when the product reached the desired concentration in active principles.

4. Conclusions

In the current study a manufacturing method for obtaining dry *A. genevensis* extracts by adsorbing the fluid extract onto an inert powder support, in a fluid bed process, was described. Moreover, an innovative NIR *in-line* polyphenolic content quantification technique was successfully developed.

The obtained results showed that the liquid extract's adsorption progressed smoothly throughout the process and that an increased quantity of active principles could be incorporated in the final product. Based on the obtained results, the dry extract proved to be three times more concentrated in polyphenolic compounds compared to the liquid one.

The method described in our study enables the monitoring of the technological process through the real-time quantification of the TPC, thus allowing the establishment of the process's end point exactly at the moment when the desired TPC concentration has been reached. The novelty of this study resides in the fact that, to the best of our knowledge, no such method has been previously described.

Author Contributions: Conceptualization, A.G., A.T. and I.T.; Investigation, A.G. and A.T.; Methodology, A.G., A.T. and I.T.; Supervision, I.T.; Writing-original draft, A.G. and L.C.; Writing-review and editing, A.G., L.C., A.M., A.T. and I.T.

Funding: This research was funded by UEFISCDI, Romania, project no. PNII-RU-TE-2014-4-1247.

Conflicts of Interest: The authors declare no conflict of interest.

References

1. Sivanesan, I.; Jeong, B.R. Silicon promotes adventitious shoot regeneration and enhances salinity tolerance of Ajuga multiflora bunge by altering activity of antioxidant enzyme. *Sci. World J.* **2014**, *2014*, 521703. [CrossRef] [PubMed]
2. Park, H.Y.; Kim, D.H.; Sivanesan, I. Micropropagation of Ajuga species: a mini review. *Biotechnol. Lett.* **2017**, *39*, 1291–1298. [CrossRef] [PubMed]
3. Toiu, A.; Vlase, L.; Arsene, A.L.; Vodnar, D.C.; Oniga, I. LC/UV/MS profile of polyphenols, antioxidant and antimicrobial effects of Ajuga genevensis L. extracts. *Farmacia* **2016**, *64*, 53–57.
4. Toiu, A.; Vlase, L.; Gheldiu, A.M.; Vodnar, D.; Oniga, I. Evaluation of the antioxidant and antibacterial potential of bioactive compounds from Ajuga Reptans extracts. *Farmacia* **2017**, *65*, 351–355.
5. Ni, B.; Dong, X.; Fu, J.; Yin, X.; Lin, L.; Xia, Z.; Zhao, Y.; Xue, D.; Yang, C.; Ni, J. Phytochemical and biological properties of ajuga decumbens (labiatae): A review. *Trop. J. Pharm. Res.* **2015**, *14*, 1525. [CrossRef]
6. Sivanesan, I.; Saini, R.K.; Noorzai, R.; Zamany, A.J.; Kim, D.H. In vitro propagation, carotenoid, fatty acid and tocopherol content of Ajuga multiflora Bunge. *3 Biotech* **2016**, *6*, 91. [CrossRef] [PubMed]

7. Toiu, A.; Mocan, A.; Vlase, L.; Pârvu, A.E.; Vodnar, D.C.; Gheldiu, A.M.; Moldovan, C.; Oniga, I. Phytochemical composition, antioxidant, antimicrobial and in vivo anti-inflammatory activity of traditionally used Romanian Ajuga laxmannii (Murray) Benth. ("nobleman's beard" - barba împăratului). *Front. Pharmacol.* **2018**, *9*, 1–15. [CrossRef] [PubMed]

8. Ghita, G.; Cioanca, O.; Gille, E.; Necula, R.; Zamfirache, M.M.; Stanescu, U. Contributions to the phytochemical study of some samples of *Ajuga reptans* L. and *Ajuga genevensis* L. *Bull. Transilv. Univ. Braşov Ser. VI* **2011**, *4*, 7–14.

9. Muzaffar, K.; Kumar, P. Parameter optimization for spray drying of tamarind pulp using response surface methodology. *Powder Technol.* **2015**, *279*, 179–184. [CrossRef]

10. Patil, V.; Chauhan, A.K.; Singh, R.P. Optimization of the spray-drying process for developing guava powder using response surface methodology. *Powder Technol.* **2014**, *253*, 230–236. [CrossRef]

11. Teixeira, C.C.C.; de Freitas Cabral, T.P.; Tacon, L.A.; Villardi, I.L.; Lanchote, A.D.; Freitas, L.A.P. de Solid state stability of polyphenols from a plant extract after fluid bed atmospheric spray-freeze-drying. *Powder Technol.* **2017**, *319*, 494–504. [CrossRef]

12. Cantor, S.; Augsburger, L.; Hoag, S.; Gerhardt, A. Wet granulation. In *Pharmaceutical Dosage Forms: Tablets*; Informa Healthcare: New York, NY, USA, 2008; pp. 1932–1944.

13. Găvan, A.; Toiu, A.M.; Tamaş, M.; Tomuţă, I. Dry rose petal extracts and compressed lozenges: Formulation and NIR—Chemometric quantification of active principles. *Farmacia* **2017**, *65*, 577–583.

14. McGoverin, C.M.; Weeranantanaphan, J.; Downey, G.; Manley, M. Review: The Application of near Infrared Spectroscopy to the Measurement of Bioactive Compounds in Food Commodities. *J. Near Infrared Spectrosc.* **2010**, *18*, 87–111. [CrossRef]

15. Mocan, A.; Crişan, G.; Vlase, L.; Crişan, O.; Vodnar, D.; Raita, O.; Gheldiu, A.-M.; Toiu, A.; Oprean, R.; Tilea, I. Comparative Studies on Polyphenolic Composition, Antioxidant and Antimicrobial Activities of Schisandra chinensis Leaves and Fruits. *Molecules* **2014**, *19*, 15162–15179. [CrossRef] [PubMed]

16. Andriamadio, J.H.; Rasoanaivo, L.H.; Benedec, D.; Vlase, L.; Gheldiu, A.-M.; Duma, M.; Toiu, A.; Raharisololalao, A.; Oniga, I. HPLC/MS analysis of polyphenols, antioxidant and antimicrobial activities of *Artabotrys hildebrandtii* O. Hffm. extracts. *Nat. Prod. Res.* **2015**, *29*, 2188–2196. [CrossRef] [PubMed]

17. Ali, L.; Mabood, F.; Rizvi, T.S.; Rehman, N.U.; Arman, M.; Al-Shidani, S.; Al-Abri, Z.; Hussain, J.; Al-Harrasi, A. Total polyphenols quantification in Acridocarpus orientalis and Moringa peregrina by using NIR spectroscopy coupled with PLS regression. *Chem. Data Collect.* **2018**, *13–14*, 104–112. [CrossRef]

18. Pandey, K.B.; Rizvi, S.I. Plant polyphenols as dietary antioxidants in human health and disease. *Oxid. Med. Cell. Longev.* **2009**, *2*, 270–278. [CrossRef] [PubMed]

19. Berté, K.A.S.; Beux, M.R.; Spada, P.K.W.D.S.; Salvador, M.; Hoffmann-Ribani, R. Chemical Composition and Antioxidant Activity of Yerba-Mate (*Ilex paraguariensis* A.St.-Hil., Aquifoliaceae) Extract as Obtained by Spray Drying. *J. Agric. Food Chem.* **2011**, *59*, 5523–5527. [CrossRef] [PubMed]

20. Jacques, R.A.; Santos, J.G.; Dariva, C.; Oliveira, J.V.; Caramão, E.B. GC/MS characterization of mate tea leaves extracts obtained from high-pressure CO_2 extraction. *J. Supercrit. Fluids* **2007**, *40*, 354–359. [CrossRef]

21. Gini, T.G.; Jeya Jothi, G. Column chromatography and HPLC analysis of phenolic compounds in the fractions of Salvinia molesta mitchell. *Egypt. J. Basic Appl. Sci.* **2018**. [CrossRef]

22. Frizon, C.N.T.; Oliveira, G.A.; Perussello, C.A.; Peralta-Zamora, P.G.; Camlofski, A.M.O.; Rossa, Ü.B.; Hoffmann-Ribani, R. Determination of total phenolic compounds in yerba mate (Ilex paraguariensis) combining near infrared spectroscopy (NIR) and multivariate analysis. *LWT Food Sci. Technol.* **2015**, *60*, 795–801. [CrossRef]

23. Liu, R.; Li, L.; Yin, W.; Xu, D.; Zang, H. Near-infrared spectroscopy monitoring and control of the fluidized bed granulation and coating processes—A review. *Int. J. Pharm.* **2017**, *530*, 308–315. [CrossRef] [PubMed]

24. U.S. Department of Health and Human Services; Food and Drug Administration; Center for Drug Evaluation and Research (CDER); Center for Veterinary Medicine (CVM); Office of Regulatory Affairs (ORA); Pharmaceutical CGMPs. FDA Guidance for Industry: PAT—A Framework for Innovative Pharmaceutical, Development, Manufacturing, and Quality Assurance. Food and Drug Administration: Rockville, MD, USA, 2004. Available online: http://www.fda.gov/CDER/guidance/6419fnl.pdf (accessed on September 2004).

25. Frake, P.; Greenhalgh, D.; Grierson, S.M.; Hempenstall, J.M.; Rudd, D.R. Process control and end-point determination of a fluid bed granulation by application of near infra-red spectroscopy. *Int. J. Pharm.* **1997**, *151*, 75–80. [CrossRef]

26. Watano, S.; Sato, Y.; Miyanami, K. Scale-Up of Agitation Fluidized Bed Granulation. IV. Scale-Up Theory Based on the Kinetic Energy Similarity. *Chem. Pharm. Bull.* **1995**, *43*, 1227–1230. [CrossRef]

27. Da Silva, C.A.M.; Butzge, J.; Nitz, M.; Taranto, O.P. Monitoring and control of coating and granulation processes in fluidized beds—A review. *Adv. Powder Technol.* **2014**, *25*, 195–210. [CrossRef]

28. Lee, M.J.; Seo, D.Y.; Lee, H.E.; Wang, I.C.; Kim, W.S.; Jeong, M.Y.; Choi, G.J. In line NIR quantification of film thickness on pharmaceutical pellets during a fluid bed coating process. *Int. J. Pharm.* **2011**, *403*, 66–72. [CrossRef] [PubMed]

29. Rizvi, T.S.; Mabood, F.; Ali, L.; Al-Broumi, M.; Al Rabani, H.K.M.; Hussain, J.; Jabeen, F.; Manzoor, S.; Al-Harrasi, A. Application of NIR Spectroscopy Coupled with PLS Regression for Quantification of Total Polyphenol Contents from the Fruit and Aerial Parts of *Citrullus colocynthis*. *Phytochem. Anal.* **2018**, *29*, 16–22. [CrossRef] [PubMed]

30. Zhan, H.; Fang, J.; Wu, H.; Yang, H.; Li, H.; Wang, Z.; Yang, B.; Tang, L.; Fu, M. Rapid Determination of Total Content of Five Major Anthraquinones in Rhei Radix et Rhizoma by NIR Spectroscopy. *Chin. Herb. Med.* **2017**, *9*, 250–257. [CrossRef]

31. Wang, J.; Wang, Y.; Cheng, J.; Wang, J.; Sun, X.; Sun, S.; Zhang, Z. *Enhanced Cross-Category Models for Predicting the Total Polyphenols, Caffeine and Free Amino Acids Contents in Chinese Tea Using NIR Spectroscopy*; Elsevier Ltd.: New York, NY, USA, 2018; Volume 96.

32. Revilla, I.; Vivar-Quintana, A.M.; González-Martín, I.; Escuredo, O.; Seijo, C. The potential of near infrared spectroscopy for determining the phenolic, antioxidant, color and bactericide characteristics of raw propolis. *Microchem. J.* **2017**, *134*, 211–217. [CrossRef]

33. Paul Findlay, W.; Peck, G.R.; Morris, K.R. Determination of fluidized bed granulation end point using near-infrared spectroscopy and phenomenological analysis. *J. Pharm. Sci.* **2005**, *94*, 604–612. [CrossRef] [PubMed]

34. Nieuwmeyer, F.J.S.; Damen, M.; Gerich, A.; Rusmini, F.; van der Voort Maarschalk, K.; Vromans, H. Granule Characterization During Fluid Bed Drying by Development of a Near Infrared Method to Determine Water Content and Median Granule Size. *Pharm. Res.* **2007**, *24*, 1854–1861. [CrossRef] [PubMed]

35. Tomuta, I.; Porfire, A.; Casian, T.; Gavan, A. Multivariate Calibration for the Development of Vibrational Spectroscopic Methods. In *Calibration and Validation of Analytical Methods*; Stauffer, M., Ed.; IntechOpen: London, UK, 2018; pp. 35–58.

36. European Directorate for the Quality of Medicines and HealthCare. *European Pharmacopoeia*; 9.0.; European Directorate for the Quality of Medicines and HealthCare: Strasbourg, France, 2017.

37. Malegori, C.; Nascimento Marques, E.J.; de Freitas, S.T.; Pimentel, M.F.; Pasquini, C.; Casiraghi, E. Comparing the analytical performances of Micro-NIR and FT-NIR spectrometers in the evaluation of acerola fruit quality, using PLS and SVM regression algorithms. *Talanta* **2017**, *165*, 112–116. [CrossRef] [PubMed]

38. Alcalà, M.; Blanco, M.; Moyano, D.; Broad, N.W.; O'Brien, N.; Friedrich, D.; Pfeifer, F.; Siesler, H.W. Qualitative and Quantitative Pharmaceutical Analysis with a Novel Hand-Held Miniature near Infrared Spectrometer. *J. Near Infrared Spectrosc.* **2013**, *21*, 445–457. [CrossRef]

39. Rohwedder, J.J.R.; Pasquini, C.; Fortes, P.R.; Raimundo, I.M.; Wilk, A.; Mizaikoff, B. iHWG-μNIR: a miniaturised near-infrared gas sensor based on substrate-integrated hollow waveguides coupled to a micro-NIR-spectrophotometer. *Analyst* **2014**, *139*, 3572. [CrossRef] [PubMed]

40. Friedrich, D.M.; Hulse, C.A.; von Gunten, M.; Williamson, E.P.; Pederson, C.G.; O'Brien, N.A. Miniature near-infrared spectrometer for point-of-use chemical analysis. In *Photonic Instrumentation Engineering*; Soskind, Y.G., Olson, C., Eds.; SPIE: Bellingham, WA, USA, 2014; p. 899203.

41. Sun, L.; Hsiung, C.; Pederson, C.G.; Zou, P.; Smith, V.; von Gunten, M.; O'Brien, N.A. Pharmaceutical Raw Material Identification Using Miniature Near-Infrared (MicroNIR) Spectroscopy and Supervised Pattern Recognition Using Support Vector Machine. *Appl. Spectrosc.* **2016**, *70*, 816–825. [CrossRef] [PubMed]

42. Eriksson, L.; Byrne, T.; Johansson, E.; Trygg, J.; Vikstrom, C. *Multi-and Megavariate Data Analysis. Basic Principles and Applications*, 3rd ed.; MKS Umetrics AB: Umea, Sweden, 2013.

Sample Availability: Samples of the compounds are not available from the authors.

molecules

MDPI

Article

Insights into Tissue-specific Specialized Metabolism in *Tieguanyin* Tea Cultivar by Untargeted Metabolomics

Si Chen [1,2,†], Jun Lin [2,†], Huihui Liu [1,2], Zhihong Gong [2], Xiaxia Wang [2], Meihong Li [1,2], Asaph Aharoni [3], Zhenbiao Yang [2,4] and Xiaomin Yu [2,*]

[1] College of Horticulture, Fujian Agriculture and Forestry University, Fuzhou 350002, China; cstc1990@hotmail.com (S.C.); liuhh42194@foxmail.com (H.L.); limei123home@gmail.com (M.L.)
[2] FAFU-UCR Joint Center for Horticultural Biology and Metabolomics, Fujian Provincial Key Laboratory of Haixia Applied Plant Systems Biology, Fujian Agriculture and Forestry University, Fuzhou 350002, China; realnadal@163.com (J.L.); zhihong_gong@sina.com (Z.G.); wangxiaxia530@126.com (X.W.); yang@ucr.edu (Z.Y.)
[3] Department of Plant & Environmental Sciences, Weizmann Institute of Science, P. O. Box 26, Rehovot 7610001, Israel; asaph.aharoni@weizmann.ac.il
[4] Center for Plant Cell Biology, Institute for Integrative Genome Biology, and Department of Botany and Plant Sciences, University of California, Riverside, CA 92521, USA
* Correspondence: xmyu0616@fafu.edu.cn; Tel.: +86-591-8639-1591
† These authors contributed equally to this work.

Academic Editors: Marcello Locatelli, Simone Carradori and Andrei Mocan
Received: 3 July 2018; Accepted: 19 July 2018; Published: 21 July 2018

Abstract: Tea plants produce extremely diverse and abundant specialized metabolites, the types and levels of which are developmentally and environmentally regulated. However, little is known about how developmental cues affect the synthesis of many of these molecules. In this study, we conducted a comparative profiling of specialized metabolites from six different tissues in a premium oolong tea cultivar, *Tieguanyin*, which is gaining worldwide popularity due to its uniquely rich flavors and health benefits. UPLC-QTOF MS combined with multivariate analyses tentatively identified 68 metabolites belonging to 11 metabolite classes, which exhibited sharp variations among tissues. Several metabolite classes, such as flavonoids, alkaloids, and hydroxycinnamic acid amides were detected predominantly in certain plant tissues. In particular, tricoumaroyl spermidine and dicoumaroyl putrescine were discovered as unique tea flower metabolites. This study offers novel insights into tissue-specific specialized metabolism in *Tieguanyin*, which provides a good reference point to explore gene-metabolite relationships in this cultivar.

Keywords: oolong tea; *Tieguanyin* tea cultivar; metabolite profiling; UPLC-QTOF MS; metabolomics

1. Introduction

Tea is the world's most consumed beverage, second only to water. The popularity of tea can be partly accounted for by the diversity of its taste and aroma, owing to the diversity and abundance of specialized metabolites in tea. Tea consumption has also been linked to a number of medicinal and nutritional properties resulting from a wide array of phytochemicals in tea plants (*Camellia sinensis*) [1]. Great efforts have been made by the tea research community to functionally characterize the bioactive components in tea. In particular, catechins, caffeine, and theanine, three of the most characteristic metabolites known to be closely associated with tea flavor and quality, have been extensively studied molecularly and biochemically [2–8]. With the recent release of genome sequences for *C. sinensis* var.

sinensis [9] and *C. sinensis* var. *assamica* [7], new insights into the molecular basis for the rich production of bioactive metabolites in tea plants will likely emerge.

The production of specialized metabolites is believed to be employed by plants mostly for the purpose of chemical defense or communication, increasing the overall fitness of the given plant producing them [10]. As with many other plants, specialized metabolism in tea plants varies in a tissue and species-specific manner, and is sensitive to both biotic and abiotic cues [7,11]. The biosynthesis of catechins, caffeine, and theanine in response to developmental cues has been most studied in tea plants [7,9]. Catechins are derived from the phenylpropanoid and flavonoid pathways [12]. The biosynthesis of caffeine involves three methylation steps to sequentially convert xanthosine to 7-methylxanthine to theobromine, and then finally to caffeine [13]. Theanine biosynthesis is catalyzed by theanine synthase acting on glutamate and ethylamine as substrates [14]. By RNA sequencing of various tissues from different developmental stages of cultivar *Longjing 43*, Li et al. analyzed the expression patterns of genes involved in the biosynthesis of flavonoids, caffeine, and theanine, and built a possible transcription factor network for the regulation of these three pathways [12]. Developmental changes in the abundance of catechins [3,11], caffeine [5,15], and theanine [5,6], and the differential expression of relevant genes in respective pathways were also documented in other tea cultivars. Nevertheless, most studies have largely focused on one or several classes of target metabolites, questions about how gene expression affects the metabolic make-up, and the distribution patterns of specialized metabolites in different tissues are not fully understood. As transcriptomics alone could not reflect the actual biochemical status (and hence the real physiology) of tea plants, non-targeted metabolomics, which involve the qualitative and semi-quantitative detection of a high number of metabolites participating in various cellular activities, is required for a more direct and comprehensive measurement of biological activities in the individual tissues of tea plants. The same approach has been successfully applied to profile many plant species during development or in response to changing environmental stress [16–19]. However, the application of untargeted metabolomics to examine the overall difference in the metabolic profiles among tea plant tissues has not been thoroughly performed in any tea cultivars.

Based on the methods of tea leaf processing, tea has been categorized into six major types: green, yellow, oolong, white, black, and dark tea [20]. *Tieguanyin* tea, originating from Anxi County, Fujian province of China, is a premium variety of oolong tea renowned for its uniquely rich flavors and various health benefits [21–23]. Due to its increasing popularity among consumers, the plantation of *Tieguanyin* tea cultivar has been greatly expanded, spreading from Anxi and surrounding areas in the Fujian province to many other regions in China. A thorough understanding of the biology and metabolism of *Tieguanyin* tea plants would facilitate the development of high-quality tea products but remains underexplored. To date, a limited number of available studies on this cultivar have only focused on the geographic origin discrimination and targeted analyses of chemical changes during processing using processed tea [7,21,24,25].

In the present study, we comprehensively investigated the phytochemical profile of *Tieguanyin* cultivar by applying a non-targeted metabolomics workflow, with the aim of revealing the differences and similarities in the metabolite composition among different tissues and identifying the tissue-specific distribution patterns of specialized metabolites. Our results provide novel insights into the developmental regulation of the specialized metabolism in *Tieguanyin* and reveal intriguing variations in the diverse classes of metabolites besides known compounds. It likely offers a valuable reference for future characterizations of the gene–metabolite relationships of metabolites uncovered in the current study.

2. Results

2.1. Prominent Metabolite Variations Observed between Tea Plant Tissues

To assess metabolite compositional differences between tissues of *Tieguanyin* tea plants, non-targeted analysis based on UPLC-QTOF MS (ultra-performance liquid chromatography-quadrupole time-of-flight mass spectrometry) was carried out to profile methanol-soluble extracts of buds, young leaves, mature leaves, new stems, flowers, and lateral roots (Figure 1 and Figure S1). Metabolite profiles were presented as PCA score plots, PCA loading plots, and the heat map (Figures 2 and 3). A total of 68 differential compounds (VIP > 1 and $p < 0.05$) were tentatively identified on the basis of their accurate masses, MS/MS fragmentation patterns, and UV absorbance, in comparison to standard compounds and references (Table 1). They were classified into 11 major classes including flavan-3-ols, proanthocyanidins, flavonol glycosides, flavone glycosides, phenolic acids, hydrolysable tannins, alkaloids, hydroxycinnamic acid amides, amino acids, aromatic alcohol glycosides, and terpenoid glycosides.

Figure 1. Phenotypic characterization of six tissues of *Tieguanyin* tea plants used in the current study. (1) bud, (2) young leaf, (3) mature leaf, (4) new stem, (5) flower, and (6) lateral root.

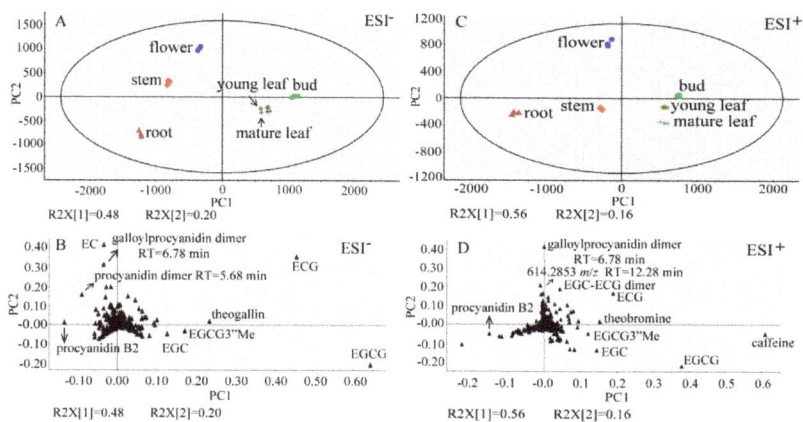

Figure 2. Metabolic profiles of tea tissue samples demonstrated by PCA score plots and PCA loading plots. (**A**) PCA score plot for tea tissue samples based on 732 single molecular features detected in ESI⁻. (**B**) PCA loading plot showing major metabolites that contribute to group separation in ESI⁻. (**C**) PCA score plot for tea tissue samples based on 821 single molecular features detected in ESI⁺. (**D**) PCA loading plot showing major metabolites that contribute to group separation in ESI⁺. R2X, explained variation. PC1, the first principal component. PC2, the second principal component. For each tissue type, three biological replicates were prepared, where one replicate was a pool of collected materials from three tea plants.

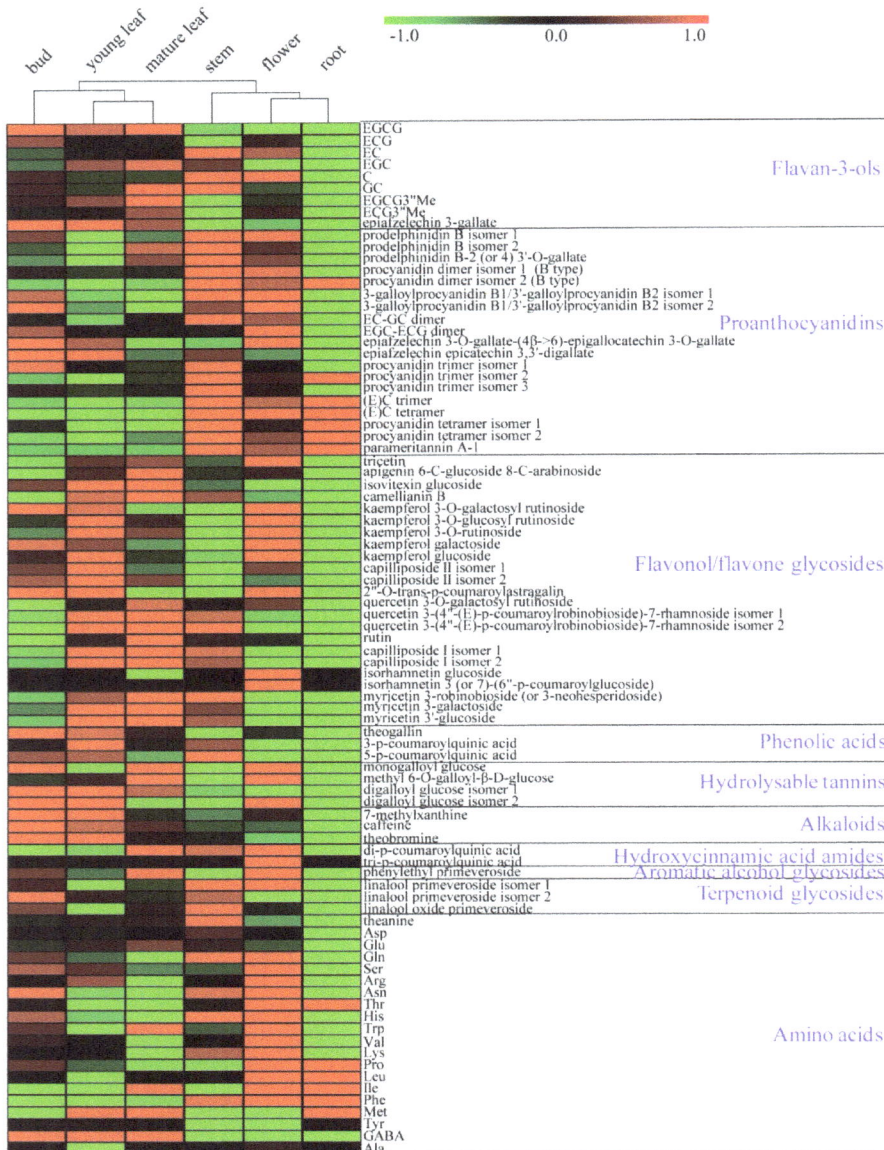

Figure 3. Comparisons of metabolite levels in six tissues. The analysis is based on the normalized average signal abundance from three biological replicates for each tissue type. Normalized values are shown on a color scale proportional to the content of each metabolite, and are expressed as log2 using the MultiExperiment Viewer software (MeV v4.9.0, J. Craig Venter Institute, La Jolla, CA, USA).

Table 1. Metabolites putatively identified in six tissues of *Tieguanyin* tea plants by UPLC-QTOF MS.

Compound	Tentative Assignments	Rt (min)	Detected [M − H]⁻ (m/z)	Theoretical [M − H]⁻ (m/z)	Mass Error (ppm)	Formula	MS/MS Fragments	Ref.
				Flavan-3-ols				
1	GC	3.84	305.0670	305.0661	2.95	$C_{15}H_{14}O_7$	219.0664, 179.0353, 167.0351, 139.0399, 125.0245	Authentic standard [b]
2	EGC	4.93	305.0677	305.0661	5.24	$C_{15}H_{14}O_7$	219.0667, 179.0349, 167.0351, 139.0402, 125.0245	Authentic standard [b]
3	C	5.36	289.0719	289.0712	2.42	$C_{15}H_{14}O_6$	245.0822, 203.0714, 125.0246	Authentic standard [b]
4	EC	6.27	289.0721	289.0712	3.11	$C_{15}H_{14}O_6$	245.0819, 203.0714, 123.0451	Authentic standard [b]
5	EGCG	6.35	457.0797	457.0771	5.69	$C_{22}H_{18}O_{11}$	305.0669, 169.0160, 125.0247	Authentic standard [b]
6	EGCG3″Me	7.42	471.0934	471.0927	1.49	$C_{23}H_{20}O_{11}$	305.0674, 287.0568, 183.0304, 161.0251, 125.0247	Authentic standard [b]
7	ECG	7.86	441.0828	441.0822	1.36	$C_{22}H_{18}O_{10}$	331.0462, 289.0720, 245.0819, 169.0147, 125.0245	Authentic standard [b]
8	ECG3″Me	8.92	455.0960	455.0978	−3.96	$C_{23}H_{20}O_{10}$	289.0721, 183.0302	[26]
9	epiafzelechin 3-gallate	8.97	425.0881	425.0873	1.88	$C_{22}H_{18}O_9$	273.0761, 169.0140, 151.0029, 137.0245, 125.0243	[27]
				Proanthocyanidins				
10	prodelphinidin B isomer 1	3.34	609.1246	609.1244	0.33	$C_{30}H_{26}O_{14}$	483.0947, 441.0827, 423.0717, 305.0667	[26]
11	prodelphinidin B isomer 2	4.11	609.1249	609.1244	0.82	$C_{30}H_{26}O_{14}$	483.0932, 441.0822, 423.0716, 305.0668	[26]
12	EC-GC dimer	4.80	593.1300	593.1295	0.84	$C_{30}H_{26}O_{13}$	425.0857, 423.0707, 305.0670, 289.0717, 125.0245	[27]
13	prodelphinidin B-2 (or 4) 3′-O-gallate	5.11	761.1352	761.1354	−0.26	$C_{37}H_{30}O_{18}$	609.1236, 591.1135, 577.1348, 423.0718	[26]
14	procyanidin trimer (B type) isomer 1	5.25	865.1962	865.1980	−2.08	$C_{45}H_{38}O_{18}$	695.1369, 577.1319, 451.1034, 287.0553	[16]
15	procyanidin trimer (B type) isomer 2	5.52	865.1966	865.1980	−1.62	$C_{45}H_{38}O_{18}$	695.1389, 575.1181, 451.0982, 287.0561	[16]
16	procyanidin dimer (B type) isomer 1	5.68	577.1349	577.1346	0.52	$C_{30}H_{26}O_{12}$	451.1031, 425.0873, 407.0766, 289.0717, 125.0243	[28]
17	procyanidin B2	5.78	577.1326	577.1346	−3.47	$C_{30}H_{26}O_{12}$	451.1022, 425.0864, 407.0763, 289.0713, 125.0243	Authentic standard [b]
18	procyanidin tetramer (B type) isomer 1	5.88	1153.2599	1153.2614	−1.30	$C_{60}H_{50}O_{24}$	1027.2271, 865.1966, 576.1259, 575.1178, 287.0546	[29]
19	procyanidin tetramer (B type) isomer 2	5.97	1153.2599	1153.2614	−1.30	$C_{60}H_{50}O_{24}$	1027.2234, 577.1329, 575.1175, 287.0557	[29]
20	procyanidin trimer (B type) isomer 3	5.99	865.1957	865.1980	−2.66	$C_{45}H_{38}O_{18}$	739.1646, 713.1482, 695.1387, 577.1292, 451.1020, 423.0711, 407.0760, 287.0557	[16]
21	EGC-ECG dimer	6.04	745.1394	745.1405	−1.48	$C_{37}H_{30}O_{17}$	593.1265, 423.0709, 407.0763, 169.0137	[30]
22	(E)C-(4→8)-(E)C-(2→7, 4→8)-(E)C	6.49	863.1814	863.1823	−1.04	$C_{45}H_{36}O_{18}$	711.1324, 693.1232, 575.1180, 573.1035, 287.0559, 285.0392	[31]

Table 1. *Cont.*

Compound	Tentative Assignments	Rt (min)	Detected $[M-H]^-$ (m/z)	Theoretical $[M-H]^-$ (m/z)	Mass Error (ppm)	Formula	MS/MS Fragments	Ref.
					Proanthocyanidins			
23	procyanidin tetramer isomer 3 (B type)	6.72	1153.2589	1153.2614	−2.17	$C_{60}H_{50}O_{24}$	865.1940, 575.1193, 287.0553	[29]
24	3-galloylprocyanidin B1/3'-galloylprocyanidin B2 isomer 1	6.78	729.1455	729.1456	−0.14	$C_{37}H_{30}O_{16}$	603.1136, 441.0826, 407.0768, 289.0716, 125.0244	[30]
25	parameritannin A-1	6.92	1153.2572 [a]	1153.2614 [a]	−3.64	$C_{60}H_{48}O_{24}$	1001.2155, 866.2023, 579.1450, 577.1265	[32,33]
26	epiafzelechin 3-O-gallate-(4β→6)-epigallocatechin 3-O-gallate	7.50	883.1722 [a]	883.1722 [a]	0.00	$C_{44}H_{34}O_{20}$	409.0919, 271.0606, 153.0190	[27]
27	epiafzelechin epicatechin 3,3'-digallate	8.14	867.1774 [a]	867.1773 [a]	0.12	$C_{44}H_{34}O_{19}$	547.1236, 393.0977, 299.0561, 267.0661, 255.0660, 243.0659, 231.0663	[27]
28	3-galloylprocyanidin B1/3'-galloylprocyanidin B2 isomer 2	8.86	729.1455	729.1456	−0.14	$C_{37}H_{30}O_{16}$	603.1140, 577.1115, 441.0829, 417.1560, 407.0777	[30]
					Flavonol/Flavone Glycosides			
29	isovitexin glucoside	6.08	595.1653 [a]	595.1663 [a]	−1.68	$C_{27}H_{30}O_{15}$	473.1142, 433.1129, 313.0711	[34,35]
30	apigenin 6-C-glucoside 8-C-arabinoside	6.91	563.1402	563.1401	0.18	$C_{26}H_{28}O_{14}$	545.1306, 503.1185, 473.1089, 443.1089, 383.0767, 353.0663	[35]
31	myricetin 3-robinobioside (or 3-neohesperidoside)	6.93	627.1556 [a]	627.1561 [a]	−0.80	$C_{27}H_{30}O_{17}$	481.1028, 319.0457	[27]
32	myricetin 3-galactoside	7.02	479.0828	479.0826	0.42	$C_{21}H_{20}O_{13}$	317.0284, 316.0232, 271.0249	[26]
33	myricetin 3'-glucoside	7.11	479.0828	479.0826	0.42	$C_{21}H_{20}O_{13}$	317.0283, 316.0232, 271.0250	[26]
34	quercetin 3-O-galactosyl rutinoside	7.21	771.1981	771.1984	−0.39	$C_{33}H_{40}O_{21}$	611.1627, 465.1064, 301.0348, 300.0270	[27]
35	camelliannin B	7.68	577.1551	577.1557	−1.04	$C_{27}H_{30}O_{14}$	433.1134, 313.0717, 269.0445	[36]
36	rutin	7.70	609.1450	609.1456	−1.85	$C_{27}H_{30}O_{16}$	301.0343, 300.0280	Authentic standard [b]
37	kaempferol 3-O-galactosyl rutinoside	7.72	757.2177 [a]	757.2191 [a]	0.00	$C_{33}H_{40}O_{20}$	595.1652, 449.1080, 287.0557	[26]
38	tricetin	7.89	303.0506 [a]	303.0505 [a]	0.33	$C_{15}H_{10}O_7$	285.0410	[27]
39	kaempferol 3-O-glucosyl rutinoside	8.00	757.2187 [a]	757.2191 [a]	−0.53	$C_{33}H_{40}O_{20}$	595.1661, 449.1079, 287.0563	[26]
40	kaempferol 3-O-rutinoside	8.43	595.1667 [a]	595.1663 [a]	0.67	$C_{27}H_{30}O_{15}$	503.0271, 449.1084, 287.0562	[27]
41	kaempferol galactoside	8.51	447.0928	447.0927	0.22	$C_{21}H_{20}O_{11}$	285.0376, 284.0328	[26]

Table 1. *Cont.*

Compound	Tentative Assignments	Rt (min)	Detected [M − H]⁻ (m/z)	Theoretical [M − H]⁻ (m/z)	Mass Error (ppm)	Formula	MS/MS Fragments	Ref.
						Flavonol/Flavone Glycosides		
42	isorhamnetin glucoside	8.65	477.1038	477.1033	1.12	$C_{22}H_{22}O_{12}$	357.1347, 315.0504, 314.0435, 300.0271, 299.0203	[28]
43	kaempferol glucoside	8.78	447.0930	447.0927	0.67	$C_{21}H_{20}O_{11}$	285.0393, 284.0333	Authentic standard [b]
44	capilliposide I isomer 1	9.93	1063.2920	1063.2931	−1.03	$C_{48}H_{56}O_{27}$	917.2346, 771.1968, 753.1868, 615.1923, 609.1423, 531.1428, 458.1134, 447.0933, 301.0351, 300.0273	[37]
45	capilliposide II isomer 1	10.18	1049.3125 [a]	1049.3138 [a]	−1.24	$C_{48}H_{56}O_{26}$	887.2597, 741.2037, 595.1495, 287.0557	[37]
46	quercetin 3-(4''-(E)-p-coumaroylrobinobioside)-7-rhamnoside isomer 1	10.24	903.2554 [a]	903.2559 [a]	−0.55	$C_{42}H_{46}O_{22}$	757.1984, 449.1078, 303.0508, 147.0448	[27]
47	capilliposide I isomer 2	10.59	1065.3074 [a]	1065.3087 [a]	−1.22	$C_{48}H_{56}O_{27}$	919.2526, 617.2090, 449.1088, 303.0505	[37]
48	capilliposide II isomer 2	10.87	1049.3136 [a]	1049.3138 [a]	−0.19	$C_{48}H_{56}O_{26}$	887.2601, 741.2042, 595.1545, 287.0559	[37]
49	quercetin 3-(4''-(E)-p-coumaroyl robinobioside)-7-rhamnoside isomer 2	10.91	903.2546 [a]	903.2559 [a]	−1.44	$C_{42}H_{46}O_{22}$	757.1981, 449.1070, 303.0505, 147.0449	[27]
50	isorhamnetin 3 (or 7)-(6''-p-coumaroylglucoside)	10.98	623.1400	623.1401	−0.16	$C_{31}H_{28}O_{14}$	477.1068, 315.0508, 300.0271, 299.0200	[27]
51	2''-O-trans-p-coumaroylastragalin	11.68	593.1296	593.1295	0.17	$C_{30}H_{26}O_{13}$	447.0938, 285.0407, 284.0325	[28]
						Phenolic Acids		
52	theogallin	2.90	343.0679	343.0665	4.08	$C_{14}H_{16}O_{10}$	191.0564	Authentic standard [b]
53	3-p-coumaroylquinic acid	5.18	337.0931	337.0923	2.37	$C_{16}H_{18}O_8$	163.0403	[26]
54	5-p-coumaroylquinic acid	6.41	337.0930	337.0923	2.08	$C_{16}H_{18}O_8$	173.0459	[26]
						Hydrolysable Tannins		
55	monogalloyl glucose	2.45	331.0672	331.0665	2.11	$C_{13}H_{16}O_{10}$	271.0461, 211.0248, 169.0144, 151.0040, 125.0244	[16]
56	methyl 6-O-galloyl-β-D-glucose	3.66	345.0827	345.0822	1.45	$C_{14}H_{18}O_{10}$	225.0406, 183.0299	[27]
57	digalloyl glucose isomer 1	4.76	483.0780	483.0775	1.04	$C_{20}H_{20}O_{14}$	313.0578, 169.0139	[38]
58	digalloyl glucose isomer 2	5.01	483.0779	483.0775	0.83	$C_{20}H_{20}O_{14}$	313.0559, 169.0142	[38]
						Alkaloids		
59	7-methylxanthine	2.84	167.0570 [a]	167.0569 [a]	0.60	$C_6H_6N_4O_2$	124.0514	[39]
60	theobromine	3.80	181.0729 [a]	181.0726 [a]	1.66	$C_7H_8N_4O_2$	163.0622, 138.0674	Authentic standard [b]
61	caffeine	5.60	195.0885 [a]	195.0882 [a]	1.54	$C_8H_{10}N_4O_2$	138.0673	Authentic standard [b]

Table 1. *Cont.*

Compound	Tentative Assignments	Rt (min)	Detected [M − H]⁻ (m/z)	Theoretical [M − H]⁻ (m/z)	Mass Error (ppm)	Formula	MS/MS Fragments	Ref.
					Hydroxycinnamic Acid Amides			
62	di-*p*-coumaroylputrescine	10.33	381.1816 [a]	381.1814 [a]	0.53	$C_{22}H_{24}N_2O_4$	235.1412, 218.1179, 147.0448, 119.0653, 91.0701	[27]
63	tri-*p*-coumaroylspermidine	12.08	584.2750 [a]	584.2761 [a]	−1.88	$C_{34}H_{37}N_3O$	438.2451, 420.2353, 292.2150, 275.1895, 205.1208, 204.1178, 147.0609, 119.0655, 91.0702	[40]
					Amino Acids			
64	theanine	1.43	173.0935	173.0926	5.20	$C_7H_{14}N_2O_3$	155.0830, 128.0354	Authentic standard [b]
					Aromatic Alcohol Glycosides			
65	phenylethyl primeveroside	7.10	415.1599	415.1604	−1.20	$C_{19}H_{28}O_{10}$	283.1177, 149.0448	[27]
					Terpenoid Glycosides			
66	linalool oxide primeveroside	8.59	463.2166	463.2179	−2.81	$C_{21}H_{36}O_{11}$	331.1761	[27]
67	linalool primeveroside isomer 1	11.25	447.2234	447.2230	0.89	$C_{21}H_{36}O_{10}$	315.1805	[27]
68	linalool primeveroside isomer 2	11.53	447.2233	447.2230	0.67	$C_{21}H_{36}O_{10}$	421.1703	[27]

[a] [M + H]⁺. [b] This letter indicates that identification of the compound was confirmed by the authentic standard.

In the PCA score plot in ESI$^-$ (electrospray ionization in the negative ion mode), the first principle component (PC1) and the second principal component (PC2) explained 48.0% and 20.0% of the variation, respectively (Figure 2A). Except for leaves from different developmental stages that were clustered, the remaining samples were clearly separated from each other at both of the PC1 and PC2 axis, suggesting distinct metabolic profiles among the tea tissues. To further depict major differential metabolites, the PCA loading plot was applied. Along PC1, (−)-epigallocatechin gallate (EGCG), (−)-epigallocatechin 3-(3-*O*-methylgallate) (EGCG3"Me), theogallin, and procyanidin B2 were observed as the main contributors toward the discrimination of buds and leaves from other tissues (Figure 2B). Along PC2, (−)-epicatechin (EC), one galloyl procyanidin dimer, and a second procyanidin dimer were responsible for the separation of stems and flowers with the remaining tissues. The separation pattern observed in the PCA score plot in ESI$^+$ was similar to that in ESI$^-$, where PC1 was 56.0% and PC2 was 16.0% (Figure 2C). Buds and leaves were grouped together, while the remaining samples were separated at the PC1 axis. Flowers were segregated from other tissues at the PC2 axis. In addition to the compounds observed in ESI$^-$, caffeine and theobromine contributed significantly ($p < 0.05$) to the separation of buds and leaves from other tissues. An EGC-ECG dimer and an unknown metabolite (m/z = 614.2853, RT = 12.28 min) were found to occur more abundantly ($p < 0.05$) in flowers (Figure 2D).

2.2. Structural Compositions of Oligomeric Proanthocyanidins Varied by Tissue Types

Proanthocyanidins (PAs) are a group of structurally complex oligomeric (degrees of polymerization or DP = 2–10) or polymeric (DP > 10) flavan-3-ols linked by interflavan C-C bonds. PAs are remarkably diverse as a result of the diversity of monomeric units, types of linkages, and variations in chain lengths [30]. Authentic standards for most PAs are not commercially available. Moreover, reports on the purification, identification, and distribution of PAs in tea plants are limited [29,30]. Therefore, unambiguous structural assignments for PAs are quite challenging. Nonetheless, according to the MS/MS fragmentation patterns previously described [41], we tentatively identified **19** oligomeric PAs by UPLC-QTOF MS, among which 10 were procyanidins (compounds **14–20, 22, 23, 25**), two were galloylated procyanidins (compounds **24** and **28**), two were prodelphinidins (compounds **10** and **11**), one was a galloylated prodelphinidin (compound **13**), two were procyanidin/prodelphinidin dimers (compounds **12** and **21**), and two were propelargonidins (compounds **26** and **27**) (Table 1, Figure S2). PAs with DPs higher than four were outside of our detection window (50–1200 Da), and hence not included in the analysis. B-type PAs, which are characteristic of C4→8 or C4→6 interflavan bonds, were predominant in tea plants and were found to exist as dimers, trimers, and tetramers. Several isomers of procyanidin oligomers were observed to elute at different times. For example, compounds **14, 15**, and **20** were all assigned as B-type procyanidin trimers (Figure S3), and compounds **18, 19**, and **23** were all assigned as B-type procyanidin tetramers (Figure S4).

In addition, two less common A-type PAs, which were characterized with an additional ether linkage between C2→7, were detected in roots and stems. A-type PAs are readily recognizable because their m/z values are two Da less than corresponding B-type PAs [41]. Compound **22**, with m/z 863.1814 in ESI$^-$, was two Da lower in mass than B-type procyanidin trimers. The fragmentation of compound **22** yielded a fragment at m/z 711.1324, as a result of retro Diels-Alder (RDA) cleavage (Figure 4 and Figure S5). Subsequent water elimination generated a fragment at m/z 693.1232. Other key fragments such as m/z 575.1180 and 287.0559 were derived from quinone methide (QM) fission (Figure 4 and Figure S5). As a result, compound **22** was speculated as (E)C-(4→8)-(E)C-(2→7, 4→8)-(E)C. Similarly, the protonated ion of compound **25**, with formula $C_{60}H_{48}O_{24}$, had m/z at 1153.2572. This was two Da less compared with B-type procyanidin tetramers, which is suggestive of a tetrameric PA containing one additional A-type linkage. Low signal intensity made it difficult to compare the spectrum with known compounds. Nevertheless, the fragment ion we observed at m/z 1001.2155 (Table 1) may arise from the RDA cleavage. One compound described previously in the barks of other plant species had the same formula and was identified as parameritannin A1, namely, EC-(2β→O7, 4β→8)-[EC-(4β→6)]-EC-(4β→8)-EC [32,33]. To the best of our knowledge, this is the first description of these two A-type PAs in tea plants.

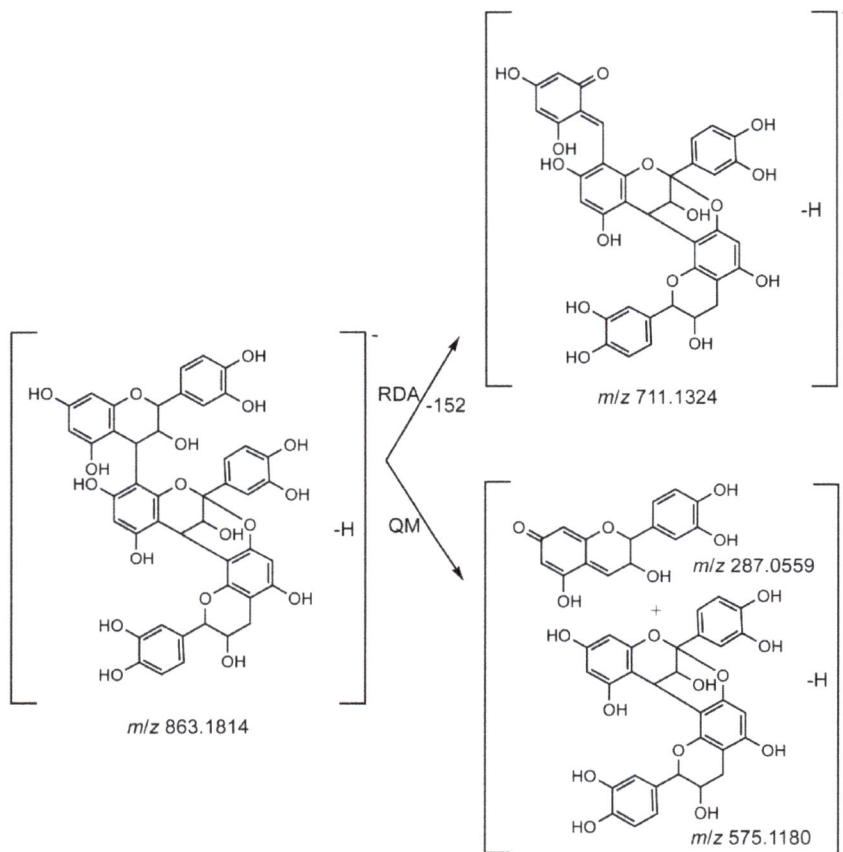

Figure 4. Proposed fragmentation pathways for compound **22**, a possible A-type procyanidin trimer, based on generated fragment ions.

Semi-quantitative comparisons of PAs and monomeric catechin units among tissues revealed that the DP of PAs increased from the upper part of tea plants to the lower part (Figure 3 and Figure S2), which is in line with the results from other studies [11,30]. The upper part of tea plants, particularly buds and leaves, was rich in monomeric catechins, including EGCG, (−)-epicatechin gallate (ECG), EC, (−)-epigallocatechin (EGC), (+)-catechin (C), (−)-gallocatechin (GC), and methylated catechins. However, most monomeric catechins were non-detectable in roots except for EC (Figure 3). PA dimers and trimers comprised of different extension units were found in higher amounts in stems relative to other tissues (Figure S2A–E). They were either not detected or present in very low levels in roots. In contrast, all four identified procyanidin tetramers occurred at the highest level in roots (Figure S2F). A similar finding was reported by Wei et al. in cultivar *Shuchazao*, where they observed a higher accumulation of more condensed PAs in fruits, flowers, and roots. In contrast, young buds and leaves contained more monomeric galloylated catechins [7].

2.3. Flavonol Glycosides with Different Aglycone Moieties Displayed Spatial Distribution

Based on UPLC-QTOF MS-based metabolite profiling, we found that *Tieguanyin* tea plants accumulated at least 21 flavonol glycosides, most of which were derivatives of kaempferol (eight compounds), quercetin (six compounds), myricetin (three compounds), and isorhamnetin (two compounds) (Table 1). Some structures were unequivocally identified by comparing with authentic standards, while others were assigned according to MS/MS fragmentation patterns, the neutral loss patterns of specific sugars, UV absorbance, and chromatographic behaviors [42,43], as exemplified in Figures S6 and S7. Among them, the sugar moieties of a few (compounds **44–51**) were further acylated to coumaric acid.

The distribution of flavonol glycosides showed intriguing patterns depending on the aglycone moiety (Figure 3 and Figure S8). For example, most kaempferol glycosides were abundant in flowers and young leaves, but scarce in stems and roots (Figure S8A). Quercetin glycosides were detected invariably at the highest level in leaves, and peaked in mature leaves. They were below detection in roots in most cases (Figure S8B). The distribution of myricetin glycosides mirrored that of quercetin glycosides, occurring mainly in the green parts of tea plants and in particular, exhibiting the highest level in mature leaves. They were barely detectable in flowers and roots (Figure S8C). Finally, isorhamnetin glucoside and isorhamnetin coumaroylglucoside were exclusively found in flowers (Figure S8D).

2.4. Distribution of Purine Alkaloids and Hydroxycinnamic Acid Amides, Two Classes of Nitrogenous Compounds, Displayed Tissue Specificity

Nitrogen-containing compounds have higher ionization efficiency in ESI$^+$. Therefore, two classes of nitrogen-containing metabolites, namely, purine alkaloids and hydroxycinnamic acid amides (HCCAs), were specifically analyzed in this mode. Three major purine alkaloids from the caffeine biosynthetic pathway, including caffeine (compound **61**), theobromine (compound **60**), and 7-methylxanthine (compound **59**), three major purine alkaloids from the caffeine biosynthetic pathway, were detected. The concentrations for all three compounds declined in the same order: buds > young leaves > mature leaves > stems > flowers > roots (Figure 3). A reduction in the caffeine content with the increased leaf age was also noted in other tea cultivars [5,44]. In each tissue, the concentration of caffeine was highest, followed by theobromine and 7-methylxanthine. Roots contained trace amounts of caffeine, theobromine, and almost no 7-methylxanthine.

Two HCCAs, including one coumaric-conjugated putrescine (compound **62**) and one coumaric-conjugated spermidine (compound **63**), with the latter being more abundant, were detected almost exclusively in tea floral organs (Table 1). Compound **63** was identified as tricoumaroyl spermidine on the basis of the fragmentation pattern and UV absorbance (Figure 5) in comparison with data available in the literature [40,45]. In MS^2 analysis, compound **63** with m/z 584.2750 generated a fragment ion at m/z 147.0452, corresponding to the coumaric moiety retaining the charge. Major ions at m/z 438.2451 and 292.2150 could arise from the loss of one coumaric acid and two coumaric acids, respectively, from the molecular ion (Figure 5A). The characteristic UV spectrum showed λ_{max} at 293 nm (Figure 5B), which was in accord with the previous report that the hydroxycinnamoyl-spermidines had a high absorption in the range of 270 nm to 330 nm [46]. Therefore, compound **63** was tentatively assigned as tri-*p*-coumaroylspermidine. Interestingly, this compound was also detected from the tea flowers of cultivar *Yabukita*, and was found to decrease during floral development [40]. Likewise, the fragmentation of compound **62** yielded the diagnostic fragment at m/z 147.0448, which also corresponded to the coumaric moiety retaining the charge. The fragment ion at m/z 235.1412 was most likely due to the cleavage of one coumaric acid from the molecular ion, and thus supported the assignment of compound **62** as a putative di-*p*-coumaroylputrescine (Figure S9). As far as we know, this is the first report of the occurrence of this compound in tea flowers.

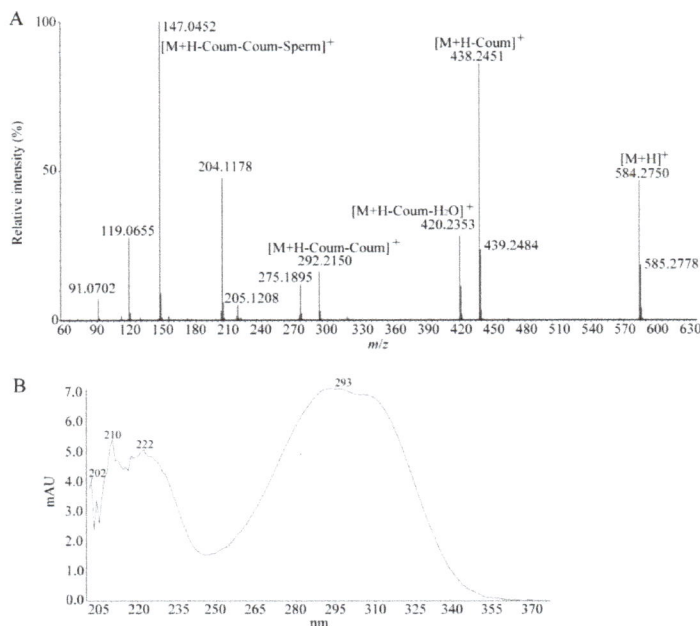

Figure 5. A spermidine derivative (compound **63**) detected in tea flowers. (**A**) CID-MS/MS spectrum of compound **63** in the ESI$^+$ mode. (**B**) The UV spectrum of compound **63** extracted from the UPLC-PDA-QTOF MS experiment. mAU, milli absorption unit.

2.5. Differential Amino Acid Profiles among Tea Plant Tissues

To compare amino acid abundance across tea tissues, hydrophilic interaction liquid chromatography (HILIC) tandem mass spectrometry was applied, with the absolute quantification results shown in Table 2. The total amino acid concentration, calculated from the sum of individual amino acid concentrations, was as follows: stems > flowers > mature leaves > young leaves > buds > roots. Theanine, aspartate, glutamate, glutamine, serine, and arginine altogether accounted for 95.9%, 98.6%, 99.2%, and 98.9% of the total amino acids in buds, young leaves, mature leaves, and stems, respectively, and thus were the major amino acids in the green parts of the tea plant.

Theanine, as the most abundant non-protein amino acid in tea, was detected in all of the tissues, but its concentration varied widely, ranging between 19.1–72.9% of the total amino acids. The highest theanine concentration was found in stems, reaching 33.36 mg/g dry weight, followed by mature leaves (10.58 mg/g dry weight). Concentration of theanine in young leaves (10.41 mg/g dry weight) was slightly lower than mature leaves, but higher ($p < 0.05$) than buds (8.36 mg/g dry weight) and flowers (6.32 mg/g dry weight). Roots contained the lowest level of theanine (1.41 mg/g dry weight).

The amino acid profile in the floral organ was distinct from the green parts of tea plants. Notably, concentrations of serine, arginine, asparagine, threonine, histidine, tryptophan, valine, lysine, proline, leucine, phenylalanine, methionine, tyrosine, γ-aminobutyric acid, and alanine were significantly higher ($p < 0.05$) in flowers than other parts, suggesting an overall up-regulation of amino acid biosynthesis in flowers (Table 2). Polyamines and CoA-activated hydroxycinnamic acids are two substrates for synthesizing HCCAs [47], in which arginine provides the substrate for the former, while phenylalanine is involved in the production of the latter [47]. Whether the highest occurrence of arginine and phenylalanine is related to the unique occurrence of coumaric-conjugated HCCAs in tea flowers is currently unknown.

Table 2. Abundance (mg/g dry weight) of amino acids in tea plant tissues.

Amino Acids	Bud	Young Leaf	Mature Leaf	Stem	Flower	Root
Theanine	8.36 ± 0.41 c	10.41 ± 1.07 b	10.58 ± 0.91 b	33.36 ± 0.25 a	6.32 ± 0.81 d	1.41 ± 0.08 e
Aspartate	8.27 ± 0.37 ab	7.08 ± 0.55 b	7.81 ± 0.25 ab	9.71 ± 1.60 a	7.69 ± 0.74 ab	ND
Glutamate	1.88 ± 0.06 c	2.55 ± 0.14 b	3.01 ± 0.10 a	2.55 ± 0.13 b	1.74 ± 0.07 c	0.26 ± 0.01 d
Glutamine	0.99 ± 0.02 c	0.58 ± 0.04 d	0.35 ± 0.02 de	6.15 ± 0.27 a	2.77 ± 0.14 b	0.03 ± 0.00 e
Serine	0.96 ± 0.03 b	0.75 ± 0.06 bc	0.42 ± 0.03 cd	0.46 ± 0.07 c	2.79 ± 0.37 a	0.03 ± 0.01 d
Arginine	0.47 ± 0.02 c	0.70 ± 0.05 b	0.09 ± 0.01 d	0.45 ± 0.01 c	1.55 ± 0.10 a	0.05 ± 0.00 d
Asparagine	0.41 ± 0.01 b	0.03 ± 0.00 d	0.01 ± 0.00 d	0.10 ± 0.02 c	0.98 ± 0.04 a	0.01 ± 0.01 d
Threonine	0.09 ± 0.00 b	0.03 ± 0.01 cd	0.02 ± 0.00 c	0.08 ± 0.01 bc	0.34 ± 0.05 a	ND
Histidine	0.08 ± 0.00 b	0.03 ± 0.00 d	0.01 ± 0.00 de	0.10 ± 0.01 b	0.34 ± 0.02 a	0.01 ± 0.00 e
Tryptophan	0.08 ± 0.00 b	0.02 ± 0.00 e	ND	0.05 ± 0.00 c	0.26 ± 0.00 a	0.03 ± 0.00 d
Valine	0.07 ± 0.00 b	0.06 ± 0.00 b	0.03 ± 0.00 c	0.06 ± 0.00 b	0.35 ± 0.02 a	0.01 ± 0.00 d
Lysine	0.06 ± 0.00 c	0.06 ± 0.00 c	0.03 ± 0.00 d	0.10 ± 0.01 b	0.28 ± 0.01 a	0.01 ± 0.00 e
Proline	0.05 ± 0.00 b	0.03 ± 0.00 b	0.02 ± 0.00 b	0.02 ± 0.00 b	4.36 ± 0.06 a	ND
Leucine	0.02 ± 0.00 b	0.01 ± 0.00 b	0.02 ± 0.00 b	0.02 ± 0.00 b	0.14 ± 0.01 a	ND
Isoleucine	0.02 ± 0.00 b	0.01 ± 0.00 bc	ND	0.01 ± 0.00 bc	0.36 ± 0.01 a	ND
Phenylalanine	0.01 ± 0.00 c	0.01 ± 0.00 c	0.02 ± 0.00 c	0.07 ± 0.00 b	1.30 ± 0.03 a	ND
Methionine	0.01 ± 0.00 b	ND	ND	0.01 ± 0.00 b	0.08 ± 0.01 a	ND
Tyrosine	ND	ND	ND	0.01 ± 0.00 b	0.13 ± 0.01 a	ND
γ-Aminobutyric acid	ND	ND	ND	0.03 ± 0.00 c	0.20 ± 0.01 a	0.06 ± 0.00 b
Alanine	ND	0.01 ± 0.01 b	ND	ND	1.15 ± 0.25 a	ND
total	21.82 ± 0.94 c	22.38 ± 1.94 c	22.43 ± 1.32 c	53.35 ± 2.40 a	33.14 ± 2.76 b	1.92 ± 0.11 d

Results are expressed as mean ± standard deviation (n = 3). Means with different letters in row are significantly different according to Tukey's HSD (honestly significant difference) test ($p < 0.05$). ND = non-detectable.

3. Discussion

The plant kingdom is predicted to produce at least 1,000,000 metabolites [48]. The production, translocation, and hydrolysis of these diverse metabolites are regulated by both intrinsic genetic programs and environmental factors. The limitations imposed by the sensitivity and resolution of analytical techniques, along with rapid metabolite turnovers, have challenged the detection of the majority of metabolites and the systematic studies of their biochemical and biological functions in any single plant species, even in model plants such as Arabidopsis and rice [49,50]. As an economically important beverage crop, tea plants produce arsenals of structurally and biologically diverse nutraceuticals to high levels, among which flavonoids, caffeine, and theanine are best known [51]. Previous studies, which typically targeted one or several classes of target metabolites, reveal the tissue-specific regulation of specialized metabolism in tea plants, as found in other plant species [3,5,6,11,15]. However, the application of untargeted LC-MS-based metabolomics to examine the overall difference in metabolic profiles among tea plant tissues has not been thoroughly performed. With the goal of understanding developmental changes in specialized metabolism, we performed a thorough comparative analysis of six different tissue types in Tieguanyin tea cultivar, which highlighted differences in tissue-specific metabolic features in tea plants. In the present study, thousands of molecular features were simultaneously detected, from which a total of 68 major specialized metabolites belonging to 11 metabolite classes were found to differentially accumulate in different tissues. The comparative results reveal the remarkable diversity of specialized metabolites in tea plants, and provide valuable information to further understand the developmental regulation of their biosynthesis.

3.1. The Abundance of Flavonol Glycosides Demonstrates Tissue-specific Variations in Different Plants

Flavonols are among the most abundant flavonoids in plants. Decorative enzymes catalyzing glycosylation, acylation, hydroxylation, and methylation provide important modifications to flavonols, conferring increased structural complexity, enhanced biological activity, as well as improved molecule solubility and stability in *Arabidopsis* and several crop species [52].

In the current study, 21 flavonol glycosides were found to be differentially distributed in different tissues, providing the first insight into the developmental regulation of flavonol glycosides in tea plants. Although the current study is the first report of the tissue-specific distribution of flavonol glycosides in tea plants, it appears as a common trait shared by many plants. Flavonol glycosides have been most thoroughly profiled in *Arabidopsis* tissues, which were discovered to be tightly regulated developmentally in different tissues [42,45]. By UPLC-QTOF MS-based profiling, Yonekura-Sakakibara et al. showed that kaempferol glycosides accounted for 97% of the total flavonoids in leaves, while quercetin glycosides took up 25% of the total flavonoids in floral buds and flowers [53]. In the same study, a higher accumulation of *C*-7 rhamnosylated flavonols in floral buds, in comparison to leaves, roots, and siliques, was found to be well coordinated with the higher expression of a flavonol 7-*O*-rhamnosyltransferase [53]. A more comprehensive flavonol profiling by the same research group revealed that kaempferol 3-*O*-rhamnoside-7-*O*-rhamnoside was one of the major flavonols in leaves, stems, and flowers. In contrast, roots contained very little of this compound, but possessed a high level of quercetin 3-*O*-glucoside-7-*O*-rhamnoside [54]. Moreover, in seeds, quercetin-3-*O*-rhamnoside and a dimer of quercetin-rhamnoside accumulated in the seed coat, while diglycosylated flavonols were only found in the embryo [55]. Significant differences in flavonol compositions were also reported in strawberry and *Compositae* plants [16,56]. For example, quercetin neohesperidoside, kaempferide neohesperidoside, and kaempferol acetylglucoside were detected in the leaf, but not in the flower of *Chrysanthemum morifolium* [56]. In strawberry flowers, dihexose derivatives of kaempferol and quercetin were present mainly in the stamen, but the malonylhexose derivatives of both flavonols were mainly detected in the pistil [16]. The differential production of flavonol glycosides were presumably caused by tissue-specific expression of genes encoding for the synthesis of different flavonols as well as decorative enzymes [57]. Detailed analysis of flavonol levels and gene coexpression is necessary to gain more knowledge on the timing expression of flavonoid biosynthetic genes in *Tieguanyin* tea plant.

Diverse biological roles of flavonols, such as maintaining plant fertility, protecting against UV stress, serving as signaling molecules, functioning as co-pigments, and modulating auxin transport, were documented in a wide range of plant species [58]. Despite their importance, the structure–function relationship for many of these molecules remains largely unknown. In tea plants, different flavonols, along with their different conjugates, presumably play specific roles in the developmental and physiological functions of different tissues, although their exact biological functions are yet to be elucidated. Moreover, flavonols and their glycosyl derivatives could confer astringency to tea infusions at much lower thresholds than catechins, making them important contributors to the flavor property of tea [59]. This, along with flavonol glycosides changing only slightly during tea processing [51], renders the understanding of the tissue-specific distribution of flavonol glycosides among unprocessed tea plant tissues important.

3.2. Coumaroyl-Conjugated Hydroxycinnamic Acid Amides (HCCAs) are Unique Flower Metabolites

Widely distributed in the plant kingdom, HCCAs are reported to have important functions in plant adaptation to biotic and abiotic stresses [60]. They are also implicated in some plant growth and developmental processes, including flower formation, sexual differentiation, tuberization, and so on, although the causal relationship is still not conclusive [61]. A wide variety of acylated polyamines have been isolated and identified in the floral parts of different plants. Neutral HCCAs such as di-*p*-coumaroylputrescine, di-*p*-coumaroylspermidine, and tri-*p*-coumaroylspermidine, have been reported in plant reproductive organs, i.e., anthers of fertile maize, male flowers of some *Araceae* species, bee pollen samples, the stamen and pistil of strawberry flowers, and the inflorescence tissues of *Arabidopsis* [16,45,61,62]. Interestingly, in an earlier study, four spermidine derivatives, namely, tricoumaroyl spermidine, feruoyl dicoumaroyl spermidine, coumaroyl diferuoyl spermidine, and triferuoyl spermidine, were found as tea flower constituents in cultivar *Yabukita*. Although we did not dissect the flowers in the current study, spermidine derivatives were previously found to mainly accumulate in the anthers of tea flowers, and as such were presumed to participate in pollen formation [40].

Enzymes that have been identified to date as responsible for synthesizing HCCAs are all acyltransferases belonging to the BAHD family, which utilizes CoA-activated hydroxycinnamic acids and polyamines as substrates [47]. Interestingly, through transcriptome analysis, we identified two unigenes for BAHD acyltransferase, which showed the highest expressions in flowers, but only basal expressions in other tea plant tissues. Functional experimentation is needed to dissect their roles in the HCCA formation in tea flowers.

3.3. Occurrence of A-Type PAs is Rare in Tea Plants but Warrants Further Analysis

PAs can be widely found in different parts of various plants, protecting plants against pathogens and herbivores [63]. Similar to monomeric flavan-3-ols, PAs also exhibit a wide array of bioactivities, including antimicrobial, antioxidative, anti-inflammatory, and antihypertensive effects, to name just a few [64]. The most commonly occurring monomeric units of PAs in tea plants are procyanidins, prodelphinidins, and their mixtures, although propelargonidins and galloylated forms of the aforementioned monomeric units also occur [29].

A total of 19 oligomeric PAs, comprising 17 B-type PAs and two A-type PAs, were identified in *Tieguanyin* tea plants. The latter included (E)C-(4→8)-(E)C-(2→7, 4→8)-(E)C and EC-(2β→O7, 4β→8)-[EC-(4β→6)]-EC-(4β→8)-EC, the structures of which were tentatively assigned based on fragmentation patterns. Both compounds were found to predominate in roots. Information on A-type PAs detected from tea plants is quite limited. Reported examples only included a dimeric A-type PA isolated from a commercial oolong tea, and a tetrameric A-type PA isolated from fresh tea leaves [65,66]. According to Kumar et al., the occurrence of A-type PAs in tea plants is rare, but is of considerable interest, because they have been implicated to contribute to the beneficial effects of cranberry juice for preventing urinary tract infections [66]. Further chemical analysis of the identified two A-type PAs is required to confirm their structures.

4. Materials and Methods

4.1. Plant Materials and Sampling

Cuttings of five-year-old cloned tea plants of *C. sinensis* cv. *Tieguanyin* were planted at the tea farm at Anxi Tea Research Institute, Anxi, Fujian Province, China (118°13′ E, 25°08′ N) under the natural environment, where the annual average temperature was 18 °C, and the annual average rainfall was between 1700–1800 mm. The *Tieguanyin* tea plant is an evergreen and perennial shrub with small leaf size, which starts to flower in late October and reaches the full-bloom stage in mid-November. In October 2015, buds, young leaves, mature leaves, new stems (no lignification), flowers, and lateral roots (Figure 1) were harvested with sterile gloves at approximately 10 o′clock in the morning from nine of such tea plants grown under the same cultivation practice. Samples were randomly divided into three groups, with each group containing plant materials collected from three tea plants. Tea plant tissues were washed with tap water to remove attached clay, immediately frozen in liquid nitrogen, brought back to the lab, and stored at −80 °C until analysis. Tissue samples were subjected to UPLC-QTOF MS and UPLC-QqQ MS analyses.

4.2. Extraction and UPLC-QTOF MS Analysis

Metabolite extraction was performed according to our previously published protocol [67]. Three biological sample replicates were prepared for each tissue type. One microliter of the metabolite extract was injected into an Acquity UPLC system coupled in tandem to a photodiode array (PDA) detector and a SYNAPT G2-Si HDMS QTOF mass spectrometer (Waters, Milford, MA, USA). Separation was achieved on a Waters Acquity UPLC HSS T3 column (2.1 × 100 mm, 1.8 μm) thermostatted at 40 °C using a gradient from solvent A (water with 0.1% formic acid) to solvent B (acetonitrile with 0.1% formic acid), as previously described [67]. The flow rate was set at 0.3 mL/min. Data were collected in the electrospray ionization (ESI) mode (both ESI$^+$ and ESI$^-$), scanning from 50–1200 Da. The instrument setup was the same as previously described [67]. Quality control (QC) samples were prepared by mixing an equal amount of each sample to become a combined sample, and were injected every five samples throughout the runs to monitor the instrument performance. The MassLynx software (version 4.1, Waters, Milford, MA, USA) was used to control all of the instruments. Each triplicate tea sample was analyzed once.

4.3. Amino Acid Quantitation by UPLC-QqQ MS

To quantify amino acid contents, two microliters of the metabolite extract, with appropriate dilutions within the range of the calibration curve, were injected into an Acquity UPLC system coupled in tandem to a PDA detector and a XEVO TQ-S MS triple quadrupole mass spectrometer (Waters, Milford, MA, USA). Separation was achieved on a Merck SeQuant ZIC-HILIC column (2.1 × 100 mm, 5 μm) thermostatted at 40 °C using a gradient from solvent A (5 mM ammonium acetate) to solvent B (acetonitrile with 0.1% formic acid), as previously described [67]. The flow rate was set at 0.4 mL/min. The instrument setup was same as previously described [67]. Calibration curves generated by injecting increasing concentrations of authentic standards were used to measure the absolute concentrations of amino acids. The MassLynx software (version 4.1, Waters, Milford, MA, USA) was used for instrument control and data acquisition. Each triplicate tea sample was analyzed once.

4.4. Data Processing, Metabolite Identification, and Statistical Analysis

Resulting chromatograms from UPLC-QTOF MS were processed using Progenesis QI software (version 2.1, Nonlinear Dynamics, Newcastle upon Tyne, UK) with default settings for peak alignment, normalization, signal integration, and initial compound assignments. Only chromatograms with an elution time between 1–14 min were included in the analysis. Thus, annotation was obtained was used for manual peak identification. Metabolites were identified by comparing accurate masses, MS/MS fragmentation patterns and isotope patterns with authentic standards, online metabolite databases of Metlin [27], MassBank [39], ReSpect [34], KNApSAcK [48] and literature references [26,28–30,43]. Each mass spectrum was manually inspected to verify if software-predicted fragments were derived from a single metabolite. UV spectra were used for identification whenever possible.

Samples were acquired in both ESI⁺ and ESI⁻ modes, and therefore, data for each ionization mode were processed in Progenesis QI separately. The software detected 2798 molecular features in ESI⁻ and 3811 molecular features in ESI⁺, which were filtered to include only 732 and 821 single molecular features in respective modes. For comparing the abundances of molecular features, the data matrix consisting of mass features and peak area values was exported from Progenesis QI to Excel. The mean peak area abundance values from three biological replicates of the same tissue type were calculated. Similarities and differences in metabolite signal abundances were compared across tissues. Single molecular features were used as inputs for principal component analysis (PCA) to observe intrinsic metabolite variance between tissues using Progenesis QI extension EZinfo after Pareto scaling. Supervised partial least squared discriminant analysis (PLS-DA) was performed to identify the metabolites that are important for group separation. The data matrix used for PCA and PLS-DA analyses was listed in Supplemental Table S1 (for ESI⁻) and S2 (for ESI⁺). One-way analysis of variance (ANOVA) was carried out using SPSS (version 13.0, Chicago, IL, USA), and differences between means were determined by Tukey's HSD test. Variable importance in projection (VIP) analysis was performed to evaluate the importance of metabolites. Significantly different metabolites between tissues were selected with VIP > 1 and a p value < 0.05. A heat map with hierarchical clustering (Pearson's correlation, average linkage), after being log2 transformed and normalized to the median level of individual compounds, was generated using MultiExperiment Viewer software (version 4.9.0), which combined data from UPLC-QTOF MS and UPLC-QqQ MS.

4.5. Chemicals and Reagents

Acetonitrile (MS grade), methanol (HPLC grade), and formic acid ($\geq$ 98%) were obtained from Sigma-Aldrich (St. Louis, MO, USA). Deionized water was produced by a Milli-Q water purification system (Millipore, Billerica, MA, USA). Standards of EGCG, EGC, C, ECG, EC, GC, rutin, and L-theanine (all with purity $\geq$ 95%) were obtained from Sigma-Aldrich (St. Louis, MO, USA). EGCG3″Me ($\geq$95%) and kaempferol glucoside ($\geq$98%) were purchased from ChemFaces (Wuhan, China). Caffeine ($\geq$98%) was obtained from Yuanye Biotechnology Inc. (Shanghai, China). Theobromine ($\geq$99%) and kaempferol glucoside ($\geq$98%) were obtained from BioBioPha Co., Ltd. (Kunming, China). Theogallin ($\geq$95%) was kindly provided by Dr. Qingxi Chen of Fujian Agriculture and Forestry University, China.

5. Conclusions

In summary, an UPLC-QTOF MS-based non-targeted metabolomics strategy was applied for the first time to comprehensively compare the specialized metabolite profiles between tea plant tissues. Many metabolite classes, including catechins, PAs, flavonol glycosides, purine alkaloids, HCCAs, and amino acids were found to demonstrate sharp variations among tissue types. The upper part of tea plants abounded in monomeric catechins, whereas the lower part was more enriched in the highly polymerized forms of catechins. The abundance of flavonol glycosides demonstrated tissue specificity depending on the aglycone moiety. Metabolite contents of purine alkaloids and amino acids significantly differed among tissues. Furthermore, two neutral HCCAs, namely, tricoumaroyl spermidine and dicoumaroyl putrescine, were discovered as unique flower metabolites. All of these results suggest that the spatial changes in metabolite levels in tea plants are likely to be developmentally regulated. It also provides a good reference point for formulating a working hypothesis for the future

Molecules **2018**, *23*, 1817

characterization of metabolic functions in tea plants. An interesting aspect for future research would be to further explore gene-metabolite relationships to pinpoint important genes/enzymes and decipher regulatory elements responsible for tissue-specific accumulations of certain metabolites (e.g., flavonol glycosides and HCCAs).

Supplementary Materials: The Supplementary Materials are available online. Figure S1 UPLC-QTOF MS total ion chromatograms in ESI$^-$ of six tea tissues. Figure S2 Mean peak area abundance values ($\pm$SD) of (A) procyanidin dimers, (B) prodelphinidin dimers, (C) procyanidin-prodelphinidin dimers, (D) propelargonidin dimers, (E) procyanidin trimers and (F) procyanidin tetramers in tea plant tissues. Figure S3 Reconstructed ion chromatograms and MS/MS fragmentation of putative procyanidin trimers. Figure S4 Reconstructed ion chromatograms and MS/MS fragmentation of putative procyanidin tetramers. Figure S5 CID-MS/MS spectrum of compound **22** in the ESI$^-$ mode. Figure S6 Reconstructed ion chromatograms and MS/MS fragmentation of kaempferol hexose-deoxyhexose-hexose. Figure S7 Reconstructed ion chromatograms and MS/MS fragmentation of capilliposide I. Figure S8 Mean peak area abundance values ($\pm$ SD) of (A) kaempferol glycosides, (B) quercetin glycosides, (C) myricetin glycosides and (D) isorhamnetin glycosides in tea plant tissues. Figure S9 CID-MS/MS spectrum of compound **62**, a putative di-*p*-coumaroylputrescine, detected from tea flowers. Table S1 Filtered and normalized PCA data matrix generated from UPLC-QTOF MS in ESI$^-$. Table S2 Filtered and normalized PCA data matrix generated from UPLC-QTOF MS in ESI$^+$.

Author Contributions: Z.Y. and X.Y. conceived and designed the experiments. S.C., J.L., H.L., Z.G., X.W. and M.L. performed the experiments. S.C., J.L. and X.Y. analyzed the data. S.C., J.L., A.A., Z.Y. and X.Y. interpreted the results. S.C., J.L., Z.Y. and X.Y. wrote the manuscript.

Funding: This work was funded by the Natural Science Foundation of Fujian (2016J01108), the Distinguished Young Scholar Program of Fujian Agriculture and Forestry University (xjq201610) and the startup fund from Fujian Agriculture and Forestry University.

Acknowledgments: We thank Ilana Rogachev (Weizmann Institute of Science) for useful discussions on the instrument setup and compound measurements. We thank the metabolomics core facility of FAFU-UCR Joint Center for Horticultural Biology and Metabolomics (Fujian Agriculture and Forestry University) for technical support.

Conflicts of Interest: The authors declare no conflict of interests.

References

1. Hayat, K.; Iqbal, H.; Malik, U.; Bilal, U.; Mushtaq, S. Tea and its consumption: Benefits and risks. *Crit. Rev. Food Sci. Nutr.* **2015**, *55*, 939–954. [CrossRef] [PubMed]

2. Deng, W.W.; Ogita, S.; Ashihara, H. Ethylamine content and theanine biosynthesis in different organs of *Camellia sinensis* seedlings. *Naturforsch. C* **2009**, *64*, 387–390. [CrossRef]

3. Ashihara, H.; Deng, W.W.; Mullen, W.; Crozier, A. Distribution and biosynthesis of flavan-3-ols in *Camellia sinensis* seedlings and expression of genes encoding biosynthetic enzymes. *Phytochemistry* **2010**, *71*, 559–566. [CrossRef] [PubMed]

4. Wu, Z.J.; Li, X.H.; Liu, Z.W.; Xu, Z.S.; Zhuang, J. *De novo* assembly and transcriptome characterization: Novel insights into catechins biosynthesis in *Camellia sinensis*. *BMC Plant Biol.* **2014**, *14*, 277. [CrossRef] [PubMed]

5. Deng, W.W.; Ashihara, H. Occurrence and *de novo* biosynthesis of caffeine and theanine in seedlings of tea (*Camellia sinensis*). *Nat. Prod. Commun.* **2015**, *10*, 703–706. [PubMed]

6. Liu, Z.W.; Wu, Z.J.; Li, H.; Wang, Y.X.; Zhuang, J. L-Theanine content and related gene expression: Novel insights into theanine biosynthesis and hydrolysis among different tea plant (*Camellia sinensis* L.) tissues and cultivars. *Front. Plant. Sci.* **2017**, *8*. [CrossRef] [PubMed]

7. Wei, C.; Yang, H.; Wang, S.; Zhao, J.; Liu, C.; Gao, L.; Xia, E.; Lu, Y.; Tai, Y.; She, G.; et al. Draft genome sequence of *Camellia sinensis* var. *sinensis* provides insights into the evolution of the tea genome and tea quality. *Proc. Natl. Acad. Sci.* **2018**, *20*. [CrossRef]

8. Wang, W.; Zhou, Y.; Wu, Y.; Dai, X.; Liu, Y.; Qian, Y.; Li, M.; Jiang, X.; Wang, Y.; Gao, L.; et al. Insight into catechins metabolic pathways of *Camellia sinensis* based on genome and transcriptome analysis. *J. Agric. Food Chem.* **2018**, *1*. [CrossRef] [PubMed]

9. Xia, E.H.; Zhang, H.B.; Sheng, J.; Li, K.; Zhang, Q.J.; Kim, C.; Zhang, Y.; Liu, Y.; Zhu, T.; Li, W.; et al. The tea tree genome provides insights into tea flavor and independent evolution of caffeine biosynthesis. *Mol. Plant* **2017**, *10*, 866–877. [CrossRef] [PubMed]

10. Wink, M. Evolution of secondary metabolites from an ecological and molecular phylogenetic perspective. *Phytochemistry* **2003**, *64*, 3–19. [CrossRef]

11. Jiang, X.; Liu, Y.; Li, W.; Zhao, L.; Meng, F.; Wang, Y.; Tan, H.; Yang, H.; Wei, C.; Wan, X.; et al. Tissue-specific, development-dependent phenolic compounds accumulation profile and gene expression pattern in tea plant *Camellia sinensis*. *PLoS ONE* **2013**, *8*, e62315. [CrossRef] [PubMed]

12. Li, C.F.; Zhu, Y.; Yu, Y.; Zhao, Q.Y.; Wang, S.J.; Wang, X.C.; Yao, M.Z.; Luo, D.; Li, X.; Chen, L.; et al. Global transcriptome and gene regulation network for secondary metabolite biosynthesis of tea plant (*Camellia sinensis*). *BMC Genom.* **2015**, *16*, 560. [CrossRef] [PubMed]

13. Suzuki, T.; Takahashi, E. Biosynthesis of caffeine by tea-leaf extracts. Enzymic formation of theobromine from 7-methylxanthine and of caffeine from theobromine. *Biochem. J.* **1975**, *146*, 87–96. [CrossRef] [PubMed]

14. Deng, W.-W.; Ogita, S.; Ashihara, H. Biosynthesis of theanine (γ-ethylamino-L-glutamic acid) in seedlings of *Camellia sinensis*. *Phytochem. Lett.* **2008**, *1*, 115–119. [CrossRef]

15. Deng, W.W.; Ashihara, H. Profiles of purine metabolism in leaves and roots of *Camellia sinensis* seedlings. *Plant Cell Physiol.* **2010**, *51*, 2105–2118. [CrossRef] [PubMed]

16. Hanhineva, K.; Rogachev, I.; Kokko, H.; Mintz-Oron, S.; Venger, I.; Karenlampi, S.; Aharoni, A. Non-targeted analysis of spatial metabolite composition in strawberry (*Fragariaxananassa*) flowers. *Phytochemistry* **2008**, *69*, 2463–2481. [CrossRef] [PubMed]

17. Asiago, V.M.; Hazebroek, J.; Harp, T.; Zhong, C. Effects of genetics and environment on the metabolome of commercial maize hybrids: A multisite study. *J. Agric. Food Chem.* **2012**, *60*, 11498–11508. [CrossRef] [PubMed]

18. Jang, Y.K.; Jung, E.S.; Lee, H.-A.; Choi, D.; Lee, C.H. Metabolomic characterization of hot pepper (*Capsicum annuum* "CM334") during fruit development. *J. Agric. Food Chem.* **2015**, *63*, 9452–9460. [CrossRef] [PubMed]

19. Wu, S.; Tohge, T.; Cuadros-Inostroza, A.; Tong, H.; Tenenboim, H.; Kooke, R.; Meret, M.; Keurentjes, J.B.; Nikoloski, Z.; Fernie, A.R.; et al. Mapping the *Arabidopsis* metabolic landscape by untargeted metabolomics at different environmental conditions. *Mol. Plant* **2018**, *11*, 118–134. [CrossRef] [PubMed]

20. Liu, T. *Chinese Tea*; China Intercontinental Press: Beijing, China, 2005.

21. Chen, Y.J.; Kuo, P.C.; Yang, M.L.; Li, F.Y.; Tzen, J.T.C. Effects of baking and aging on the changes of phenolic and volatile compounds in the preparation of old Tieguanyin oolong teas. *Food Res. Int.* **2013**, *53*, 732–743. [CrossRef]

22. Meng, W.; Xu, X.; Cheng, K.-K.; Xu, J.; Shen, G.; Wu, Z.; Dong, J. Geographical origin discrimination of oolong Tea (TieGuanYin, *Camellia sinensis* (L.) O. Kuntze) using proton nuclear magnetic resonance spectroscopy and near-infrared spectroscopy. *Food. Anal. Method* **2017**, *10*, 3508–3522. [CrossRef]

23. Li, Y.; Lei, J.; Yang, J.; Liu, R. Classification of Tieguanyin tea with an electronic tongue and pattern recognition. *Anal. Lett.* **2014**, *47*, 2361–2369. [CrossRef]

24. Yan, S.M.; Liu, J.P.; Xu, L.; Fu, X.S.; Cui, H.F.; Yun, Z.Y.; Yu, X.P.; Ye, Z.H. Rapid discrimination of the geographical origins of an oolong tea (anxi-tieguanyin) by near-infrared spectroscopy and partial least squares discriminant analysis. *J. Anal. Methods. Chem.* **2014**, *2014*. [CrossRef] [PubMed]

25. Xu, Y.Q.; Liu, P.P.; Shi, J.; Gao, Y.; Wang, Q.S.; Yin, J.F. Quality development and main chemical components of Tieguanyin oolong teas processed from different parts of fresh shoots. *Food Chem.* **2018**, *249*, 176–183. [CrossRef] [PubMed]

26. Dou, J.; Lee, V.S.; Tzen, J.T.; Lee, M.R. Identification and comparison of phenolic compounds in the preparation of oolong tea manufactured by semifermentation and drying processes. *J. Agric. Food Chem.* **2007**, *55*, 7462–7468. [CrossRef] [PubMed]

27. Tautenhahn, R.; Cho, K.; Uritboonthai, W.; Zhu, Z.; Patti, G.J.; Siuzdak, G. An accelerated workflow for untargeted metabolomics using the METLIN database. *Nat. Biotechnol.* **2012**, *30*, 826–828. [CrossRef] [PubMed]

28. Dai, W.D.; Qi, D.D.; Yang, T.; Lv, H.P.; Guo, L.; Zhang, Y.; Zhu, Y.; Peng, Q.H.; Xie, D.C.; Tan, J.F.; et al. Nontargeted analysis using ultraperformance liquid chromatography-quadrupole time-of-flight mass spectrometry uncovers the effects of harvest season on the metabolites and taste quality of tea (*Camellia sinensis* L.). *J. Agric. Food Chem.* **2015**, *63*, 9869–9878. [CrossRef] [PubMed]

29. Fraser, K.; Harrison, S.J.; Lane, G.A.; Otter, D.E.; Hemar, Y.; Quek, S.-Y.; Rasmussen, S. HPLC–MS/MS profiling of proanthocyanidins in teas: A comparative study. *J. Food Compos. Anal.* **2012**, *26*, 43–51. [CrossRef]

30. Jiang, X.; Liu, Y.; Wu, Y.; Tan, H.; Meng, F.; Wang, Y.S.; Li, M.; Zhao, L.; Liu, L.; Qian, Y.; et al. Analysis of accumulation patterns and preliminary study on the condensation mechanism of proanthocyanidins in the tea plant [*Camellia sinensis*]. *Sci. Rep.* **2015**, *5*. [CrossRef] [PubMed]

31. Yan, T.; Hu, G.S.; Wang, A.H.; Hong, Y.; Jia, J.M. Characterisation of proanthocyanidins from *Schisandra chinensis* seed coats by UPLC-QTOF/MS. *Nat. Prod. Res.* **2014**, *28*, 1834–1842. [CrossRef] [PubMed]

32. Lin, G.M.; Lin, H.Y.; Hsu, C.Y.; Chang, S.T. Structural characterization and bioactivity of proanthocyanidins from indigenous cinnamon (*Cinnamomum osmophloeum*). *J. Sci. Food Agric.* **2016**, *96*, 4749–4759. [CrossRef] [PubMed]

33. Kamiya, K.; Watanabe, C.; Endang, H.; Umar, M.; Satake, T. Studies on the constituents of bark of *Parameria laevigata* Moldenke. *Chem. Pharm. Bull.* **2001**, *49*, 551–557. [CrossRef] [PubMed]

34. Sawada, Y.; Nakabayashi, R.; Yamada, Y.; Suzuki, M.; Sato, M.; Sakata, A.; Akiyama, K.; Sakurai, T.; Matsuda, F.; Aoki, T.; et al. RIKEN tandem mass spectral database (ReSpect) for phytochemicals: A plant-specific MS/MS-based data resource and database. *Phytochemistry* **2012**, *82*, 38–45. [CrossRef] [PubMed]

35. Ferreres, F.; Silva, B.M.; Andrade, P.B.; Seabra, R.M.; Ferreira, M.A. Approach to the study of C-glycosyl flavones by ion trap HPLC-PAD-ESI/MS/MS: Application to seeds of quince (*Cydonia oblonga*). *Phytochem. Anal.* **2003**, *14*, 352–359. [CrossRef] [PubMed]

36. Zheng, Y.; Hu, X.; Zhai, Y.; Liu, J.; Wu, G.; Wu, L.; Shen, T.J. Pharmacokinetics and tissue distribution study of camellianin A and its major metabolite in rats by liquid chromatography with tandem mass spectrometry. *J. Chromatogr. B* **2015**, *997*, 200–209. [CrossRef] [PubMed]

37. Xie, C.; Xu, L.Z.; Luo, X.Z.; Zhong, Z.; Yang, S.L. Flavonol glycosides from *Lysimachia capillipes*. *J. Asian Nat. Prod. Res.* **2002**, *4*, 17–23. [CrossRef] [PubMed]

38. Kashiwada, Y.; Nonaka, G.-I.; Nishioka, I.; Yamagishi, T. Galloyl and hydroxycinnamoylglucoses from rhubarb. *Phytochemistry* **1988**, *27*, 1473–1477. [CrossRef]

39. Horai, H.; Arita, M.; Kanaya, S.; Nihei, Y.; Ikeda, T.; Suwa, K.; Ojima, Y.; Tanaka, K.; Tanaka, S.; Aoshima, K.; et al. MassBank: A public repository for sharing mass spectral data for life sciences. *J. Mass Spectrom.* **2010**, *45*, 703–714. [CrossRef] [PubMed]

40. Yang, Z.; Dong, F.; Baldermann, S.; Murata, A.; Tu, Y.; Asai, T.; Watanabe, N. Isolation and identification of spermidine derivatives in tea (*Camellia sinensis*) flowers and their distribution in floral organs. *J. Sci. Food Agric.* **2012**, *92*, 2128–2132. [CrossRef] [PubMed]

41. Li, H.J.; Deinzer, M.L. Tandem mass spectrometry for sequencing proanthocyanidins. *Anal. Chem.* **2007**, *79*, 1739–1748. [CrossRef] [PubMed]

42. Saito, K.; Yonekura-Sakakibara, K.; Nakabayashi, R.; Higashi, Y.; Yamazaki, M.; Tohge, T.; Fernie, A.R. The flavonoid biosynthetic pathway in *Arabidopsis*: Structural and genetic diversity. *Plant. Physiol. Biochem.* **2013**, *72*, 21–34. [CrossRef] [PubMed]

43. Dai, W.; Tan, J.; Lu, M.; Xie, D.; Li, P.; Lv, H.; Zhu, Y.; Guo, L.; Zhang, Y.; Peng, Q.; et al. Nontargeted modification-specific metabolomics investigation of glycosylated secondary metabolites in tea (*Camellia sinensis* L.) based on liquid chromatography-high-resolution mass spectrometry. *J. Agric. Food Chem.* **2016**, *64*, 6783–6790. [CrossRef] [PubMed]

44. Li, Z.X.; Yang, W.J.; Ahammed, G.J.; Shen, C.; Yan, P.; Li, X.; Han, W.Y. Developmental changes in carbon and nitrogen metabolism affect tea quality in different leaf position. *Plant Physiol. Biochem.* **2016**, *106*, 327–335. [CrossRef] [PubMed]

45. Matsuda, F.; Yonekura-Sakakibara, K.; Niida, R.; Kuromori, T.; Shinozaki, K.; Saito, K. MS/MS spectral tag-based annotation of non-targeted profile of plant secondary metabolites. *Plant J.* **2009**, *57*, 555–577. [CrossRef] [PubMed]

46. Youhnovski, N.; Werner, C.; Hesse, M. N,N′,N″-Triferuloylspermidine, a new UV absorbing polyamine derivative from pollen of *Hippeastrum* x hortorum. *Naturforsch. C* **2001**, *56*, 526–530. [CrossRef]

47. Michael, A.J. Biosynthesis of polyamines and polyamine-containing molecules. *Biochem. J.* **2016**, *473*, 2315–2329. [CrossRef] [PubMed]

48. Afendi, F.M.; Okada, T.; Yamazaki, M.; Hirai-Morita, A.; Nakamura, Y.; Nakamura, K.; Ikeda, S.; Takahashi, H.; Altaf-Ul-Amin, M.; Darusman, L.K.; et al. KNApSAcK family databases: Integrated metabolite-plant species databases for multifaceted plant research. *Plant Cell Physiol.* **2012**, *53*. [CrossRef] [PubMed]

49. Fiehn, O. Metabolomics–the link between genotypes and phenotypes. *Plant Mol. Biol.* **2002**, *48*, 155–171. [CrossRef] [PubMed]

50. Dixon, R.A.; Strack, D. Phytochemistry meets genome analysis, and beyond. *Phytochemistry* **2003**, *62*, 815–816. [CrossRef]

51. Engelhardt, U.H. Chemistry of Tea. In *Reference Module in Chemistry, Molecular Sciences and Chemical Engineering*; Elsevier: Braunschweig, Germany, 2013.

52. Tohge, T.; de Souza, L.P.; Fernie, A.R. Current understanding of the pathways of flavonoid biosynthesis in model and crop plants. *J. Exp. Bot.* **2017**, *68*, 4013–4028. [CrossRef] [PubMed]

53. Yonekura-Sakakibara, K.; Tohge, T.; Niida, R.; Saito, K. Identification of a flavonol 7-O-rhamnosyltransferase gene determining flavonoid pattern in *Arabidopsis* by transcriptome coexpression analysis and reverse genetics. *J. Biol. Chem.* **2007**, *282*, 14932–14941. [CrossRef] [PubMed]

54. Yonekura-Sakakibara, K.; Tohge, T.; Matsuda, F.; Nakabayashi, R.; Takayama, H.; Niida, R.; Watanabe-Takahashi, A.; Inoue, E.; Saito, K. Comprehensive flavonol profiling and transcriptome coexpression analysis leading to decoding gene–metabolite correlations in *Arabidopsis*. *Plant Cell* **2008**, *20*, 2160–2176. [CrossRef] [PubMed]

55. Routaboul, J.M.; Kerhoas, L.; Debeaujon, I.; Pourcel, L.; Caboche, M.; Einhorn, J.; Lepiniec, L. Flavonoid diversity and biosynthesis in seed of *Arabidopsis thaliana*. *Planta* **2006**, *224*, 96–107. [CrossRef] [PubMed]

56. Lai, J.P.; Lim, Y.H.; Su, J.; Shen, H.M.; Ong, C.N. Identification and characterization of major flavonoids and caffeoylquinic acids in three *Compositae* plants by LC/DAD-APCI/MS. *J. Chromatogr. B* **2007**, *848*, 215–225. [CrossRef] [PubMed]

57. Schmid, M.; Davison, T.S.; Henz, S.R.; Pape, U.J.; Demar, M.; Vingron, M.; Scholkopf, B.; Weigel, D.; Lohmann, J.U. A gene expression map of *Arabidopsis thaliana* development. *Nat. Genet.* **2005**, *37*, 501–506. [CrossRef] [PubMed]

58. Ishihara, H.; Tohge, T.; Viehoever, P.; Fernie, A.R.; Weisshaar, B.; Stracke, R. Natural variation in flavonol accumulation in *Arabidopsis* is determined by the flavonol glucosyltransferase BGLU6. *J. Exp. Bot.* **2016**, *67*, 1505–1517. [CrossRef] [PubMed]

59. Scharbert, S.; Hofmann, T. Molecular definition of black tea taste by means of quantitative studies, taste reconstitution, and omission experiments. *J. Agric. Food Chem.* **2005**, *53*, 5377–5384. [CrossRef] [PubMed]

60. Macoy, D.M.; Kim, W.Y.; Lee, S.Y.; Kim, M.G. Biosynthesis, physiology, and functions of hydroxycinnamic acid amides in plants. *Plant Biotechnol. Rep.* **2015**, *9*, 269–278. [CrossRef]

61. Facchini, P.J.; Hagel, J.; Zulak, K.G. Hydroxycinnamic acid amide metabolism: Physiology and biochemistry. *Can. J. Bot.* **2002**, *80*, 577–589. [CrossRef]

62. Negri, G.; Teixeira, E.W.; Florêncio Alves, M.L.T.M.; Moreti, A.C.d.C.C.; Otsuk, I.P.; Borguini, R.G.; Salatino, A. Hydroxycinnamic acid amide derivatives, phenolic compounds and antioxidant activities of extracts of pollen samples from Southeast Brazil. *J. Agric. Food Chem.* **2011**, *59*, 5516–5522. [CrossRef] [PubMed]

63. Pang, Y.Z.; Abeysinghe, I.S.B.; He, J.; He, X.Z.; Huhman, D.; Mewan, K.M.; Sumner, L.W.; Yun, J.F.; Dixon, R.A. Functional characterization of proanthocyanidin pathway enzymes from tea and their application for metabolic engineering. *Plant Physiol.* **2013**, *161*, 1103–1116. [CrossRef] [PubMed]

64. Gu, L.; Kelm, M.A.; Hammerstone, J.F.; Beecher, G.; Holden, J.; Haytowitz, D.; Prior, R.L. Screening of foods containing proanthocyanidins and their structural characterization using LC-MS/MS and thiolytic degradation. *J. Agric. Food Chem.* **2003**, *51*, 7513–7521. [CrossRef] [PubMed]

65. Hashimoto, F.; Nonaka, G.; Nishioka, I. Tannins and related compounds. XC. 8-C-Ascorbyl (-)-epigallocatechin 3-O-gallate and novel dimeric flavan-3-ols, oolonghomobisflavans A and B, from oolong tea. *Chem. Pharm. Bull.* **1989**, *37*, 3255–3263. [CrossRef]

66. Kumar, N.S.; Bandara, B.M.R.; Hettihewa, S.K. Isolation of a tetrameric A-type proanthocyanidin containing fraction from fresh tea (*Camellia sinensis*) leaves using high-speed counter-current chromatography. *J. Liq. Chromatogr. Relat. Technol.* **2015**, *38*, 1571–1575. [CrossRef]

67. Chen, S.; Li, M.; Zheng, G.; Wang, T.; Lin, J.; Wang, S.; Wang, X.; Chao, Q.; Cao, S.; Yang, Z.; et al. Metabolite profiling of 14 Wuyi Rock tea cultivars using UPLC-QTOF MS and UPLC-QqQ MS combined with chemometrics. *Molecules* **2018**, *23*, 104. [CrossRef] [PubMed]

Sample Availability: Samples of the compounds are not available from the authors.

molecules

MDPI

Article

Hydrophilic Interaction Liquid Chromatography-Electrospray Ionization Mass Spectrometry for Therapeutic Drug Monitoring of Metformin and Rosuvastatin in Human Plasma

Nikolaos Antonopoulos [1], Giorgos Machairas [1], George Migias [2], Ariadni Vonaparti [3], Vasiliki Brakoulia [1], Constantinos Pistos [4], Dimitra Gennimata [5] and Irene Panderi [1,*]

[1] Laboratory of Pharmaceutical Analysis, School of Pharmacy, Division of Pharmaceutical Chemistry, National and Kapodistrian University of Athens, Panepistimiopolis, Zografou, 15771 Athens, Greece; nikos_antono@windowslive.com (N.A.); giorgosmachairas@yahoo.com (G.M.); vasilikibrakoulia@gmail.com (V.B.)

[2] General Hospital of Athens Alexandra, 80 Vas. Sofias Avenue, 11528 Athens, Greece; g.migias@gmail.com

[3] Qatar Doping Analysis Laboratory, Sports City Road, Aspire Zone, P.O. Box 27775, Doha, Qatar; avonaparti@adlqatar.qa

[4] Department of Chemistry, West Chester University, West Chester, PA 19383, USA; cpistos@wcupa.edu

[5] General Hospital Korgialenio-Benakio National Red Cross, Erithrou Stavrou 1, 11526 Athens, Greece; dimigenn@gmail.com

* Correspondence: irenepanderi@gmail.com or ipanderi@pharm.uoa.gr; Tel.: +30-697-401-5798

Academic Editors: Marcello Locatelli, Simone Carradori and Andrei Mocan
Received: 10 June 2018; Accepted: 25 June 2018; Published: 27 June 2018

Abstract: In this work a hydrophilic interaction liquid chromatography/positive ion electrospray mass spectrometric assay (HILIC/ESI-MS) has been developed and fully validated for the quantitation of metformin and rosuvastatin in human plasma. Sample preparation involved the use of 100 μL of human plasma, following protein precipitation and filtration. Metformin, rosuvastatin and 4-[2-(propylamino) ethyl] indoline 2 one hydrochloride (internal standard) were separated by using an X-Bridge-HILIC BEH analytical column (150.0 × 2.1 mm i.d., particle size 3.5 μm) with isocratic elution. A mobile phase consisting of 12% (v/v) 15 mM ammonium formate water solution in acetonitrile was used for the separation and pumped at a flow rate of 0.25 mL min^{-1}. The linear range of the assay was 100 to 5000 ng mL^{-1} and 2 to 100 ng mL^{-1} for metformin and rosuvastatin, respectively. The current HILIC-ESI/MS method allows for the accurate and precise quantitation of metformin and rosuvastatin in human plasma with a simple sample preparation and a short a chromatographic run time (less than 15 min). Plasma samples from eight patients were further analysed proving the capability of the proposed method to support a wide range of clinical studies.

Keywords: rosuvastatin; metformin; HILIC; LC-MS; therapeutic drug monitoring

---

1. Introduction

Diabetes mellitus is a metabolic disorder caused by failures in the action and/or secretion of insulin. During the last decades, incidents of diabetes have increased from 108 million to 422 million corresponding mainly in type 2 diabetes mellitus [1]. Diabetes mellitus is a risk factor for a wide range of vascular diseases like ischaemic stroke and coronary heart disease [2]. Metformin belongs to a class of biguanides and it is usually administered for non-insulin-dependent type II diabetes mellitus as the first line of treatment when lifestyle modification alone is not enough [3]. Several studies in the literature report the beneficial protective effects of metformin administration in cardiac function, resulting in the deterioration of mortality caused from cardiovascular deceases as a result of type 2 diabetes mellitus [4]. For the treatment of patients with type 2 diabetes mellitus it is often required drug therapy with beneficial effects on the lowering of blood glucose and on dyslipidemia. Statins are administered for the prevention of cardiovascular diseases and are also effective to prevent vascular events in diabetic patients [5]. Rosuvastatin is a 3-hydroxy-3-methylglutaryl-coenzyme A (HMG-CoA) reductase inhibitor that is administered for hypercholesterolemia in patients with a high risk of developing atherosclerosis and in patients with established cardiovascular disease [6].

Metformin and rosuvastatin are frequently prescribed in combination due to the high comorbidity of these two diseases [7]. The development of novel analytical methods for therapeutic drug monitoring is crucial to support the model of personalized medicine and the introduction of new drug combinations in therapy. Thus, the purpose of this study was on the development of a new method to detect and quantify those two commonly prescribed drugs in human plasma.

Several methods have been reported in literature for the quantitation of metformin alone [8–12] or in combination with other drugs [13–21] in biofluids. A number of other methods have also been reported in literature [22] for the quantitation of rosuvastatin alone [23–30] or in combination with other drugs [31,32] in human plasma. The pharmacokinetic parameters of metformin and rosuvastatin in healthy volunteers have been studied by the use of two different reversed-phase liquid chromatography tandem mass spectrometric methods procedures for each analyte and complicated sample preparation procedures involving liquid-liquid extraction [7]. A study on the pharmacokinetics of rosuvastatin after concomitant administration of metformin and/or furosemide was conducted by using two different RP-LC tandem mass spectrometric methods. The first method was used in combination with a solid phase extraction procedure for the determination of metformin alone and the second one in combination with a liquid-liquid extraction for the determination of rosuvastatin and furosemide combination [33]. Recently, a reversed-phase tandem mass spectrometric method was developed and fully validated for the simultaneous determination of metformin and rosuvastatin in human plasma in combination with a simple sample preparation procedure [34]. Most recently a HILIC method coupled to diode array detection has been published by our group for the quantitation of rosuvastatin and metformin impurities in fixed-dose combination tablets containing rosuvastatin and metformin [35]. To the extent of our knowledge, no bioanalytical assay has been previously developed for the simultaneous quantitation of metformin and rosuvastatin in human plasma by hydrophilic interaction liquid chromatography (HILIC) and by using a single quadrupole mass spectrometer. HILIC offers improved sensitivity when combined with mass spectrometric detection due to the high percentage of organic solvents used in the mobile phase in order to elute polar compounds. The proposed method is suitable for therapeutic drug monitoring of metformin and rosuvastatin and it has been successfully applied to the analysis of clinical samples obtained from patients that have been treated with the analysed drugs.

2. Results and Discussion

2.1. Method Development

2.1.1. MS Detection Optimization

The parameters for electrospray ionization (ESI) have been optimized so as to allow maximum abundance of the molecular ions of metformin, rosuvastatin and the ISTD. The mass spectrometer was operated in positive ESI ion mode and selected ion monitoring (SIM) was used to analyze the protonated molecules $[M + H]^+$ at m/z 130, 482 and 219 for rosuvastatin, metformin and N-despropyl ropinirole (ISTD), respectively. The maximum abundance of the selected ions was achieved when capillary voltage, mass span and dwell time were set at 4.8 kV, 0.1 and 0.3 s, respectively. Mass spectra of metformin, rosuvastatin and N-despropyl ropinirole (ISTD) are presented in Figure 1.

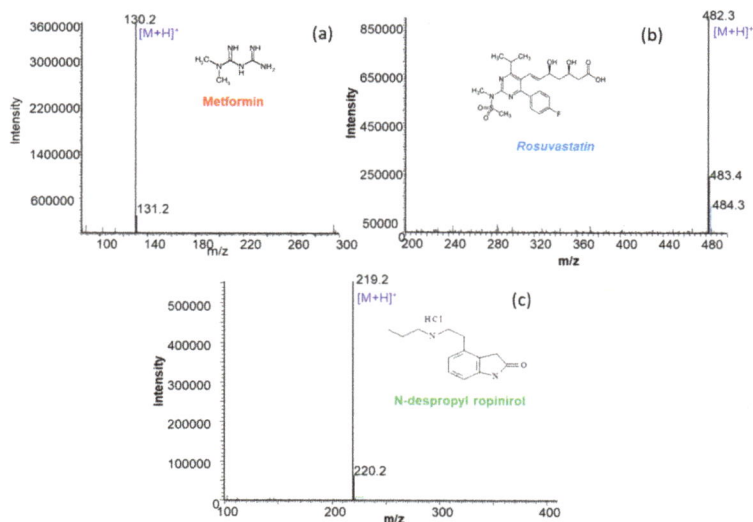

Figure 1. ESI mass spectra in positive ion mode of a 10 μg mL^{-1} solution prepared in mobile phase of (**a**) metformin (**b**) rosuvastatin and (**c**) N-despropyl ropinirole.

2.1.2. Chromatography

The chromatographic separation of compounds with a broad band of polarity can be easily achieved by using HILIC chromatography [36]. The separation mechanism in HILIC combines partition, reversed-phase interactions, normal phase/adsorption, electrostatic interactions and hydrogen bonding and thus it is ideal to separate compounds with diverse physicochemical properties [37]. Since metformin and rosuvastatin have sufficiently different physicochemical properties a zwitterionic ZIC®-pHILIC analytical column has been used in preliminary experiments to separate these compounds. In most of the mobile phases that have been tested rosuvastatin was eluted very close to the solvent front even when the aqueous content of the mobile phase was reduced below 6%. Thus, a hydrophilic XBridge®-HILIC BEH analytical column was chosen in this work as the most suitable for the chromatography. The packing material of this column consists of two kind of monomers, bis-triethoxysilylethane and tetraethoxysilane, which are joined with ethylene bridges. Some accessible silanols that remain on the surface of these BEH particles are responsible for electrostatic interactions. Metformin is 100% positively-charged over a pH range from 1 to 12, while rosuvastatin is 100% negatively-charged at pH values above 6. The internal standard, N-despropyl ropinirol is 100% positively-charged over a pH range from 1 to 7.5. Previous findings, of our research group [35], pointed out that the mechanism of separation for rosuvastatin and metformin in HILIC comprises both secondary electrostatic interactions and hydrophilic partition.

Several combinations of the mobile phase constituents, acetonitrile asorganic modifier and aqueous ammonium formate buffer, have been tested to optimize the chromatographic parameters in order to achieve adequate separation of the analytes from matrix interferences and reach the best sensitivity in ESI-MS detection. With a constant water content of the mobile phase eluent at 12%, the concentration of the aqueous ammonium formate buffer was varied from 2.5 to 25 mM. Figure 2a shows that the retention of both metformin and N-despropyl ropinirol is decreased by increasing the concentration of ammonium formate up to 25 mM, while the retention of rosuvastatin is increased. Based on the above findings it can be concluded for metformin and ISTD, that by increasing the buffering salt concentration the electrostatic attraction between the positively-charged compounds and the negatively-charged silanol groups of the XBridge®-HILIC BEH particle is reduced. On the contrary, for rosuvastatin the slight increase in retention by increasing the buffering salt concentration reduces the electrostatic repulsion of the negatively charged rosuvastatin molecule and the negatively charged silanol groups of the XBridge®-HILIC BEH particles. These results have been obtained from the analysis of human plasma samples spiked with 2500 ng mL^{-1} metformin, 50 ng mL^{-1} rosuvastatin and 190 ng mL^{-1} of N-despropyl ropinirol (ISTD). It was also observed that by reducing the concentration of ammonium formate below 10 mM the peak shape of metformin and N-despropyl ropinirole was distorted and split. Thus, a 15 mM aqueous ammonium formate concentration was chosen as the optimum.

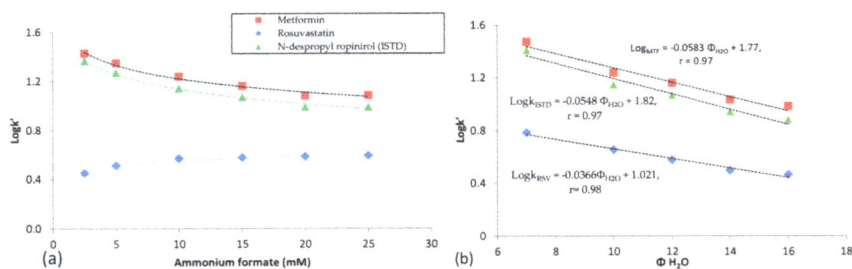

Figure 2. Plots of logk values for rosuvastatin, metformin and N-despropyl ropinirole versus (**a**) the concentration of ammonium formate, and (**b**) the percentage of the water of the mobile phase on an XBridge®-HILIC BEH analytical column. Data have been obtained from the analysis of human plasma samples spiked with 2500 ng mL^{-1} metformin, 50 ng mL^{-1} rosuvastatin and 190 ng mL^{-1} N-despropyl ropinioro (ISTD).

The effect of the percentage of water, ϕ_{water}, was also evaluated in experiments where ϕ_{water} was varied from 7 to 16% while the concentration of ammonium formate was kept constant at the optimum value of 1.8 mM (pH = 6.5) in whole mobile phase. As it can be observed in Figure 2b, the retention of the analytes decreases linearly with increasing the percentage of water, implying a partition mechanism in HILIC separation. It was also observed an improvement in ESI/MS sensitivity for metformin (Figure 3a) and rosuvastatin (Figure 3b) by decreasing the concentration of the buffering salt in the mobile phase, while the percentage of water, ϕ_{water}, was kept constant at 12% (*v/v*).

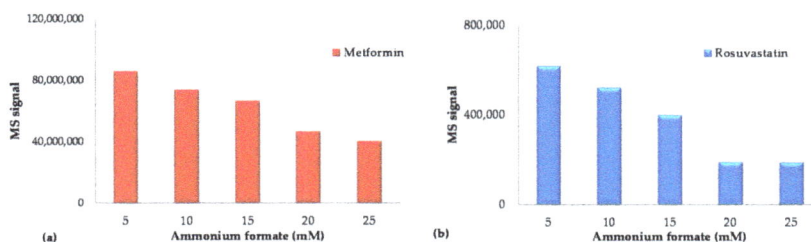

Figure 3. Impact of the concentration of ammonium formate in the mobile phase on the MS signal of (**a**) metformin and (**b**) rosuvastatin. Data have been obtained from the analysis of human plasma samples spiked with 2500 ng mL^{-1} metformin, 50 ng mL^{-1} rosuvastatin and 190 ng mL^{-1} N-despropyl ropinioro (ISTD).

Based on the above studies, the optimum mobile phase composition consists of 12% 15 mM ammonium formate water solution pH 6.5 in acetonitrile. The proposed HILIC-ESI/MS method allows the isocratic separation of the analytes with 15 min. After column equilibration time of 1 h a great number of samples can be analysed within one analytical daily batch. The internal standard,

N-despropyl ropinirol, exhibits adequate ion intensity and good chromatographic peak shape under the selected chromatographic and MS conditions. This compound is an impurity of ropinirol, thus, it cannot be found in human plasma of patients.

The selectivity of the proposed method is illustrated in Figure 4, where a representative ion chromatogram of a blank plasma sample is overlaid with ion chromatograms of calibration plasma samples spiked with 100 and 1000 ng mL^{-1} of metformin and 2 and 20 ng mL^{-1} of rosuvastatin, respectively and 190 ng mL^{-1} of N-despropyl ropinirol (ISTD). Rosuvastatin, metformin and N-despropyl ropinirol were eluted at 2.4, 7.7 and 6.3 min, respectively.

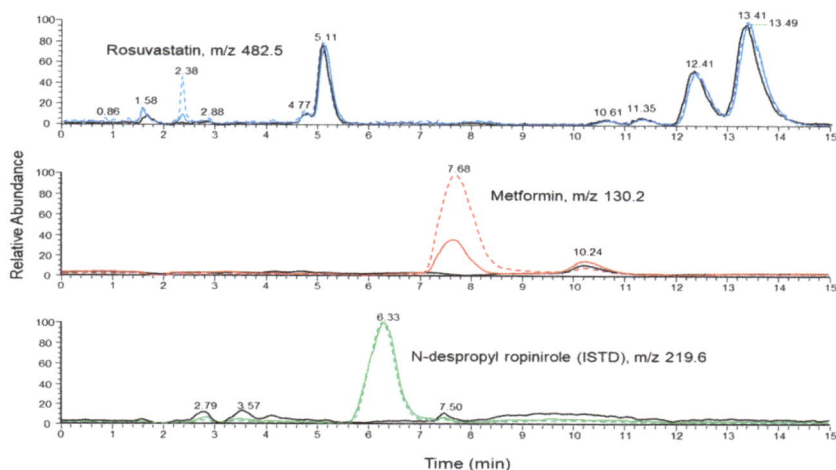

Figure 4. Ion chromatogram (SIM mode) of a blank human plasma sample (black line) overlaid with ion chromatograms (SIM mode) of calibration spiked plasma samples at 2 ng mL^{-1} (solid line) and 20 ng mL^{-1} (dashed line) of rosuvastatin (blue line), 100 ng mL^{-1} (solid line) and 1000 ng mL^{-1} (dashed line) of metformin (red line) and 190 ng mL^{-1} of ISTD (green line).

2.1.3. Optimization of the Preparation of the Biological Sample

Protein precipitation is a simple sample preparation procedure that allows for the analysis of a large number of samples in short time. In this work protein precipitation gave adequate recovery for metformin, rosuvastatin and the ISTD, while further filtration was applied to minimize the matrix interference, and to prolong the column life time and prevent system blockages [38]. In HILIC the nature of the dissolution solvent is more crucial than in reversed-phase HPLC on the chromatographic peak shape. It was observed during method development that the peak shapes of metformin and rosuvastatin were split and distorted in different dissolution solvent mixtures and by changing the percentage of the biological sample in the injection sample.

Various proportions of human plasma have been tested to determine the optimum amount of biological material for the analyses. Figure 5 shows that by increasing the amount of human plasma up to 100 µL, peak area signal of metformin (Figure 5a) and rosuvastatin (Figure 5b) was increased. A further increase of human plasma above 100 µL caused serious distortion of the peak shapes of the analytes. Thus, analysis was performed by using 100 µL of biological sample. To further improve the chromatographic peak shapes, buffered eluents have been used. Therefore, we tested different concentrations of 30% (*v/v*) ammonium formate water solution in the dissolution solvent. Results presented in Figure 5c,d, for metformin and rosuvastatin, respectively, indicate that peak area signal of rosuvastatin decreases by increasing the concentration of ammonium formate. We concluded that the best peak shapes were obtained by using 100 µL of biological sample and processed with a dissolution solvent mixture consisting of 30% (*v/v*) of an ammonium formate water solution at 5 mM (pH 6.5) in acetonitrile. A membrane syringe filter was also used for the filtration of the samples prior to HILIC-ESI/MS analysis. Filtration was essential to remove any additional matrix interferences and increase the sensitivity in HILIC-ESI/MS system. To evaluate whether filtering of the processed biological samples would cause loss of the analytes, eight different types of syringe filters were selected for testing, five 0.45 μm pore size and three 0.22 μm pore size, all 13 mm diameter.

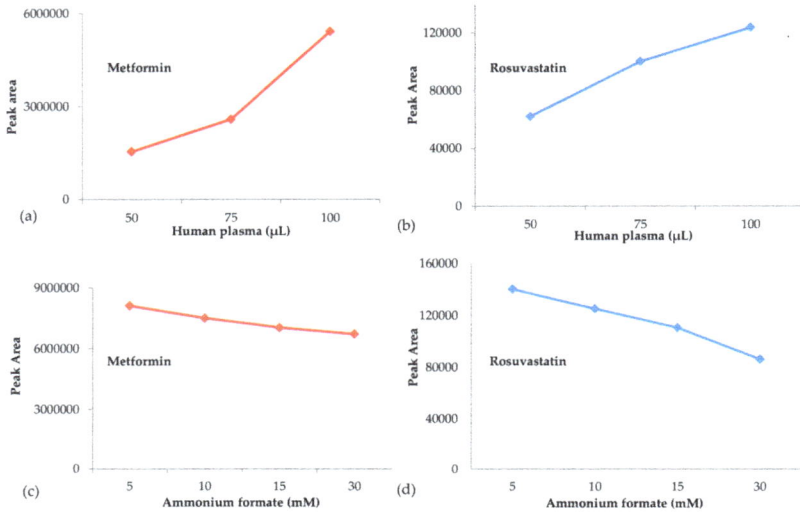

Figure 5. Effect of the proportion of the biological sample (**a** and **b**) and the concentration of a 30% (*v/v*) ammonium formate water solution in acetonitrile (**c** and **d**) on the peak area signal of metformin and (**b,d**) rosuvastatin.

To evaluate the % recovery, the peak area of spiked plasma samples was compared with the analytical response of blank plasma samples spiked with the equivalent concentration of the analytes after the sample preparation and filtration procedure. Data presented in Figure 6, shows the % recovery of the analytes under the various types of the filters that have been tested.

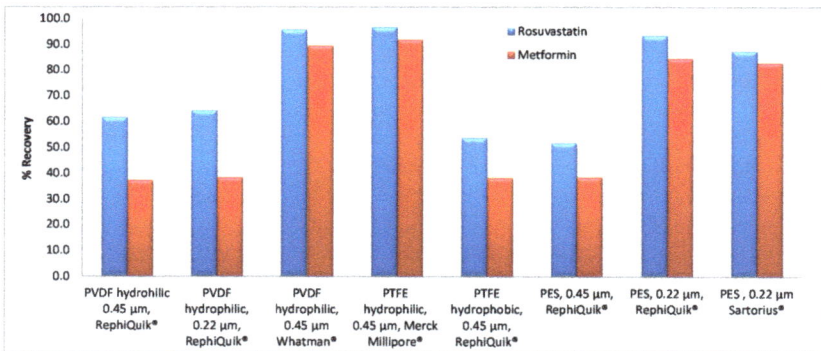

Figure 6. Effect of the type of the filters used for sample preparation of the biological samples on the % recovery of the analytes.

Optimum % recovery for metformin and rosuvastatin was achieved when a 100 μL aliquot of each human plasma sample was processed by the addition of 50 μL of ISTD solution (3.8 μg mL^{-1}), 30 μL of 5 mM ammonium formate solution in water and 820 μL of acetonitrile. After vortex mixing and centrifugation, the supernatant was filtered through a 13 mm PTFE membrane syringe filter (hydrophilic) with a pore size of 0.45 μm.

2.2. Statistical Analysis of Data

2.2.1. Selectivity and Specificity

The selectivity test met the pre-defined criteria as co-eluting peaks where less than 5% of the area of metformin and rosuvastatin at the LOQ level, and less than 5% of the area of N-despropyl ropinirole in the ion chromatograms (SIM mode) obtained from the analysis of six batches of human plasma. The carry-over test has been performed by analysing blank human plasma samples after the injection of the highest concentration calibration and showed no interfering peaks with peak areas greater than 5% of the peak areas at the LOQ level of metformin and rosuvastatin.

2.2.2. Linearity, Precision and Accuracy

A weighting factor of $1/y^2$ was used for the linear regression analysis and data presented in Table 1, indicate that linear relationships have been achieved between the measured signal ratios of the analytes and the corresponding concentrations. In agreement with international guidelines [39], the back-calculated concentrations in the calibration curves were less than 11.5% and 14.5% of the nominal concentration of metformin and rosuvastatin.

Table 1. Analytical concentration parameters of the calibration equations for the determination of metformin and rosuvastatin in human plasma by HILIC-ESI/MS.

Compound	Concentration Range, ng mL^{-1}	Regression Equations [a]	r [b]	Standard Deviation		S$_r$ [c]
				Slope	Intercept	
Mean of three calibration curves over a period of one month						
Metformin	62.5–5000	$R_{Mtf} = 0.00715 \times C_{Mtf} + 0.066$	≥0.996	2.2×10^{-4}	4.3×10^{-4}	≤0.10
Rosuvastatin	2–100	$R_{Rsv} = 0.01775 \times C_{Rsv} - 0.0129$	≥0.998	5.2×10^{-4}	7.2×10^{-3}	≤0.062

[a] Ratios of the peak areas signals of metformin, R_{Mtf} and rosuvastatin, R_{Rsv}, to that to the internal standard vs. the corresponding concentration of metformin, C_{Mtf} and rosuvastatin, C_{Rsv}; [b] Correlation coefficient; [c] Standard error of the estimate.

The LOD and LLOQ were estimated from the signal-to-noise ratio (S/N) and by analysing human plasma samples spiked with metformin and rosuvastatin at low concentrations. The LOD was estimated at a S/N ratio of 3:1 and the LLOQ was estimated at a S/N ratio of at least 10:1, until a % CV of less than 20% was obtained. The LODs were found to be at 11 and 0.3 ng mL^{-1}, and the LLOQs at 35 and 1 ng mL^{-1}, for metformin and rosuvastatin, respectively.

Precision and accuracy data are presented in Table 2 and indicate that intra-assay coefficients of variations, % CV, were between 5.3% and 9.1% for metformin and between 6.2% and 8.5% for rosuvastatin. The inter-assay % CVs were lower than 3.0% and 7.1% for metformin and rosuvastatin, respectively. The overall accuracy, which was assessed by the percentage relative error, was ranged from −3.3 to 0.4% for metformin and from −1.1 to 7.0% for rosuvastatin.

Table 2. Data on accuracy and precision obtained from the analysis of quality control samples containing rosuvastatin and metformin (n = 3 runs; five replicates per run).

Compound	Concentration (ng mL^{-1})		
	Metformin		
Added Concentration	**62.5**	**500**	**5000**
Run 1 (mean ± s.d.)	63.9 ± 6.1	489 ± 29	5045 ± 352
Run 2 (mean ± s.d.)	64.3 ± 5.1	42.9 ± 8.3	4698 ± 143
Run 3 (mean ± s.d.)	60.1 ± 6.0	488 ± 39	4768 ± 232
Overall mean	62.7	497.9	4837
Intra-day CV (%) [a]	9.1	7.6	5.3
Inter-day CV (%) [a]	1.8	1.0	3.0
Overall accuracy Er% [b]	0.4	−0.4	−3.3
	Rosuvastatin		
Added Concentration	**2**	**10**	**100**
Run 1 (mean ± s.d.)	2.31 ± 0.21	9.41 ± 0.33	104.1 ± 6.6
Run 2 (mean ± s.d.)	2.12 ± 0.22	9.61 ± 0.89	96.5 ± 5.7
Run 3 (mean ± s.d.)	2.01 ± 0.21	10.81 ± 0.46	96.2 ± 9.1
Overall mean	2.1	9.9	98.9
Intra-day CV (%) [a]	8.5	6.2	7.3
Inter-day CV (%) [a]	4.4	7.1	3.0
Overall accuracy Er% [b]	7.0	−1.0	−1.1

[a] Intra- and inter-assay coefficient of variations as calculated by ANOVA. [b] Relative percentage error.

2.2.3. Recovery and Matrix Effect

Recovery data are presented in Table 3 and indicate average recovery of more than 92.6% and 95.8% for metformin and rosuvastatin, respectively. The average recovery for N-despropyl ropinirol (ISTD) was found to be 91.7 ± 3.8%.

Table 3. Recovery and ion suppression data for the quantitation of metformin and rosuvastatin by HILIC-ESI/MS.

Compound	Concentration Levels (ng mL^{-1})	
Metformin	**1000**	**5000**
% Recovery (mean ± s.d.)$_{n = 3}$	92.6 ± 2.5	93.9 ± 1.4
% Matrix Factor (mean ± s.d.)$_{n = 3}$	59.9 ± 1.0	63.7 ± 1.3
Rosuvastatin	**20**	**100**
% Recovery (mean ± s.d.)$_{n = 3}$	96.6 ± 2.1	95.8 ± 4.1
% Matrix Factor (mean ± s.d.)$_{n = 3}$	84.5 ± 7.1	81.3 ± 2.0
N-despropyl ropinirole (ISTD)	3800	
% Recovery (mean ± s.d.)$_{n = 3}$	91.7 ± 3.8	
% Matrix Factor (mean ± s.d.)$_{n = 3}$	61.8 ± 1.8	

The method described by Matuszewski et al. [40] was used to evaluate the matrix effect for metfromin, rosuvastatin and the ISTD. The matrix effect was expressed by the percentage matrix factor which is defined as the percentage of the ratio of the analyte peak response in the presence of matrix ions to the analyte peak response in the absence of matrix ions. As shown by the results presented in Table 4, the % matrix factor was greater than 59.9% and 81.3% for metformin and rosuvastatin, respectively indicating low ion suppression. The % matrix factor for N-despropyl ropinirol (ISTD) was found to be 61.8 ± 1.8%.

Table 4. Stability data for metformin and rosuvastatin in human plasma under various storage conditions.

Compound	Calculated Concentration (ng mL^{-1})					
Metformin	50		1250		2500	
	Mean ± S.D.$_{(n = 3)}$	%E$_r$ [a]	Mean ± S.D.$_{(n = 3)}$	%E$_r$ [a]	Mean ± S.D.$_{(n = 3)}$	%E$_r$ [a]
Ambient temperature/4 h	50.4 ± 1.9	0.9	1271.5 ± 8.2	1.7	2522 ± 138	0.01
−20 °C/4 weeks	49.7 ± 2.1	−0.6	1260.1 ± 0.80	0.8	2412 ± 18	−0.04
−20 °C/4 Freeze-thaw cycles	51.21 ± 0.81	2.4	1282.3 ± 2.6	2.6	2386 ± 24	−0.05
Rosuvastatin	2.5		25		50	
Ambient temperature/4 h	2.509 ± 0.0089	0.4	24.72 ± 0.11	−1.1	49.4 ± 3.4	−0.9
−20 °C/4 weeks	2.457 ± 0.045	−1.7	24.53 ± 0.15	−1.9	50.2 ± 3.2	0.4
−20 °C/4 Freeze-thaw cycles	2.435 ± 0.031	−2.6	24.94 ± 0.42	−0.2	49.5 ± 2.4	−1.0

[a] %E$_r$: relative percentage error = (overall mean assayed concentration − added concentration)/(added concentration) × 100.

2.2.4. Stability

The stability of the analytes in spiked human plasma samples was evaluated under various storage conditions presented in detail in Table 4. To evaluate stability the results of the stored samples were compared to the results of the freshly prepared samples spiked with the analytes in human plasma at equivalent concentrations.

Stability results are presented in Table 4 and indicate that the concentrations of metformin and rosuvastatin deviate by no more than −2.6% relative to the reference for any of the analytes. Thus, human plasma samples containing metformin and rosuvastatin may be kept without any substantial degradation under the various tested storage conditions.

2.3. Application to the Analysis of Real Samples

Human plasma samples collected from patients that were treated with the analytes have been analysed by the current method so as to demonstrate the applicability of the method to support several clinical studies related to metformin and rosuvastatin therapy. Human plasma samples collection was approved by the local Ethics Committee of clinical settings, although not demanded by national legislation, as this study is not regarded as a clinical trial.Blood samples were collected in tubes containing sodium heparin as anticoagulant and immediately after drawing, the samples were shaken gently and centrifuged at 4000 rpm for 10 min at 4 °C. Human plasma samples were stored at −20 °C and analysed within two weeks after storage by the proposed HILIC-ESI/MS method. In particular, plasma samples obtained from eight human patients (five female and three male) with ages ranging from 57 to 88 years old have been analysed by the proposed method. During the period of sample collection, all of the patients had receiving other medication described in detail in Table 5.

Table 5. Clinical data in patients treated with rosuvastatin and metformin.

Patient Sex/Age (years)	Time Post Dose (h)	Drug/Dose Peros (mg)	C (ng mL^{-1}) Mean ± S.D.$_{(n=3)}$	Co-Administered Drugs
♀/57	$2_{1/2}$	Metformin/1000 × 1 × 30 days Rosuvastatin/10 × 1 × 30 days	1956 ± 93 22.2 ± 0.4	Acetylsalicylic acid 100 mg tb 1 × 1
♂/64	11	Rosuvastatin/20 × 1 × 30 days	16.2 ± 1.4	Furosemide 40 mg tb 1 × 1, ramipril 2.5 mg tb 1 × 1, carvedilol 6.35 mg tb 1 × 1.
♂/63	16	Rosuvastatin/20 × 1 × 30 days	9.21 ± 0.39	Nadroparin 5700 anti-ha 1 × 1 subcutaneous, omeprazole caps 20 mg 1 × 1, metoprolol tb 25 mg 1 × 2, sulbactam 1 g & ampicillin i.v. 1 × 3 × 5 days, acetyl salicylic acid 100 mg E.C.tb 1 × 1, atalopram 20 mg 1 × 1.
♀/80	$2_{1/2}$	Metformin/1000 × 1 × 30 days	2635 ± 67	Acetylsalicylic acid 100 mg tb 1 × 1, insuline glarine 100 iu/mL 10IU ×1 night subcutaneous, Ramipril 5 mg tb 1 × 1, nadropanin 2850 Anti-xa 1 × 1, piracetam i.v. 1 × 2.
♀/83	10	Metformin/1000 × 2 × 30 days	1850 ± 55	Furosemide 20 mg tb 1 × 1, 1 × 1, insuline glarine 100 iu/mL 14IU ×1 night subcutaneous, pantoprazole 40 mg caps 1 × 1, T4-50 1 × 1, carvedilol 12.5 mg tb 1 × 2, apixaban 2.5 mg tb 1 × 2, Lutein/Zeaxanthine/Mesozeaxanthine caps 1 × 1, VITAMIN D3 2000 IU 1 × 1.
♂/70	$2_{1/2}$	Metformin/850 × 2 × 30 days	1856 ± 84	Ezetimibe/simvastatin 10/40 mg tb 1 × 1, gliclazide 30 mg tb 1 × 1, sertraline 50 mg tb 1 × 2, galantamine 8 mg caps 1 × 1, levodopa/carvidopa/entacapone 100/25/200 tb 1 × 3, clopidogrel 75 mg tb 1 × 1.
♀/88	12	Metformin/850 × 1 × 30 days	1676 ± 60	Atorvastatin tb 40 mg 1 × 1, furosemide 20 mg i.v. 2 × 3, bisoprostol 10 mg 1 × 1, insuline glarine 100 iu/mL 10IU ×1 night subcutaneous, thyrromone 100 mg 1 × 1, acetyl salicylic acid 100 mg E.C.tb 1 × 1, digoxin tb 0.25 mg $\frac{1}{2}$ × 1.
♀/88	16	Metformin/850 × 1 × 30 days	226 ± 12	Atorvastatin tb 40 mg 1 × 1, furosemide 20 mg i.v. 2 × 3, insuline glarine 100 iu/mL 10IU ×1 night subcutaneous, thyrromone 100 mg 1 × 1, digoxin tb 0.25 mg $\frac{1}{2}$ × 1, acetyl salicylic acid 100 mg E.C. tb 1 × 1.

As shown by the clinical data presented in Table 5, rosuvastatin was quantitated in human plasma samples obtained from two male patients who had receiving 20 mg of rosuvastatin once daily in the morning (Crestor® 20, AstraZeneca A.E., Cambridge, UK) for a long-term treatment and blood samples were collected at 11 and 16 h after dosing. The concentrations of rosuvastatin ranged from 9.21 ± 0.39 to 16.2 ± 1.4 ng mL^{-1}.

Metformin was quantitated in human plasma samples obtained from five patients (four female and one male) who had receiving either 1000 mg of metformin (Glucophage® 1000, Merck A.E., Kenilworth, NJ, USA) once daily in the morning or 850 mg of metformin twice daily (Glucophage® 850, Merck A.E.) and for a long-term treatment. The concentrations of metformin ranged from 226 ± 12 to 2635 ± 67 ng mL^{-1}.

Rosuvastatin and metformin were quantitated in one female patient who had receiving 1000 mg of metformin once daily (Glucophage® 1000, Merck A.E.) and 10 mg of rosuvastatin once daily (Crestor® 10, AstraZeneca A.E.). Plasma sample obtained from this patient exhibited metformin and rosuvastatin plasma concentrations 1956 ± 93 ng mL^{-1} and 22.2 ± 0.4 ng mL^{-1}, respectively.

3. Materials and Methods

3.1. Chemicals and Reagents

Metformin hydrochloride of pharmaceutical purity grade was purchased from Wanbury Limited (BSEL Tech Park, Maharashtra, India). Rosuvastatin calcium of pharmaceutical purity grade was obtained from MSN Laboratories Private Limited (Rudraram, Telengana, India). The internal standard, N-despropyl ropinirole was purchased from Vitalife Chemipharma Pvt. Ltd. (Mumbai, Maharashtra,

India). HPLC grade solvents were used and purchased from E. Merck (Darmstadt, Germany). A Synergy UV water purification system (Merck Millipore, Danvers, MA, USA) was used to provide water HPLC grade. Hydrophilic polytetrafluorethylene membrane syringe filters (PTFE, 13 mm, pore size 0.45 μm) were purchased from Merck Millipore (Danvers, MA, USA). Pooled drug-free human plasma was obtained from National Red Cross General Hospital, Athens, Greece.

3.2. Instrumentation

A single quadrupole mass spectrometer model Finnigan AQA (Thermo, Manchester, UK) equipped with an electrospray ionisation interface and an isocratic pump model Spectra Series P100 (ThermoSeparation, Piscataway, NJ, USA) LC system were used to perform the HILIC-ESI/MS experiments. Highly pure nitrogen was produced by using a Model Nitrox-N2, Domnick hunter (Gateshead, UK) nitrogen generator. Xcalibur software (v. 1.2, ThermoQuest, Austin, UK) was used for data acquisition and analysis. Metformin, rosuvastatin and N-despropyl ropinirol were separated using a Xbridge HILIC BEH analytical column (150.0 × 2.1 mm i.d., particle size 3.5 μm, 135 Å). To prolong column lifetime, a XBridge HILIC BEH guard cartridge (20 × 2.1 mm, 3.5 μm) was also used. A mobile phase comprising of 12% 15 mM ammonium formate water solution in acetonitrile and pumped at a flow rate of 0.25 mL min^{-1} was used to run the experiments. The mobile phase was always filtered prior to use through a 0.45 μm nylon-membrane filter, GelmanSciences (Northampton, UK) and degassed under vacuum. Chromatography was performed at 25 ± 2 °C with a chromatographic run time of 15 min. All analytes and the ISTD have been detected in electrospray positive ionization mode and quantitation was achieved in SIM mode. Source parameters were adjusted to the following settings: temperature 260 °C, capillary voltage 4.8 kV, cone voltage 20 V.

3.3. Stock and Working Standard Solutions

Stock standard solutions of metformin and rosuvastatin were prepared at 500 μg mL^{-1} in acetonitrile-water mixture (90:10, *v/v*). These stock standard solutions were diluted in acetonitrile to prepare the mixed working standard solutions over the concentration ranges of 62.5 to 5000 ng mL^{-1} and 2 to 100 ng mL^{-1} for metformin and rosuvastatin, respectively. Stock standard solution of N-despropyl ropinirole (ISTD) was prepared at 190 μg mL^{-1} in acetonitrile. A working standard solution of ISTD at 3.8 μg mL^{-1} was prepared after further dilution in acetonitrile. The stock standard solutions were found to be stable for several months when stored at −20 °C. The working standard solutions were stored at 4 °C and prepared every month.

3.4. Calibration Standards and Quality Control Samples

Calibration standards spiked in human plasma were prepared at seven different concentration levels 62.5, 100, 250, 500, 1000, 2500 and 5000 ng mL^{-1} for metformin and at six different concentration levels 2, 5, 10, 20, 50 and 100 ng mL^{-1} for rosuvastatin. Each calibration sample contained 190 ng mL^{-1} of the internal standard. Quality control (QC) samples were prepared independently, in an analogous manner as the calibration standards, using separate stock solutions of the analytes. QC samples were prepared at three different concentration levels (62.5, 500 and 5000 ng mL^{-1}) for metformin and (2, 10 and 100 ng mL^{-1}) for rosuvastatin in human plasma.

3.5. Sample Preparation Procedure

Cleanup of biological samples is carried out by protein precipitation. On the day of extraction the samples are thawed at room temperature followed by vortex mixing to ensure homogeneity. Consequently, a 100 μL aliquot of each human plasma sample is transferred to a 2 mL Eppendorf tube, followed by addition of 50 μL of ISTD solution (3.8 μg mL^{-1}), 30 μL of 5 mM ammonium formate solution in water and 820 μL of acetonitrile. The mixture is vortexed for 1.0 min and centrifuged at 16,000× *g* for 20 min at 25 °C. A 13 mm hydrophilic PTFE membrane syringe filter (pore size 0.45 μm) is then used to filter the samples prior to HILIC-ESI/MS analysis.

4. Conclusions

In this work the advantages of HILIC, on the separation of compounds with diverse physicochemical properties and on the improvement in ESI/MS sensitivity, were demonstrated through the development of a HILIC-ESI/MS for the quantitation of metformin and rosuvastatin in human plasma. For the first time metformin and rosuvastatin have been adequately retained and separated from matrix interferences with an analytical run time of 15 min by using a hydrophilic Xbridge®-HILIC BEH analytical column. The method is combined with a fast and simple procedure for sample preparation based on protein precipitation and filtration that requires only 100 μL of biological sample. Validation results demonstrate that under the optimum conditions metformin and rosuvastatin can be quantified accurately and precisely in human plasma. The method was successfully applied to the analysis of real samples and it can be used to support various clinical studies.

Author Contributions: I.P., G.M. (George Migias) and D.M. participated in designing the study. N.A., G.M. (Giorgos Machairas), I.P. and V.B. conducted the study. Data was collected and analysed by N.A., I.P., A.V. and C.P., G.M. (George Migias) and D.M. organized the collection of clinical samples. The manuscript was written by I.P. and A.V.

Funding: This research received no external funding. Article Processing Charge (APC) was sponsored by MDPI.

Conflicts of Interest: The authors declare no conflict of interest.

References

1. Mathers, C.D.; Loncar, D. Projections of global mortality and burden of disease from 2002 to 2030. *PLoS Med.* **2006**, *3*, e442. [CrossRef] [PubMed]
2. The Emerging Risk Factors Collaboration. Diabetes mellitus, fasting blood glucose concentration, and risk of vascular disease: A collaborative meta-analysis of 102 prospective studies. *Lancet* **2010**, *375*, 2215–2222.
3. Bretnall, A.E.; Clarke, G.S. Metformin hydrochloride. *Anal. Profil. Drug Sub. Excip.* **1998**, *25*, 243–293.
4. Zilinyi, R.; Czompa, A.; Czegledi, A.; Gajtko, A.; Pituk, D.; Lekli, I.; Tosaki, A. The Cardioprotective Effect of Metformin in Doxorubicin-Induced Cardiotoxicity: The Role of Autophagy. *Molecules* **2018**, *23*, 1184. [CrossRef] [PubMed]
5. Cui, J.Y.; Zhou, R.R.; Han, S.; Wang, T.S.; Wang, L.Q.; Xie, X.H. Statin therapy on glycemic control in type 2 diabetic patients: A network meta-analysis. *J. Clin. Pharm. Ther.* **2018**. [CrossRef] [PubMed]
6. McTaggart, F. Comparative pharmacology of rosuvastatin. *Atheroscl. Suppl.* **2003**, *4*, 9–14. [CrossRef]
7. Lee, D.; Roh, H.; Son, H.; Jang, S.B.; Lee, S.; Nam, S.Y.; Park, K. Pharmacokinetic interaction between rosuvastatin and metformin in healthy Korean male volunteers: A randomized, open-label, 3-period, crossover, multiple-dose study. *Clin. Ther.* **2014**, *36*, 1171–1181. [CrossRef] [PubMed]
8. Hsieh, Y.; Galviz, G.; Hwa, J.J. Ultra-performance hydrophilic interaction LC–MS/MS for the determination of metformin in mouse plasma. *Bioanalysis* **2009**, *1*, 1073–1079. [CrossRef] [PubMed]
9. Zhang, W.; Han, F.; Zhao, H.; Lin, Z.J.; Huang, Q.M.; Weng, N. Determination of metformin in rat plasma by HILIC-MS/MS combined with Tecan automation and direct injection. *Biomed. Chromatogr.* **2012**, *26*, 1163–1169. [CrossRef] [PubMed]
10. Ben-Hander, G.M.; Makahleh, A.; Saad, B.; Saleh, M.I. Hollow fiber liquid phase microextraction with in situ derivatization for the determination of trace amounts of metformin hydrochloride (antidiabetic drug) in biological fluids. *J. Chromatogr. B* **2013**, *941*, 123–130. [CrossRef] [PubMed]
11. Nielsen, F.; Christensen, M.M.H.; Brosen, K. Quantitation of metformin in human plasma and urine by hydrophilic interaction liquid chromatography and application to a pharmacokinetic study. *Therap. Drug Monitor.* **2014**, *36*, 211–217. [CrossRef] [PubMed]
12. Michel, D.; Gaunt, M.C.; Arnason, T.; El-Aneed, A. Development and validation of fast and simple flow injection analysis–tandem mass spectrometry (FIA-MS/MS) for the determination of metformin in dog serum. *J. Pharm. Biomed. Anal.* **2015**, *107*, 229–235. [CrossRef] [PubMed]
13. Zhong, G.; Bi, H.; Zhou, S.; Chen, X.; Huang, M. Simultaneous determination of metformin and gliclazide in human plasma by liquid chromatography-tandem mass spectrometry: Application to a bioequivalence study of two formulations in healthy volunteers. *J. Mass Spectrom.* **2005**, *40*, 1462–1471. [CrossRef] [PubMed]

14. Mistri, H.N.; Jangid, A.G.; Shrivastav, P.S. Liquid chromatography tandem mass spectrometry method for simultaneous determination of antidiabetic drugs metformin and glyburide in human plasma. *J. Pharm. Biomed. Anal.* **2007**, *45*, 97–106. [CrossRef] [PubMed]

15. Georgita, C.; Albu, F.; David, V.; Medvedovici, A. Simultaneous assay of metformin and glibenclamide in human plasma based on extraction-less sample preparation procedure and LC/(APCI)MS. *J. Chromatogr. B* **2007**, *854*, 211–218. [CrossRef] [PubMed]

16. Pontarolo, R.; Gimenez, A.C.; Francisco, T.M.G.; Ribeiro, R.P.; Pontes, F.L.D.; Gasparetto, J.C. Simultaneous determination of metformin and vildagliptin in human plasma by a HILIC-MS/MS method. *J. Chromatogr. B* **2014**, *965*, 133–141. [CrossRef] [PubMed]

17. Ben-Hander, G.M.; Makahleh, A.; Saad, B.; Saleh, M.I.; Cheng, K.W. Sequential hollow-fiber liquid phase microextraction for the determination of rosiglitazone and metformin hydrochloride (anti-diabetic drugs) in biological fluids. *Talanta* **2015**, *131*, 590–596. [CrossRef] [PubMed]

18. Mohamed, A.M.I.; Mohamed, F.A.F.; Ahmed, S.; Mohamed, Y.A.S. An efficient hydrophilic interaction liquid chromatographic method for the simultaneous determination of metformin and pioglitazone using high-purity silica column. *J. Chromatogr. B* **2015**, *997*, 16–22. [CrossRef] [PubMed]

19. Zhang, X.; Wang, X.; Vernikovskaya, D.I.; Fokina, V.M.; Nanovskaya, T.N.; Hankins, G.D.V.; Ahmed, M.S. Quantitative determination of metformin, glyburide and its metabolites in plasma and urine of pregnant patients by LC-MS/MS. *Biomed. Chromatogr.* **2015**, *29*, 560–569. [CrossRef] [PubMed]

20. Fachi, M.M.; Cerqueira, L.B.; Leonart, L.P.; Guimarães de Francisco, T.M.; Pontarolo, R. Simultaneous quantification of antidiabetic agents in human plasma by a UPLC QToF-MS method. *PLoS ONE* **2016**, *11*, e0167107. [CrossRef] [PubMed]

21. Shah, P.A.; Shah, J.V.; Sanyal, M.; Shrivastav, P.S. LC-MS/MS analysis of metformin, saxagliptin and 5-hydroxy saxagliptin in human plasma and its pharmacokinetic study with a fixed-dose formulation in healthy Indian subjects. *Biomed. Chromatogr.* **2017**, *31*, 1–11. [CrossRef] [PubMed]

22. Ângelo, M.L.; Moreira, F.L.; Ruela, A.L.M.; Santos, A.L.A.; Salgado, H.R.N.; Araújo, M.B. Analytical methods for the determination of rosuvastatin in pharmaceutical formulations and biological fluids: A critical review. *Crit. Rev. Anal. Chem.* **2018**. [CrossRef] [PubMed]

23. Hull, C.K.; Penman, A.D.; Smith, C.K.; Martin, P.D. Quantification of rosuvastatin in human plasma by automated solid phase extraction using tandem mass spectrometric detection. *J. Chromatogr. B* **2002**, *772*, 219–228. [CrossRef]

24. Xu, D.H.; Ruan, Z.R.; Zhou, Q.; Yuan, H.; Jiang, B. Quantitative determination of rosuvastatin in human plasma by liquid chromatography with electrospray ionization tandem mass spectrometry. *Rapid Commun. Mass Spectrom.* **2006**, *20*, 2369–2375. [CrossRef] [PubMed]

25. Oudhoff, K.A.; Sangster, T.; Thomas, E.; Wilson, I.D. Application of microbore HPLC in combination with tandem MS for the quantification of rosuvastatin in human plasma. *J. Chromatogr. B* **2006**, *832*, 191–196. [CrossRef] [PubMed]

26. Gao, J.; Zhong, D.; Duan, X.; Chen, X. Liquid chromatography/negative ion electrospray tandem mass spectrometry method for the quantification of rosuvastatin in human plasma: Application to a pharmacokinetic study. *J. Chromatogr. B* **2007**, *856*, 35–40. [CrossRef] [PubMed]

27. Lan, K.; Jiang, X.; Li, Y.; Wang, L.; Zhou, J.; Jiang, O.; Ye, L. Quantitative determination of rosuvastatin in human plasma by ion pair liquid-liquid extraction using liquid chromatography with electrospray ionization tandem mass spectrometry. *J. Pharm. Biomed. Anal.* **2007**, *44*, 540–546. [CrossRef] [PubMed]

28. Lee, H.K.; Ho, C.S.; Hu, M.; Tomlinson, B.; Wong, C.K. Development and validation of a sensitive method for simultaneous determination of rosuvastatin and N-desmethyl rosuvastatin in human plasma using liquid chromatography/negative electrosprayionization/tandem mass spectrometry. *Biomed. Chromatogr.* **2013**, *17*, 1369–1374. [CrossRef] [PubMed]

29. Macwan, J.S.; Ionita, I.A.; Akhlaghi, F. A simple assay for the simultaneous determination of rosuvastatin acid, rosuvastatin-5S-lactone, and N-desmethyl rosuvastatin in human plasma using liquid chromatography–tandem mass spectrometry (LC/MS/MS). *Anal. Bioanal. Chem.* **2012**, *402*, 1217–1227. [CrossRef] [PubMed]

30. Shah, Y.; Iqbal, Z.; Ahmad, L.; Nazir, S.; Watson, D.G.; Khuda, F.; Khan, A.; Khan, M.I.; Khan, A.; Nasir, F. Determination of rosuvastatin and its metabolite N-desmethyl rosuvastatin in human plasma by liquid chromatography-high resolution mass spectrometry: Method development, validation, and application to pharmacokinetic study. *J. Liq. Chromatogr. Rel. Technol.* **2015**, *38*, 863–873. [CrossRef]

31. Vittal, S.; Shitut, N.R.; Kumar, T.R.; Vinu, M.C.A.; Mullangi, R.; Srinivas, N.R. Simultaneous quantitation of rosuvastatin and gemfibrozil in human plasma by high-performance liquid chromatography and its application to a pharmacokinetic study. *Biomed. Chromatogr.* **2006**, *20*, 1252–1259. [CrossRef] [PubMed]

32. Narapusetti, A.; Bethanabhatla, S.S.; Sockalingam, A.; Repaka, N.; Saritha, V. Simultaneous determination of rosuvastatin and amlodipine in human plasma using tandem mass spectrometry: Application to disposition kinetics. *J. Adv. Res.* **2015**, *6*, 931–940. [CrossRef] [PubMed]

33. Stopfer, P.; Giessmann, T.; Hohl, K.; Sharma, A.; Ishiguro, N.; Taub, M.E.; Jungnik, A.; Gansser, D.; Ebner, T.; Muller, F. Effects of metformin and furosemide on rosuvastatin pharmacokinetics in healthy volunteers: Implications for their use as probe drugs in a transporter cocktail. *Eur. J. Drug Metab. Pharmacokinet.* **2018**, *43*, 69–80. [CrossRef] [PubMed]

34. Kumar, P.P.; Murthy, T.E.G.K.; Rao, M.V.B. Development, validation of liquid chromatography-tandem mass spectrometry method for simultaneous determination of rosuvastatin and metformin in human plasma and its application to a pharmacokinetic study. *J. Adv. Pharm. Technol. Res.* **2015**, *6*, 118–124. [CrossRef] [PubMed]

35. Machairas, G.; Panderi, I.; Geballa-Koukoula, A.; Rozou, S.; Antonopoulos, N.; Charitos, C.H.; Vonaparti, A. Development and validation of a hydrophilic interaction liquid chromatography method for the quantitation of impurities in fixed-dose combination tablets containing rosuvastatin and metformin. *Talanta* **2018**, *183*, 131–141. [CrossRef] [PubMed]

36. Pedrali, A.; Tengattini, S.; Marrubini, G.; Bavaro, T.; Hemström, P.; Massolini, G.; Terreni, M.; Temporini, C. Characterization of Intact Neo-Glycoproteins by Hydrophilic Interaction Liquid Chromatography. *Molecules* **2014**, *19*, 9070–9088. [CrossRef] [PubMed]

37. Johnsen, E.; Leknes, S.; Wilson, S.R.; Lundanes, E. Liquid chromatography-mass spectrometry platform for both small neurotransmitters and neuropeptides in blood, with automatic and robust solid phase extraction. *Sci. Rep.* **2015**, *5*, 9308. [CrossRef] [PubMed]

38. Kiriazopoulos, E.; Zaharaki, S.; Vonaparti, A.; Vourna, P.; Panteri-Petratou, E.; Gennimata, D.; Lombardo, K.; Panderi, I. Quantification of three beta-lactam antibiotics in breast milk and human plasma by hydrophilic interaction liquid chromatography/positive-ion electrospray ionization mass spectrometry. *Drug Test. Anal.* **2017**, *9*, 1062–1072. [CrossRef] [PubMed]

39. Committee for Medicinal Products for Human Use (CHMP). *Guideline on Bioanalytical Method Validation*; EMEA/CHMP/EWP/192217/2009 Rev. 1; European Medicines Agency: London, UK, 2015.

40. Matuszewski, B.K.; Constanzer, M.L.; Chavez-Eng, C.M. Strategies for the assessment of matrix effect in quantitative bioanalytical methods based on HPLC-MS/MS. *Anal. Chem.* **2003**, *75*, 3019–3030. [CrossRef] [PubMed]

Sample Availability: Samples of the metformin and rosuvastatin and N-despropyl ropinirole are not available from the authors but it can be purchased from the manufacturers mentioned in the materials section.

molecules

MDPI

Article

Production of Minor Ginenosides from *Panax notoginseng* by Microwave Processing Method and Evaluation of Their Blood-Enriching and Hemostatic Activity

Huiying Liu [1], Jun Pan [2], Ye Yang [1], Xiuming Cui [1] and Yuan Qu [1,*]

[1] Faculty of Life Science and Technology, and Yunnan Provincial Key Laboratory of Panax Notoginseng, Kunming University of Science and Technology, Kunming 650500, China; lhying7836@163.com (H.L.); q270228@126.com (Y.Y.); sanqi37@vip.sina.com (X.C.)

[2] Yunnan Provincal Academy of Agricultural Sciences, Kunming 650231, China; Pjun2017@126.com

* Correspondence: quyuan2001@126.com; Tel: +86-136-6970-6827

Received: 10 May 2018; Accepted: 20 May 2018; Published: 23 May 2018

Abstract: A green solvent extraction technology involving a microwave processing method was used to increase the content of minor ginsenosides from *Panax notoginseng*. This article aims to investigate the optimization of preparation of the minor ginsenosides by this microwave processing method using single-factor experiments and response surface methodology (RSM), and discuss the blood-enriching activity and hemostatic activity of the extract of microwave processed *P. notoginseng* (EMPN) The RSM for production of the minor ginsenosides was based on a three-factor and three-level Box-Behnken design. When the optimum conditions of microwave power, temperature and time were 495.03 W, 150.68 °C and 20.32 min, respectively, results predicted that the yield of total minor ginsenosides (Y_9) would be 93.13%. The actual value of Y_9 was very similar to the predicted value. In addition, the pharmacological results of EMPN in vivo showed that EMPN had the effect of enriching blood in *N*-acetylphenylhydrazine (APH) and cyclophosphamide (CTX)-induced blood deficient mice because of the increasing content of white blood cells (WBCs) and hemoglobin (HGB) in blood. Hemostatic activity in vitro of EMPN showed that it had significantly shortened the clotting time in PT testing ($p < 0.05$). The hemostatic effect of EMPN was mainly caused by its components of Rh_4, 20(S)-Rg_3 and 20(R)-Rg_3. This microwave processing method is simple and suitable to mass-produce the minor ginsenosides from *P. notoginseng*.

Keywords: microwave processing; response surface methodology; minor ginsenosides; blood-enriching activity; hemostatic activity

---

1. Introduction

The root of *Panax notoginseng* (Burk.) F. H. Chen is a well-known traditional Chinese medicine in China and around the world. Traditionally, raw notoginseng is commonly used in the treatment of trauma and blood circulation, while the steamed notoginseng can enrich the blood [1,2]. Both of them are not only different in treatment function, but also different in chemical composition. So far, more than 100 dammarane-type saponins have been isolated and identified from raw notoginseng, five major components of which are respectively ginsenosides Rg_1, Re, Rb_1, Rd and notoginsenoside R_1 [3–5]. However, more than 50 new components are produced by dehydration, hydroxylation or epoxidation of the main saponins in the raw notoginseng after hydrolysis of the C-20 sugar moiety by the steaming process [6–8]. Eight minor ginsenosides of the steamed notoginseng—20(S)-Rh_1, 20(R)-Rh_1, Rh_4, Rk_1, Rk_3, 20(S)-Rg_3, 20(R)-Rg_3 and Rg_5—are less polar compounds [9–11]. These minor ginsenosides cause the functional differences between raw and steamed notoginseng.

So far, the minor ginsenosides have been difficult to obtain because of their low content, but some of them are good potential drug candidates based on their biological activity, such as the anti-tumor activity of ginsenoside compound K, ginsenoside Rg_3 and ginsenoside Rk_1 [12–15], anti-apoptotic activity of ginsenoside Rg_5 [16], and antiviral activity of 20(*R*)-ginsenoside Rh_2 [17]. Methods for obtaining minor saponins mainly include acid-base degradation [18], enzymatic degradation [19,20] and microbial transformation [21,22]. Some of the main ginsenosides can be transformed into minor ginsenosides by the steaming process. It is reported that the *P. notoginseng* steaming method is usually performed for a long time at a high temperature (48 h at 120 °C) [23]. However, there is a new method to increase the content of minor ginsenosides Rg_3, Rg_5, and Rk_1 from the ginseng extract using a microwave processing method [24,25]. The microwave technique is widely used as a "green" solvent extraction technology [26,27]. Studies have showed that the microwave extraction technique had many advantages such as a faster extraction rate, reduced organic solvent consumption, and sample preparation at lower costs. Compared with the conventional steaming method, the microwave processing method is highly efficient process because of the shorter processing time. However, there are no reports in the production of minor saponins from *P. notoginseng* using the microwave processing method. The chemical components and pharmacological activities of transformed minor saponins from *P. notoginseng* with this processing method have not been studied.

In our study, the extracts of *P. notoginseng* containing minor ginsenosides were prepared by different microwave processing methods. As many factors affect the response, response surface methodology (RSM) was used to optimize the reaction conditions for production of the minor ginsenosides. The main effecting factors of microwave power, temperature and time were discussed in RSM based on single factor experiments. In addition, the blood-enriching activity and hemostatic activity of the extract of microwave processed *P. notoginseng* (EMPN) containing minor ginsenosides were analyzed in order to evaluate their potential medicinal value.

2. Results and Discussion

2.1. Structural Changes during the Microwave Processing of Saponins

The chemical ingredient changes of raw notoginseng and processed notoginseng are shown in the HPLC chromatograms in Figure 1. The main components of raw notoginseng are ginsenosides Rg_1, Re, Rb_1, Rd and notoginsenoside R_1, which are obviously different from those of processed notoginseng. After steaming and microwave processing, the main constituents of raw notoginseng are decreased and transformed into eight minor ginsenosides, which were identified as 20(*S*)-Rh_1, 20(*R*)-Rh_1, Rh_4, Rk_1, Rk_3, 20(*S*)-Rg_3, 20(*R*)-Rg_3 and Rg_5 by comparing their retention times with standard ginsenosides [23]. When notoginseng was steamed at 120 °C for 4 h, most of saponins were transformed into minor ginsenosides except ginsenoside Rg_1. Under the condition of microwave treatment of 500 W, 150 °C and 20 min, all the saponins were degraded into minor ginsenosides.

The principle and application of microwave dielectric heating in chemistry has been reported by Galema [28]. Microwave radiation induces molecular dipoles to orientate in the direction of electromagnetic field, and generates heat. The heating process occurs within the molecule, and is hence described as "inside heating". If the microwave energy matches the rotation energy of polar molecules, the reactivity of polar bonds such as the glycosidic bond is increased and it becomes easy to break [29]. The structural changes of saponins in the microwave processing are shown in Scheme 1. When the microwave power and temperature increased, the glycosyl residues of major saponins were easily decreased. The protopanaxadiol (PPD) group, e.g., in ginsenosides Rb_1 and Rd, was hydrolyzed at the glucosyl residue of C-20 to produce 20(*S*)/(*R*)-Rg_3, which was then dehydrated at C-20 to yield Rk_1 and Rg_5. Similarly, the protopanaxatriol (PPT) group, such as found in ginsenosides Rg_1, Re and notoginsenoside R_1 easily lost the glycosyl residue at C-6 or C-20 to produce 20(*S*)/(*R*)-Rh_1, and then Rh_1 formed Rk_3 and Rh_4 through dehydration at C-20. During this process, it was evident that C-3 sugar moiety has higher temperature stability than the C-6 or C-20 sugar moieties. After elimination of the glycosyl residue at C-20, 20(S) and 20(R) epimers were produced by the selective attack of the OH group like 20(S)/(R)-Rg$_3$ and 20(S)/(R)-Rh$_1$ [30]. The 20(S) saponin content was higher than that of 20(R) saponin in this process.

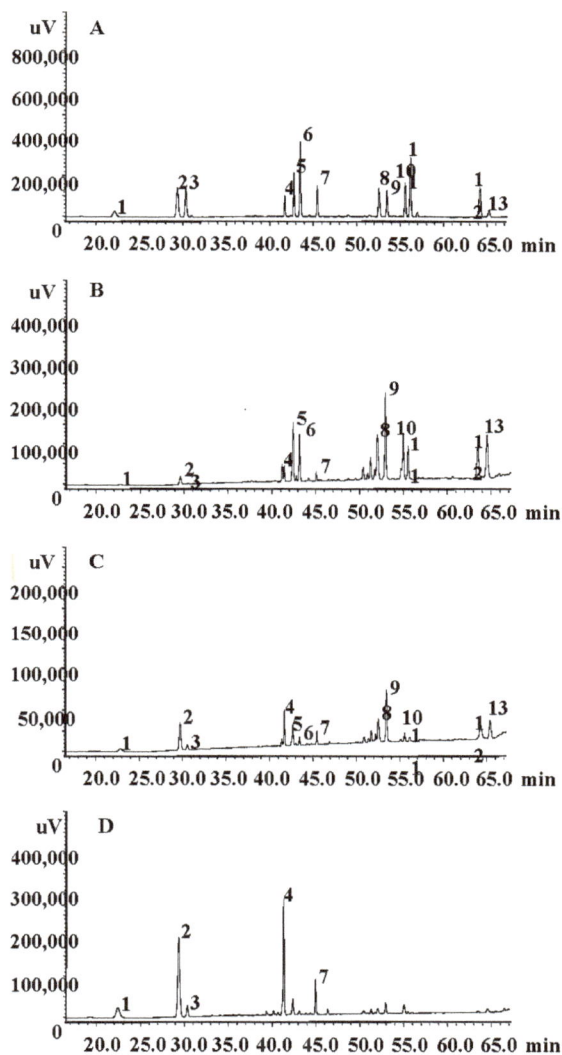

Figure 1. HPLC chromatograms of mixed standards (**A**); notoginseng processed by the microwave processing method at 500 W and 150 °C for 20 min (**B**); notoginseng steamed at 120 °C for 4 h (**C**) and raw notoginseng (**D**). Peaks: 1, notoginsenoside R$_1$; 2, ginsenoside Rg$_1$; 3, ginsenoside Re; 4, ginsenoside Rb$_1$; 5, 20(S)-ginsenoside Rh$_1$; 6, 20(R)-ginsenoside Rh$_1$; 7, ginsenoside Rd; 8, ginsenoside Rk$_3$; 9, ginsenoside Rh$_4$; 10, 20(S)-ginsenoside Rg$_3$; 11, 20(R)-ginsenoside Rg$_3$; 12, ginsenoside Rk$_1$; 13, ginsenoside Rg$_5$.

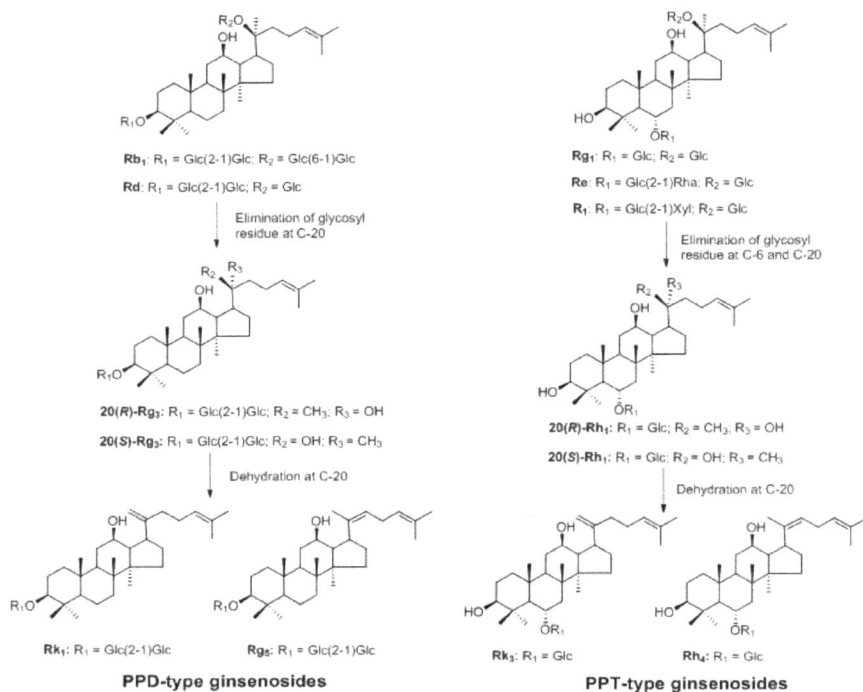

Scheme 1. Proposed structural changes of saponins during the microwave processing of *P. notoginseng*. -Glc, D-glucopyranosyl; -Rha, L-rhamnopyranosyl; -Xyl, D-xylopyranosyl.

2.2. Effects of Single Factors

2.2.1. Effect of Solvent

In this paper, firstly the solvent used in the microwave processing method was studied. When different concentrations of ethanol or methanol were used, the total yield of the minor ginsenosides (sum of eight saponins $20(S)$-Rh$_1$, $20(R)$-Rh$_1$, Rk$_3$, Rh$_4$, $20(S)$-Rg$_3$, $20(R)$-Rg$_3$, Rk$_1$ and Rg$_5$) was different (Figure 2A,B). The results showed that the yield decreased with the increase of ethanol or methanol ratio, when other conditions of microwave power, temperature, time and solid-to-liquid ratio were 500 W, 135 °C, 15 min and 1:60, respectively, so water was thought to be the best solvent in the green microwave extraction. The preparation of the minor ginsenosides without any organic solvent in microwave processing method reduces the environmental burden.

2.2.2. Effect of Microwave Power

Microwave power was an important factor in any microwave processing method. When the other conditions of temperature, time and solid-to-liquid ratio were 150 °C, 15 min and 1:60, respectively, the microwave power was varied from 300 W to 1000 W (Figure 2C). When the microwave power was lower than 300 W or higher than 600 W, the conversion yield of the total minor ginsenosides was less than 65%. Finally, the optimal microwave power was set at 500 W.

2.2.3. Effect of Temperature

In order to further study the temperature of microwave processing method, the temperature range was from 60 °C to 180 °C (Figure 2D). The other conditions of microwave power, time and solid-to-liquid ratio were set at 500 W, 15 min and 1:60, respectively. The yield of the total minor ginsenosides was obviously increased when temperature was increased to 150 °C, so it is concluded that temperature had an important influence on the transformation of ginsenosides and 150 °C was the appropriate temperature for the transformation method.

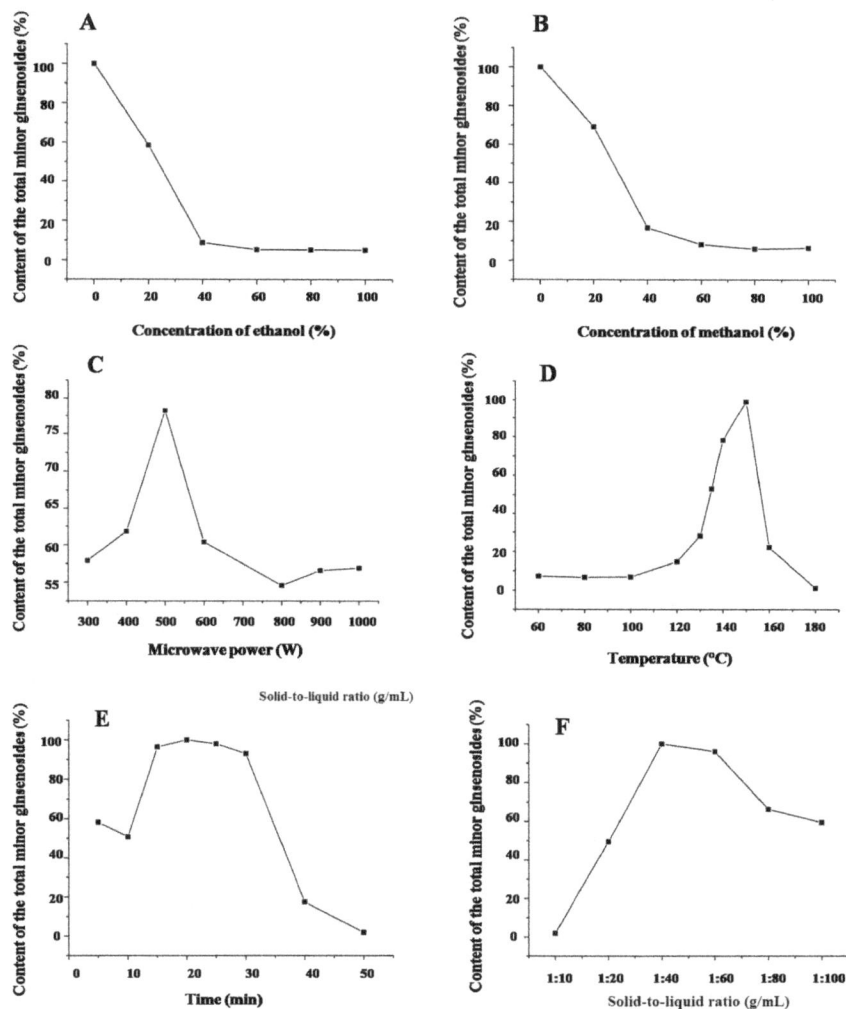

Figure 2. Effect of solvent, microwave power, temperature, time and solid-to-liquid ratio on production of the total minor ginsenosides. (**A**) Ethanol concentration; (**B**) Methanol concentration; (**C**) Microwave power; (**D**) Microwave temperature; (**E**) Microwave time; (**F**) Solid-to-liquid ratio.

2.2.4. Effect of Microwave Time

When the microwave power, temperature and solid-to-liquid ratio were 500 W, 150 °C and 1:60, respectively, the microwave time was varied from 5 min to 50 min (Figure 2E). According to the yield of the total minor ginsenosides, a suitable time for this method ranged from 15 min to 35 min. When the microwave exposure time was 15 min, the yield of the total minor ginsenosides was more than 90%. If the conventional steaming method was used, it took at least 4 h to achieve the same yield. The steaming method is usually a highly time and energy consuming process. Compared with the steaming method, the minor ginsenosides were produced by the microwave method with less energy and less waste.

2.2.5. Effect of Solid-to-Liquid Ratio

While the other optimized conditions were microwave power of 500 W, temperature 150 °C and 20 min time, t solid-to-liquid ratios of 1:10, 1:20, 1:40, 1:60, 1:80, and 1:100 (w/v) were examined. When the volume of water was 10 mL (ratio 1:40 or 1:60), the total minor ginsenosides had a higher transformation yield (Figure 2F).

In conclusion, considering a variety of factors, we obtained that optimal conditions for the microwave processing method. Water was selected as the solvent, and other conditions of microwave power, temperature, time and solid-to-liquid ratio were 500 W, 150 °C, 15 min–35 min and 1:40, respectively. Next, based on the single factor analysis for production of the minor ginsenosides by microwave processing, the three most important variables such as microwave power, temperature and time were further investigated by RSM.

2.3. Response Surface Optimization of the Minor Ginsenosides Production

2.3.1. Fitting the Model

Response Surface Methodology (RSM) is widely used for optimization of minor ginsenosides production [31,32] and understanding the relationships among variables. Table 1 shows the different combinations of variables and the nine responses in the Box-Behnken design used. Nine responses in the experimental design, 20(S)-Rh$_1$ content (Y_1), 20(R)-Rh$_1$ content (Y_2), Rk$_3$ content (Y_3), Rh$_4$ content (Y_4), 20(S)-Rg$_3$ content (Y_5), 20(R)-Rg$_3$ content (Y_6), Rk$_1$ content (Y_7), Rg$_5$ content (Y_8) and total minor ginsenosides content (Y_9) were measured. In these treatments, treatment 10 (500 W, 150 °C, 20 min) achieved the highest yield of total minor ginsenosides (95.74%) and treatment 3 (600 W, 160 °C, 20 min) achieved the lowest yield (8.14%). The nine responses in the design were analyzed using analysis of variance (ANOVA). The regression coefficients of second order polynomial models for response variables (Yi) are listed in Table 2. For example, among the coefficients for total minor ginsenosides (Y_9), some factors (β_{12}, β_{11}, β_{22} and β_{33}) were considered as important factors because their p-value was less than 0.05. The following final regression Equation (1) for production of total minor ginsenosides (Y_9) was obtained in terms of coded factors:

$$Y = 92.8 - 2.16X_1 + 5.82X_2 + 2.46X_3 - 9.79X_1X_2 + 3.45X_1X_3 - 0.88X_2X_3 - 26.25X_1{}^2 - 46.02X_2{}^2 - 17.31X_3{}^2 \quad (1)$$

The ANOVA results (Table 3) showed that the total model for production of the minor ginsenosides (Y_1–Y_9) was highly significant, with a p value of <0.001, while the lack of fit is not significant relative to the pure error, with a p value of >0.05. Hence, this model could be used for predicting all points. Figure 3 further shows that the actual values of total minor ginsenosides (Y_9) were very similar to the predicted values. The fitting degree of the model was checked by regression coefficient (R^2 = 0.9766), and this is in agreement with the adjusted R^2 value of 0.9466.

According to the regression coefficient results in Table 2, the second-order variables microwave power (X_{11}) with significance ($p < 0.05$) and temperature (X_{22}) with high significance ($p < 0.001$) showed a negative effect on yield of the minor ginsenosides (Y_1–Y_9), while the interaction variables X_{12} between microwave power and temperature had a significantly negative effect on Rh$_4$ yield (Y_4), Rg$_5$ yield (Y_8) and yield total minor ginsenosides (Y_9) ($p < 0.05$), X_{13} between microwave power and time had a positive effect on 20(S)-Rh$_1$ yield (Y_1), 20(R)-Rh$_1$ yield (Y_2), 20(S)-Rg$_3$ yield (Y_5), 20(R)-Rg$_3$ yield (Y_6) and Rk$_1$ yield (Y_7).

Table 1. Experimental design factors and response values in Box-Behnken Design.

No.	Independent Variables			Response Variables								
	Microwave Power (X_1, W)	Temperature (X_2, °C)	Time (X_3, min)	20(S)-Rh1 (Y_1, %)	20(R)-Rh1 (Y_2, %)	Rk3 (Y_3, %)	Rh4 (Y_4, %)	20(S)-Rg3 (Y_5, %)	20(R)-Rg3 (Y_6, %)	Rk1 (Y_7, %)	Rg5 (Y_8, %)	Total Minor Ginsenosides (Y_9 [a], %)
1	600 (1)	150 (0)	25 (1)	8.49	4.23	3.89	13.31	2.87	1.17	7.63	10.85	52.45
2	500 (0)	160 (1)	25 (1)	7.63	2.04	2.74	12.59	1.92	0.90	6.29	9.66	43.76
3	600 (1)	160 (1)	20 (0)	0.86	1.29	0.87	1.07	1.35	0.90	0.77	1.04	8.14
4	400 (−1)	140 (−1)	20 (0)	1.51	2.27	1.03	3.13	2.45	0.79	0.90	1.26	13.34
5	400 (−1)	160 (1)	20 (0)	4.24	1.72	6.03	11.86	1.75	0.79	3.02	7.41	36.83
6	500 (0)	160 (1)	15 (−1)	2.95	2.23	7.42	10.93	1.93	1.09	1.03	6.97	34.53
7	400 (−1)	150 (0)	25 (1)	5.97	0.62	6.14	14.42	0.96	0.29	4.68	11.80	44.88
8	600 (1)	140 (−1)	20 (0)	5.41	1.54	2.84	5.85	1.24	0.40	2.69	3.82	23.80
9	500 (0)	150 (0)	20 (0)	14.25	7.00	14.76	24.22	6.51	3.65	12.20	13.01	95.61
10	500 (0)	150 (0)	20 (0)	14.17	7.77	13.52	22.05	5.80	4.03	12.53	15.88	95.74
11	500 (0)	150 (0)	20 (0)	15.76	7.17	15.80	20.73	6.12	4.28	9.85	14.97	94.67
12	500 (0)	150 (0)	20 (0)	14.85	8.55	11.52	21.62	4.46	4.61	13.17	16.28	95.05
13	600 (1)	150 (0)	15 (−1)	2.30	1.31	9.54	17.76	1.03	0.84	1.46	12.27	46.51
14	400 (−1)	150 (0)	15 (−1)	7.69	3.75	7.26	12.55	2.81	3.79	6.44	8.82	53.12
15	500 (0)	150 (0)	25 (1)	1.63	0.83	6.07	8.91	1.05	1.24	0.73	5.69	26.15
16	500 (0)	140 (−1)	15 (−1)	1.13	1.02	0.84	5.45	0.93	0.22	0.52	3.28	13.41
17	500 (0)	150 (0)	20 (0)	13.06	5.21	10.17	17.74	4.92	4.92	11.48	15.44	82.93

[a] Total minor ginsenosides content (Y_9), Sum of eight saponins 20(S)-Rh1, 20(R)-Rh1, Rk3, Rh4, 20(S)-Rg3, 20(R)-Rg3, Rk1 and Rg5.

Table 2. Regression coefficients of second order polynomial for response variables.

Variable	Coefficient (β)									p-Value [a]								
	Y_1	Y_2	Y_3	Y_4	Y_5	Y_6	Y_7	Y_8	Y_9	Y_1	Y_2	Y_3	Y_4	Y_5	Y_6	Y_7	Y_8	Y_9
Intercept	14.42	7.14	13.15	21.27	5.56	4.29	11.85	15.12	92.8	0.0003	0.0027	0.004	0.0025	0.0023	0.0028	0.0002	0.0003	<0.0001
X_1	−0.29	0.0018	−0.41	−0.50	−0.19	−0.29	−0.31	−0.16	−2.16	0.6100	0.9963	0.5849	0.6218	0.5094	0.2698	0.5068	0.7649	0.4236
X_2	0.75	0.20	0.78	1.64	0.16	0.13	0.78	1.38	5.82	0.2156	0.6135	0.3150	0.1316	0.5736	0.6211	0.1233	0.0338	0.0558
X_3	1.21	−0.075	−0.78	0.32	0.013	−0.29	1.23	0.83	2.46	0.0645	0.8490	0.3187	0.7503	0.9628	0.2733	0.0283	0.1565	0.3654
X_{12}	−1.82	0.075	−1.74	−3.38	0.20	0.12	−1.01	−2.23	−9.79	0.0520	0.8924	0.1327	0.0417	0.6120	0.7345	0.1546	0.0194	0.0296
X_{13}	1.98	1.51	−1.13	−1.58	0.92	0.96	1.98	−1.10	3.54	0.0386	0.0261	0.3052	0.2831	0.0457	0.0279	0.0166	0.1810	0.3567
X_{23}	1.04	0.0012	−2.48	−0.45	−0.032	−0.30	1.26	0.071	−0.88	0.2223	0.9982	0.0463	0.7506	0.9362	0.4108	0.0860	0.9259	0.8141
X_{11}	−4.32	−2.24	−4.01	−5.38	−1.70	−1.46	−3.55	−3.60	−26.25	0.0007	0.0037	0.0051	0.0048	0.0025	0.0036	0.0007	0.0016	0.0001
X_{22}	−7.09	−3.19	−6.45	−10.42	−2.16	−2.12	−6.45	−8.13	−46.03	<0.0001	0.0005	0.0003	0.0001	0.0006	0.0004	<0.0001	<0.0001	<0.0001
X_{33}	−3.99	−2.42	−2.44	−1.38	−1.94	−1.31	−3.25	−0.58	−17.31	0.0012	0.0024	0.0448	0.3303	0.0012	0.0060	0.0012	0.4467	0.0017

[a] $p < 0.01$, highly significant; $0.01 < p < 0.05$ significant; $p > 0.05$, not significant.

Table 3. ANOVA for response surface quadratic model analysis.

Variable	Degree of Freedom	Sum of Squares								
		Y_1	Y_2	Y_3	Y_4	Y_5	Y_6	Y_7	Y_8	Y_9
Model [a]	9	445.35	108.39	346.55	707.60	57.37	44.58	344.40	399.30	15,109.32
Residual	7	17.00	8.10	29.39	51.59	4.04	3.38	11.22	15.35	361.53
Lack of fit [b]	3	13.09	1.97	8.12	29.36	1.14	2.40	4.77	8.85	239.01
Pure error	4	3.91	6.13	21.27	22.23	2.90	0.97	6.45	6.49	122.52
Correct total	16	462.35	116.49	375.94	759.19	61.40	47.95	355.62	414.65	15,470.85
R-Squared		0.9632	0.9305	0.9218	0.9320	0.9342	0.9296	0.9684	0.9630	0.9766
Adj *R*-Squared		0.9160	0.8411	0.8213	0.8447	0.8497	0.8391	0.9279	0.9154	0.9466

[a] Model for Y_1–Y_9 is highly significant ($p < 0.001$); [b] Lack of fit for Y_1–Y_9 is not significant ($p > 0.05$).

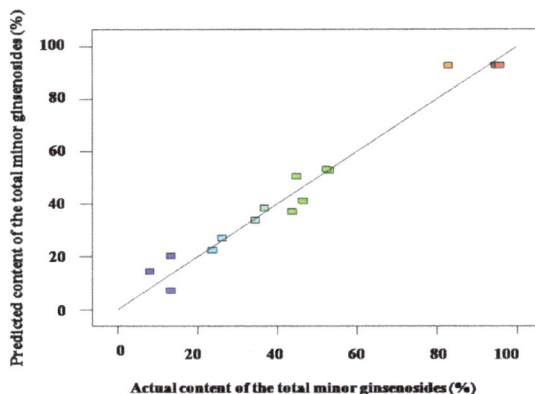

Figure 3. Correlation between actual and predicted values of response (Y_9).

2.3.2. Response Surface Method Analysis

Response surface and contour plots for production of the total minor ginsenosides based on the interaction between the variables are shown in Figure 4. The relationships between variables and response can be better understood by response surface plots (3D). The content of total minor ginsenosides firstly increased and then decreased with the increase of microwave power, process temperature and time. The shape of contour plots indicated the effect of the interaction between the variables. From Figure 4A, the ellipse-shaped contour plots showed that the interaction of microwave power and processing temperature had a strong influence on the production of the total minor ginsenosides.

2.3.3. Prediction and Verification of the Optimum Conditions

Microwave processing conditions were optimized using response surface methodology (Table 4). For example, the predicted maximum yield of the total minor ginsenosides was 93.13%, when the optimum conditions of microwave power, temperature and time were 495.03 W, 150.68 °C and 20.32 min, respectively. The verification test was carried out three times and compared with the predicted values obtained from the model (Equation (2)). In the optimum conditions, the actual yield of total minor ginsenosides was 94.15% that agree with the predicted values. Therefore, RSM can effectively optimize the production of the minor ginsenosides.

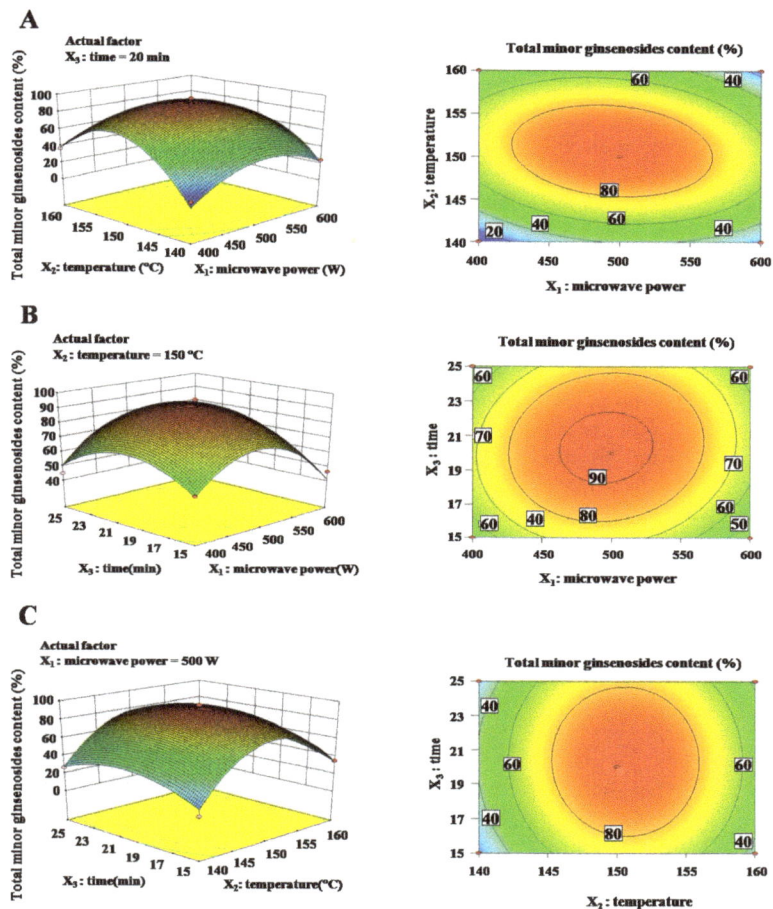

Figure 4. Response surface plots and contour plots for production of the total minor ginsenosides using interaction of variables. (**A**) Effect of microwave power (X_1) and temperature (X_2) on total minor ginsenosides content (Y_9) at time 20 min; (**B**) Effect of microwave power (X_1) and time (X_3) on total minor ginsenosides content (Y_9) at temperature 150 °C; (**C**) Effect of temperature (X_2) and time (X_3) on total minor ginsenosides content (Y_9) at microwave power 500 W.

Table 4. Optimum conditions for maximum 20(S)-Rh$_1$ content, 20(R)-Rh$_1$ content, Rk$_3$ content, Rh$_4$ content, 20(S)-Rg$_3$ content, 20(R)-Rg$_3$ content, Rk$_1$ content, Rg$_5$ content and total minor ginsenosides content.

Response Variables (Y_i)	Microwave Power (X_1, W)	Temperature (X_2, °C)	Time (X_3, min)	Content (%)
20(S)-Rh$_1$ (Y_1)	498.80	150.66	20.79	14.54
20(R)-Rh$_1$ (Y_2)	499.52	150.31	19.91	7.14
Rk$_3$ (Y_3)	495.36	151.07	18.98	13.28
Rh$_4$ (Y_4)	490.33	150.91	20.78	21.39
20(S)-Rg$_3$ (Y_5)	494.40	150.34	19.95	5.57
20(R)-Rg$_3$ (Y_6)	484.35	150.38	19.13	4.35
Rk$_1$ (Y_7)	500.13	150.81	21.03	12.00
Rg$_5$ (Y_8)	480.22	151.16	24.55	15.59
Total minor ginsenosides (Y_9)	495.03	150.68	20.32	93.13

2.3.4. Industrial Feasibility of Production of Minor Ginenosides with the Microwave Processing Method

Microwave technology is widely applied for improving extraction of plant secondary metabolites from leaves, flowers and seeds [33]. The first important contribution of microwave technology from lab to industrialization was the application of the extraction of volatile organic compounds from Boldo leaves [34]. Comparing the microwave method to conventional hydrodistillation, the volatile oil was extracted with less time and energy. In our work, ginsenoside transformation using the microwave method was completed in 20 min, while the conventional steam processing took 4 h. Using this method, the energy efficiency is higher and time consumption is less. The sample thickness is the most important factor in the process of large-scale extraction, because uniform heating is rarely achievable in the microwave method [35]. Under the optimized solid-to-liquid ratio conditions, we found that the crude extract of *P. notoginseng* was completely dissolved in the solvent water before microwave treatment. This result overcame the problem of uneven heating resulting from low penetration depth of microwaves. We have also established HPLC systems suitable for quality control of the transformed products (Table 5), and solved the problem of poor quality control in industrial microwave equipment. Therefore, the scale up of production of minor ginenosides from *Panax notoginseng* in microwave processing method to industrial scale is feasible.

Table 5. Method validation.

Compounds	Calibration Curve	R^2	Linear Range (mg/mL)	Precision (RSD, %)	Stability (RSD, %)	Repeatability (RSD, %)	Recovery (mean $\pm$ s, %)
20(S)-ginsenoside Rh$_1$	y = 1,000,000x + 52,373	0.9998	0.02–3.20	0.89	3.65	3.58	102.16 $\pm$ 2.46
20(R)-ginsenoside Rh$_1$	y = 2,000,000x + 154,738	0.9994	0.02–3.20	0.79	2.58	3.26	101.58 $\pm$ 3.11
ginsenoside Rk$_3$	y = 1,000,000x + 22,910	0.9999	0.08–6.40	0.36	2.45	3.31	105.25 $\pm$ 3.24
ginsenoside Rh$_4$	y = 826,150x − 23,734	0.9996	0.16–9.60	0.45	2.69	3.24	101.78 $\pm$ 2.58
20(S)-ginsenoside Rg$_3$	y = 1,000,000x + 9671.9	0.9999	0.01–3.20	0.48	2.16	3.83	108.48 $\pm$ 3.64
20(R)-ginsenoside Rg$_3$	y = 2,000,000x + 65,498	0.9996	0.005–1.60	0.46	2.28	3.34	104.15 $\pm$ 2.64
ginsenoside Rk$_1$	y = 1,000,000x + 14881	0.9997	0.08–6.40	0.52	3.97	3.86	103.78 $\pm$ 2.89
ginsenoside Rg$_5$	y = 301,295x − 8602.8	0.9997	0.08–6.40	0.85	3.41	4.46	105.89 $\pm$ 3.66

2.4. Effect of the Extract of Microwave Processed P. notoginseng on Blood Deficient Mice In Vivo

As previously stated, raw notoginseng has the effect of promoting blood circulation, and steaming notoginseng has the effect of enriching blood [1]. In our work, the effects of raw notoginseng and processed notoginseng prepared by steaming and microwave transformation methods on the peripheral blood index of *N*-acetyl phenylhydrazine (APH) and cyclophosphamide (CTX)-induced blood deficient mice were investigated (Table 6).

The concentration of white blood cells (WBC), red blood cells (RBC) and hemoglobin (HGB) and hematocrit (HCT) are usually used in the clinical diagnosis of anemia [36]. Compared with normal control group, the content of WBC ($p < 0.001$), RBC ($p < 0.01$) and HGB ($p < 0.001$) and hematocrit (HCT) ($p < 0.05$) in model group decreased significantly, which suggested that establishment of this blood deficiency model was effect. WBC and HGB in extract of steamed *P. notoginseng* (ESPN) and extract of microwave processed *P. notoginseng* (EMPN) groups showed a significant increase trend compared with model group, suggesting processed notoginseng could improve the symptoms of blood deficiency. And in the high dose group (800 mg/kg), the results showed the best effect, which were similar to the positive sample of Fufang E'jiao Jiang (FEJ). Fufang E'jiao Jiang in the Chinese Pharmacopoeia (2015 edition) is used to treat qi-blood deficiency, dizziness, loss of appetite, leucopenia and anemia. However, raw notoginseng (EPN) had no effect on reduced WBC, RBC, HGB, and HCT. The combination of APH and CTX reduced the number of RBC in the blood, and damaged the immune organs of mice [37]. Furthermore, we studied the effect of raw notoginseng (EPN) and processed notoginseng (ESPN and EMPN) on immune organs of blood deficient mice (Figure 5). Compared with the normal control group, the thymus index in the model group decreased significantly ($p < 0.05$), and the spleen index increased significantly ($p < 0.01$). The results showed that there was no effect on the thymus index and spleen index in the EPN group. But ESPN at the dose of 400 mg/kg and

800 mg/kg reduced the increasing spleen index ($p < 0.05$). EMPN at the dose of 800 mg/kg significantly improved thymus index ($p < 0.05$) and decreased spleen index ($p < 0.05$) of blood deficiency mice. EMPN had protective effect on the immune organs of mice.

Table 6. Effect of raw notoginseng and processed notoginseng on peripheral blood index of blood deficiency mice ($x \pm s$, $n = 10$).

Group	Dose (mg/kg)	WBC (10^9/L)	RBC (10^{12}/L)	HGB (g/L)	HCT (%)
Control		7.62 ± 0.66	7.11 ± 0.68	146.8 ± 4.98	33.77 ± 2.36
Model		4.38 ± 0.66 $^{\triangle\triangle\triangle}$	5.77 ± 0.58 $^{\triangle\triangle}$	96.2 ± 4.11 $^{\triangle\triangle\triangle}$	30.4 ± 1.37 $^{\triangle}$
FEJ	10	5.45 ± 0.42 **	6.81 ± 0.63 *	128.67 ± 8.35 ***	31.07 ± 1.56
	200	4.2 ± 0.18	4.54 ± 1.18	103.18 ± 8.18	30.33 ± 3.78
EPN	400	4.12 ± 0.08	4.76 ± 0.85	107.52 ± 6.96	30.92 ± 1.31
	800	4.07 ± 0.42	5.03 ± 1.61	111.47 ± 14.87	31.63 ± 1.11
	200	4.35 ± 0.34	4.97 ± 0.69	102.47 ± 18.16	30.12 ± 2.12
ESPN	400	5.38 ± 0.26 **	5.93 ± 0.66	124.08 ± 10.7 ***	30.58 ± 2.62
	800	7.10 ± 0.97 ***	6.63 ± 0.83	128.86 ± 16.01 ***	32.3 ± 3.14
	200	4.16 ± 0.09	4.65 ± 0.43	102.94 ± 6.66	30.74 ± 0.82
EMPN	400	5.65 ± 0.23 **	5.24 ± 0.76	105.68 ± 14.11	30.27 ± 1.48
	800	5.55 ± 0.39 **	5.59 ± 0.35	115.63 ± 6.52 ***	32.22 ± 2.67

Control group, *N*-acetyl phenylhydrazine (APH) and cyclophosphamide (CTX) were not induced. Model group, APH and CTX were induced. FEJ positive group, FEJ was administered to APH and CTX induced mice in a dose of 10 mg/kg. EPN group, EPN (200, 400 and 800 mg/kg of body weight) was respectively administered to APH and CTX induced mice. ESPN group, ESPN (200, 400 and 800 mg/kg of body weight) was respectively administered to APH and CTX induced mice. EMPN group, EMPN (200, 400 and 800 mg/kg of body weight) was respectively administered to APH and CTX induced mice. FEJ, Fufang E'jiao Jiang; EPN, extract of *P. notoginseng*; ESPN, extract of steamed *P. notoginseng*; EMPN, extract of microwave processed *P. notoginseng*. WBC, white blood cell; RBC, red blood cell; HGB, hemoglobin; HCT, hematocrit. Compared with normal control group: $^{\triangle}$ $p < 0.05$, $^{\triangle\triangle}$ $p < 0.01$, $^{\triangle\triangle\triangle}$ $p < 0.001$; compared with blood deficiency model group: * $p < 0.05$, ** $p < 0.01$, *** $p < 0.001$.

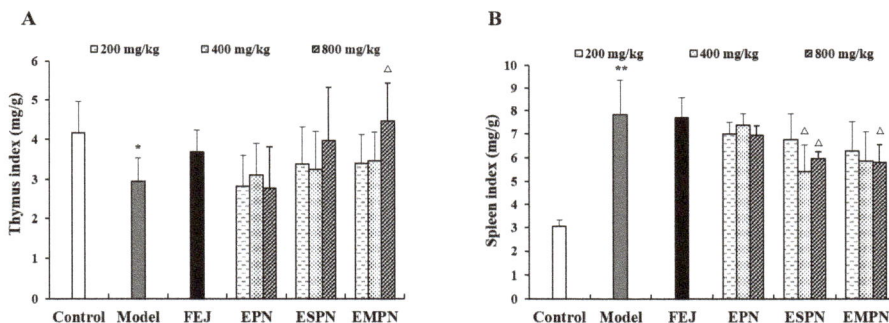

Figure 5. Effect of raw notoginseng and processed notoginseng on thymus index (**A**) and spleen index (**B**) of blood deficiency model mice. FEJ, Fufang E'jiao Jiang; EPN, the extract of *P. notoginseng*; ESPN, the extract of steamed *P. notoginseng*; EMPN, the extract of microwave processed *P. notoginseng*. Data are expressed as mean $\pm s$ ($n = 10$ in group). Compared with the negative control group: * $p < 0.05$, ** $p < 0.01$; compared with the model group: $^{\triangle}$ $p < 0.05$.

2.5. Effect of the Extract of Microwave Processed P. notoginseng on Hemostatic Activity In Vitro

PT value reflects the activity of extrinsic coagulation system [38]. The hemostatic activity of EPN, ESPN, EMPN and their main saponins is shown in Figure 6. Compared with the negative control, EPN and ESPN had a two-way regulatory effect of hemostasis and activating blood circulation. EPN and ESPN had hemostatic effect significantly ($p < 0.001$) at the low concentration of 1 mg/mL and 3 mg/mL. EPN and ESPN has significant anticoagulant effect ($p < 0.001$), when using the high concentration of 20 mg/mL (Figure 6A). However, EMPN showed the hemostatic effect at a different concentration. The main components of raw notoginseng (EPN) were ginsenosides Rg$_1$, Re, Rb$_1$, Rd and notoginsenoside R$_1$. PT testing results showed that ginsenosides R$_1$ and

Rd significantly prolonged the clotting time ($p < 0.01$), and ginsenoside Rb_1 shortened the clotting time at the concentration of 2 mg/L (Figure 6B). It is reported that Rb_1 has the effect of antiplatelet aggregation [39]. No statistically significant effect on PT value was observed for the other compounds of Rg_1 and Re. The main components in ESPN or EMPN are ginsenosides $20(S)$-Rh_1, $20(R)$-Rh_1, Rk_3, Rh_4, $20(S)$-Rg_3, $20(R)$-Rg_3, Rk_1 and Rg_5. The PT values of these eight compounds are shown in Figure 6C. When the concentration of ginsenosides was 2 mg/mL, the PT time of Rk_1 and Rg_5 was prolonged ($p < 0.01$), and the PT time of Rh_4 ($p < 0.05$), $20(S)$-Rg_3 ($p < 0.05$), $20(R)$-Rg_3 ($p < 0.001$) was shortened. The results showed that the hemostatic activity of raw notoginseng (EPN) and processed notoginseng (ESPN and EMPN) were significantly different, which was related to their different chemical composition. The hemostatic effect of EMPN is mainly caused by its components of Rh_4, $20(S)$-Rg_3, $20(R)$-Rg_3.

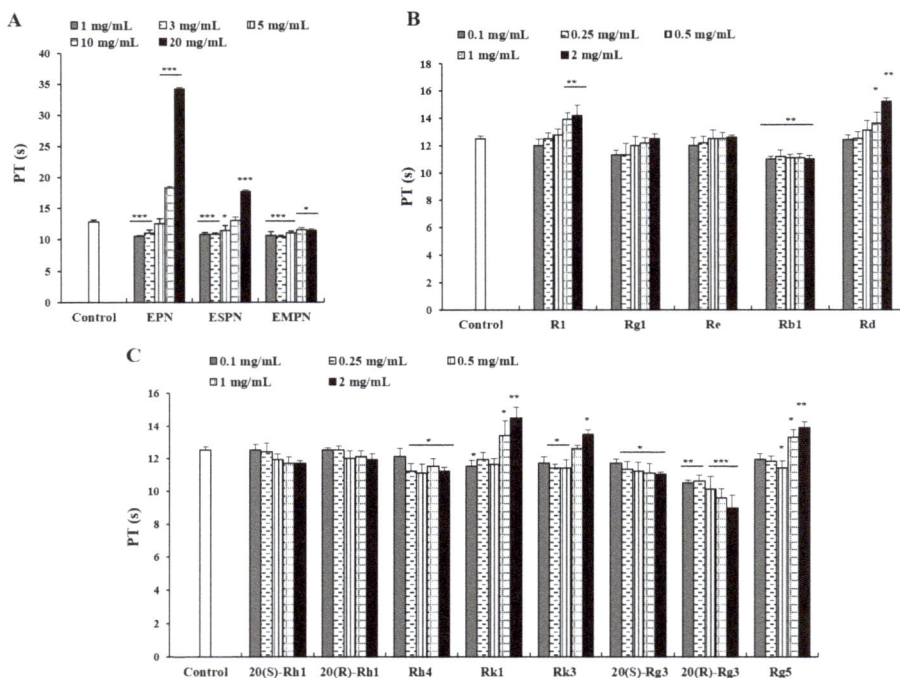

Figure 6. Hemostatic activities of raw notoginseng and processed notoginseng. (**A**) PT assays of the extract of *P. notoginseng* (EPN), the extract of steamed *P. notoginseng* (ESPN) and the extract of microwave processed *P. notoginseng* (EMPN); (**B**) PT assays of ginsenosides Rg_1, Re, Rb_1, Rd and notoginsenoside R_1; (**C**) PT assays of ginsenosides $20(S)$-Rh_1, $20(R)$-Rh_1, Rh_4, Rk_1, Rk_3, $20(S)$-Rg_3, $20(R)$-Rg_3 and Rg_5. Data are expressed as mean $\pm$ s ($n = 3$). Compared with the negative control group: * $p < 0.05$, ** $p < 0.01$, *** $p < 0.001$.

3. Materials and Methods

3.1. Chemicals and Materials

Standard ginsenosides Re (lot number: 20141015), Rg_1 (lot number: 20140809), Rb_1 (lot number: 20140828), Rd (lot number: 20141105), $20(S)$-Rg_3 (lot number: 20141109), $20(R)$-Rg_3 (lot number: 20141205) and notoginsenoside R_1 (lot number: 20140925) were purchased from Jinsui Bio-technology Co. Ltd. (Shanghai, China). Standard ginsenoside $20(S)$-Rh_1 (lot number: MUST-13052205) was purchased from Must Bio-technology Co. Ltd. (Chengdu, China). Standard ginsenoside 20(*R*)-Rh$_1$ (lot number: JBZ-1065) was purchased from Nanjing Jin Yibai Biological Technology Co. Ltd. (Nanjing, China). Standard ginsenosides Rk$_3$ (lot number: GR-133-140724) and Rh$_4$ (lot number: GR-133-130721) were purchased from Guangrun Bio-technology Co. Ltd. (Nanjing, China). Standard ginsenosides Rk$_1$ (lot number: P20N6F6254), Rg$_5$ (lot number: P20N6F6253), cyclophosphamide (lot number: SJ0121RA14) and *N*-acetylphenylhydrazine (lot number: AA14446) were purchased from Yuanye Bio-technology Co. Ltd. (Shanghai, China). Fufang E'jiao Jiang (lot number: 160519) were purchased from Shandong Fujiao Group Co., Ltd. (Jinan, China). The PT kit was purchased from Wuhan Zhongtai Biotech Co. Ltd. (Wuhan, China). Acetonitrile (HPLC grade) was purchased from Sigma Aldrich (Saint Louis, MO, USA) and other solvents (A.R.) were purchased from Fengchuan Chemical Reagent Technology Co. Ltd. (Tianjin, China).

3.2. Preparation of Processed notoginseng Using the Microwave Processing Method

The dried roots of *P. notoginseng* were purchased from the Wenshan market in Yunnan Province, China. The roots were ground to pass through a 40-mesh sieve and extracted ultrasonically with 70% aqueous MeOH, three times for 45 min. After filtration, filtrate was evaporated to give the extract of *P. notoginseng* (32% of dried weight). The extract (250 mg) was added to the solvent (10 mL) in a microwave extraction system (model no.: MDS-6G) manufactured by Sineo Microwave Chemical Technology Co. Ltd. (Shanghai, China). This microwave instrument has a high precision platinum electric resistance sensor, which controls the temperature and displays temperature curve in real time. The temperature range is 0–300 °C. The solution was irradiated by microwave in sealed containers for the different microwave power, temperature and time. The solution was centrifuged at 500× *g* for 5 min and the supernatant was added the solvent to 10 mL. This sample solution was prepared for HPLC analysis.

3.3. HPLC Analysis of Ginenosides

By comparing with the retention time of standard ginsenosides in HPLC, eight new transformed ginsenosides of *P. notoginseng* were identified as 20(*S*)-Rh$_1$, 20(*R*)-Rh$_1$, Rh$_4$, Rk$_1$, Rk$_3$, 20(*S*)-Rg$_3$, 20(*R*)-Rg$_3$ and Rg$_5$ (Figure 1). Quantitative analysis of the eight minor ginsenosides processed by microwave method was determined by HPLC. HPLC was performed by a Shimadzu Analytical Instrument (Shimadzu, Kyoto, Japan), equipped with a Shimadzu, DGU-20A3R(C) solvent degasser, Shimadzu, LC-20AB binary pump, a Shimadzu SIL-20A auto sampler and a Shimadzu SPD-20A UV detector. The detection wavelength was 203 nm, and the column temperature was at 35 °C. The column was the Vision HT C18 (250 mm × 4.6 mm, 5 μm). The mobile phase consisted of distilled water (solvent A) and acetonitrile (solvent B) at the flow rate of 1.0 mL/min. The solvent B ratios were as follows: 20–20%, 20–46%, 46–55%, 55–55% with retention times of 0 min–20 min, 20 min–45 min, 45 min–50 min, 50 min–60 min, 60 min–65min respectively. The method was validated for the determination of eight minor ginsenosides (Table 5). The calibration curves for 20(*S*)-Rh$_1$, 20(*R*)-Rh$_1$, Rh$_4$, Rk$_1$, Rk$_3$, 20(*S*)-Rg$_3$, 20(*R*)-Rg$_3$ and Rg$_5$ showed good linearity (R^2 > 0.9990) in the concentration ranges.

3.4. Experimental Design

3.4.1. Single-Factor Experiments

Next microwave processed products were obtained in different conditions. The single factor conditions for the minor ginsenosides production were studied. Solvents with different concentrations of ethanol or methanol (from 0% to 100%) were used. Considering the influence of the ratio of material to volume, the ratios of 1:10, 1:20, 1:40, 1:60, 1:80, and 1:100 (*w*/*v*) were selected. When the solvent is water and material ratio is 1:60, the temperature ranged from 60 °C to 180 °C, microwave power ranged from 300 W to 1000 W, and time ranged from 5 min to 50 min. The obtained samples were filtered through a 0.45 μm filter membrane for HPLC analysis.

3.4.2. Response Surface Methodology

According to the results of single factor analysis, the three most important independent variables, X_1 (microwave power), X_2 (temperature) and X_3 (time) were chosen for further evaluation in response surface methodology (RSM). Table 1 shows different combinations of the independent factors (X_i) and actual experimental responses (Y_i) using Box-Behnken Design. The responses included 20(S)-Rh$_1$ content (Y_1), 20(R)-Rh$_1$ content (Y_2), Rk$_3$ content (Y_3), Rh$_4$ content (Y_4), 20(S)-Rg$_3$ content (Y_5), 20(R)-Rg$_3$ content (Y_6), Rk$_1$ content (Y_7), Rg$_5$ content (Y_8) and total minor ginsenosides content (Y_9). The predicted responses were calculated by the second degree polynomial Equation (2):

$$Y = \beta_0 + \beta_1 X_1 + \beta_2 X_2 + \beta_3 X_3 + \beta_{11} X_1^2 + \beta_{22} X_2^2 + \beta_{33} X_3^2 + \beta_{12} X_1 X_2 + \beta_{13} X_1 X_3 + \beta_{23} X_2 X_3 + \varepsilon \quad (2)$$

The coefficients of the polynomial were intercept (β_0), linear coefficients (β_1, β_2, β_3), squared coefficients (β_{11}, β_{22}, β_{33}) and interaction coefficients (β_{12}, β_{13}, β_{23}). Graphical analysis of data was completed by Design Expert trial 8.0.5 (Stat-Ease, Minneapolis, MN, USA). Response surface quadratic model was analyzed by variance analysis (ANOVA). The significance of model was checked by R^2, adjusted R^2 and goodness of fit. *P* values of less than 0.05 were considered significant.

3.5. Pharmacological Assays

3.5.1. Animals

Kunming mice (18–22 g) of both sexes were supplied by Changsha Tianqin Biotechnology Co., (Changsh, China). The approval number is SCXK 2014-0011. The mice were kept at $25 \pm 2\,^{\circ}\text{C}$ and a 12 h dark/light cycle condition. All experimental protocols were approved by the Animal Ethical Committee of Laboratory Animals of Kunming University of Science and Technology.

3.5.2. Extract Preparation

The roots of *P. notoginseng* (500 g) were powdered and ultrasonically extracted with 70% aqueous MeOH, three times for 45 min. After filtration, the solvent was evaporated to give the extract of *P. notoginseng* (EPN, 162.3 g) with a yield of 32.5%.

Powdered roots of *P. notoginseng* (500 g) were steamed at 120 °C for 4 h [23] and ultrasonically extracted with 70% aqueous MeOH, three times for 45 min. After filtration and concentration, the extract of steamed *P. notoginseng* (ESPN, 125.4 g) was obtained with a yield of 25.1%.

The methnal extract of *P. notoginseng* was processed using optimal microwave conversion conditions. The EPN (50 g) was irradiated with microwave at a power of 500 W and a temperature of 150°C for 20 min to give the extract of microwave processed *P. notoginseng* (EMPN, 47.1 g) with a yield of 94.2%.

3.5.3. *N*-Acetyl Phenylhydrazine (APH) and Cyclophosphamide (CTX) Induced Blood Deficiency in Mice

The establishment of the blood deficiency model was used according to the method described by Li et al. [36]. Mice (120) were randomly divided into 12 groups with 10 mice in each group, i.e., normal control group (Control); blood deficiency model group (Model); positive group of Fufang E'jiao Jiang (FEJ, 10 mg/kg); low dose group of EPN (200 mg/kg), middle dose group of EPN (400 mg/kg), high dose group of EPN (800 mg/kg); low dose group of ESPN (200 mg/kg), middle dose group of ESPN (400 mg/kg), high dose group of ESPN (800 mg/kg); low dose group of EMPN (200 mg/kg), middle dose group of EMPN (400 mg/kg), high dose group of EMPN (800 mg/kg). The FEJ, EPN, ESPN and EMPN were intragastrically administered in each group, once a day for two weeks. The normal control group and model group were given an equal volume of normal saline. Except for the normal control group, mice in each group were hypodermically injected with 2% APH saline solution (20 mg/kg) on the 1st day, and intraperitoneally injected with CTX saline solution (70 mg/kg) on the 2st day. After 7 days of continuous injection, the blood deficiency model was established. Orbit blood samples of mice was collected on day 7 for detecting the white blood cell (WBC), red blood cell (RBC), hemoglobin (HGB), and hematocrit (HCT). After collection of blood, the mice were weighed and killed by cervical dislocation. The thymus and spleen were removed and weighed. The spleen or thymus index was calculated according to the following formula:

Spleen or thymus index (mg/g) = (spleen weight or thymus weight/body weight)

3.5.4. Blood Plasma Clotting Analysis

Prothrombin time (PT) was measured using a coagulation analyzer (XN06-IV, Aierfu, Wuhan, China). The plasma was obtained from mice whole blood added with 0.109 mol/L sodium citrate (9:1 ratio of blood to citrate) by centrifugation at $1000 \times g$ for 15 min. The mixtures of the plasma (50 μL) with EPN, ESPN, EMPN or their compounds (50 μL) were incubated at 37 °C for 3 min. The samples of EPN, ESPN and EMPN were diluted with purified water to give the five concentrations: 20, 10, 5, 3, 1 mg/mL. The ginsenosides Rg_1, Rb_1, Re, Rd, 20(S)-Rh_1, 20(R)-Rh_1, Rh_4, Rk_1, Rk_3, 20(S)-Rg_3, 20(R)-Rg_3 and Rg_5 and notoginsenoside R_1 were diluted to other five concentrations: 2, 1, 0.5, 0.25, 0.1 mg/mL. The PT assay reagent (100 μL) was added to the mixed samples and the clotting time was recorded. Purified water was used as negative control.

3.5.5. Statistical Analysis

Statistical analyses were performed with SPSS 18.0 software (SPSS Inc., Chicago, IL, USA), and the data were expressed as mean ± s. Single factor analysis of variance was used to compare between groups. The values of $p < 0.05$ were considered to be statistically significant, and $p < 0.01$ and $p < 0.001$ being very significant.

4. Conclusions

In this study, the generated minor ginsenosides from *P. notoginseng* can be used to illustrate the transformation of ginsenosides during microwave processing. Eight minor ginsenosides from *P. notoginseng* were identified as ginsenosides 20(S)-Rh_1, 20(R)-Rh_1, Rh_4, Rk_1, Rk_3, 20(S)-Rg_3, 20(R)-Rg_3 and Rg_5. When the microwave power and temperature were increased, the glycosyl residues of major saponins were easily decreased.

Eight minor ginsenosides was successfully prepared from *P. notogingseng* by microwave processing and the transformation conditions were optimized by RSM. The experimental results showed that the yield of the total minor gisenosides was 93.13% when the optimum conditions of microwave power, temperature and time were 495.03 W, 150.68 °C and 20.32 min, respectively. Under the optimum conditions, the actual yield of total minor ginsenosides was 94.15% that agreed with the predicted values. The RSM in this study was effective to optimize the production of the minor ginsenosides.

The pharmacological results of the extract of microwave processed *P. notoginseng* (EMPN) in vivo showed that EMPN had the effect of enriching blood, similar to the extract of steamed *P. notoginseng* (ESPN) in APH and CTX induced blood deficient mice. WBC and HGB in EMPN group showed a significant increase trend compared with the model group. EMPN had a protective effect on the immune organs of mice. Meanwhile, the in vitro hemostatic activity of EMPN showed that it had significantly shortened the clotting time on PT testing. The hemostatic effect of EMPN is mainly caused by its components of Rh_4, 20(S)-Rg_3, 20(R)-Rg_3.

Therefore, this microwave transformed method was simple and suitable to mass-produce the minor ginsenosides on an industrial scale and conducive to studying their blood-enriching and hemostatic activity.

Author Contributions: Y.Q. and X.C. conceived and designed the experiments; H.L. and J.P. performed the experiments; Y.Y. performed the animal experiments and analyzed the data; H.L. and Y.Q. drafted the manuscript.

Acknowledgments: This work was supported by key special project of national key research and development program (2017YFC1702506) and analysis and testing fund of Kunming University of Science and Technology (2016T20120047).

Conflicts of Interest: The authors declare no conflicts of interest in this work.

References

1. Wan, X.Q.; Chen, S.H.; Peng, Y.S.; Lou, Z.H.; Yang, M.C. Comparative of Notoginseng Radix Et Rhizome and its processed products on enriching blood and supplementing qi of rats with blood-deficiency. *Chin. J. Modern Appl. Pharm.* **2014**, *13*, 881–892.

2. Wang, T.; Guo, R.G.; Zhou, G.H.; Zhou, X.D.; Kou, Z.Z.; Sui, F.; Li, C.; Tang, L.Y.; Wang, Z.J. Traditional uses, botany, phytochemistry, pharmacology and toxicology of *Panax notoginseng* (Burk.) F.H. Chen: A review. *J. Ethnopharmacol.* **2016**, *188*, 234–258. [CrossRef] [PubMed]

3. Wang, C.Z.; McEntee, E.; Wicks, S.; Wu, J.A.; Yuan, C.S. Phytochemical and analytical studies of *Panax notoginseng* (Burk.) F.H. Chen. *J. Nat. Med.* **2006**, *60*, 97–106. [CrossRef]

4. Song, M.; Zhang, S.Y.; Xu, X.Y.; Hang, T.J.; Lee, J. Simultaneous determination of three *Panax notoginseng* saponins at sub-nanograms by LC-MS/MS in dog plasma for pharmacokinetics of compound Danshen tablets. *J. Chromatogr. B* **2010**, *878*, 3331–3337. [CrossRef] [PubMed]

5. Wan, J.B.; Zhang, Q.W.; Hong, S.J.; Li, P.; Li, S.P.; Wang, Y.T. Chemical investigation of saponins in different parts of *Panax notoginseng* by pressurized liquid extraction and liquid chromatography-electrospray ionization-tandem mass spectrometry. *Molecules* **2012**, *17*, 5836–5853. [CrossRef] [PubMed]

6. Gu, C.Z.; Lv, J.J.; Zhang, X.X.; Qiao, Y.J.; Yan, H.; Li, Y.; Wang, D.; Zhu, H.T.; Luo, H.R.; Yang, C.R.; et al. Triterpenoids with promoting effects on the differentiation of PC12 cells from the steamed roots of *Panax notoginseng*. *J. Nat. Prod.* **2015**, *78*, 1829–1840. [CrossRef] [PubMed]

7. Gu, C.Z.; Lv, J.J.; Zhang, X.X.; Yan, H.; Zhou, H.T.; Luo, H.R.; Wang, D.W.; Yang, C.R.; Xu, M.; Zhang, Y.J. Minor dehydrogenated and cleavaged dammarane-type saponins from the steamed roots of *Panax notoginseng*. *Fitoterapia* **2015**, *103*, 97–105. [CrossRef] [PubMed]

8. Liao, P.Y.; Wang, D.; Zhang, Y.J.; Yang, C.R. Dammarane-type glycosides from steamed notoginseng. *J. Agric. Food Chem.* **2008**, *56*, 1751–1756. [CrossRef] [PubMed]

9. Sun, S.; Wang, C.Z.; Tong, R.; Li, X.L.; Fishbein, X.; Wang, Q.; He, T.C.; Du, W.; Yuan, C.S. Effects of steaming the root of *Panax notoginseng* on chemical composition and anticancer activities. *Food Chem.* **2010**, *118*, 307–314. [CrossRef]

10. Lau, A.J.; Woo, S.O.; Koh, H.L. Analysis of saponins in raw and steamed *Panax notoginseng* using high-performance liquid chromatography with diode array detection. *J. Chromatogr. A* **2003**, *1011*, 77–87. [CrossRef]

11. Lau, A.J.; Seo, B.H.; Woo, S.O.; Koh, H.L. High-performance liquid chromatographic method with quantitative comparisons of whole chromatograms of raw and steamed *Panax notoginseng*. *J. Chromatogr. A* **2004**, *1057*, 141–149. [CrossRef] [PubMed]

12. Chae, S.; Kang, K.A.; Chang, W.Y.; Kim, M.J.; Lee, S.J.; Lee, Y.S.; Kim, H.S.; Kim, D.H.; Hyun, J.W. Effect of compound K, a metabolite of ginseng saponin, combined with gamma-ray radiation in human lung cancer cells in vitro and in vivo. *J. Agric. Food Chem.* **2007**, *57*, 5777–5782. [CrossRef] [PubMed]

13. Shinkai, K.; Akedo, H.; Mukai, M.; Imamura, F.; Isoai, A.; Kobayashi, M.; Kitagawa, I. Inhibition of in vitro tumor cell invasion by ginsenoside Rg$_3$. *Jpn. J. Cancer Res.* **1996**, *87*, 357–362. [CrossRef] [PubMed]

14. Li, L.; Ni, J.Y.; Li, M.; Chen, J.R.; Han, L.F.; Zhu, Y.; Kong, D.L.; Mao, J.Y.; Wang, Y.; Zhang, B.L. Ginsenoside Rg$_3$ micelles mitigate doxorubicin-induced cardiotoxicity and enhance its anticancer efficacy. *Drug Deliv.* **2017**, *24*, 1617–1630. [CrossRef] [PubMed]

15. Ko, H.; Kim, Y.J.; Park, J.S.; Park, J.H.; Yang, H.O. Autophagy inhibition enhances apoptosis induced by ginsenoside Rk$_1$ in hepatocellular carcinoma cells. *Biosci. Biotechnol. Biochem.* **2009**, *73*, 2183–2189. [CrossRef] [PubMed]

16. Wang, Z.; Hu, J.N.; Yan, M.H.; Xing, J.J.; Liu, W.C.; Li, W. Caspase-mediated anti-apoptotic effect of ginsenoside Rg$_5$, a main rare ginsenoside, on acetaminophen-induced hepatotoxicity in mice. *J. Agric. Food Chem.* **2017**, *65*, 9226–9236. [CrossRef] [PubMed]

17. Kang, S.; Im, K.; Kim, G.; Min, H. Antiviral activity of 20(*R*)-ginsenoside Rh$_2$ against murine gammaherpesvirus. *J. Ginseng Res.* **2017**, *41*, 496–502. [CrossRef] [PubMed]
18. Wang, R.F.; Li, J.; Hu, H.J.; Li, J.; Yang, Y.B.; Yang, L.; Wang, Z.T. Chemical transformation and target preparation of saponins in stems and leaves of *Panax notoginseng*. *J. Ginseng Res.* **2016**, *40*, 1–7. [CrossRef]
19. Choi, H.S.; Kim, S.Y.; Park, Y.; Jung, E.Y.; Suh, H.J. Enzymatic transformation of ginsenosides in Korean Red Ginseng (*Panax ginseng* Meyer) extract prepared by Spezyme and Optidex. *J. Ginseng Res.* **2014**, *38*, 264–269. [CrossRef] [PubMed]
20. Upadhyaya, J.; Kim, M.J.; Kim, Y.H.; Ko, S.R.; Park, H.W.; Kim, M.K. Enzymatic formation of compound-K from ginsenoside Rb$_1$ by enzyme preparation from cultured mycelia of *Armillaria mellea*. *J. Ginseng Res.* **2016**, *40*, 105–112. [CrossRef] [PubMed]
21. Lee, S.J.; Kim, Y.J.; Kim, M.G. Changes in the ginsenoside content during the fermentation process using microbial strains. *J. Ginseng Res.* **2015**, *39*, 392–397. [CrossRef] [PubMed]
22. Cui, L.; Wu, S.Q.; Zhao, C.A.; Yin, C.R. Microbial conversion of major ginsenosides in ginseng total saponins by *Platycodon grandiflorum* endophytes. *J. Ginseng Res.* **2016**, *40*, 366–374. [CrossRef] [PubMed]
23. Wang, D.; Liao, P.Y.; Zhu, H.T.; Chen, K.K.; Xu, M.; Zhang, Y.J.; Yang, C.R. The processing of *Panax notoginseng* and the transformation of its saponin components. *Food Chem.* **2012**, *132*, 1808–1813. [CrossRef]
24. Choi, P.; Park, J.Y.; Kim, T.; Park, S.H.; Kim, H.K.; Kang, K.S.; Ham, J. Improved anticancer effect of ginseng extract by microwave-assisted processing through the generation of ginsenosides Rg$_3$, Rg$_5$ and Rk$_1$. *J. Funct. Foods* **2015**, *14*, 613–622. [CrossRef]
25. Park, J.Y.; Choi, P.; Kim, H.K.; Kang, K.S.; Ham, J. Increase in apoptotic effect of *Panax ginseng* by microwave processing in human prostate cancer cells: In vitro and in vivo studies. *J. Ginseng Res.* **2016**, *40*, 62–67. [CrossRef] [PubMed]
26. Chemat, F.; Rombaut, N.; Meullemiestre, A.; Turk, M.; Perino, S.; Fabiano-Tixier, A.S.; Abert-Vian, M. Review of Green Food Processing techniques. Preservation, transformation, and extraction. *Innov. Food Sci. Emerg.* **2017**, *41*, 357–377. [CrossRef]
27. Jacotet-Navarro, M.; Rombaut, N.; Deslis, S.; Fabiano-Tixier, A.S.; Pierre, F.X.; Bily, A.; Chemat, F. Towards a "dry" bio-refinery without solvents or added water using microwaves and ultrasound for total valorization of fruits and vegetables by-products. *Green Chem.* **2016**, *18*, 3106–3115. [CrossRef]
28. Galema, S.A. Microwave chemistry. *Chem. Soc. Rev.* **1997**, *26*, 233–238. [CrossRef]
29. Shao, J.; Yang, Y.; Zhong, Q. Studies on preparation of oligoglucosamine by oxidative degradation under microwave irradiation. *Polym. Degrad. Stabil.* **2003**, *82*, 395–398. [CrossRef]
30. Kang, K.S.; Yamabe, N.; Kim, H.Y.; Okamoto, T.; Sei, Y.; Yokozawa, T. Increase in the free radical scavenging activities of American ginseng by heat processing and its safety evaluation. *J. Ethnopharmacol.* **2007**, *113*, 225–232. [CrossRef] [PubMed]
31. Wang, Z.; Zhao, L.C.; Li, W.; Liu, S.L.; Yang, G.; Liu, Z.; Zhang, L.X. Optimization of preparation process of ginsenosides Rh$_4$ and Rk$_3$ by response surface methodology. *Chin. Tradit. Herbal Drugs* **2017**, *48*, 2207–2211.
32. Sun, C.P.; Gao, W.P.; Zhao, B.Z.; Cheng, L.Q. Optimization of the selective preparation of 20(*R*)-ginsenoside Rg$_3$ catalyzed by D, L-tartaric acid using response surface methodology. *Fitoterapia* **2013**, *84*, 213–221. [CrossRef] [PubMed]
33. Chemat, F.; Cravotto, G. *Microwave-Assisted Extraction for Bioactive Compounds: Theory and Practice*; Springer: New York, NY, USA, 2013.
34. Petigny, L.; Périno, S.; Minuti, M.; Visinoni, F.; Wajsman, J.; Chemat, F. Simultaneous microwave extraction and separation of volatile and non-volatile organic compounds of Boldo leaves from lab to industrial scale. *Int. J. Mol. Sci.* **2014**, *15*, 7183–7198. [CrossRef] [PubMed]
35. Ciriminna, R.; Carnaroglio, D.; Delisi, R.; Arvati, S.; Tamburino, A.; Pagliaro, M. Industrial Feasibility of Natural Products Extraction with Microwave Technology. *Chemistryselect* **2016**, *1*, 549–555. [CrossRef]
36. Li, P.L.; Sun, H.G.; Hua, Y.L.; Ji, P.; Zhang, L.; Li, J.X.; Wei, Y.M. Metabolomics study of hematopoietic function of Angelica sinensis on blood deficiency mice model. *J. Ethnopharmacol.* **2015**, *166*, 261–269. [CrossRef] [PubMed]
37. Li, W.X.; Tang, Y.P.; Guo, J.M.; Huang, M.Y.; Li, W.; Qian, D.W.; Duan, J.A. Enriching blood effect comparison in three kinds of blood deficiency model after oral administration of drug pair of Angelicae Sinensis Radix and Chuanxiong Rhizoma and each single herb. *China J. Chin. Mater. Med.* **2011**, *36*, 1808–1814.

38. Li, C.T.; Wang, H.B.; Xu, B.J. A comparative study on anticoagulant activities of three Chinese herbal medicines from the genus *Panax* and anticoagulant activities of ginsenosides Rg_1 and Rg_2. *Pharm. Biol.* **2013**, *51*, 1077–1080. [CrossRef] [PubMed]
39. Kim, E.S.; Lee, J.S.; Lee, H.G. Nanoencapsulation of red ginseng extracts using chitosan with polyglutamic acid or fucoidan for improving antithrombotic activities. *J. Agric. Food Chem.* **2016**, *64*, 4765–4771. [CrossRef] [PubMed]

Sample Availability: Samples of the extracts and compounds are available from the authors.

molecules

MDPI

Article

Optimisation of Ethanol-Reflux Extraction of Saponins from Steamed *Panax notoginseng* by Response Surface Methodology and Evaluation of Hematopoiesis Effect

Yupiao Hu [1], Xiuming Cui [1,2,3], Zejun Zhang [1], Lijuan Chen [1], Yiming Zhang [1], Chengxiao Wang [1], Xiaoyan Yang [1], Yuan Qu [1] and Yin Xiong [1,2,3,*]

[1] Faculty of Life Science and Technology, Kunming University of Science and Technology, Kunming 650500, China; hypflygo@163.com (Y.H.); cuisanqi37@163.com (X.C.); 18380802826@163.com (Z.Z.); chen13990207@163.com (L.C.); jr93586@163.com (Y.Z.); wcx1192002@126.com (C.W.); yangxiaoyan9999@163.com (X.Y.); quyuan2001@126.com (Y.Q.)

[2] Yunnan Key Laboratory of *Panax notoginseng*, Kunming University of Science and Technology, Kunming 650500, China

[3] Laboratory of Sustainable Utilization of *Panax notoginseng* Resources, State Administration of Traditional Chinese Medicine, Kunming University of Science and Technology, Kunming 650500, China

* Correspondence: yhsiung@163.com; Tel.: +86-0871-5915818

Received: 17 April 2018; Accepted: 9 May 2018; Published: 17 May 2018

Abstract: The present study aims to optimize the ethanol-reflux extraction conditions for extracting saponins from steamed *Panax notoginseng* (SPN). Four variables including the extraction time (0.5–2.5 h), ethanol concentration (50–90%), water to solid ratio (W/S, 8–16), and times of extraction (1–5) were investigated by using the Box-Behnken design response surface methodology (BBD-RSM). For each response, a second-order polynomial model with high R^2 values (>0.9690) was developed using multiple linear regression analysis and the optimum conditions to maximize the yield (31.96%), content (70.49 mg/g), and antioxidant activity (EC_{50} value of 0.0421 mg/mL) for saponins extracted from SPN were obtained with a extraction time of 1.51 h, ethanol concentration of 60%, extraction done 3 times, and a W/S of 10. The experimental values were in good consistency with the predicted ones. In addition, the extracted SPN saponins could significantly increase the levels of blood routine parameters compared with the model group ($p < 0.01$) and there was no significant difference in the hematopoiesis effect between the SPN group and the SPN saponins group, of which the dose was 15 times lower than the former one. It is suggested that the SPN saponins extracted by the optimized method had similar functions of "blood tonifying" at a much lower dose.

Keywords: steamed *Panax notoginseng*; saponins; extraction; optimization; antioxidant activity; response surface methodology; hematopoiesis

1. Introduction

Panax notoginseng (PN) (Burk.) F. H. Chen, a highly valued Chinese medicinal herb, has been used in Asia to treat blood disorders for thousands of years [1]. Numerous studies have shown that saponins are the major active components of PN, with pharmacologic effects such as dilating blood vessels, lowering the blood pressure, anti-thrombosis, anti-inflammation, anti-vascular aging, anti-cancer, and antioxidant activities [2–4]. Therefore, a large number of studies have focused on the technology of extraction and purification of saponins. Related products with PN saponins as the main ingredients have even been developed. For example, Xuesaitong, one of the bestselling prescriptions of herbal medicine, consists of over 85% ginsenosides attributed to the extract obtained from PN, showing good therapeutic effects on the cardiovascular and cerebrovascular system, blood system, and nervous system [5].

There was a saying for PN that "the raw materials eliminate and the steamed ones tonify". The so-called "eliminate" means raw PN can stop bleeding, promote blood circulation, diminish swelling, and ease pain [2]. The "tonify" means that steamed PN (SPN) performs better efficacies on improving the immunity and nourishing the blood [6,7]. The variation in the types and contents of saponins has been reported to be the reason resulting in the efficacy difference between raw and SPN [8–10]. According to our previous studies [11,12], the contents of notoginsenoside R_1, ginsenosides Rg_1, Rb_1, Re, and Rd in raw PN were decreased along with the duration of steaming, whereas those of ginsenosides Rh_1, Rk_3, Rh_4, 20(R)-Rg_3, and 20(S)-Rg_3 were increased, which were also found to be closely related to the tonifying functions of SPN. However, the current studies on the total saponins of PN were generally focused on those from raw materials and there were few reports on the processing of extracts or total saponins from SPN, not to speak of their activities, which hinders the development of this valuable medicine. Therefore, the ethanol-reflux extraction process of saponins from SPN was optimized in this research. Constituents of notoginsenoside R_1, ginsenosides Rg_1, Rb_1, Re, Rd, Rh_1, Rk_3, Rh_4, 20(R)-Rg_3, 20(S)-Rg_3 were included as the indices of saponins from SPN based on our previous study. The yield, content, and antioxidant activity of saponins from SPN were evaluated for the process optimization.

The Box–Behnken design response surface methodology (BBD-RSM) is a collection of mathematical and statistical techniques useful for establishing models which can be used to evaluate multiple parameters and their interactions with quantitative data, and effectively optimizing complex extraction procedures in a statistical way [13–15]. It reduces not only the number of experimental trials, the development time, and overall cost determines, but also the interactions among the independent variables [16–19]. Therefore, in this research, BBD-RSM was used to determine the optimal conditions of ethanol-reflux extraction of saponins from SPN. Besides, SPN is traditionally used as a tonic to enrich blood and tonify the body, which can improve the blood deficiency syndrome by increasing the production of various blood cells in anemic conditions [6,7]. Thus, to verify the pharmacologic activity of extracted saponins from SPN, we also investigated the hematopoiesis effect of SPN saponins on the levels of blood routine parameters in anemic mice induced by acetylphenylhydrazine (APH) and cyclophosphamide (CTX).

2. Results and Discussion

2.1. HPLC Analyses

The results of the HPLC analyses for the standards solution and sample solution were shown in Figure 1a,b. By comparing the chromatograms of SPN saponins to that of the mixed standards solution, constituents corresponding to major peaks were identified as notoginsenoside R_1, ginsenosides Rg_1, Re, Rb_1, Rd, Rk_3, Rh_4, Rh_1, 20(S)-Rg_3, and 20(R)-Rg_3. Among which, notoginsenoside R_1, ginsenosides Rg_1, and Rb_1 were marker constituents for the quality control of PN in the Chinese Pharmacopoeia (2015 edition) [20]. Ginsenosides Re and Rd were active constituents related to the antioxidation effect of SPN [12]. The rest of the five ginsenosides were characteristic and active constituents of SPN, of which the contents were increased along with the increase of steaming temperature and duration of time. Therefore, the ten constituents were determined as the evaluation markers for the preparation of SPN saponins.

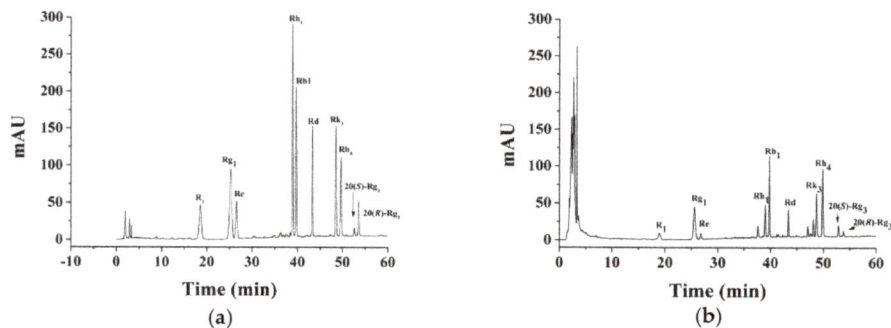

Figure 1. The HPLC chromatograms of the mixed standards solution (**a**) and SPN saponins (**b**). HPLC, high performance liquid chromatography; SPN, steamed *Panax notoginseng*.

2.2. Single-Factor Experimental Analysis

2.2.1. The Effect of Extraction Time on the Content and Yield of Saponins

Extraction time is one of the most important factors that could affect the content of saponins and the yield. The content and yield of saponins under different extraction times are shown in Figure 2a,b. These results were obtained by, firstly setting the ethanol concentration, water to solid ratio, and extraction times to 70%, 10, and 2, respectively. The effect of extraction time on the content and yield of saponins was then investigated by sequentially setting the extraction time at 0.5, 1, 1.5, 2, and 2.5 h. According to the results, with the extension of extraction time, the content and yield of saponins both increased firstly and then decreased. The content of saponins and dry extract yield were firstly increased along with the extraction time, which took 1.5 h and 1 h to reach the maximum, respectively. After that, the levels of the two indexes decreased. The reason might be that at the initial stage of extraction, a relatively longer extraction time was beneficial for dissolving saponins and other polar compounds in the extract. However, if they were kept at high temperature for a long period of time, triterpenoids saponins could be easily decomposed [21]. The concentration of other ingredients in the extract reached the equilibrium with little change of dissolution and there might also be negative reactions causing the yield decline [22]. In addition, with the duration of steaming time, the increase of suspension viscosity was disadvantageous to the extraction efficiency [23] and some saponins might not be separated from the cell debris [24]. Based on the results, the extraction time of 1 h, 1.5 h, and 2 h could achieve saponins of higher content and yield. Therefore, 1 h, 1.5 h, and 2 h were selected as optimal conditions in the following BBD-RSM experiments.

2.2.2. The Effect of Ethanol Concentration on the Content and Yield of Saponins

Another significant factor that would affect the content and yield of saponins is the ethanol concentration. Thus, in the present study, the effect of ethanol concentration on the content and yield of saponins was evaluated. The extraction time, water to solid ratio, and the times of extraction were set as 1.5, 10, and 2 h, respectively. The ethanol concentration was then sequentially set as 50%, 60%, 70%, 80%, and 90%. As shown in Figure 2c,d, when the ethanol concentration was varied between 50% and 90%, the content and yield of saponins increased firstly and then decreased. While the higher or lower ethanol concentration led to the decrease of the content and yield of saponins. A lower ethanol content means more water in the extract, which can result in the increased swelling of *Panax notoginseng*. Saponins can be extracted in a shorter time when the swelling of *Panax notoginseng* is greater. However, more saponins may hydrolyze when the water content in the mixed solvent is higher because of the higher boiling temperature [25]. According to the results, the ethanol concentrations of 60%, 70%, and 80%, were chosen to extract the saponins in the optimization step.

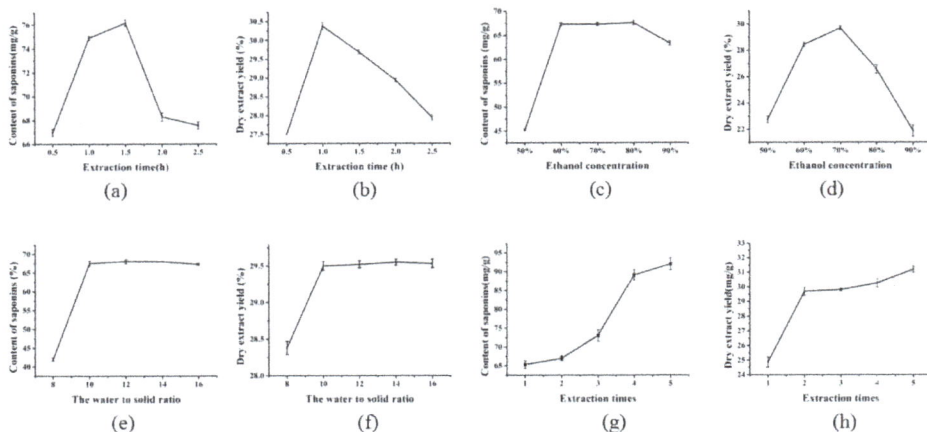

Figure 2. The result of the single-factor experimental analyses.

2.2.3. The Effect of the Water to Solid Ratio on the Content and Yield of Saponins

Increasing the water to solid ratio can improve the content and yield of saponins by affecting the concentration gradient inside and outside cells in the extract. In the research, the water to solid ratios of 8, 10, 12, 14, and 16 were investigated. As shown in Figure 2e,f, as the water to solid ratio was increased from 8 to 10, the content and yield of saponins increased accordingly. This might be due to the increasing water to solid increasing the diffusivity of the solvent into the cells and enhancing the desorption of the saponins from the cells [26]. However, when the water to solid ratios were set at 10, 12, 14, and 16, the variation in the content and yield of saponins was not significant ($p > 0.05$). This might be due to the full dissolution of the components in the extract. Thus, in order to save the costs, the water to solid ratio was determined as 10.

2.2.4. The Effect of the Times of Extraction on the Content and Yield of Saponins

Increasing the times of extraction can significantly improve the content of saponins and the dry extract yield. The effect of the times of extraction on the content and yield of saponins was shown in Figure 2g,h. The extraction time, ethanol concentration, and water to solid ratio were set as 1.5 h, 70%, and 10, respectively. The times of extraction was then sequentially set as 1, 2, 3, 4, and 5. It could be seen that with the increase of the times of extraction, the content and yield of saponins had different degrees of increase. At the same time, the production costs were constantly rising. Therefore, we selected extraction 2, 3, and 4 times as the optimal conditions in the following experiments.

In general, according to the single-factor experimental analysis, our study adopted the extraction time of 1 h, 1.5 h, and 2 h; the ethanol concentration of 60%, 70%, and 80%; the water to solid ratio of 10; and the times of extraction of 2, 3, and 4 for the BBD-RSM.

2.3. *Fitting the Response Surface Models*

As shown in Tables 1 and 2, in the BBD-RSM experimental design, a total of 17 tests of different conditions were performed and the corresponding result of each test was also shown in Table 2. Meanwhile, the contents of ten saponins were shown in Table 3. According to the experimental data, the multiple linear regression equation between the response value and the experimental condition was calculated by using Design-Expert, version 8.6 (Stat-Ease Inc., Minneapolis, MN, USA). At the same time, the regression coefficients for each value were also determined. The fitted equations to predict the yield, content, and antioxidant activity of saponins from the SPN are given below regardless of the significance of the coefficients:

$$\text{Yield of saponins} = 31.13 + 0.52X_1 - 1.36X_2 + 0.81X_3 + 0.45X_{12} - 0.5X_{12} - 0.54X_{22} - 0.52X_{32} \quad (1)$$

$$\text{Content of saponins} = 70.14 - 1.65X_1 + 6.09X_3 - 1.91X_{12} - 1.50X_{23} - 2.32X_{12} - 2.97X_{32} \quad (2)$$

$$EC_{50} = 0.042 + (5.713E - 003)X_1 - 0.020X_3 + (3.580E - 003)X_{12} + (4.550E - 003)X_{23} + (7.154E - 003)X_{12} + 0.012X_{32} \quad (3)$$

Table 1. The experimental domain of BBD-RSM. BBD-RSM, Box-Behnken design response surface methodology.

Independent Variables	Unit	Symbol	Coded Levels		
			−1	0	+1
Extraction time	min	X_1	60	90	120
Ethanol concentration	%	X_2	60	70	80
Times of extraction		X_3	2	3	4

Table 2. The BBD matrix and the experimental data for the responses. BBD-RSM, Box-Behnken design response surface methodology.

Treatment Number	Extraction Time (h)	Ethanol Concentration (%)	Times of Extraction	Content of Saponins (mg/g)	Dry Extract Yield (%)	EC_{50} Value (mg/mL)
1	1.00	80.00	3.00	73.16 ± 0.058	28.08 ± 0.038	0.0393 ± 0.0006
2	1.50	80.00	2.00	62.81 ± 0.069	28.13 ± 0.026	0.0638 ± 0.0005
3	1.50	70.00	3.00	68.96 ± 0.072	31.13 ± 0.019	0.0407 ± 0.0008
4	1.50	70.00	3.00	69.81 ± 0.135	31.50 ± 0.0017	0.0429 ± 0.0006
5	2.00	80.00	3.00	64.97 ± 0.046	29.58 ± 0.023	0.0563 ± 0.0005
6	2.00	70.00	2.00	57.83 ± 0.083	29.96 ± 0.026	0.0903 ± 0.0009
7	1.50	70.00	3.00	71.05 ± 0.122	31.04 ± 0.057	0.0429 ± 0.0004
8	2.00	60.00	3.00	67.55 ± 0.096	31.71 ± 0.049	0.0484 ± 0.0007
9	1.50	60.00	4.00	75.76 ± 0.073	32.19 ± 0.025	0.0324 ± 0.0005
10	1.00	70.00	4.00	71.62 ± 0.108	30.58 ± 0.027	0.0380 ± 0.0005
11	1.50	80.00	4.00	72.62 ± 0.125	30.38 ± 0.018	0.0336 ± 0.0006
12	1.00	60.00	3.00	68.08 ± 0.098	31.49 ± 0.043	0.0457 ± 0.0008
13	2.00	70.00	4.00	69.6 ± 0.062	31.25 ± 0.025	0.0450 ± 0.0009
14	1.00	70.00	2.00	60.32 ± 0.085	28.67 ± 0.036	0.0713 ± 0.0011
15	1.50	70.00	3.00	70.65 ± 0.094	31.50 ± 0.054	0.0429 ± 0.0006
16	1.50	60.00	2.00	59.96 ± 0.067	30.38 ± 0.075	0.0808 ± 0.0012
17	1.50	70.00	3.00	70.22 ± 0.083	30.50 ± 0.026	0.0387 ± 0.0008

Table 3. The contents of the ten markers for SPN saponins.

Treatment Number	R_1 (mg/g)	Rg_1 (mg)	Rb_1 (mg/g)	Re (mg/g)	Rd (mg/g)	Rh_1 (mg/g)	Rk_3 (mg/g)	Rh_4 (mg/g)	20(R)-Rg_3 (mg/g)	20(S)-Rg_3 (mg/g)
1	4.71 ± 0.075	18.51 ± 0.163	17.29 ± 0.132	2.57 ± 0.036	6.76 ± 0.079	0.12 ± 0.025	6.78 ± 0.098	11.41 ± 0.125	1.87 ± 0.028	3.14 ± 0.053
2	3.72 ± 0.036	14.53 ± 0.172	15.54 ± 0.195	1.92 ± 0.063	5.63 ± 0.138	0.05 ± 0.009	6.70 ± 0.084	9.50 ± 0.112	2.87 ± 0.096	2.35 ± 0.107
3	5.25 ± 0.052	21.52 ± 0.164	17.94 ± 0.118	3.30 ± 0.082	6.55 ± 0.125	0.15 ± 0.019	4.02 ± 0.043	7.64 ± 0.124	0.77 ± 0.086	1.84 ± 0.097
4	5.55 ± 0.068	21.79 ± 0.155	17.41 ± 0.184	3.11 ± 0.065	5.78 ± 0.153	0.14 ± 0.008	4.57 ± 0.052	7.61 ± 0.098	1.34 ± 0.052	2.53 ± 0.025
5	3.91 ± 0.034	17.17 ± 0.125	16.28 ± 0.185	2.28 ± 0.083	5.64 ± 0.126	0.11 ± 0.007	5.69 ± 0.086	9.76 ± 0.153	1.47 ± 0.036	2.65 ± 0.026
6	1.12 ± 0.025	10.52 ± 0.126	14.09 ± 0.208	1.69 ± 0.057	5.17 ± 0.134	0.03 ± 0.002	6.85 ± 0.068	12.53 ± 0.209	2.39 ± 0.058	3.43 ± 0.053
7	5.46 ± 0.086	23.35 ± 0.126	18.85 ± 0.152	3.43 ± 0.084	6.10 ± 0.275	0.12 ± 0.011	4.12 ± 0.045	6.86 ± 0.095	0.83 ± 0.063	1.93 ± 0.087
8	4.83 ± 0.047	19.78 ± 0.165	15.98 ± 0.126	3.16 ± 0.096	5.15 ± 0.159	0.13 ± 0.023	5.40 ± 0.058	9.36 ± 0.089	1.27 ± 0.086	2.48 ± 0.082
9	5.63 ± 0.058	20.62 ± 0.396	18.05 ± 0.151	3.15 ± 0.113	6.72 ± 0.223	0.13 ± 0.008	6.25 ± 0.066	10.85 ± 0.103	1.58 ± 0.032	2.78 ± 0.096
10	4.85 ± 0.063	18.99 ± 0.154	17.51 ± 0.185	3.06 ± 0.079	6.53 ± 0.128	0.12 ± 0.007	6.04 ± 0.059	10.54 ± 0.108	1.25 ± 0.025	2.71 ± 0.057
11	4.15 ± 0.035	21.09 ± 0.157	17.87 ± 0.182	3.30 ± 0.102	5.56 ± 0.135	0.16 ± 0.005	5.88 ± 0.048	10.23 ± 0.099	1.53 ± 0.041	2.85 ± 0.035
12	4.93 ± 0.028	20.19 ± 0.065	17.37 ± 0.093	2.52 ± 0.094	5.76 ± 0.124	0.13 ± 0.009	5.00 ± 0.038	8.64 ± 0.075	1.22 ± 0.086	2.32 ± 0.026
13	5.06 ± 0.096	21.69 ± 0.182	17.87 ± 0.133	3.32 ± 0.087	5.84 ± 0.119	0.12 ± 0.006	4.73 ± 0.086	7.98 ± 0.082	0.94 ± 0.023	2.05 ± 0.345
14	2.22 ± 0.025	11.19 ± 0.168	14.28 ± 0.182	1.85 ± 0.058	5.25 ± 0.132	0.06 ± 0.002	6.93 ± 0.073	12.63 ± 0.096	2.45 ± 0.055	3.47 ± 0.182
15	5.50 ± 0.125	23.04 ± 0.154	18.40 ± 0.121	3.37 ± 0.106	6.48 ± 0.157	0.12 ± 0.011	4.17 ± 0.021	6.82 ± 0.135	0.81 ± 0.062	1.92 ± 0.069
16	1.90 ± 0.019	10.94 ± 0.096	14.60 ± 0.086	1.13 ± 0.048	5.15 ± 0.122	0.04 ± 0.002	7.27 ± 0.102	13.08 ± 0.152	2.39 ± 0.096	3.45 ± 0.122
17	5.15 ± 0.097	21.82 ± 0.193	18.14 ± 0.063	3.31 ± 0.115	5.90 ± 0.112	0.11 ± 0.005	4.65 ± 0.083	7.81 ± 0.096	0.96 ± 0.025	2.06 ± 0.083

The analysis of variance (ANOVA) was used to evaluate the significance of the quadratic polynomial models [27]. For each term in the models, a large *F*-value and a small *P*-value would imply a more significant effect on the respective response variable [28]. The ANOVA results of some important terms in the models were summarized in Tables 4–6. To verify the adequacy of a model, the coefficient of determination (R^2), lack of fit, R^2_{adj}, AP, and CV tests were typically used. R^2 represents a percentage of the variables that can be explained by the model. Commonly, the higher R^2 not only represents the majority of the variables that can be explained by the model, but also represents that the experimental data are very consistent with the second-order polynomial equation. As shown in Tables 4–6, the R^2 values were 0.9690, 0.9840, and 0.9874, respectively. This indicated that there was only 1.60–3.10% of the total variation which was not explained by the models. In these models, the values of R^2 were high, which met our requirements. However, R^2 is not a decisive factor and R^2_{adj} is also important. R^2_{adj} is a modification of R^2 that adjusts for the number of explanatory terms in a model. Unlike R^2, the R^2_{adj} increases only if the new term improves the model more than would be expected by chance [29]. For the values of R^2 and R^2_{adj}, the greater the better, the closer the better. From Tables 3–5, the values of R^2_{adj} of the models were 0.9292, 0.9635, and 0.9713, respectively. The high values of R^2_{adj} indicated that the model was significant. The significance of the lack of fit test indicated that the points were not properly distributed around the model; as a result, the model could not be applied to predict the values of the independent variables. Therefore, the insignificance of the lack of fit test implied that the model was able to fit the data properly [13]. In our models, all the values of "p-value prob $> F$" of the lack of fit were greater than 10% and they were insignificant. This indicated that the model was able to fit the data properly. "AP" measures the signal to noise ratio. A ratio greater than 4 is desirable. In these models, the values of AP were 16.965, 24.331, and 27.165, respectively. This indicated an adequate signal. As a general rule, the CV should not be greater than 10% [30]. The values of the CV of the models were 1.16, 1.43, and 5.67, respectively. In addition to the above important parameters which could verify the suitability of the model, the figure of the predicted value versus the measured one can also be used to prove it. As shown in Figure 3, the predicted value of the model and the actual value of the experiment were fitted almost in a straight line. It proved that the second-order polynomial regression model was in good agreement with the experimental results and indicated that the models applied in this study were able to identify the operating conditions for selective extraction of saponins from SPN.

Table 4. The ANOVA results for the response surface quadratic models for the yield of saponins extracted from the SPN.

Source (Yield of Saponins)	Sum of Squares	DF	Mean Square	F Value	p-Value
Model	27.03	9	3.00	24.33	0.0002
X_1	2.17	1	2.17	17.61	0.0041
X_2	14.88	1	14.88	120.54	<0.0001
X_3	5.22	1	5.22	42.26	0.0003
X_{12}	0.80	1	6.49	6.49	0.0382
X_{13}	0.096	1	0.78	0.78	0.4060
X_{23}	0.032	1	0.26	0.26	0.6242
X_{12}	1.05	1	8.47	8.47	0.0227
X_{22}	1.24	1	10.07	10.07	0.0156
X_{32}	1.14	1	9.25	9.25	0.0188
Residual	0.86	7	0.12		
Lack of fit	0.19	3	0.062	0.36	0.7837
Pure error	0.68	4	0.17		
Cor total	27.89	16			
R^2			0.9690		
R^2_{adj}			0.9292		
CV			1.16		
AP			16.965		

Table 5. The ANOVA results for the response surface quadratic models for the content of saponins extracted from the SPN.

Source (Content of Saponins)	Sum of Squares	DF	Mean Square	F-Value	p-Value
Model	406.03	9	45.11	47.91	<0.0001
X_1	21.88	1	21.88	23.24	0.0019
X_2	0.61	1	0.61	0.65	0.4472
X_3	296.22	1	296.22	314.61	<0.0001
X_{12}	14.67	1	14.67	15.58	0.0056
X_{13}	0.055	1	0.055	0.059	0.8156
X_{23}	8.97	1	8.97	9.53	0.0177
X_{12}	22.69	1	22.69	24.10	0.0017
X_{22}	1.64	1	1.64	1.74	0.2288
X_{32}	37.24	1	37.24	39.55	0.0004
Residual	6.59	7	0.94		
Lack of fit	3.99	3	1.33	2.05	0.2493
Pure error	2.60	4	0.65		
Cor total	412.62	16			
R^2			0.9840		
R^2_{adj}			0.9635		
CV			1.43		
AP			24.331		

Table 6. The ANOVA results for the response surface quadratic models for the antioxidant activity of saponins extracted from the SPN.

Source (Antioxidant Activity)	Sum of Squares	DF	Mean Square	F-Value	p-Value
Model	4.448×10^{-3}	9	4.942×10^{-4}	61.06	<0.0001
X_1	2.611×10^{-4}	1	2.611×10^{-4}	32.25	0.0008
X_2	2.556×10^{-5}	1	2.556×10^{-5}	3.16	0.1188
X_3	3.089×10^{-3}	1	3.089×10^{-3}	381.62	<0.0001
X_{12}	5.112×10^{-5}	1	5.112×10^{-5}	6.32	0.0402
X_{13}	3.600×10^{-5}	1	3.600×10^{-5}	4.45	0.0729
X_{23}	8.281×10^{-5}	1	8.281×10^{-5}	10.23	0.0151
X_{12}	2.154×10^{-4}	1	2.154×10^{-4}	26.61	0.0013
X_{22}	7.645×10^{-6}	1	7.645×10^{-6}	0.94	0.3635
X_{32}	6.451×10^{-4}	1	6.451×10^{-4}	79.69	<0.0001
Residual	5.666×10^{-5}	7	8.094×10^{-6}		
Lack of fit	4.237×10^{-5}	3	1.412×10^{-5}	3.95	0.1087
Pure error	1.429×10^{-5}	4	3.572×10^{-6}		
Cor total	4.505×10^{-3}	16			
R^2			0.9874		
R^2_{adj}			0.9713		
CV			5.67		
AP			27.165		

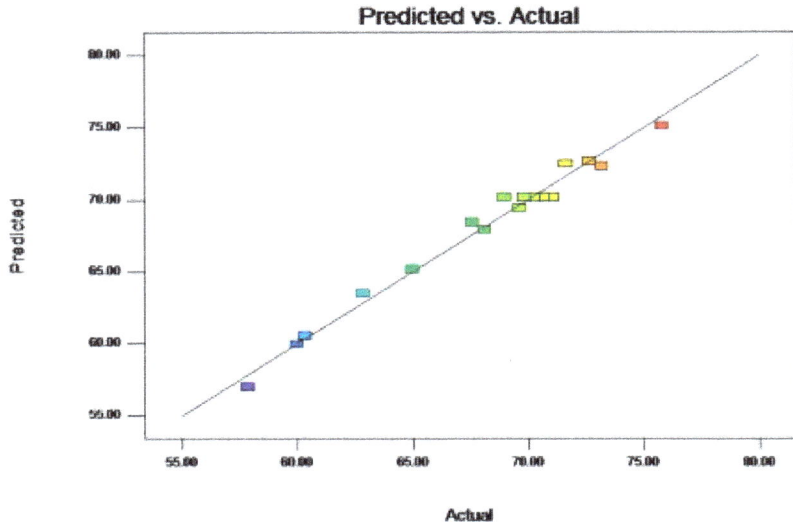

(a)

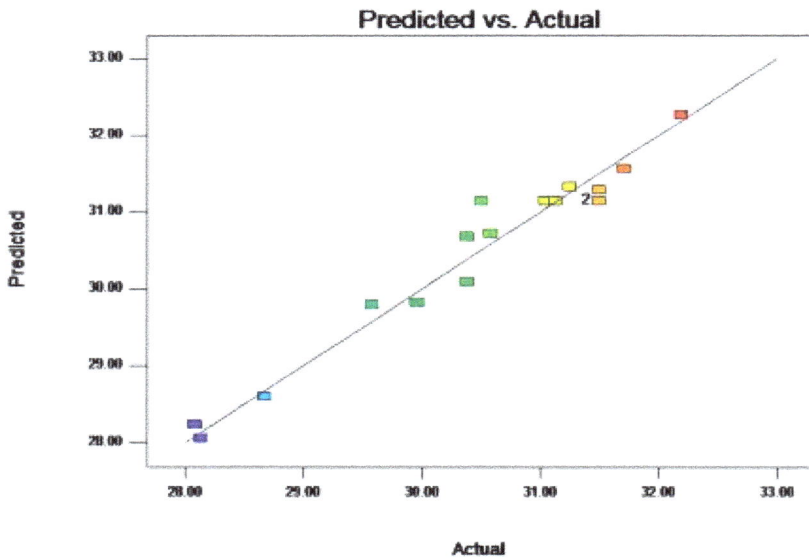

(b)

Figure 3. *Cont.*

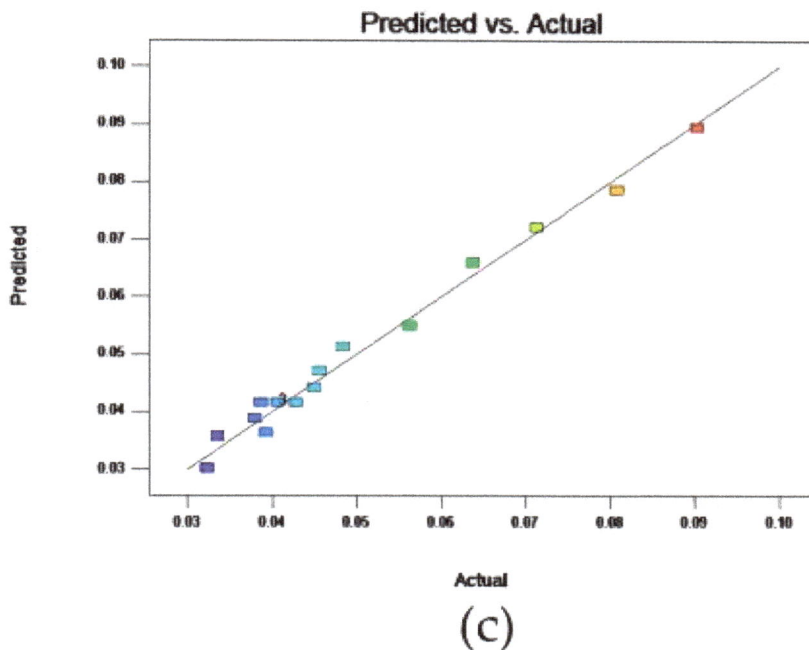

Figure 3. The comparison between the predicted and measured values of the content (**a**), yield (**b**), and antioxidant activity (**c**) of the saponins from the SPN.

2.4. Analysis of Influence of Variables on the Yield of Saponins

In Table 4, the liner and quadratic effects of the extraction time, ethanol concentration, and times of extraction were significant ($p < 0.05$). The most significant effect on the yield of saponins was shown to be the linear effect of ethanol concentration ($p < 0.05$). Among the different interaction effects, there was only one interaction of extraction time with ethanol concentration which was significant ($p < 0.05$). Figure 4a presented the interaction between the extraction time and ethanol concentration. The yield of saponins was initially increased along with the duration of the extraction, following the decrease which might be due to the providence of the time requirement of the exposure of the ingredients of SPN to the release medium where the liquid penetrated into the cell wall of dried raw materials, dissolved the saponins, and subsequently diffused out from the raw materials [31]. However, with the extension of the extraction time, the components in the solvent were transformed and even degraded. At the same time, the yield of the saponins was significantly decreased by increasing the ethanol concentration, which was probably due to the decreased solubility of the weakly-polar and non-polar components, such as carbohydrates contained in SPN, induced by the increase of the solvent polarity. As shown in Figure 4a,b, the yield of the saponins reached the maximum when the extraction time and ethanol concentration were approximately 1.56 h and 60%, respectively.

(a)

(b)

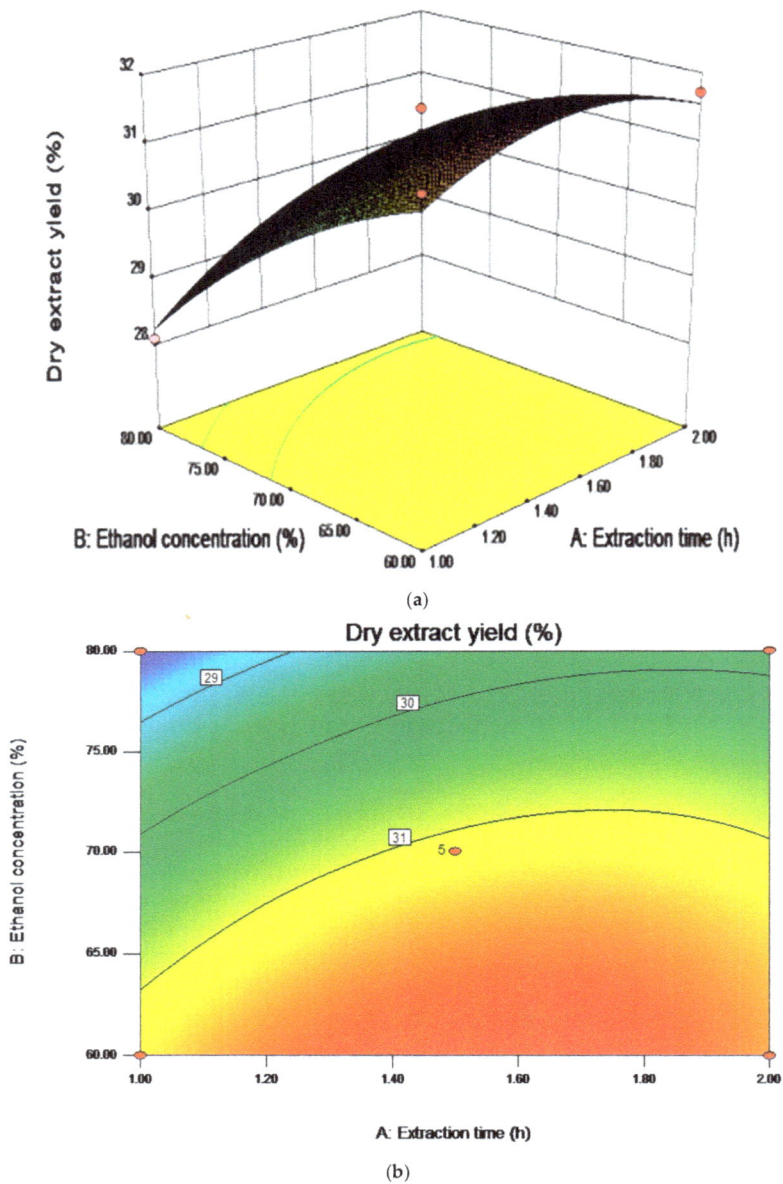

Figure 4. The response surface (**a**) and contour (**b**) plots showing the significant ($p < 0.05$) interaction effects of ethanol concentration with extraction time on the yield of saponins.

2.5. Analysis of Influence of Variables on the Content of Saponins

From Table 5, the linear and quadratic effects of the extraction time and extraction times were found to be significant on the content of saponins extracted from SPN ($p < 0.01$). However, the linear and quadratic effects of the ethanol concentration were not significant ($p > 0.05$). Among the different interaction effects, only the interactions of extraction time with ethanol concentration, and ethanol concentration with extraction times were significant ($p < 0.05$). As shown in Table 5, the most significant ($p < 0.05$) effect on the content of saponins was shown to be the linear one of the extraction times followed by the quadratic effect of extraction time.

Figure 5a presents the interaction between the extraction time and ethanol concentration. Initially, the content of saponins was increased along with the duration of the extraction time, then following the decrease which might be due to the time requirement of the exposure of the saponins from the SPN to the release medium. However, with the extension of the extraction time, the saponins in the solvent were transformed and even degraded. As shown in Figure 5a,c, the content of saponins reached the maximum when the extraction time and ethanol concentration were approximately 1.50 h and 70%, respectively.

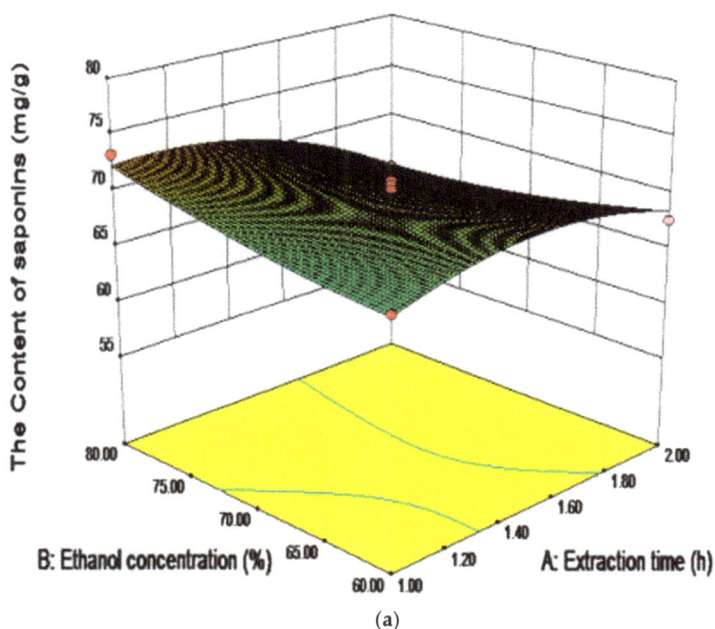

(a)

Figure 5. *Cont.*

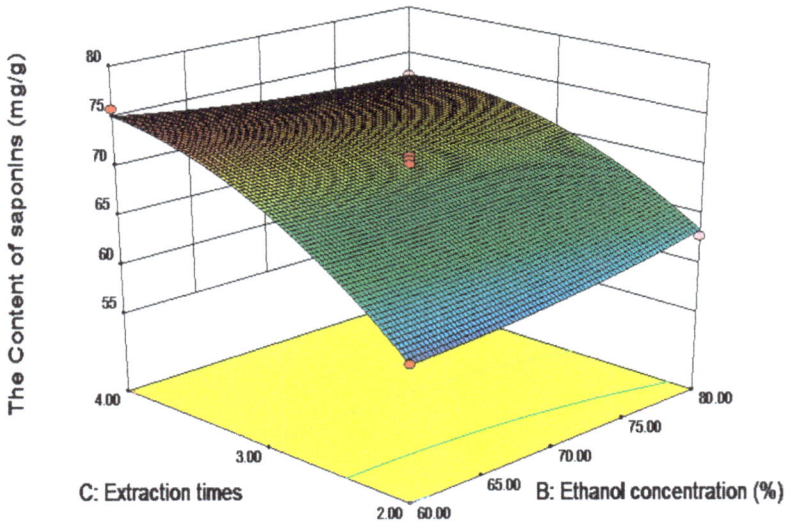

(b)

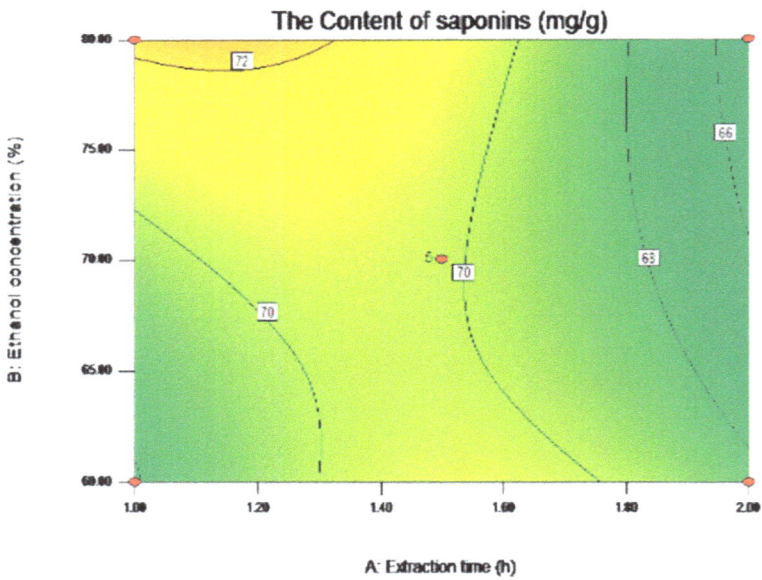

(c)

Figure 5. *Cont.*

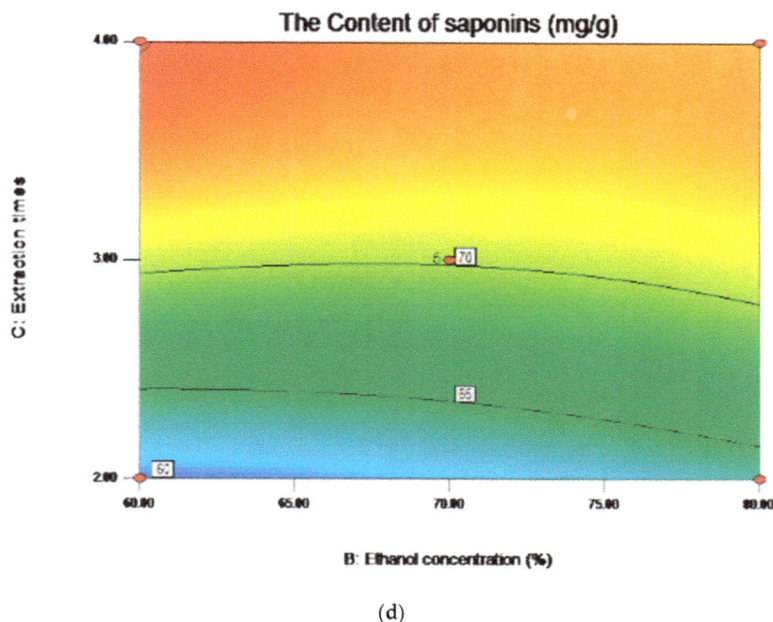

(d)

Figure 5. The response surface (**a,b**) and contour (**c,d**) plots showing the significant ($p < 0.05$) interaction effects of the ethanol concentration with the extraction time and extraction times on the content of saponins.

Figure 5b shows the interaction between the times of extraction and ethanol concentration. The content of saponins was significantly increased with the increase of the times of extraction, which might be due to the full degradation of saponins from SPN. As shown in Figure 5b,d, the content of saponins reached the maximum when the times of extraction and ethanol concentration were approximately 3% and 70%, respectively.

2.6. Analysis of Influence of Variables on the Antioxidant Activity

In our study, the effect of the ethanol-reflux extraction variables on the antioxidant activity of saponins from SPN was determined based on the hydroxyl radicals scavenging activity because, in our previous study, we found that SPN exhibited a stronger activity of scavenging hydroxyl radicals than scavenging DPPH free radicals [11]. Hydroxyl radicals are strong free radicals. The excessive hydroxyl radicals have a close relationship with various diseases and health problems, such as aging arthritis, cancer, inflammation, and heart diseases [32]. In the research, the antioxidant activity of SPN saponins was evaluated by determining the EC_{50} value of the hydroxyl radical scavenging capacity. The lower value of EC_{50} indicated a stronger clearance ability.

According to Table 6, the linear and quadratic effects of the extraction time and times of extraction on the hydroxyl radicals scavenging activity were significant ($p < 0.05$). However, the linear and quadratic effects of the ethanol concentration on the antioxidation were not significant ($p > 0.05$). Among different interaction effects, the interactions of the extraction time with ethanol concentration, and the ethanol concentration with the times of extraction were significant ($p < 0.05$). As shown in Figure 6a,b, the EC_{50} value decreased along with the increase of the extraction time and times of extraction. The results indicated that the EC_{50} value was minimized when the extraction time and times of extraction were approximately 1.5 h and 3 times, respectively.

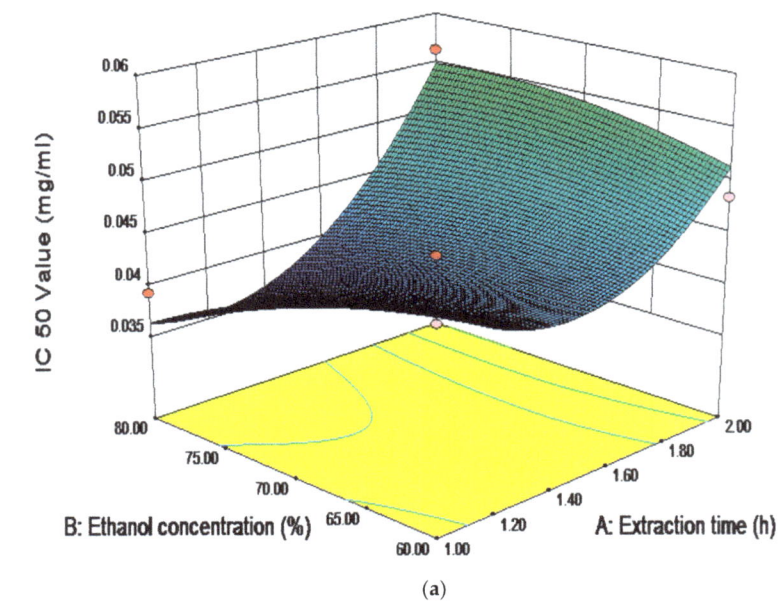

(a)

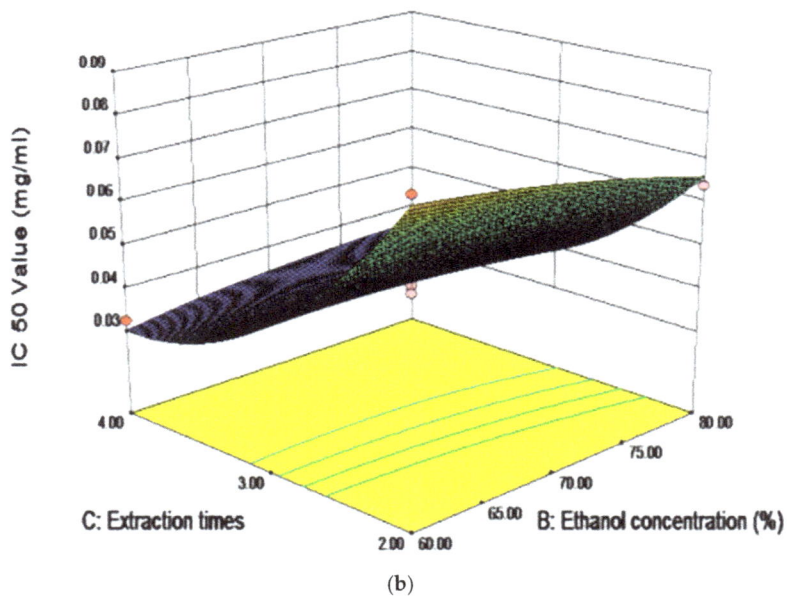

(b)

Figure 6. *Cont.*

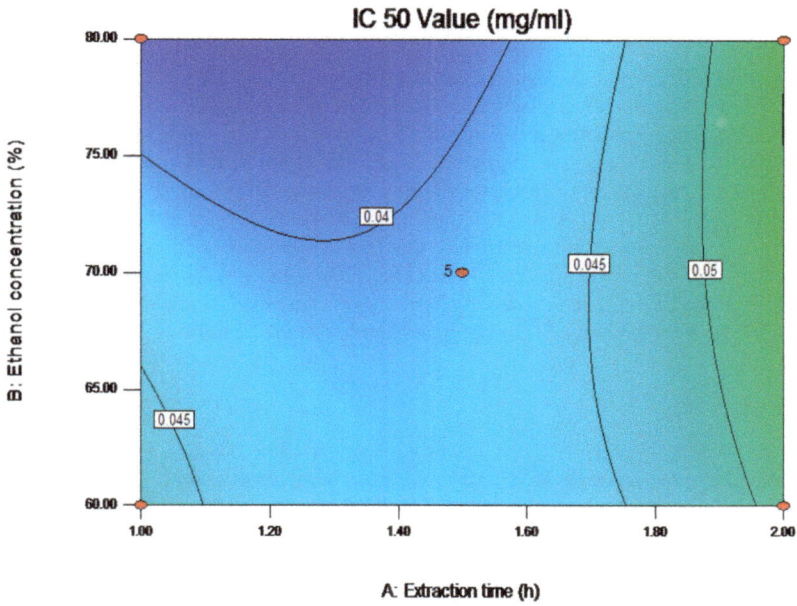

(c)

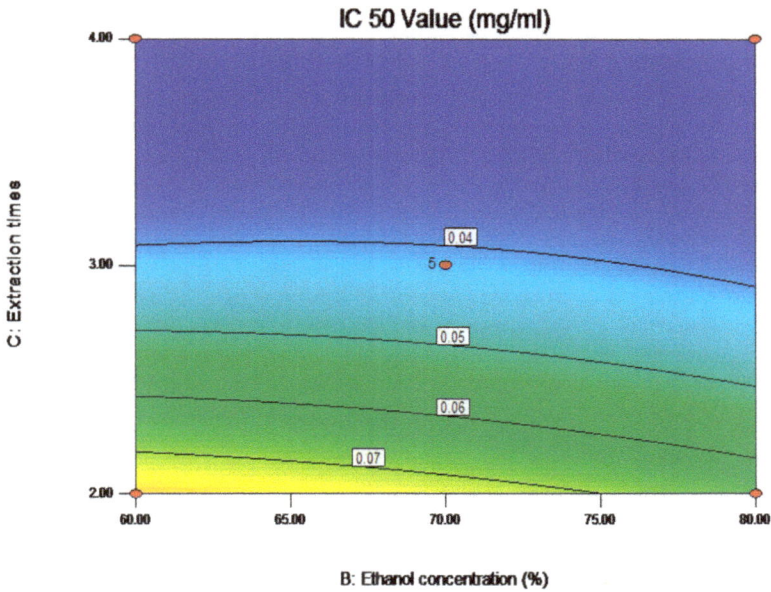

(d)

Figure 6. Response surface (**a**,**b**) and contour (**c**,**d**) plots showing the significant ($p < 0.05$) interaction effects of ethanol concentration with extraction time and extraction times on the value of EC $_{50}$.

2.7. Optimization and Validation Procedures

To maximize the yield, content, and antioxidant activity of SPN saponins, the quadratic models within the studied experimental range of various process variables were established. The predicted optimal conditions were shown as follows: the extraction time of 1.51 h, ethanol concentration of 60%, and extraction times of 3. Under the recommended optimum extraction condition, the predicted values of the yield, the content of saponins, and the EC_{50} of antioxidant activity were 31.95%, 70.49 mg/g, and 0.0421 mg/mL, respectively. The corresponding experimental values of the above indices were determined as 31.94 ± 0.02%, 70.46 ± 0.0971 mg/g, and 0.0419 ± 0.0005 mg/mL, respectively, which were very close to the values predicted by the constructed models. In addition, compared with our previous study [12] on the antioxidant activity of SPN (EC_{50} value of 1.197 ± 0.11 mg/mL), the EC_{50} value of 0.0419 ± 0.005 mg/mL of SPN saponins was significantly lower, indicating that the antioxidant activity of SPN saponins extracted by the optimized method was much stronger than the unprocessed SPN powder.

2.8. Blood Routine Test

After the administration for 12 days, the quantities of white blood cell (WBC), red blood cell (RBC), platelet (PLT) and hemoglobin (Hb) from the peripheral blood of mice were shown in Figure 7. Compared with the control group, the levels of WBC, RBC, PLT, and Hb in the model group were significantly decreased ($p < 0.01$), indicating the anemia model was successfully established. Compared with the model group, WBC, RBC, PLT, and Hb levels in the *Fufang E'jiao Jiang* (FEJ) and three doses of SPN and SPN saponins groups were increased at different degrees. Besides, the levels of the above four parameters were increased with the added dose of SPN and SPN saponins in a dose-dependent manner. For the same type of dose, there was no significant difference in the hematopoiesis effect between SPN and SPN saponins, although the dose of SPN saponins was 15 times lower than the former one. It suggested that taking a much smaller amount of SPN saponins could achieve the similar hematopoiesis effect as SPN.

Traditionally, SPN is used as a tonic to enrich blood and tonify the body, which can improve anemia by increasing the production of various blood cells in anemic conditions [6,7]. Besides, the body tonifying function of herbal medicines is partly attributed to their antioxidant effect by modern pharmacological research [33], which can be evaluated by investigating the hydroxyl radical scavenging activity [32]. Therefore, the changed levels of blood parameters and EC_{50} value of the hydroxyl radical scavenging activity were determined to evaluate the tonifying efficacies of SPN. The lower the value of EC_{50}, the stronger the antioxidant ability. According to the results, the crude saponins of SPN obtained by the optimized extraction method did not only exhibit strong antioxidant activity, but also show good hematopoiesis effect on increasing the levels of blood parameters in a dose-dependent way. The consistency in improving the functions of antioxidation in vitro and hematopoiesis in vivo suggested that the crude extract of saponins with strong activities related to the clinic efficacies of SPN could be obtained by the optimized extraction method.

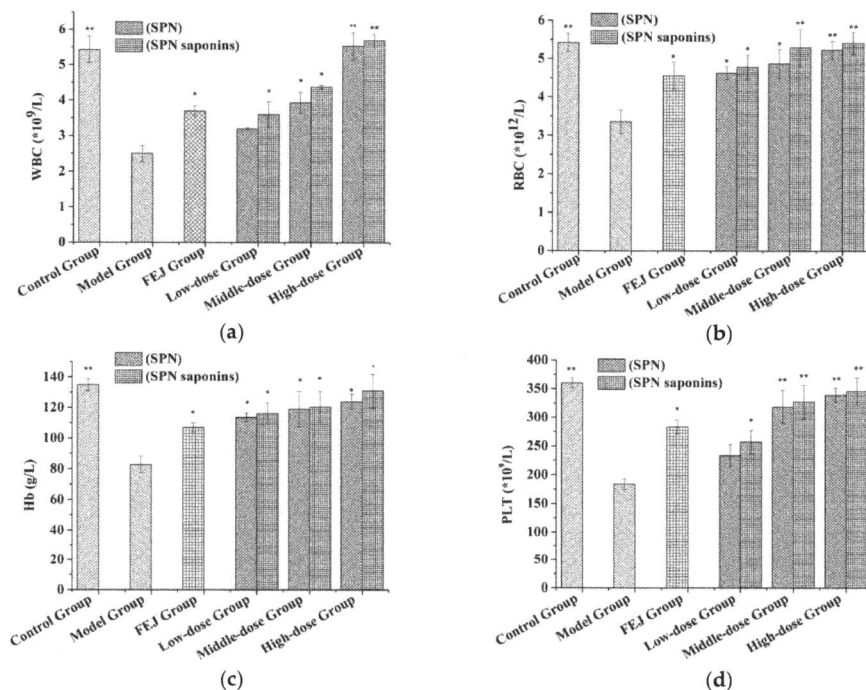

Figure 7. The blood parameters after treating SPN and SPN saponins, where (**a**) is the content of WBC, (**b**) is the content of RBC, (**c**) is the content of Hb and (**d**) is the content of platelet. Each value represents means $\pm$ SD ($n = 10$); *$p < 0.05$ and ** $p < 0.01$ compared to the control group.

3. Materials and Methods

3.1. Plant Material and Chemicals

SPN samples were prepared by steaming the crushed raw PN in an autoclave (Shanghai, China) at 120 °C for 2 h. The steamed powder was then dried in a heating-air drying oven at about 45 °C until constant weight and then sieved through a 40 mesh sieve. Ethanol was of analytical grade and purchased from Tianjin Feng Chuan Chemical Reagent Technologies Co, Ltd. (Tianjin, China). Acetonitrile of chromatographic grade and ferrous chloride and hydrogen peroxide of analytical grade were purchased from Merck Chemical Co. (Darmstadt, Germany). Notoginsenoside R_1, ginsenosides Rg_1, Re, Rb_1, Rd, Rh_1, Rk_3, Rh_4, 20(*S*)-Rg_3, and 20(*R*)-Rg_3 (Sichuan Weikeqi Biological Technology Co., Ltd. Chengdu, China) with a purity $\geq$ 98% were used as the standard compounds.

3.2. Animals

Animal experimental procedures in the study strictly conformed to the Guide for the Care and Use of Laboratory Animals and related ethics regulations of Kunming University of Science and Technology. The protocol was approved by the Experimental Animal Welfare and Ethics Committee, Kunming University of Science and Technology (project number: 81660661; code: KKGD201626039; date of approval: 9 January 2017). The experimental method refers to our previous study [11], in which "Kunming mice, male and female, weighing 18–22 g, were purchased from Tianqin Biotechnology Co., Ltd., Changsha, China [SCXK (Xiang) 2014-0011]. Before the experiments, the mice were given a one-week acclimation period in a laboratory at room temperature (20–25 °C) and constant humidity (40–70%), and fed with standard rodent chow and tap water freely."

3.3. Ethanol-Reflux Extraction Process

In order to obtain saponins from SPN, ethanol-reflux extraction was used. Ethanol-reflux extraction was performed by using ethanol as the extraction solvent at the given extraction time, ethanol concentration, water to solid ratio, and times of extraction. That is to say, a 5.0 g sample powder was extracted by 50 mL of 70% ethanol at 85 °C in a water bath for 1.5 h. After extraction three times, the extraction of ethanol-reflux was combined. Subsequently, the extract solution was centrifuged, filtered, concentrated, and dried to obtain the crude saponins of SPN.

3.4. Determination of the Yield of Saponins.

The saponin extraction yield (%) was obtained by dividing the dried crude saponins weight (g) to powder weight (g) and it was calculated by the following equation:

$$\text{Yield (\%)} = \text{dried crude saponins weight (g)}/\text{powder weight (g)} \times 100\% \qquad (4)$$

3.5. Determination of the Content of Saponins

HPLC analyses were performed according to the previous method [11]. The crude SPN saponins were dissolved with 20 mL of ultra-pure water. The supernatant after the filtration of the solution was used as the sample solution. "A mixed standards solution containing (in mg/mL) 0.40 notoginsenoside R_1, 0.55 ginsenosides Rg_1, 0.50 Re, 0.60 Rb_1, 0.50 Rd, 0.60 Rh_1, 1.00 Rk_3, 1.00 Rh_4, 0.45 20(S)-Rg_3 and 0.55 20(R)-Rg_3, was prepared by adding each standard into a volumetric flask and dissolving with methanol. A series of standards solutions of seven concentrations were prepared by diluting the mixed standard solution with methanol for the determination of the standard curves. HPLC analyses were done on a 1260 series system (Agilent Technologies, Santa Clara, CA, USA) consisting of a G1311B Pump, a G4212B DAD detector, and a G1329B autosampler. A Vision HT C_{18} column (250 mm × 4.6 mm, 5 μm) (welch Co., Ltd., Shanghai, China) was adopted for the analyses. The mobile phase was comprised of A (ultra-pure water) and B (methyl cyanide). The gradient mode was as follows: 0–20 min, 80% A; 20–45 min, 54% A; 45–55 min, 45% A; 55–60 min, 45% A; 60–65 min, 100% B; 65–70 min, 80% A; 70–90 min, 80% A. The flow rate was set at 1.0 mL/min. The detection wavelength was set at 203 nm. The column temperature was set at 30 °C and the sample volume was set at 10 μL."

3.6. Measurement of Antioxidant Activity

The antioxidant activity was determined based on the hydroxyl radicals scavenging activity. The scavenging activity for hydroxyl radicals was measured with the Fenton reaction [34]. The reaction mixture contained 60 μL of 1.0 mM $FeCl_2$, 90 μL of 1 mM 1,10-phenanthroline, 2.4 mL of 0.2 M phosphate buffer (pH 7.8), 150 μL of 0.17 M H_2O_2, and 1.5 mL of extracts at various concentrations. The reaction was started by adding H_2O_2. After incubation at room temperature for 5 min, the absorbance of the mixture at 560 nm was measured with a Beckman spectrophotometer(Shimadzu corporation, kyoyo, Japan). The hydroxyl radicals scavenging activity was calculated according to the following equation:

$$\text{Scavenging rate} = [1 - (A_1 - A_2)/A_0] \times 100\% \qquad (5)$$

where A_0 was the absorbance of the control (blank, without extract), A_1 was the absorbance in the presence of the extract, and A_2 was the absorbance without 1,10-phenanthroline [32].

3.7. BBD-RSM Experimental Design

BBD-RSM was used to optimize the effect of the extraction parameters including the extraction time (1–2 h), ethanol concentration (60–80%), and times of extraction (2–4) on the yield, content, and antioxidant activity of the saponins from the SPN. Based on the preliminary range of the process variables in a single factor test, the BBD-RSM with three mentioned independent variables at three levels was carried out (Tables 1 and 2). In this design, the randomized run order was created by considering six factorial points, six axial points, and five center points. For the model analysis, the results of the experimental design were fitted by a polynomial equation to correlate the response to the independent variables. The general equation to predict the optimal point was explained as follows:

$$Y = b_0 + b_1X_1 + b_2X_2 + b_3X_3 + b_{11}X_{11} + b_{22}X_{22} + b_{33}X_{33} + b_{12}X_{12} + b_{13}X_{13} + b_{23}X_{23} \tag{6}$$

where Y is the predicted response; b_0, b_1, b_2, b_3, b_{11}, b_{22}, b_{33}, b_{12}, b_{13}, and b_{23} represent the regression coefficients; and X_1, X_2, X_3, X_4 are the coded independent factors. The regression coefficient (R^2), adjusted-R^2 (R^2_{adj}), the prediction error sum of squares (PRESS), and adequate precision (AP) were used to determine the goodness-of-fit of the constructed polynomial models.

3.8. Blood Routine Test

One hundred and eight Kunming mice, half male and half female, were randomly divided into nine groups, namely the control group, model group, FEJ group, high, moderate, low-dose SPN groups, and the high, moderate, and low-dose SPN saponins groups; 12 mice in each group. The APH and CTX-induced anemia model was applied to evaluate the "blood tonifying" function of SPN saponins combined with previous methods [35]. The anemia model was established by intraperitoneal injection of CTX of 0.07 g/kg for the first three days and a hypodermic injection of APH of 0.02 g/kg on the fourth day. Mice in the control group were administered with 0.9% normal saline, whereas the other groups were administered with FEJ (8 g/kg), SPN (0.45 g/kg, 0.90 g/kg, and 1.8 g/kg, respectively), and SPN saponins (0.03 g/kg, 0.06 g/kg, and 0.12 g/kg, respectively), respectively, by gavage for 12 days. After 30 minutes of the last administration, the blood was collected from the tail vein of mice and then removed from a centrifuge tube containing heparin sodium. Then the collected blood was used for the routing analysis by an automatic blood analyzer (Healife, Taian, China), including levels of WBC, RBC, Hb, and PLT after 30 min of the last administration.

3.9. Statistical Analysis

The software Design-Expert (version 8.6.0, Stat-Ease Inc., Minneapolis, MN, USA) was used to calculate the relationship between the independent variables and responses. And all the data were expressed as means ± SD. The SPSS 21.0 software (Statistical Program for Social Sciences, SPSS Inc., Chicago, IL, USA) was applied to carry out the two-tailed unpaired t-test. A value of $p < 0.05$ was considered to be a significant difference. A value of $p < 0.01$ was considered to be a highly significant difference. The EC_{50} value was fitted by probit regression with the Origin 7.5 software for Windows (OriginLab Corporation, Northampton, MA, USA).

4. Conclusions

In the research, the effects of different ethanol-reflux extraction variables (extraction time, ethanol concentration, the water to solid ratio, and extraction times) on the yield, content, and antioxidant activity of saponins from SPN were studied and the extraction condition to obtain saponins from SPNs was successfully optimized by using BBD-RSM. The second-order polynomial models for all the response variables were found to be statistically significant. The optimal extraction conditions were as follows: an extraction time of 1.51 h, an ethanol concentration of 60%, and extraction 3 times. Under this condition, the yield, content, and EC_{50} of antioxidant activity of SPN saponins were 31.94 ± 0.02%, 70.46 ± 0.0971 mg/g, and 0.0419 ± 0.0005 mg/mL, respectively, which were corresponding well with the predicted values (31.95%, 70.49 mg/g, and 0.0421 mg/mL) by the models. Compared with the unprocessed SPN powder, the extracted SPN saponins also showed significantly stronger antioxidant activity and similar hematopoiesis effects at a 15-times lower concentration level, indicating that the optimized extracted method could be beneficial for producing SPN saponins with good therapeutic functions.

Author Contributions: Y.H. and X.C. participated in the design of the study and performed the experiments; Z.Z., L.C., and Y.Z. collected samples and analyzed the data; C.W., X.Y., and Y.Q. participated in the experimental design; Y.X. supervised the project and wrote this paper.

Acknowledgments: The study was supported by National Natural Science Foundation of China (81660661), Yunnan Applied Basic Research Project (2016FD040), and Basic Research Program of the Ministry of S&T of China (2015FY111500-070).

Conflicts of Interest: The authors have declared no conflicts of interest.

Abbreviations

ANOVA	analysis of variance
AP	adequate precision
APH	acetylphenylhydrazine
BBD-RSM	Box-Behnken design response surface methodology
CTX	cyclophosphamide
Hb	hemoglobin
PLT	platelet
PN	*Panax notoginseng*
PRESS	prediction error sum of squares
R^2	regression coefficient
RBC	red blood cell
SPN	steamed *Panax notoginseng*
WBC	white blood cess

References

1. Wang, C.Z.; Eryn, M.E.; Sheila, W.; Wu, J.A.; Yuan, C.S. Phytochemical and analytical studies of *Panax notoginseng* (Burk.) F.H. Chen. *J. Nat. Med.* **2006**, *60*, 97–106. [CrossRef]
2. Ng, T.B. Pharmacological activity of sanchi ginseng (*Panax notoginseng*). *J. Pharm. Pharmacol.* **2010**, *58*, 1007–1019. [CrossRef] [PubMed]
3. Zhao, H.P.; Han, Z.P.; Li, G.W.; Zhang, S.J.; Luo, Y.M. Therapeutic potential and cellular mechanisms of *Panax notoginseng* on prevention of aging and cell senescence-associated diseases. *A&D* **2017**, *8*, 721–739. [CrossRef]
4. Li, X.; Wang, G.; Sun, J.; Hao, H.; Xiong, Y.; Yan, B.; Zheng, Y.; Sheng, L. Pharmacokinetic and absolute bioavailability study of total *Panax notoginsenoside*, a typical multiple constituent traditional Chinese medicine (TCM) in rats. *Biol. Pharm. Bull.* **2007**, *30*, 847–851. [CrossRef] [PubMed]
5. Liao, J.Q.; Wei, B.J.; Chen, H.W.; Liu, Y.M.; Wang, J. Bioinformatics investigation of therapeutic mechanisms of Xuesaitong capsule treating ischemic cerebrovascular rat model with comparative transcriptome analysis. *Am. J. Transl. Res.* **2016**, *8*, 2438–2449. [PubMed]
6. Gu, C.; Lv, J.; Zhang, X.; Qiao, Y.; Yan, H.; Li, Y.; Wang, D.; Zhu, H.; Luo, H.; Yang, C.; Xu, M.; Zhang, Y. Triterpenoids with promoting effects on the differentiation cells from the steamed roots of *Panax notoginseng*. *J. Nat. Prod.* **2015**, *78*, 1829–1840. [CrossRef] [PubMed]
7. Ge, F.; Huang, Z.J.; Yu, H.; Wang, Y.; Liu, D.Q. Transformation of *Panax notoginseng* saponins by steaming and *Trichoderma longibrachiatum*. *Biotechnol. Biotechnol. Equip.* **2015**, *30*, 165–172. [CrossRef]
8. Sun, S.; Wang, C.Z.; Tong, R.B.; Li, X.L.; Anna, F.B.; Wang, Q.; He, T.C.; Du, W.; Yuan, C.S. Effects of steaming the root of *Panax notoginseng* on chemical composition and anticancer activities. *Food Chem.* **2010**, *118*, 307–314. [CrossRef]
9. Toh, D.F.; Patel, D.N.; Chan, E.C.; Teo, A.; Neo, S.Y.; Koh, H.L. Anti-proliferative effects of raw and steamed extracts of *Panax notoginseng* and its ginsenoside constituents on human liver cancer cells. *Chin. Med.* **2011**, *6*. [CrossRef] [PubMed]

10. Lau, A.J.; Toh, D.F.; Chua, T.K.; Pang, Y.K.; Woo, S.O.; Koh, H.L. Antiplatelet and anticoagulant effects of *Panax notoginseng*: comparison of raw and steamed *Panax notoginseng* with *Panax ginseng* and *Panax quinquefolium*. *Ethnopharmacology* **2009**, *125*, 380–386. [CrossRef] [PubMed]

11. Xiong, Y.; Chen, L.J.; Man, J.H.; Hu, Y.P.; Cui, X.M. Chemical and bioactive comparison of *Panax notoginseng* root and rhizome in raw and steamed forms. *J. Gins. Res.* **2017**, 1–9. [CrossRef]

12. Xiong, Y.; Chen, L.J.; Hu, Y.P.; Cui, X.M. Uncovering active constituents responsible for different activities of raw and steamed *Panax notoginseng* roots. *Front. Pharmacol.* **2017**, *8*, 1–11. [CrossRef] [PubMed]

13. Heydarian, M.; Jooyandeh, H.; Nasehi, B.; Noshad, M. Characterization of Hypericum perforatum, polysaccharides with antioxidant and antimicrobial activities: optimization based statistical modeling. *Int. J. Biol. Macromol.* **2017**, *104*, 287–293. [CrossRef] [PubMed]

14. Li, W.; Zhao, L.C.; Sun, Y.S.; Lei, F.J.; Wang, Z.; Gui, X.B.; Wang, H. Optimization of pressurized liquid extraction of three major acetophenones from cynanchum bungei using a box-behnken design. *Int. J. Mol. Sci.* **2012**, *13*, 14533–14544. [CrossRef] [PubMed]

15. Yolmeh, M.; Habibi Najafi, M.B.; Farhoosh, R. Optimisation of ultrasound-assisted extraction of natural pigment from *annatto* seeds by response surface methodology (RSM). *Food. Chem.* **2014**, *155*, 319–324. [CrossRef] [PubMed]

16. Noshad, M.; Shahidi, F.; Mortazavi, S.A. Multi-Objective Optimization of osmotic-ultrasonic pretreatments and hot-air drying of quince using response surface methodology. *Food Bioprocess Technol.* **2012**, *5*, 2098–2110. [CrossRef]

17. Wang, X.; Wu, Q.; Wu, Y.; Chen, G.; Yue, W.; Liang, Q. Response surface optimized ultrasonic-assisted extraction of flavonoids from Sparganii Rhizoma and evaluation of their in vitro antioxidant activities. *Molecules* **2012**, *17*, 6769–6783. [CrossRef] [PubMed]

18. Gharibzahedi, S.M.T.; Mousavi, S.M.; Hamedi, M.; Khodaiyan, F.; Razavi, S.H. Development of an optimal formulation for oxidative stability of walnut-beverage emulsions based on gum arabic and xanthan gum using response surface methodology. *Carbohydr. Polym.* **2012**, *87*, 1611–1619. [CrossRef]

19. Zhao, L.C.; Liang, J.; Li, W.; Cheng, K.M.; Xia, X.; Deng, X.; Yang, G.L. The use of response surface methodology to optimize the ultrasound-assisted extraction of five anthraquinones from *Rheum palmatum* L. *Molecules* **2011**, *16*, 5928–5937. [CrossRef] [PubMed]

20. Chinese Pharmacopoeia Commission. *Pharmacopoeia of the People's Republic of China*; Chinese Medical Science and Technology Press: Beijing, China, 2015; pp. 11–12.

21. Chen, Y.; Xie, M.Y.; Gong, X.F. Microwave-assisted extraction used for the isolation of total triterpenoids saponins from *Ganoderma atrum*. *J. Food. Eng.* **2007**, *81*, 162–170. [CrossRef]

22. Hu, T.; Guo, Y.Y.; Zhou, Q.F.; Zhong, X.K.; Zhu, L.; Piao, J.H.; Chen, J.; Jiang, J.G. Optimization of ultrasonic-assisted extraction of total saponins from *Eclipta prostrasta* L. using response surface methodology. *J. Food Sci.* **2012**, *77*, C975–C982. [CrossRef] [PubMed]

23. Taubert, J.; Krings, U.; Berger, R.G. A comparative study on the disintegration of filamentous fungi. *J. Microbiol. Meth.* **2000**, *42*, 225–232. [CrossRef]

24. Vongsangnak, W.; Jian, G.; Chauvatcharin, S.; Zhong, J.J. Towards efficient extraction of notoginseng saponins from cultured cells of *Panax notoginseng*. *Biochem. Eng. J.* **2004**, *18*, 115–120. [CrossRef]

25. Gong, X.C.; Zhang, Y.; Pan, J.Y.; Qu, H.B. Optimization of the ethanol recycling reflux extraction process for saponins using a design space approach. *PLoS ONE* **2014**, *9*. [CrossRef] [PubMed]

26. Pakrokh, G.P. The extraction process optimization of antioxidant polysaccharides from Marshmallow (*Althaea officinalis* L.) roots. *Int. J. Biol. Macromol.* **2015**, *75*, 51–57. [CrossRef] [PubMed]

27. Yuan, Y.; Gao, Y.; Mao, L.; Zhao, J. Optimization of condition for the preparation of β-carotene nanoemulsions using response surface methodology. *Food Chem.* **2008**, *107*, 1300–1306. [CrossRef]

28. Li, Q.H.; Fu, C.L. Application of response surface methodology for extraction optimization of *germinant pumpkin* seeds protein. *Food Chem.* **2005**, *92*, 701–706. [CrossRef]

29. Ghasemlou, M.; Khodaiyan, F.; Jahanbin, K.; Gharibzahedi, S.M.; Taheri, S. Structural investigation and response surface optimisation for improvement of kefiran production yield from a low-cost culture medium. *Food. Chem.* **2002**, *133*, 383–389. [CrossRef] [PubMed]

30. Nath, A.; Chattopadhyay, P.K. Optimization of oven toasting for improving crispness and other quality attributes of ready to eat potato-soy snack using response surface methodology. *J. Food Eng.* **2007**, *80*, 1282–1292. [CrossRef]

31. Gan, C.Y.; Manaf, H.A.; Latiff, A.A. Optimization of alcohol insoluble polysaccharides (AIPS) extraction from the Parkia speciosa pod using response surface methodology (RSM). *Carbohydr. Polym.* **2010**, *79*, 825–831. [CrossRef]

32. Zhao, G.R.; Xiang, Z.J.; Ye, T.X.; Juan, Y.J.; Guo, Z.X. Antioxidant activities of *Salvia miltiorrhiza* and *Panax notoginseng*. *Food. Chem.* **2006**, *99*, 767–774. [CrossRef]

33. Yim, T.K.; Ko, K.M. Antioxidant and immunomodulatory activities of chinese tonifying herbs. *Pharm. Biol.* **2002**, *40*, 329–335. [CrossRef]

34. Roberto, M.; Samantha, R.V.; Novella, B.; Roberto, L.S. Fenton-dependent damage to carbohydrates: Free radical scavenging activity of some simple sugars. *J. Agric. Food. Chem.* **2003**, *51*, 7418–7425. [CrossRef]

35. Shi, X.Q.; Shang, E.X.; Tang, Y.P.; Zhu, H.X.; Guo, J.M.; Huang, M.Y.; Li, W.X.; Duan, J.A. Interaction of nourishing and tonifying blood effects of the combination of Angelicae sinensis Radix and Astragali Radix studied by response surface method. *Acta. Pharm. Sin.* **2012**, *47*, 1375–1383.

Sample Availability: Samples of the reference standards mentioned in the paper are available from the authors.

molecules

MDPI

Article

Compounds Identification in Semen Cuscutae by Ultra-High-Performance Liquid Chromatography (UPLCs) Coupled to Electrospray Ionization Mass Spectrometry

Ying Zhang, Hui Xiong, Xinfang Xu, Xue Xue, Mengnan Liu, Shuya Xu, Huan Liu, Yan Gao, Hui Zhang and Xiangri Li *

School of Chinese Materia Medica, Beijing University of Chinese Medicine, Beijing 102488, China; zhang0312ying@163.com (Y.Z.); xionghui@bucm.edu.cn (H.X.); xuxinfang007@163.com (X.X.); sherry.xue@bucm.edu.cn (X.X.); 20160931810@bucm.edu.cn (M.L.); xushuya11@163.com (S.X.); 20150931937@bucm.edu.cn (H.L.); 20150931752@bucm.edu.cn (Y.G.); zh19930503@sina.com (H.Z.)
* Correspondence: lixiangri@sina.com

Academic Editor: Marcello Locatelli
Received: 19 April 2018; Accepted: 15 May 2018; Published: 17 May 2018

Abstract: Semen Cuscutae is commonly used in traditional Chinese medicine and contains a series of compounds such as flavonoids, chlorogenic acids and lignans. In this study, we identified different kinds of compositions by ultra-high-performance liquid chromatography (UPLC) coupled to electrospray ionization mass spectrometry (MS). A total of 45 compounds were observed, including 20 chlorogenic acids, 23 flavonoids and 2 lignans. 23 of them are reported for the first time including 6-*O*-caffeoyl-β-glucose, 3-*O*-(4'-*O*-Caffeoylglucosyl) quinic acid, etc. Their structures were established by retention behavior, extensive analyses of their MS spectra and further determined by comparison of their MS data with those reported in the literature. As chlorogenic acids and flavonoids are phenolic compounds that are predominant in Semen Cuscutae, in conclusion, phenolic compounds are the major constituents of Semen Cuscutae.

Keywords: Semen Cuscutae; ultra-high-performance liquid chromatography coupled to electrospray ionization mass spectrometry; chlorogenic acids; flavonoids

---

1. Introduction

Semen Cuscutae is the dry mature seed of *Cuscuta australis* R.Br. or *Cuscuta chinensis* Lam., belonging to convolvulaceae family. It was first recorded in the *"Shen Nong's Herbal"* as an upper grade drug about 2000 years ago. Semen Cuscutae has been widely prescribed by Chinese medicinal practitioners to nourish the liver and kidney, improve eyesight, treat the aching and weakness of the loins and knees, prevent abortion, and treat diarrhea due to hypofunction of the kidney and the spleen [1]. Previous phytochemical investigations on Semen Cuscutae have led to the isolation of a series of natural compounds, including flavonoids, lignans, polysaccharides, alkaloids and other chemicals [2–4].

Most studies of identification and quantification of flavonoids and polysaccharide in Semen Cuscutae have been performed by HPLC-UV [5–7], but few studies have been performed by ultra-high-performance liquid chromatography (UPLC) coupled with electrospray ionization tandem mass spectrometry. This method has the advantage that it is more sensitive and selective than a HPLC-UV, leading to a more exact identification of a higher number of compounds [8].

The purpose of this work was to identify different kinds of ingredients with significant biological functions in Semen Cuscutae for further phytochemical and pharmacological study.

2. Results and Discussion

2.1. Optimization of UPLC-MS Conditions

In this study, an optimized chromatographic separation was achieved using acetonitrile-water containing 0.05% formic acid solvent system as the mobile phases. Waters ACQUITY UPLC BEH C_{18} (2.1 × 100 mm i.d., 1.7 μm) was selected for qualitative analysis due to better separation efficiency. A representative total ion chromatographic (TIC) were shown in Figure 1.

To obtain the satisfactory analytical method, chromatographic conditions, including mobile phase (methanol, acetonitrile and acetonitrile-water), flow rate (0.1, 0.2, and 0.3 mL·min^{-1}), formic acid addition (0.05% and 0.1%), and column type (Waters ACQUITY BEH C_{18}, 2.1 × 100 mm, 1.7 μm, and Agilent Eclipse Plus C_{18} column (2.1 × 100 mm i.d., 1.8 μm) were optimized after several trials. Meanwhile, in order to achieve massive fragment ions, all the factors related to MS performance, including ionization mode, sheath gas flow rate, aux gas flow rate, spray voltage of the ion source, and collision energy have been optimized.

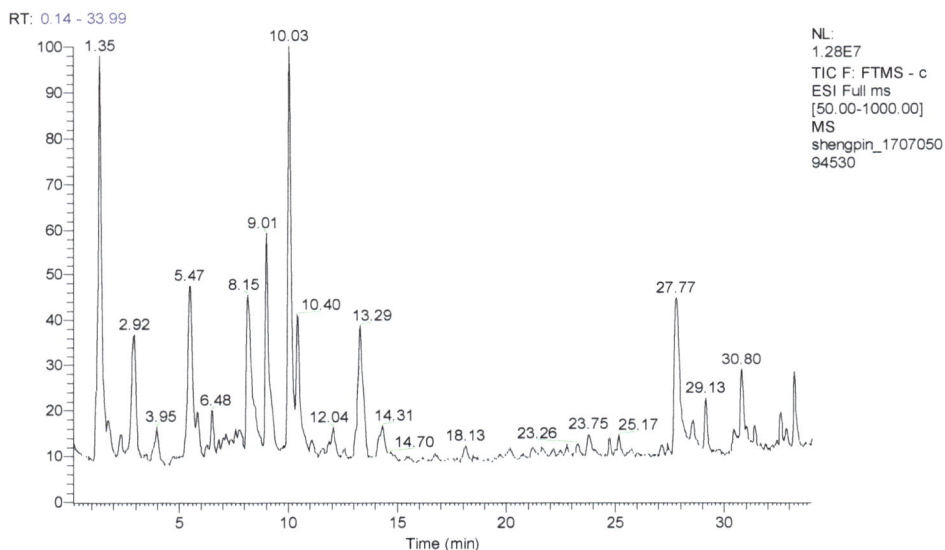

Figure 1. The total ion chromatographic (TIC) of Semen Cuscutae

2.2. Optimization for Sample Extraction

The extraction method had been established by our team [5]. The best extracted condition was established as follows: 1.0 g of sample was extracted by refluxing using 50 mL 80% methanol as solvent for 2 h. To obtain satisfactory extraction efficiency, the extraction method (refluxing and ultrasonication), extraction concentration (40%, 60% and 80%), and extraction time (0.5, 1 and 2 h) were optimized.

2.3. Structural Characterization by UPLC-MS

Due to the lack of standards for some of the compounds, their negative identification was based on the correspondence of the ion from the deprotonated molecule with literature data, fragmentation patterns of other similar compounds and database. The chromatographic of standards were shown in Figures 2 and 3. For the LC-MS measurements, negative ion mode was used to obtain the better tandem mass spectra and high-resolution mass spectra. In total of 45 compounds were identified, including 23 flavonoids, 2 lignans and 20 chlorogenic acids (Table 1). For all the compounds the high-resolution mass data was in good agreement with the theoretical molecular formulas, all displaying a mass error of below 5 ppm (Table 2) thus confirming their elemental composition.

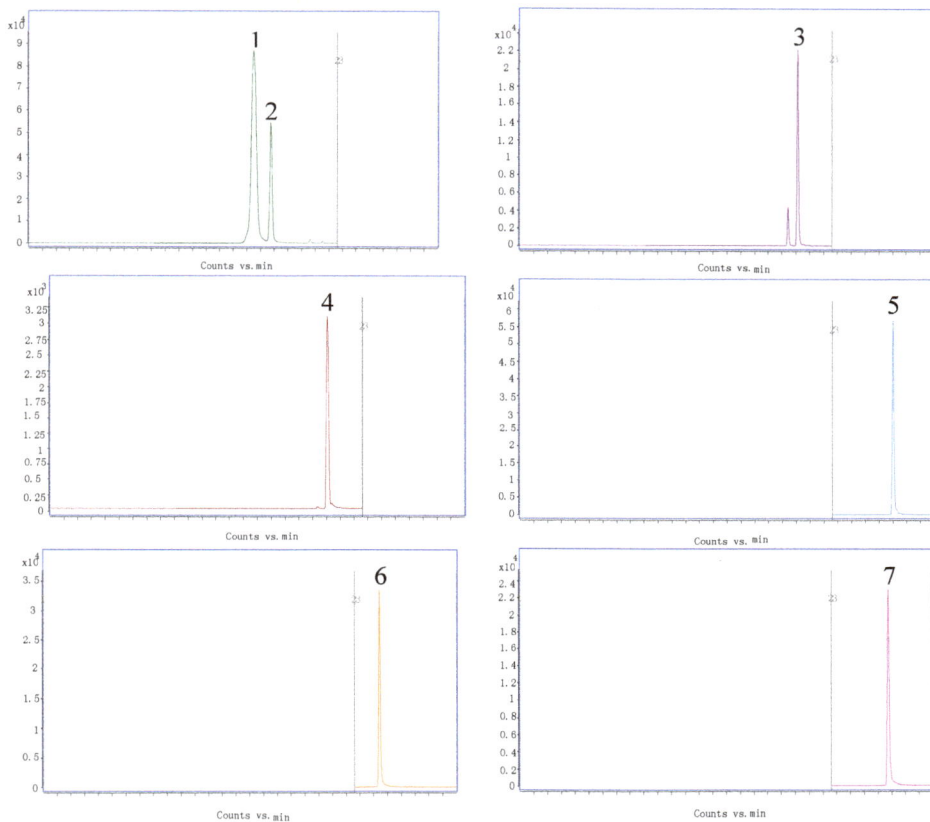

Figure 2. Chromatographic of flavonoid standards. Note: 1. Hyperoside 2. Isoquercitrin 3. Astragalin 4. Luteolin-7-O-glucoside 5. Isorhamnetin 6. Quercetin 7. Kaempferol.

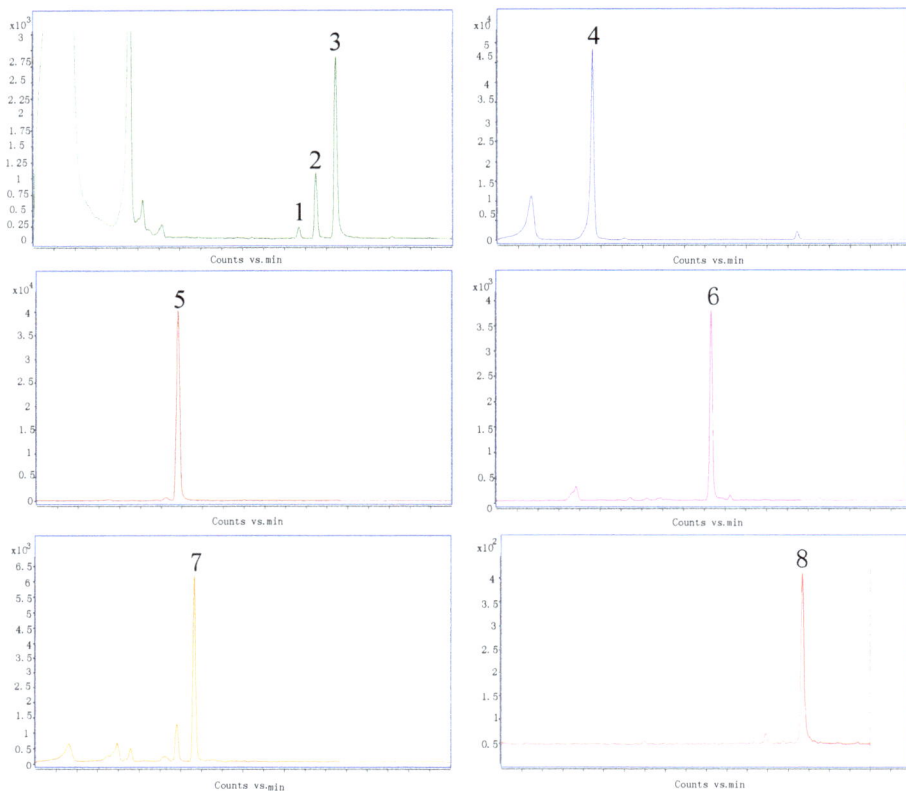

Figure 3. Chromatographic of chlorogenic acid standards. Note: 1. 3,4-dicaffeoylquinic acid 2. 3,5-dicaffeoylquinic acid 3. 4,5-dicaffeoylquinic acid 4. 3-caffeoylquinic acid 5. 4-caffeoylquinic acid 6. *p*-hydroxycinnamic acid 7. caffeic acid 8. 5-FQA.

Table 1. Compounds identified in Semen Cuscutae.

No.	t_R(min)	Identification	[M − H] or [M+FA-H]	Molecular Formula	Exact Mass	[M − H]-m/z	Characteristic m/z of Ions in Negative Ion Mode
1	1.18	Caffeoyl glucoside	M+FA-H [M+FA-H]	C15H18O9	342.0951	539.1653	MS2:503(100), 341(73) MS3:179(100), 323(25), 341(10)
2	1.31	6-O-caffeoyl-β-glucose	M − H			341.1068	MS2:179(100), 281(2), 221(1) MS3:135(100)
3	1.32	Caffeoyl diglucoside	M − H			503.1590	MS2:179(100), 221(51), 323(22), 161(15), 341(15) MS3:161(100),143(63), 131(14)
4	1.46	3,4-diCQA	M − H	C25H24O12	516.1268	515.1378	MS2:353(100), 191(62), 178(9), 173(2)
5	1.58	6-O-Caffeoyl-α-glucose	M − H	C15H18O9	342.0951	341.1075	MS2:179(100), 281(2) MS3:135(100)
6	2.34	3,5-diCQA	M − H	C25H24O12	516.1268	515.1400	MS2:179(100), 353(69), 191(22), 335(3), 173(3), MS3:191(100), 179(10), 135(4)
7	2.93	4,5-diCQA	M − H	C25H24O12	516.1479	515.1403	MS2:353(100), 191(63), 179(2) MS3:191(100), 179(5)
8	4.01	3-O-(4'-O-Caffeoylglucosyl)quinic acid	M − H	C22H28O14	516.1434	515.1397	MS2:341(100), 353(87), 191(4), 173(41), 179(57)
9	4.04	p-Hydroxycinnamic acid	M − H	C9H8O3	164.0473	163.0398	MS2:119(100) MS3:119(100), 75(46)
10	4.73	5-O-(3'-O-Caffeoylglucosyl)quinic acid	M − H	C22H28O14	516.1479	515.1396	MS2:323(100), 191(26), 353(24), 341(16), 179(3)
11	5.35	Quercetin-3-O-galactoside-7-O-glucoside	M − H	C27H30O17	626.1483	625.1390	MS2:463(100),505(3), 301(2) MS3:301(100), 179(1)
12	5.47	Quinic acid	M − H	C7H12O6	192.0634	191.0553	MS2:127(100), 85(86), 173(63), 93(58) MS3:85(100), 99(40), 109(27)
13	5.51	3-O-Caffeoylquinic acid	M − H	C16H18O9	354.0951	353.0856	MS2:191(100), 179(3) MS3:127(100), 173(76)
14	5.85	Caffeic acid	M − H	C9H8O4	180.0423	179.0347	MS2:135(100), 107(27) MS3:107(100), 78(68)
15	6.01	Quercetin-3-O-caffeoylgalactoside	M − H	C27H30O17	626.1483	625.1396	MS2:463(100), 301(33), 505(5) MS3:301(100), 343(10)
16	6.48	Coumaroyl caffeoylglycoside	M − H	C24H24O11	488.1319	487.1449	MS2:265(100), 163(85), 307(71), 235(60), 145(53), 325(30), 341(15), 323(6)
17	6.58	Kaempferol-diheoroside	M − H	C27H30O16	610.1534	609.1452	MS2:447(100), 285(16) MS3:284(100), 285(52), 327(22), 255(14)
18	7.56	Coumaroyl-tricaffeoylquinic acid	M − H			823.2269	MS2:661(100) MS3:487(100), 353(94), 515(34),
19	7.73	5-FQA	M − H	C17H20O9	368.1107	367.1025	MS2:191(100), 173(100) MS3:127(100), 85(63), 173(81), 93(46), 111(36)
20	9.01	Quercetin-3-O-apiosyl-(1→2)-galactoside	M − H	C26H28O16	596.1377	595.1292	MS2:300(100), 301(48), 463(24) MS3:271(100), 255(58), 179(1)
21	10.03	Hyperoside	M − H	C21H20O12	464.0955	463.0860	MS2:301(100), 300(33), 343(4) MS3:179(100), 151(79), 273(18), 257(12)
22	10.40	Isoquercitrin	M − H	C21H20O12	464.0955	463.0866	MS2:301(100) MS3:179(100), 151(91), 273(16), 255(9)
23	10.65	Kaempferol-3-apiosyl-(1→2)-glucoside	M − H	C26H28O15	580.1428	579.1344	MS2:285(100), 284(60), 447(24), 255(15) MS3:257(100), 151(69), 241(46)
24	11.50	Isorhamnetin-3-apiosyl-(1→2)-hexoside	M − H	C27H30O16	610.1534	609.1448	MS2:315(100), 314(29), 300(22), 459(13) MS3:300(100), 287(5)
25	11.86	Quercetin 3-(2''-acetylgalactoside)	M − H	C23H22O13	506.1060	505.0979	MS2:301(100), 300(53), 463(23) MS3:179(100), 151(89)
26	12.04	Kaempferol-3-O-galactoside	M − H	C21H20O11	448.1006	447.0926	MS2:284(100), 285(61), 327(14) MS3:255(100), 227(9)
27	13.29	Astragalin	M − H	C21H20O11	448.1006	447.0927	MS2:284(100), 285(58), 327(12) MS3:255(100), 256(17), 227(13)
28	13.42	Isorhamnetin-7-glucoside	M − H	C22H22O12	478.1111	477.1029	MS2:314(100), 315(71), 357(18), 449(8) MS3:285(100), 271(74), 300(34), 243(27)
29	14.31	Isorhamnetin-3-O-glucoside	M − H	C22H22O12	478.1111	477.1033	MS2:314(100), 315(36), 285(9), 300(3) MS3:285(100), 271(77), 300(35), 243(25),
30	14.4	Luteolin-hexoside	M − H	C21H20O11	448.1006	447.0927	MS2:285(100), 327(5) MS3:285(100), 151(32), 257(20), 229(10), 241(20)
31	16.12	Luteolin-7-O-glucoside	M − H	C21H20O11	448.1006	447.0926	MS2:285(100), 284(13), 327(5), 257(2) MS3:151(100), 257(32), 241(22)
32	16.73	4-O-Caffeoylquinic acid	M − H	C16H18O9	354.0951	353.0871	MS2:173(100), 179(47), 191(16), 135(7) MS3:93(100), 111(77)
33	21.23	Kaempferol-3-O-p-hydroxybenzoylglucoside	M − H	C28H24O13	568.1217	567.1135	MS2:284(100), 285(97), 447(37), 255(30), 429(11), 327(9) MS3:285(100), 151(6)
34	22.17	cis-4-pCoQA	M − H	C16H18O8	338.1002	337.0922	MS2:173(100), 163(9) MS3:93(100), 111(58)
35	22.76	Apigenin	M − H	C15H10O5	270.0528	269.0928	MS2:225(100)
36	23.23	Cuscutoside D	M − H	C37H46O21	826.2532	825.2439	MS2:369(100), 663(76), 323(43) MS3:219(100), 339(74), 311(54), 323(11)

Table 1. *Cont.*

No.	t$_R$(min)	Identification	[M − H] or [M+FA-H]	Molecular Formula	Exact Mass	[M − H]-*m/z*	Characteristic *m/z* of Ions in Negative Ion Mode
37	23.32	Quercetin-3-O-coumaroylgalactoside	M − H	C$_{30}$H$_{26}$O$_{14}$	610.1323	609.1242	MS^2:463(100), 301(13) MS^3:301(100), 343(3)
38	23.75	Quercetin	M − H	C$_{15}$H$_{10}$O$_7$	302.0427	301.0349	MS^2:151(100), 179(97)
39	24.69	Kaempferol-3-O-glucoside-7-rhamnoside	M − H	C$_{30}$H$_{26}$O$_{13}$	594.1373	593.1289	MS^2:447(13), 285(100), 327(1) MS^3:257(81),285(100), 151(63), 241(45), 229(3)
40	25.17	Kaempferol-3-O-coumaroylglucoside	M − H	C$_{30}$H$_{26}$O$_{13}$	594.1373	593.1290	MS^2:285(100), 447(12), 307(6) MS^3:285(100), 257(70), 151(59), 229(36), 241(35)
41	25.21	*trans*-4-*p*CoQA	M − H	C$_{16}$H$_{18}$O$_8$	338.1002	337.0923	MS^2:173(100), 163(8) MS^3:93(100), 111(57)
42	25.74	*p*CoQA isomer	M − H	C$_{16}$H$_{18}$O$_8$	338.1002	337.0923	MS^2:173(100), 322(57), 306(14) MS^3:93(100), 111(62), 155(36),
43	27.12	Cuscutoside A	M − H			663.1917	MS^2:369(100) MS^3:219(100), 339(77), 311(45)
44	27.77	Kaempferol	M − H	C$_{15}$H$_{10}$O$_6$	286.0477	285.0392	MS^2:285(100), 229(17)
45	28.54	Isorhamnetin	M − H	C$_{16}$H$_{12}$O$_7$	316.0583	315.0502	MS^2:300(100) MS^3:271(100), 151(91), 272(62), 255(30)

Table 2. The existence of each component in Semen Cuscutae.

No.	t_R(min)	Compound	Molecular Formula	New or Not	Mass Error (ppm)
1	1.18	Caffeoyl glucoside		+	4
2	1.31	6-O-caffeoyl-β-glucose	$C_{15}H_{18}O_9$	+	2
3	1.32	Caffeoyl diglucoside		+	3
4	1.46	3,4-diCQA	$C_{25}H_{24}O_{12}$	-	1
5	1.58	6-O-Caffeoyl-α-glucose	$C_{15}H_{18}O_9$	+	4
6	2.34	3,5-diCQA	$C_{25}H_{24}O_{12}$	-	1
7	2.93	4,5-diCQA	$C_{25}H_{24}O_{12}$	-	1
8	4.01	3-O-(4′-O-Caffeoylglucosyl)quinic acid	$C_{22}H_{28}O_{14}$	+	3
9	4.04	p-Hydroxycinnamic acid	$C_9H_8O_3$	-	1
10	4.73	5-O-(3′-O-Caffeoylglucosyl)quinic acid	$C_{22}H_{28}O_{14}$	+	4
11	5.35	Quercetin-3-O-galactoside-7-O-glucoside	$C_{27}H_{30}O_{17}$	-	3
12	5.47	Quinic acid	$C_7H_{12}O_6$	-	4
13	5.51	3-O-Caffeoylqunic acid	$C_{16}H_{18}O_9$	-	2
14	5.85	Caffeic acid	$C_9H_8O_4$	+	1
15	6.01	Quercetin-3-O-caffeoylgalactoside	$C_{27}H_{30}O_{17}$	-	2
16	6.48	Coumaroyl caffeoylglycoside	$C_{24}H_{24}O_{11}$	+	4
17	6.58	Kaempferol-dihexoside	$C_{27}H_{30}O_{16}$	-	1
18	7.56	Coumaroyl-tricaffeoylquinic acid		-	3
19	7.73	5-FQA	$C_{17}H_{20}O_9$	+	3
20	9.01	Quercetin-3-O-apiosyl-(1→2)-galactoside	$C_{26}H_{28}O_{16}$	-	2
21	10.03	Hyperoside	$C_{21}H_{20}O_{12}$	-	4
22	10.40	Isoquercitrin	$C_{21}H_{20}O_{12}$	-	4
23	10.65	Kaempferol-3-apiosyl-(1→2)-glucoside	$C_{26}H_{28}O_{15}$	+	1
24	11.50	Isorhamnetin-3-apiosyl-(1→2)-hexoside	$C_{27}H_{30}O_{16}$	+	2
25	11.86	Quercetin 3-(2″-acetylgalactoside)	$C_{23}H_{22}O_{13}$	+	1
26	12.04	Kaempferol-3-O-galactoside	$C_{21}H_{20}O_{11}$	-	1
27	13.29	Astragalin	$C_{21}H_{20}O_{11}$	-	1
28	13.42	Isorhamnetin 7-glucoside	$C_{22}H_{22}O_{12}$	-	1
29	14.31	Isorhamnetin-3-O-glucoside	$C_{22}H_{22}O_{12}$	+	1
30	14.40	Luteolin-hexoside	$C_{21}H_{20}O_{11}$	+	1
31	16.12	Luteolin-7-O-glucoside	$C_{21}H_{20}O_{11}$	+	1
32	16.73	4-O-Caffeoylqunic acid	$C_{16}H_{18}O_9$	-	1
33	21.23	Kaempferol-3-O-p-hydroxybenzoylglucoside		-	1
34	22.17	cis-4-pCoQA	$C_{16}H_{18}O_8$	+	2
35	22.76	Apigenin	$C_{15}H_{10}O_5$	-	1
36	23.23	Cuscutoside D	$C_{37}H_{46}O_{21}$	-	1
37	23.32	Quercetin-3-O-coumaroylgalactoside	$C_{30}H_{26}O_{14}$	-	1
38	23.75	Quercetin	$C_{15}H_{10}O_7$	-	1
39	24.69	Kaempferol-3-O-glucoside-7-rhamnoside	$C_{30}H_{26}O_{13}$	-	2
40	25.17	Kaempferol-3-O-coumaroylglucoside	$C_{30}H_{26}O_{13}$	+	1
41	25.21	trans-4-pCoQA	$C_{16}H_{18}O_8$	+	2
42	25.74	pCoQA isomer	$C_{16}H_{18}O_8$	+	2
43	27.12	Cuscutoside A		-	1
44	27.77	Kaempferol	$C_{15}H_{10}O_6$	-	4
45	28.54	Isorhamnetin	$C_{16}H_{12}O_7$	-	2

Note: + represents a newly discovered compound in Semen Cuscutae, - represents the existing compound.

2.4. Chlorogenic Acids

2.4.1. Characterization of p-Coumaroylquinic Acids (pCoQA, Mr = 338.1002)

Chlorogenic acids (CGAs) are a family of esters of trans-cinnamic acids (most commonly p-coumaroyl, caffeoyl, feruloyl and dimethoxycinnamoyl acids) with quinic acid [9,10]. The trans-cinnamic acids can be esterified at one or more of the hydroxyls at positions 1, 3, 4, and 5 of quinic acid, originating series of positional isomers. More importantly, it is easy to distinguish a 4-acyl chlorogenic acids by its "dehydrated" MS^2 base peak at m/z 173 ([quinic acid-H-H_2O]$^-$), supported by strong MS^3 ions at m/z 93 [11,12]. P-Coumaroylquinic acid (pCoQA) has a molecular weight (Mr) of 338.1002 and three peaks (peak 34, 41, 42) at m/z 337 were detected. The three peaks are all pCoQA isomers. In addition, according to previous reported literature, the retention time of a 4-position substituted cis-isomer on a reverse phase chromatographic column is obviously longer than that of a trans-isomer [13]. Based on the above analysis, compounds 34 and 41 were identified as cis-4-*p*-CoQC and trans-4-*p*-CoQA, respectively. While compound 42 is *p*CoQA isomer which is uncertain.

2.4.2. Characterization of Caffeoylquinic Acids (CQA, Mr = 354.0951)

Chromatographic peaks 13 and 32 presented m/z 353 as base peaks in negative ionization mode mass spectra, which suggested positional isomers of a quinic acid (QA) esterified with a single caffeoyl (CAF) unit. The product ion spectra obtained by negative ion MS/MS for precursor ions m/z 353 were different from each other. The product ion spectrum for peak 32 showed m/z 173 (dehydrated quinic moiety) as the base peak, m/z 191 [loss of caffeic moiety], and m/z 179 (loss of quinic moiety). As m/z 173 is a diagnostic ion that acylated at position 4, peak 32 was attributed to 4-CQA (Figure 4). As reported before, the retention times of acylated CQAs repeat the elution pattern: 3-acylquinic acid elutes first, followed by 4-acylquinic acids. So peak 13 is assumed as 3-CQA (Figure 5) [9,14].

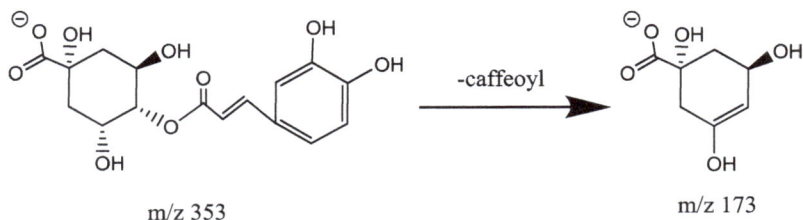

m/z 353   m/z 173

Figure 4. Fragmentation pathways of 4-*O*-caffeoylqunic acid.

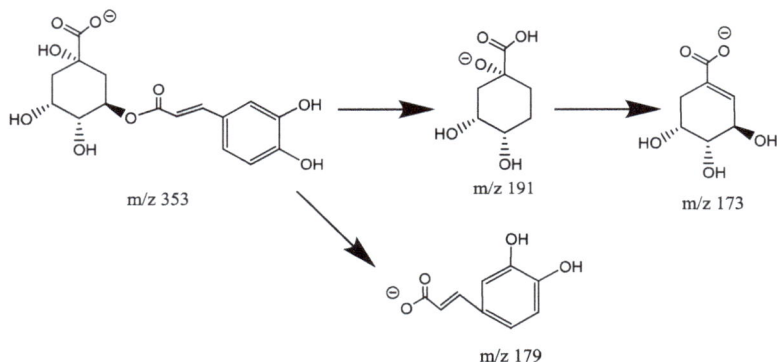

m/z 353   m/z 191   m/z 173

m/z 179

Figure 5. Fragmentation pathways of 3-*O*-caffeoylqunic acid.

2.4.3. Characterization of Feruloylquinic Acids (FQA, Mr = 368.1107)

Feruloylquinic acids has a Mr = 368.1107. Similar to *p*CoQA and CQA, molecules harboring ferulic acid moieties were also identified. The negative ionization mode fragmentation of the precursor ion m/z 367 of peak 19 produced m/z 191 as base peak. This is a diagnostic ion of acylation in position 5 of quinic acid (19) and allows the identification of the compound as 5-FQA (Figure 6), based on the chlorogenic acid identification by LEONARDO et al. [9].

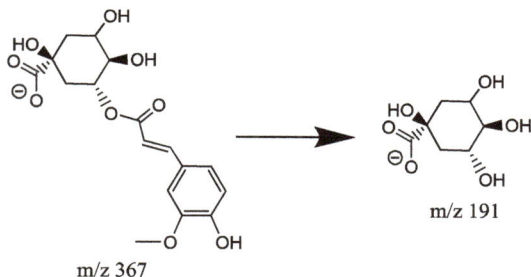

Figure 6. Fragmentation pathways of 5-FQA

2.4.4. Characterization of Di-Caffeoylquinic Acid (di-CQA, Mr = 516.1268)

As previously reported, di-CQA produce an isobaric pseudomolecular ion at m/z 515. The diCQA are isomers as their chemical structures possess the same skeleton of quinic acid, and they can be differentiated by different substitution positions (Figure 7). Peaks 4 and 7 generated $[M - H]^-$ ion at m/z 515 and $[M - H\text{-caffeoyl}]^-$ ion at m/z 353 (the deprotonated molecular ions yielded via the neutral loss of 162 ($C_9H_6O_3$)). Base peak in peak 6 at m/z 179, indicating it was 3-substituted quinic acid. This ion in 3, 4-diCQA (peaks 4) was at m/z 178 and in 4, 5-diCQA (peaks 7) was absent. 3, 5-DiCQA (peaks 6) was relatively easy to distinguish, owing to its MS^3 base peak m/z 191 and similar intensities of ions at m/z 179 with data previously published [15]. Generally, it was observed that the order of elution for the diacyl CGAs in RP columns is 3, 4 > 3, 5 > 4, 5 [16]. By comparing with reference substances, peaks 4, 6 and 7 were assigned as 3, 4-diCQA, 3, 5-diCQA, 4, 5-diCQA which is consist with data previously published.

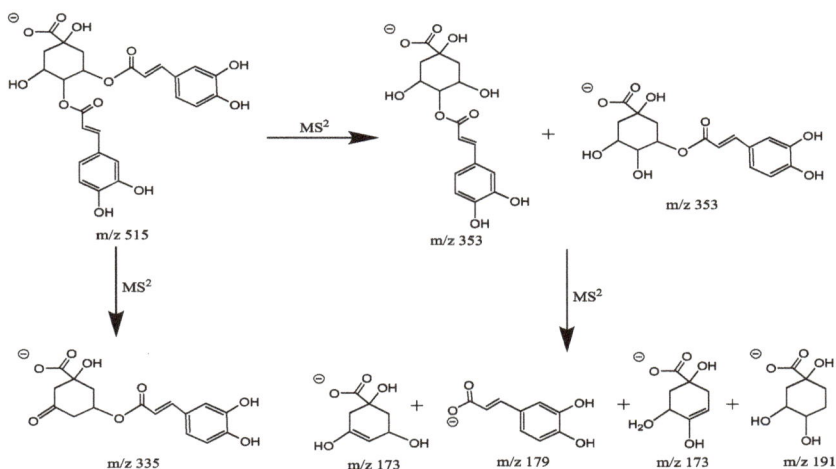

Figure 7. Fragmentation pathways of di-caffeoylquinic acid.

2.4.5. Characterization of Caffeoylglycoside (Mr = 342.0951)

Caffeoylglycoside have molecular weights of 342.0951 which indicates that the glucosyl group was linked to the caffeic acid, not quinic acid [13]. Two molecules (2 and 5) with pseudomolecular peaks at m/z 341 were assigned as isomers of caffeoylglycoside. They produced distinctive ions at m/z 179 ([caffeic acid-H]$^-$) by the loss of a glucosyl residue ($C_6H_{10}O_5$) and m/z 135 ([caffeic acid-H]$^-$). As reported before, the retention time of a 6-position substituted β-glucose isomer on a reverse phase chromatographic column is obviously longer than that of α-glucose isomer [17]. On the basis of these arguments, the first eluting isomer was assigned as 6-*O*-caffeoyl-β-glucose (2) and the later eluting isomer as 6-*O*-caffeoyl-α-glucose (5).

2.4.6. Characterization of Caffeoylquinic Acid Glucoside (CQA-Glycoside, Mr = 516.1434)

Caffeoylquinic acid glucoside has a Mr = 516.1434. Unlike the diCQA, the CQA glycosides showed a typical fragmentation pattern of chlorogenic acids. They produce distinctive ions which originated from the cinnamoyl glycoside part at m/z 341 ($C_{15}H_{17}O_9$, [caffeoyl glucoside-H]$^-$) or/and 323 ($C_{15}H_{15}O_8$, [caffeoyl glucoside-H-H_2O]$^-$) which were not present in the diCQA MS spectra. Previous studies [14] led to the conclusion that CQA forms a glycoside through an ether bond at either C-3 or C-4 on the aromatic caffeoyl ring. However, a MS^2 base peak at m/z 323 is a characteristic of glucosyl attachment at C-3 [18].

Peak 8 produced the MS^2 base peak at m/z 341 ([caffeoyl glucoside-H]$^-$) due to the loss of a quinic acid moiety (174 Da); the secondary peaks occurred as follows: the peak at m/z 353 ([caffeoyquinic acid-H]$^-$) via the loss of a glucosyl residue (162 Da), the peak at m/z 179 ([caffeic acid-H]$^-$) due to the loss of a glucosyl, the peak at m/z 191 ([quinic acid-H]$^-$) due to the loss of a caffeoyl, a glucosyl residues and the peak at m/z 173 ([quinic acid-H_2O-H]$^-$) due to the loss of a caffeoyl and a glucosyl residues followed by H_2O (18 Da) (Table 1).The MS^2 spectrum was identical to 3-*O*-caffeoylquinic acid. The MS^2 peaks at m/z 353 and 341 were absent which suggested that glucose was connected with the quinic acid moiety by an ether linkage and caffeic acid was connected with quinic acid at C-3 by an ester bond. This isomer is assigned as 3-*O*-(4′-*O*-caffeoyl glucosyl) quinic acid.

Peak 10 produced the MS^2 base peak at m/z 323 ([caffeoyl glucoside-H_2O-H]$^-$) due to the loss of a quinic acid moiety (174 Da) followed by H_2O (18 Da) and secondary peaks at m/z 341 ([caffeoyl glucoside-H]$^-$) due to the loss of a quinic acid moiety (174 Da), m/z 353 ([caffeoylquinic acid-H]$^-$) via the loss of a glucosyl residue (162 Da), m/z 191 ([quinic acid-H]$^-$) due to the loss of a caffeoyl. The MS^2 spectra were identical to the MS^2 spectra of 5-*O*-caffeoylquinic acid (Table 1). The MS^2 peaks at m/z 353, 341 and 323 suggested that glucose was connected with the caffeic acid moiety by an ether linkage and caffeic acid was connected with quinic acid at C-5 by an ester bond [19]. From the above points it was clear that peak 10 can be identified as 5-*O*-(3′-*O*-Caffeoylglucosyl) quinic acid (Figure 8).

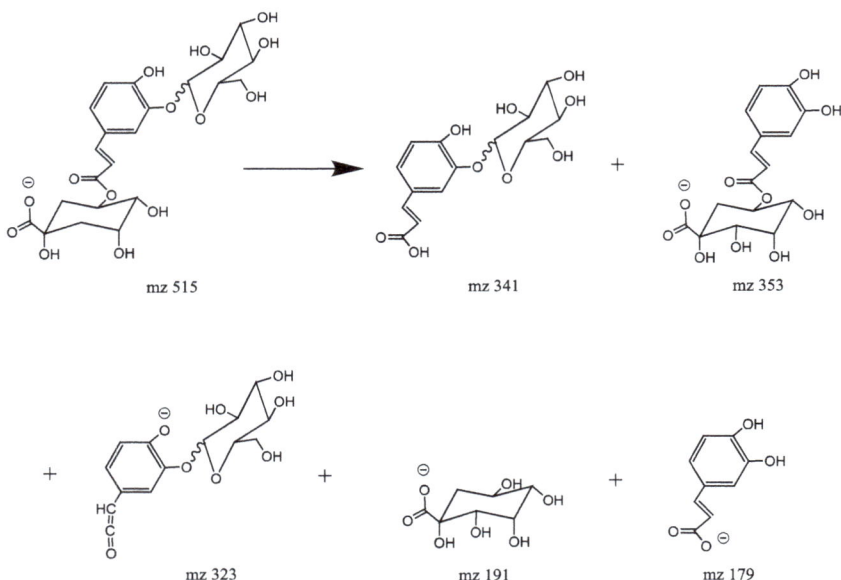

Figure 8. Fragmentation pathways of 5-*O*-(3′-*O*-Glucosylcaffeoyl) quinic acid.

2.4.7. Characterization of Coumaroyl-Tricaffeoylquinic Acid (Mr = 824)

Peak 18 exhibited $[M - H]^-$ ion at m/z 823, which revealed m/z 661 by losing a caffeoyl moiety. Compounds 18 consequently lost a coumaroyl moiety and a caffeoyl moiety to produce $[M - H-146-162]^-$ ion at m/z 353. It was tentatively identified as coumaroyl-tricaffeoylquinic acids.

2.5. Flavonoids Derivatives

An number of flavonoids have been reported from Cuscuta species previously. Most of them are flavonoids, flavonols and flavones glycoside [20].

2.5.1. Characterization of Apigenin (Mr = 270.0528)

One peak was detected at m/z 269(peak 35) in the extracted ion chromatogram. It produced the MS^2 base peak at m/z 225 corresponding to neutral loss of CO_2 (44 Da). It was identified as apigenin, consistent with the literature data [20].

2.5.2. Characterization of Isorhamnetin (Mr = 316.0583)

One peak was detected at m/z 315(peak 45) in the extracted ion chromatogram and was tentatively assigned as isorhamnetin. It produced the MS^2 base peak at m/z 300 indicating the presence of a methoxyl group (loss of a methyl radical). These data matched those previously reported for isorhamnetin [20].

2.5.3. Characterization of Kaempferol-Hexoside (Mr = 448.1006)

Four isomers were found at M/S 447(26, 27, 30 and 31). The four isomers yielded MS^2 ion at m/z 285 as the base peak, indicating the existence of hexoside. Peak 26, 27 yielded fragment at m/z 285($[A-H]^-$), m/z 255($[A-H-HCHO]^-$) and m/z 227($[A-H-CO]^-$) ion, consistent with the reported data for kaempferol [20]. As reported [20], peak 26 showed weaker retention on the RP-HPLC column than peak 27, therefore they were assigned as kaempferol-3-*O*-galactoside, astragalin respectively. For peak 31, there were no $[A-H-30]^-$ ion, the base peak for MS^3 was M/S 151 which is consist with luteolin, therefore peak 31 was identified as luteolin-7-*O*-glucoside and was confirmed by comparison with a reference standard. Peak 30 had similar fragments with peak 31, thus it was assigned as luteolin-hexoside.

2.5.4. Characterization of Kaempferol-*O*-Dihexoside, Isorhamnetin-3-Apiosyl-(1→2)-Hexoside and Quercetin-3-*O*-Coumaroylgalactoside (Mr = 610.1323)

Peak 17, 24 and 37 all produced $[M - H]^-$ ion at m/z 609. However, they produced obviously different MS^2 ions. Peak 17 MS^2 spectrum gave ions at m/z 447 and 285, originating from successive losses of 162 Da, suggesting the presence of two hexosyl residues. The $[A-H]^-$ ion at m/z 285 yielded a $[A-H-30]^-$ fragment at m/z 255, consistent with kaempferol. Thus, compound 17 was identified as kaempferol-*O*-dihexoside. Peak 24 exhibited a $[M - H]^-$ ion at m/z 609. Its MS^2 spectrum gave ions at m/z 315 and 300, originating from successive losses of 162 Da and 132 Da, suggesting the presence of hexosyl and apiosyl residues. The $[A-H]^-$ ion at m/z 315 yielded a $[A-H-15]^-$ fragment at m/z 300, consistent with isorhamnetin. Thus, peak 24 was assigned as isorhamnetin-3-apiosyl-(1→2)-hexoside. Peak 37 MS^2 spectrum gave ions at m/z 463 and 301, originating from successive losses of 146 Da and 162 Da, suggesting the presence of hexosyl residue and coumaroyl residue. The product spectrum of the m/z 301 ion was very similar to that of quercetin, though no $[A-H-30]^-$ ion was observed. Based on the fragmentation pattern, this compound was assigned as quercetin-3-*O*-coumaroylgalactoside.

2.5.5. Characterization of Isorhamnetin-Hexoside (Mr = 478.1111)

Two peaks were detected at m/z 477 in the extracted ion chromatogram and were tentatively assigned as isorhamnetin-hexoside (28 and 29). These two compounds produced base peak at m/z

314, originating from the loss of a hexose (162 Da), and MS^3 spectrum was very similar to that of isorhamnetin. These compounds were thus tentatively identified as isorhamnetin-7-glucoside and isorhamnetin-3-O-glucoside, respectively.

2.5.6. Characterization of Kaempferol-Glucoside (Mr = 568.1217)

For peak 33, a significant loss of 120 Da was also observed, but no direct loss of 162 Da from the $[M − H]^-$ ion was observed. Therefore, it is rational to assign a *p*-hydroxybenzoyl group linked to the hexose moiety rather than the aglycone in this structure. Interestingly, a second loss of 120 Da (m/z 447→327) was also observed, which presumably results from $^{1,2}X$ fragmentation of the hexose. Peak 33 produced MS^2 base peak at m/z 285 whose fragmentation was consistent with kaempferol, and therefore peak 33 was finally confirmed as kaempferol-3-O-*p*-hydroxybenzoylglucoside, which is consistent with the previous report [20].

2.5.7. Characterization of Kaempferol-O-Glucoside-7-Rhamnoside and Kaempferol-3-O-Coumaroylglucoside (Mr = 594.1373)

Peak 39, 40 displayed a $[M − H]^-$ ion at m/z 593.The MS^2 spectra of the reference substances kaempferol-3-O-glucoside-7-rhamnoside showed ion $[M − H-146]^-$ that clearly indicates the removal of rhamnosyl moiety from the hydroxyl group of C7 and showed almost the same intensity as the aglycone. Previous studies [10] described the removal of the sugar residues from the hydroxyl in position 7 as being much more favored in ESI-MS than from position 3. Due to these findings compound 39 was identified as kaempferol-3-O-glucoside-7-rhamnoside.

Peak 40 displayed a $[M − H]^-$ ion at m/z 593. The MS^2 spectrum gave a base peak at m/z 285, originating from the concurrent losses of coumaroyl (146 Da) and a hexose (162 Da), which made of a disaccharide moiety. The m/z 447 ion resulted from the cleavage of the coumaroyl, which should thus be connected directly with the hexose moiety. The product spectrum of the m/z 285 ion was very similar to that of kaempferol, though no $[A-H-30]^-$ ion was observed. Previous study [20] reported that the retention times of flavonoid diglycosides on RP-HPLC columns generally is longer than monoglycosides. Based on the above points, compound 40 was assigned as kaempferol-3-O-coumaroylglucoside.

2.5.8. Characterization of Quercetin-3-O-Apiosyl-(1→2)-Galactoside (Mr = 596.1377)

One peak 20 was detected at m/z 595 in the extracted ion chromatogram and was tentatively assigned as quercetin-3-O-apiosyl-(1→2)-galactoside. CID of the $[M − H]^-$ ion gave two major ions at m/z 463 and 301, consistent with successive losses of apiose (132 Da) and galactose (162 Da). Similar to astragalin, daughter ion at m/z 301 produced a $[A-H-30]^-$ ion at m/z 271, thus conforming it as a quercetin glycoside. The $^{1,2}A^-$ ion at m/z 179 was also observed, in agreement with the literature data [20].

2.5.9. Characterization of Kaempferol-3-O-Galactoside and Quercetin-Hexoside (Mr = 464.0955)

Peak 21 gave MS^2 and MS^3 spectra very similar to those of astragalin, and was plausibly identified as hyperoside, which has been previously reported [20]. However peak 22 had similar fraction information with 21, so it was tentatively identified as isoquercitrin.

2.5.10. Characterization of Quercetin-3-(2″-Acetylgalactoside) (Mr = 506.1060)

Peak 25 exhibited $[M − H]^-$ ion at m/z 505, which yielded an MS^2 base peak at m/z 301 by losing a m/z 204 moiety which is identified as acetylhexose [21]. The base peak at m/z 301 indicates that hexose group is connected with flavonoid. M/z 301 yielded fragment at m/z 179 and 151, consist with quercetin. Peak 25 was identified as quercetin 3-(2″-acetylgalactoside), which is consistent with the previous report [20].

2.5.11. Characterization of Quercetin-Dihexoside (Mr = 626.1483)

Quercetin-3-*O*-galactoside-7-*O*-glucoside (peak 11) exhibited base peak at *m/z* 463, originating from the loss of hexose (162 Da). Fragment at *m/z* 301 was consequently the successive loss of hexose (162 Da). The fragmentation pattern was identical to previously studied data [20]. Similarly, peak 11 was assigned as quercetin-dihexoside.

2.5.12. Characterization of Kaempferol 3-Apiosyl-(1→2)-Glucoside (Mr = 580.1428)

One peak was detected at *m/z* 579 in the extracted ion chromatogram and was tentatively assigned as kaempferol-3-apiosyl-(1→2)-glucoside (23). Apiose was the only pentose hitherto reported in flavonoid glycosides of Cuscuta species, and could be characterized by a 132 Da loss [20]. Compound 23 produced base peak at *m/z* 285 via the loss of pentose and hexose residues which indicated that hexose was connected with pentose residue. Fragment ion at *m/z* 285 yielded a [A-H-30]⁻ fragment at *m/z* 255, consistent with kaempferol. Thus, compound 23 was identified as kaempferol-3-apiosyl-(1→2)-glucoside. Furthermore, peak 23 is its geometric isomer.

Chemical constituents in Semen Cuscutae were analyzed by LC/MS. From the structural characterization by HPLC-MS, 45 compounds were identified based on their retention behavior. As a result, 45 compounds including 20 chlorogenic acids, 23 flavonoids and 2 lignans were identified based on their retention behavior. 23 of them are reported for the first time including 6-*O*-caffeoyl-*β*-glucose and 3-*O*-(4′-*O*-Caffeoylglucosyl) quinic acid and so on (Table 2).

3. Materials and Methods

3.1. Materials and Chemicals

Five flavonoid reference standards including hyperoside, quercetin, astragalin, kaempferol, and isorhamnetin were purchased from National Institutes for Food and Drug Control (Beijing, China). Two flavonoid reference standards including isoquercitrin and luteolin-7-*O*-glucoside were purchased from Shanghai Yuanye Bio-Technology Co., Ltd. (Shanghai, China). Eight chlorogenic acid reference standards including 3-caffeoylquinic acid(3-CQA), 4-caffeoylquinic acid(4-CQA), 3,4-dicaffeoylquinic acid (3,4-DiCQA), 3,5-dicaffeoylquinic acid(3,5-DiCQA), 4,5-dicaffeoylquinic acid(4,5-DiCQA), *p*-hydroxycinnamic acid, caffeic acid(CA) and 5-*O*-feruloylquinic acid (5-FQA) were purchased from Shanghai Yuanye Bio-Technology Co., Ltd. (Shanghai, China). Their structures (shown in Figure 9) were fully elucidated by spectra data (ESI-MS). The purities of all the standards were no less than 98%. All the standard were resolved in 80% methanol.

Acetonitrile (MS grade) and formic acid (MS grade) were purchased from Thermo Fisher Scientific Inc. Deionized water used throughout the experiment was purified by MilliQ50 SP Reagent Water System (Bedford, MA, USA) for preparing samples.

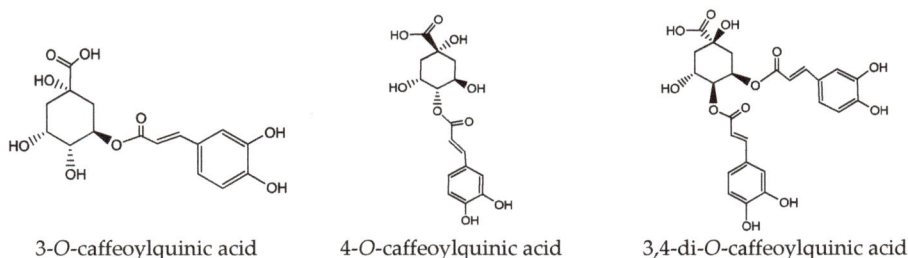

3-*O*-caffeoylquinic acid   4-*O*-caffeoylquinic acid   3,4-di-*O*-caffeoylquinic acid

Figure 9. *Cont.*

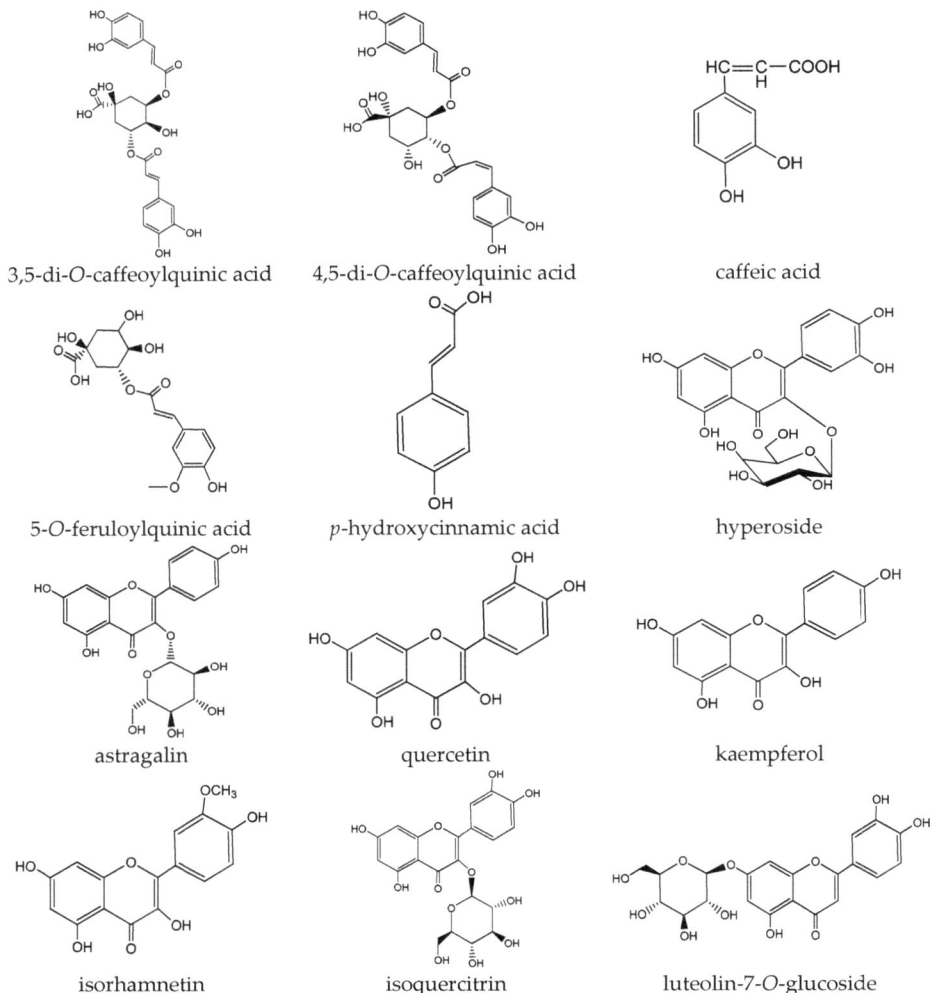

Figure 9. Chemical formula of flavonoids of Semen Cuscutae.

3.2. Sample Collection

The crude products of Semen Cuscutae (Lot number: 160161211) were purchased from Beijing Kangmei Pharmaceutical Co., Ltd. which were identified and authenticated as Semen Cuscutae by Yang Yaojun, the professor of Pharmacognosy Department in Beijing University of Chinese Medicine. Voucher specimens were retained in the School of Chinese Materia, Beijing University of Chinese Medicine.

3.3. Extraction Method

The extraction method referenced to our previous study and was set as follow [5]: powdered samples (60 mesh, 1 g) were suspended in 80% methanol (50 mL) and extracted under reflux for 2 h. After cooling, the loss of weight was replenished with 80% methanol. All solvents and samples were filtered through 0.22-μm organic-membranes prior to injection.

3.4. UPLC−MS Analysis

The extracts were chromatographically separated on an ACQUITY UPLC BEH C_{18} column (2.1 × 100 mm i.d., 1.7 μm). The mobile phase consisted of A (acetonitrile) and B (water containing 0.05% formic acid, v/v). The flow rate was 0.20 mL/min. The elution conditions applied with a linear gradient as follows: 0–4 min, 7–16% A; 4–8 min, 16–17% A; 8–15 min, 17–17% A; 15–20 min, 17–24% A; 20–27 min, 35–52% A; 27–33 min, 52–97% A. Column temperature was 35 °C.

For LC/MS analysis, an LTQ-Orbitrap mass spectrometer (Thermo Scientific, Bremen, Germany) was connected to the Ultra-High-Performance Liquid Chromatography instrument via an electrospray ionization (ESI) interface. Samples were analyzed in negative ion mode with a tune method set as follows: sheath gas (nitrogen) flow rate of 40 arb, aux gas (nitrogen) flow rate of 20 arb, source voltage, 4 kV, capillary temperature of 350 °C, capillary voltage of 25 V, and tube lens voltage of −110 V. Accurate mass analysis was calibrated according to the manufacturer's guidelines. Centroided mass spectra were acquired in mass range of m/z 50–1000 and resolution set at 30,000 using a normal scan rate detected by Orbitrap analyzer.

3.5. Data Processing

Thermo Xcaliber 2.1 (Thermo Fisher Scientific, San Jose, CA, USA) was used for qualitative data acquiring and processing. All the relevant data including peak number, retention time, accurate mass and predicted chemical formula were recorded into an Excel file.

4. Conclusions

In this study, we identified 45 compositions in Semen Cuscutae using UPLC coupled with electrospray ionization tandem mass spectrometry system. 23 of them are reported for the first time including 6-*O*-caffeoyl-*β*-glucose, 3-*O*-(4′-*O*-Caffeoylglucosyl) quinic acid, etc. As chlorogenic acids and flavonoids are phenolic compounds which are predominant compounds in Semen Cuscutae, we can conclude that phenolic compounds are the major constituents of Semen Cuscutae.

Author Contributions: Y.Z., H.X., X.X. and X.L. conceived and designed the experiment. Y.Z. and H.X. performed the experiment and data analysis. Y.Z., Y.G., H.Z. and X.L. drafted the paper. S.X., H.L., X.X. and M.L. revised the manuscript. All authors have contributed to the final version and approved the publication of the final manuscript.

Acknowledgments: This work was supported in part by grants from the National Natural Science Foundation of China (Grant no. 81573608).

Conflicts of Interest: The authors declare that there is no conflict of interests regarding the publication of this paper.

References

1. Liao, J.C.; Chang, W.T.; Lee, M.S.; Chiu, Y.J.; Chao, W.K.; Lin, Y.C.; Lin, M.K.; Peng, W.H. Antinociceptive and Anti-Inflammatory Activities of Cuscuta chinensis Seeds in Mice. *Am. J. Chin. Med.* **2014**, *42*, 223–242. [CrossRef] [PubMed]

2. Yang, S.; Xu, X.; Xu, H.; Xu, S.; Lin, Q.; Jia, Z.; Han, T.; Zhang, H.; Zhang, Y.; Liu, H.; Gao, Y.; Li, X. Purification, characterization and biological effect of reversing the kidney-yang deficiency of polysaccharides from semen cuscutae. *Carbohydr. Polym.* **2017**, *175*, 249–256. [CrossRef] [PubMed]

3. Yang, S.; Xu, H.; Zhao, B.; Li, S.; Li, T.; Xu, X.; Zhang, T.; Lin, R.; Li, J.; Li, X. The Difference of Chemical Components and Biological Activities of the Crude Products and the Salt-Processed Product from Semen Cuscutae. *Evid. Based Complement. Altern. Med.* **2016**, *2016*, 8656740. [CrossRef] [PubMed]

4. Wang, J.; Tan, D.; Wei, G.; Guo, Y.; Chen, C.; Zhu, H.; Xue, Y.; Yin, C.; Zhang, Y. Studies on the Chemical Constituents of *Cuscuta chinensis*. *Chem. Nat. Compd.* **2016**, *52*, 1133–1136. [CrossRef]

5. Yang, S.; Kuai, D.; Li, S.; Xu, H.; Li, T.; Xu, X.; Li, X. The Study on Total Flavonoids Content Comparison of Cuscuta chinensis and Three Differently Processed Products. *World Sci. Technol.-Mod. Tradit. Chin. Med. Mater. Medica* **2015**, *17*, 178–181. [CrossRef]

6. Xu, X.; Xu, L.; Guo, Z.; Pan, L.; Yuan, S.; Hao, B.; Lv, J.; Li, X. Study on simultaneous determination of five kinds of flavonoids in seeds of semen cuscutae by HPLC. *World Chin. Med.* **2014**, *9*, 491–493. [CrossRef]

7. Xu, L.; Lv, Y.; Wang, D.; Yang, L.; Li, X. HPLC Simultaneous Determination of 7 Components in Shuanghuangqudu Tablets. *Chin. J. Exp. Tradit. Med. Formulae* **2012**, *18*, 119–122. [CrossRef]

8. Torras-Claveria, L.; Jauregui, O.; Codina, C.; Tiburcio, A.F.; Bastida, J.; Viladomat, F. Analysis of phenolic compounds by high-performance liquid chromatography coupled to electrospray ionization tandem mass spectrometry in senescent and water-stressed tobacco. *Plant Sci.* **2012**, *182*, 71–78. [CrossRef] [PubMed]

9. Gobbo-Neto, L.; Lopes, N.P. Online Identification of Chlorogenic Acids, Sesquiterpene Lactones, and Flavonoids in the Brazilian Arnica *Lychnophora ericoides* Mart. (Asteraceae) Leaves by HPLC-DAD-MS and HPLC-DAD-MS/MS and a Validated HPLC-DAD Method for Their Simultaneous Analysis. *J. Agric. Food Chem.* **2008**, *56*, 1193–1204. [CrossRef] [PubMed]

10. Regos, I.; Urbanella, A.; Treutter, D. Identification and Quantification of Phenolic Compounds from the Forage Legume Sainfoin (*Onobrychis viciifolia*). *J. Agric. Food Chem.* **2009**, *57*, 5843–5852. [CrossRef] [PubMed]

11. Clifford, M.N.; Knight, S.; Kuhnert, N. Discriminating between the Six Isomers of Dicaffeoylquinic Acid by LC-MS^n. *J. Agric. Food Chem.* **2005**, *53*, 3821–3832. [CrossRef] [PubMed]

12. Clifford, M.N.; Johnston, K.L.; Knight, S.; Kuhnert, N. Hierarchical scheme for LC-MSn identification of chlorogenic acids. *J. Agric. Food Chem.* **2003**, *51*, 2900–2911. [CrossRef] [PubMed]

13. Ouyang, H.; Li, J.; Wu, B.; Zhang, X.; Li, Y.; Yang, S.; He, M.; Feng, Y. A robust platform based on ultra-high performance liquid chromatography Quadrupole time of flight tandem mass spectrometry with a two-step data mining strategy in the investigation, classification, and identification of chlorogenic acids in Ainsliaea fragrans Champ. *J. Chromatogr. A* **2017**, *1502*, 38–50. [CrossRef] [PubMed]

14. Ncube, E.N.; Mhlongo, M.I.; Piater, L.A.; Steenkamp, P.A.; Dubery, I.A.; Madala, N.E. Analyses of chlorogenic acids and related cinnamic acid derivatives from Nicotiana tabacumtissues with the aid of UPLC-QTOF-MS_MS based on the in-source collision-induced dissociation method. *Chem. Cent. J.* **2014**, *8*, 66. [CrossRef] [PubMed]

15. Parveen, I.; Threadgill, M.D.; Hauck, B.; Ḍonnison, I.; Winters, A. Isolation, identification and quantitation of hydroxycinnamic acid conjugates, potential platform chemicals, in the leaves and stems of Miscanthus × giganteus using LC-ESI-MSn. *Phytochemistry* **2011**, *72*, 2376–2384. [CrossRef] [PubMed]

16. Kuhnert, N.; Said, I.H.; Jaiswal, R. Assignment of Regio- and Stereochemistry of Natural Products Using Mass Spectrometry Chlorogenic Acids and Derivatives as a Case Study. *Stud. Nat. Prod. Chem.* **2014**, *42*, 305–339. [CrossRef]

17. Jaiswal, R.; Kuhnert, N. Identification and Characterization of the Phenolic Glycosides of *Lagenaria siceraria* Stand. (Bottle Gourd) Fruit by Liquid Chromatography–Tandem Mass Spectrometry. *J. Agric. Food Chem.* **2014**, *62*, 1261–1271. [CrossRef] [PubMed]

18. Jaiswal, R.; Halabi, E.A.; Karar, M.G.E.; Kuhnert, N. Identification and characterisation of the phenolics of Ilex glabra L. Gray (Aquifoliaceae) leaves by liquid chromatography tandem mass spectrometry. *Phytochemistry* **2014**, *106*, 141–155. [CrossRef] [PubMed]

19. Jaiswal, R.; Müller, H.; Müller, A.; Karar, M.G.; Kuhnert, N. Identification and characterization of chlorogenic acids, chlorogenic acid glycosides and flavonoids from Lonicera henryi L. (Caprifoliaceae) leaves by LC-MSn. *Phytochemistry* **2014**, *108*, 252–263. [CrossRef] [PubMed]

20. Ye, M.; Yan, Y.; Guo, D.A. Characterization of phenolic compounds in the Chinese herbal drug Tu-Si-Zi by liquid chromatography coupled to electrospray ionization mass spectrometry. *Rapid Commun. Mass Spectrom.* **2005**, *19*, 1469–1484. [CrossRef] [PubMed]

21. Cuyckens, F.; Claeys, M. Mass spectrometry in the structural analysis of flavonoids. *J. Mass Spectrom.* **2004**, *39*, 1–15. [CrossRef] [PubMed]

molecules

MDPI

Article

Effects of Ultrasound Assisted Extraction in Conjugation with Aid of Actinidin on the Molecular and Physicochemical Properties of Bovine Hide Gelatin

Tanbir Ahmad [1,2], Amin Ismail [3,4], Siti A. Ahmad [5], Khalilah A. Khalil [6], Teik K. Leo [1], Elmutaz A. Awad [7,8], Jurhamid C. Imlan [7,9] and Awis Q. Sazili [1,4,7,*]

[1] Department of Animal Science, Faculty of Agriculture, Universiti Putra Malaysia, Serdang 43400, Selangor, Malaysia; tanbirvet05@rediffmail.com (T.A.); leoteikee@gmail.com (T.K.L.)

[2] ICAR-Central Institute of Post-Harvest Engineering and Technology, Ludhiana, Punjab 141004, India

[3] Faculty of Medicine and Health Sciences, Universiti Putra Malaysia, Serdang 43400, Selangor, Malaysia; aminis@upm.edu.my

[4] Halal Products Research Institute, Putra Infoport, Universiti Putra Malaysia, Serdang, Selangor 43400, Malaysia

[5] Faculty of Biotechnology and Biomolecular Sciences, Universiti Putra Malaysia, Serdang, Selangor 43400, Malaysia; aqlima@upm.edu.my

[6] Faculty of Applied Sciences, Universiti Teknologi MARA, Shah Alam 40450, Selangor, Malaysia; khalilahabdkhalil@gmail.com

[7] Laboratory of Sustainable Animal Production and Biodiversity, Institute of Tropical Agriculture and Food Security, Universiti Putra Malaysia, Serdang, Selangor 43400, Malaysia; motazata83@gmail.com (E.A.A.); jurhamidimlan@yahoo.com.ph (J.C.I.)

[8] Department of Poultry Production, University of Khartoum, Khartoum 13314, Sudan

[9] Department of Animal Science, College of Agriculture, University of Southern Mindanao, Kabacan 9407, North Cotabato, Philippines

* Correspondence: awis@upm.edu.my; Tel.: +60-3-8947-1841

Received: 25 December 2017; Accepted: 12 February 2018; Published: 22 March 2018

Abstract: Actinidin was used to pretreat the bovine hide and ultrasonic wave (53 kHz and 500 W) was used for the time durations of 2, 4 and 6 h at 60 °C to extract gelatin samples (UA2, UA4 and UA6, respectively). Control (UAC) gelatin was extracted using ultrasound for 6 h at 60 °C without enzyme pretreatment. There was significant ($p < 0.05$) increase in gelatin yield as the time duration of ultrasound treatment increased with UA6 giving the highest yield of 19.65%. Gel strength and viscosity of UAC and UA6 extracted gelatin samples were 627.53 and 502.16 g and 16.33 and 15.60 mPa.s, respectively. Longer duration of ultrasound treatment increased amino acids content of the extracted gelatin and UAC exhibited the highest content of amino acids. Progressive degradation of polypeptide chains was observed in the protein pattern of the extracted gelatin as the time duration of ultrasound extraction increased. Fourier transform infrared (FTIR) spectroscopy depicted loss of molecular order and degradation in UA6. Scanning electron microscopy (SEM) revealed protein aggregation and network formation in the gelatin samples with increasing time of ultrasound treatment. The study indicated that ultrasound assisted gelatin extraction using actinidin exhibited high yield with good quality gelatin.

Keywords: ultrasound assisted extraction; gelatin; actinidin; bovine hide; physicochemical properties; gel strength

1. Introduction

Gelatin is a high molecular weight biopolymer obtained from collagen by thermal hydrolysis causing its denaturation. Being a versatile biomaterial, it is extensively used in preparing various food products, medicines, cosmetic items and in photography because of its film-forming capability, water binding ability and emulsifying and foaming properties [1,2].

Insoluble collagen is required to be converted into soluble form by pretreatment with either acid or alkali resulting in the loss of the triple-helical arrangement of native collagen chains which is swollen but still insoluble [3]. Finally, conversion into gelatin takes place during extraction process due to the cleavage of hydrogen and covalent bonds by heat leading to helix-to-coil transition [4]. Cleavage of covalent and non-covalent bonds in sufficient numbers releases free α chains and oligomers [5]. Additionally, few amide bonds present in the original collagen triple chains are broken down by hydrolysis [6]. Consequently, the recovered gelatin has lower molecular weight polypeptide chains compared to native collagen chain and the extracted gelatin represents a mixture of polypeptide chains having molecular weight ranging from 16 to 150 kDa [7].

There is lack of published research on ultrasound assisted extraction (UAE) of bioactive materials from animal sources [8–10]. UAE can increase extraction efficiency and extraction rate particularly for aqueous extraction and lower processing temperatures can be applied for enhanced extraction of heat sensitive bioactive food components at lower processing temperatures [11]. Its promising effect in food science has attracted attention of food industry [12]. High power ultrasound (power >1 W cm^{-2} and frequencies between 20 and 500 kHz) can be applied to aid the extraction process of different food components such as herbal, oil, protein and polysaccharides including bioactive compounds such asantioxidants from various animal and plant materials [11]. Ultrasonic irradiation increased the yield of collagen from bovine tendon and significantly shortened the extraction time in comparison to the traditional pepsin aided extraction process [13]. The extraction yield of collagen increased with the ultrasonic treatment [14]. Good quality gelatin with high yield (30.94–46.67%) was obtained from bighead carp scales by using ultrasound bath and the presence of α-and β-chains were observed in the resulting gelatin [15].

Collagen cross-links bonds are resistant to thermal and acid hydrolysis [16] resulting in a low gelatin yield [17]. Previously, some proteases capable of breaking the collagen cross-links have been used to increase the extractability of gelatin [17]. Pepsin and proctase (isolated from *Aspergillus niger*) were used to extract the gelatin from bovine hide but the gelatin yield, its gel strengths and viscosities were low [18]. Crude proteolytic enzyme from papaya latex and commercial papain were used to extract gelatin from the raw hide and higher yield was obtained but the gel strength was relatively low and complete degradation of α and β chains in the recovered gelatins were observed in both types of samples [19]. Papain was used to extract gelatin from rawhide splits but the obtained gelatin showed low gel strength and viscosity [20]. Although better gelatin yield was achieved, the functional qualities of the obtained gelatin were lowered. Gelatin with high molecular weight polymers (less degraded peptides) are reported to be better in functional properties [21–24]. Therefore, novel enzymes capable of cleaving long chains of gelatin only at few sites should be explored so that a long chain gelatin of high quality can be produced [25]. Actinidin protease was most specifically effective at hydrolysing meat myofibril proteins out of papain, bromelain, actinidin and zingibain [26]. Earlier study from this laboratory (unpublished results) showed encouraging result in term of gelatin yield and quality when bovine hide was pretreated with actinidin at level of 20 unit of enzyme per gram of hide. Therefore, actinidin has been included in this study. There is no published research work on the ultrasound assisted extraction of gelatin as well as on the ultrasound–enzyme assisted extraction of gelatin from bovine hide. Hence, the objectives of this study were to extract gelatin using ultrasound in conjugation with enzyme actinidin pretreatment and investigating their effects on the quality characteristics of the recovered gelatin.

2. Results and Discussion

2.1. Gelatin Yield

The effects of ultrasound assisted extraction in conjugation with actinidin on gelatin yield are shown in Table 1. The gelatin yield was significantly ($p < 0.05$) increased with increasing the duration of ultrasound treatment. The result was in accord with those of Arnesen and Gildberg [27] and Tu et al. [15] who reported that higher yield of gelatin was obtained with longer extraction time from Atlantic salmon skin and bighead carp scales, respectively. More energy was provided by increasing time to destroy the stabilizing bonds present in the collagen structures and peptide bonds of α-chains resulting in helix-to-coil transformation [28]. At higher temperature, conversion of collagen to gelatin is brought about by destruction of the stabilizing hydrogen bonds of collagen resulting in the transformation of helix-to-coil structure [29]. In addition, few peptides bonds are also broken down [30].

Table 1. Yields, pH, turbidity, gel strength and viscosity of gelatin extracted using ultrasound from bovine hide pretreated with enzyme actinidin. Values are presented as mean ± SE from triplicate determination.

Sample	Yield (%) of Gelatin	pH	Turbidity (ppm)	Gel Strength (g)	Viscosity (mPa.s)
UAC	18.72 ± 018 [b]	2.91 ± 0.02 [c]	53.28 ± 0.47 [b]	627.5 ± 4.48 [a]	16.33 ± 0.03 [a]
UA2	8.64 ± 0.08 [d]	2.75 ± 0.01 [d]	105.53 ± 0.15 [a]	451.5 ± 5.29 [d]	15.67 ± 0.03 [c]
UA4	15.17 ± 0.18 [c]	2.97 ± 0.01 [b]	25.98 ± 0.27 [d]	520.3 ± 4.18 [b]	15.87 ± 0.06 [b]
UA6	19.65 ± 0.19 [a]	3.03 ± 0.02 [a]	32.03 ± 0.15 [c]	502.2 ± 4.06 [c]	15.60 ± 0.04 [c]

[a, b, c, d] Means with different superscripts in the same column indicate significant difference at $p < 0.05$. UA2, UA4 and UA6 refers to ultrasound assisted gelatin extracted for the time duration of 2, 4 and 6 h, respectively using actinidin pretreatment. UAC: control gelatin extracted using ultrasound without enzymatic pretreatment.

The higher yield of gelatin with increasing duration could also be due to cavitation and mechanical effect of ultrasound [15]. Acoustic cavitation is mainly responsible for the increased extraction obtained from UAE [31] as it releases more energy to wash out the gelatin from the hide sample. Besides, ultrasound increases the contact surface area between sample matrix and solvent by producing mechanical effect and thus enabling greater penetration of liquid medium into the solid phase for extraction [32]. Thus, a greater penetration of solvent into sample matrix and improved mass transfer was facilitated by the acoustic cavitation and mechanical effects of ultrasound [33]. Li et al. [13] observed enhanced collagen extraction with the use of ultrasound due to cavitation which opened the collagen fibrils and improved the dispersal of enzyme aggregates and this assisted in carrying molecules of pepsin in the close vicinity of collagen chains affecting the hydrolysis.

In the present study, UA6 had significantly ($p < 0.05$) higher gelatin yield compared to UAC (19.65% vs. 18.72%). This result corroborated the previous findings where higher gelatin yield was obtained with the proteolytic enzymes pretreatment [34–36]. Balti et al. [34] reported extraction yield increased from 2.21% to 7.84% on wet weight basis from skin of cuttle fish (*Sepia officinalis*) when smooth hound crude acid protease at level 15 units/g was used. Bougatef et al. [36] obtained 54.61% and 15.22% gelatin from skin of smooth hound in the presence and absence of smooth hound crude acid protease (SHCAP) enzyme when citric acid was used as pretreatment agent. In addition, Lassoued et al. [35] obtained higher gelatin yield from thornback ray (*Raja clavata*) skin with pepsin pretreatment. Nalinanon et al. [17] also reported markedly higher gelatin yield when proteases were added to extract the gelatin compared to the gelatin yield without enzyme.

2.2. Colour

Colour coordinates a^* and b^* of UA2 sample were significantly ($p < 0.05$) lower than the rest of the samples (Table 2). The highest lightness (L^* value) for UA2 was consistent with the finding of Sinthusamran et al. [37] who reported highest L^* (lightness) for gelatin extracted for short time.

The higher yellowness (b^* value) for UA4, UA6 and UAC samples might be due to non-enzymatic browning reaction [37].

Table 2. Colour of gelatin extracted using ultrasound from bovine hide pretreated with enzyme actinidin. Values are presented as mean $\pm$ SE from triplicate determination.

Treatment	L^*	a^*	b^*
UAC	64.45 $\pm$ 0.29 [b]	1.91 $\pm$ 0.02 [b]	17.10 $\pm$ 0.20 [a]
UA2	73.15 $\pm$ 0.23 [a]	0.26 $\pm$ 0.05 [c]	10.27 $\pm$ 0.18 [d]
UA4	62.64 $\pm$ 0.09 [c]	2.47 $\pm$ 0.05 [a]	16.40 $\pm$ 0.21 [b]
UA6	63.43 $\pm$ 0.55b [c]	1.84 $\pm$ 0.07 [b]	14.57 $\pm$ 0.23 [c]

[a, b, c, d] Means with different superscripts in the same column indicate significant difference at $p < 0.05$. UA2, UA4 and UA6 refers to ultrasound assisted gelatin extracted for the time duration of 2, 4 and 6 h, respectively using actinidin pretreatment. UAC: control gelatin extracted using ultrasound without enzymatic pretreatment.

2.3. pH

There was a significant ($p < 0.05$) increase in pH as the extraction time increased and the highest pH (3.03) was recorded for UA6 (Table 1). Mohtar et al. [38] reported that the pH of bovine gelatin was 5.48. The HCl used for pretreatment of hide could be a possible explanation for the lower pH in our current study. The relationship between the pH of gelatin and processing method used to extract gelatin has not been established yet [39].

2.4. Amino Acid Composition of Gelatin

Gelatin properties are greatly determined by the amino acid composition and molecular weight distribution [40]. The most abundant amino acid in gelatin is glycine [41]. Repeating chains of Gly-X-Y, where X and Y usually denote proline and hydroxyproline, respectively, are present in triple peptides which make up to 50–60% of α-chains [7]. A higher content of proline and hydroxyproline (imino acid) amino acids, particularly hydroxyproline, are found in the gelatins extracted from warm-blooded animal tissues [42].

There is a dearth of published research on the effects of duration of ultrasound treatment on the amino acid content. Improved hydrophobic amino acids content of rice dreg protein (RDP) extracted from rice dreg flour (RDF) using ultrasound treatment was obtained [43]. Micro fractures, molecule unfolding and protein structure changes occurred due to high-intensity shock waves, microjets, shear forces and turbulence produced as a result of cavitation effect [44] leading to increased amino acid content [43]. In present study, glycine, proline and hydroxyproline contents for UAC, UA2, UA4 and UA6 were25.54%, 11.39% and 17.00%;16.86%, 8.33% and 10.77%;18.95%, 9.26% and 12.64%; and 20.60%, 9.78% and 13.65%, respectively (Table 3). The amino acids content increased with the increase in time duration of ultrasonic treatment and UAC exhibited the highest content of amino acids.

Ox skin and calf skin contained 27.6%, 16.5% and 13.4%, and 26.9%, 14.0% and 14.6% glycine (Gly), proline (Pro) and hydroxyproline (Hyp), respectively [45]. Furthermore, Lassoued et al. [35] and Balti et al. [34] reported the glycine, proline and hydroxyproline content of food grade halal bovine gelatin as 34.48%, 13.39% and 9.54%, and 34.1%, 12.3% and 9.6% of the total amino acids, respectively. Our amino acid results are expressed in terms of percentage of sample weight (mg of amino acid per 100 mg of sample). The observed variations in the amino acid contents might be also due to differences in manufacturing processes of gelatin [46].

Table 3. Amino acid composition (per centof gelatin sample) of gelatin samples. UA2, UA4 and UA6 refers to gelatin extracted using ultrasound for the time duration of 2, 4 and 6 h, respectively from bovine hide pretreated with enzyme actinidin. UAC: control gelatin extracted using ultrasound without enzymatic pretreatment.

Amino Acids	Gelatin Samples			
	UAC	UA2	UA4	UA6
Hydroxyproline (Hyp)	17.00	10.77	12.64	13.65
Aspartic acid (Asp)	2.99	3.02	3.29	3.28
Serine (Ser)	3.30	2.30	2.64	2.91
Glutamic acid (Glu)	8.28	6.07	6.61	6.96
Glycine (Gly)	25.54	16.86	18.95	20.60
Histidine (His)	0.96	0.63	0.71	0.74
Arginine (Arg)	8.41	5.43	6.31	6.89
Threonine (Thr)	1.91	1.29	1.49	1.58
Alanine (Ala)	7.64	5.43	6.06	6.31
Proline (Pro)	11.39	8.33	9.26	9.78
Tyrosine (Tyr)	0.66	0.40	0.46	0.50
Valine (Val)	2.18	1.60	1.78	1.87
Lysine (Lys)	3.28	2.37	2.72	2.79
Isoleucine (Ile)	1.34	0.95	1.07	1.12
Leucine (Leu)	2.81	2.02	2.26	2.35
Phenylalanine (Phe)	1.99	1.37	1.55	1.64
Imino acids (Pro + Hyp)	28.39	19.10	21.90	23.43

The amino acid (Pro + Hyp) content of UAC, UA2, UA4, UA6 and were 28.39%, 19.10%, 21.90%, and 23.43%, respectively. The imino acid content in bovine gelatin ranged between 21.90% [34,35] and 23.3% [47]. Hyp content (for UA2, UA4, UA6 and UAC were 10.77%, 12.64%, 13.65% and 17.00%, respectively) were higher than those (9.6% and 9.54%, respectively) previously reported [34,35] in halal bovine gelatin. The stability of the triple helical structure of the gelatins gel is directly dependent on the quantity of Pro and Hyp (imino acids) as nucleation zones are formed in Pro + Hyp rich areas [48]. Additionally, stability to the triple-stranded collagen helix is believed to be provided by Hyp through its ability to form hydrogen bond through its hydroxyl group [48,49]. The high imino acid content as obtained for different samples in this study was reflected in the high gel strength of the UG samples.

All data areexpressed in the unit of g/100 g of gelatin. Measurements were performed in triplicate and data correspond to mean values. Standard deviations were in all cases lower than 2%.

2.5. SDS-PAGE Analysis of Gelatin

Functional properties of gelatin are affected by the amino acid composition, the molecular weights distribution, structure and compositions of its subunits [34]. Pretreated hide samples (PS), UAC, UA2, UA4, and UA6 samples were subjected to SDS-PAGE analysis (Figure 1). Presence of α1 and α2 chains, β chains (covalently linked α-chains dimers) and γ chains (covalently linked α-chains trimers) were observed in the molecular distribution pattern of pretreated hide samples with highest intensity. UA2 sample revealed the presence of β chains, α1 and α2 chains. Progressive degradation to these chains was observed as the time duration of ultrasound treatment increased. Subsequently, there was complete absence of β and α2 chains in UA6 and very faint presence of β and α2 in UAC. α1 chain was observed in both UAC and UA6. The result showed that the ultrasonic treatment for long duration was responsible for the breakdown of the polypeptides chains. Similar molecular weight distribution pattern was observed for all replicates.

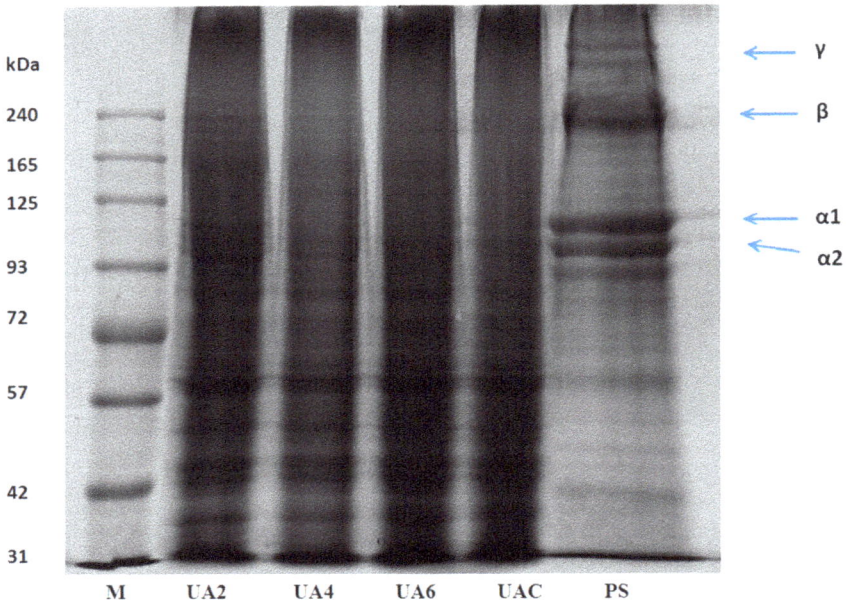

Figure 1. SDS-PAGE pattern of pretreated hide (PS) sample along with gelatin extracted using ultrasound for the time duration of 2, 4 and 6 h (UA2, UA4 and UA6, respectively) from bovine hide with actinidin pretreatment. UAC: control gelatin extracted using ultrasound without enzyme pretreatment. M denotes the marker.

Utilization of ultrasonic in various food products was reported previously. There were no differences in the protein fraction of various food products when ultrasonic treatment was applied for very short durations (i.e., minutes) [50–55]. However, there was a decrease in molecular weight when ultrasound treatment of 20 and 40 kHz was applied for 30 min in whey protein concentrate (WPC) and whey protein isolate (WPI) [56] and α-lactalbumin [57]. Degradation of α chains was observed with long duration of ultrasound assisted extraction of gelatin from bighead carp scales [15]. Degradation of protein molecular structure might be due to higher shear stress and turbulence effects of ultrasound treatment [55].

2.6. Turbidity

The higher turbidity of UA2 reflected its low quality compared to other samples [22,58]. UA4 and UA6 had significantly ($p < 0.05$) lower turbidity than the UAC (Table 1). This might be due to size reduction of the suspended insoluble aggregates by ultrasound [56]. No earlier reports could be found to compare our results.

2.7. Gel Strength

The most significant functional property of gelatin is gel strength which is function of complex interaction decided by molecular weight distribution [22]. Complicated interactions occurring between among amino acid composition and α chain ratio and quantity of β components control the gel strength [34].

The gel strength values of all the ultrasound extracted gelatin (UG) samples were high (Table 1). The highest gel strength value of 627.5 g was found for UAC. The corresponding values for UA2, UA4 and UA6 were 451.5, 520.3 and 502.2 g, respectively. The UA2 sample revealed the presence of β chains along with α1 and α2 chains which degraded progressively and very faint presence of β chain was found in UAC together with only α1 chain. Normally, high molecular weight polypeptides gelatins show high gel strength than the gelatin having low molecular weight distribution [18] because lower weight peptides could not be able to establish inter-junction zones efficiently, failing to form the gelatin chains aggregates leading to low gelling property.

The presence of cross linked two α-chains and the β-component facilitate the peptide chains to regain the triple helical structure when cooled and thereby aids in increasing coiled helix formation during gel maturation resulting in high gel strength [34]. However, the molecular weight distribution and the gelatin molecules aggregate formation could also contribute to the differences in gel strength [34]. Polypeptide chains configuration and the inter-junction zones formed during the maturation process also determine the gel strength [30].

In addition, amino acid composition and the type of extraction treatments also influence the gel strength of gelatin [34]. The imino acid (proline and hydroxyproline) content also governs the gelatin gelling property [59]. Among the two, hydroxyproline is considered the major determining factor for the stability due to its hydrogen bonding ability through the -OH groups [60]. More stable gel structures are formed by the formation of hydrogen bonds by imino acid leading to high gel strength. Significantly ($p < 0.05$) low gel strength of UA2, UA4 and UA6 compared to UAC could be explained by the low proline and hydroxyproline (imino acid) content in these samples, which could be resulted in less organized triple helix structure. Triple helices are partially recovered during maturation of gel and the stability to triple helices is provided by the regions rich in Gly-Pro-Hyp [60].

2.8. Viscosity

Viscosity is the second most important commercial physical property of gelatin [60]. Gelatin having high viscosity is commercially valuable [21]. Collagen kept in hot water gets denatured by the breakdown of the hydrogen and probably electrostatic bonds and thus destroying the triple helical structure of collagen to produce one, two or three random chain gelatin molecules that results in a solution in water of high viscosity [21]. Viscosity is partially governed by molecular weight and polydipersity of the gelatin polypeptides [61] meaning that presence of higher molecular weight components increases viscosity but polydispersity can have variable effect depending on the molecular weight distribution [62].

In this study, viscosity values were 16.33, 15.67, 15.87 and 15.60 mPa.s for UAC, UA2, UA4 and UA6, respectively (Table 1). Presence of enzyme decreased the viscosity significantly ($p < 0.05$). Viscosity of the commercial bovine gelatin was 9.80 cP [38]. Comparatively high viscosities obtained for these samples might be due to particle sized denatured collagen recovered during ultrasonic extraction attributed to cavitation which caused impingement by micro-jets that resulted in surface peeling, erosion and particle breakdown [11].

2.9. FTIR Spectra

Functional groups and secondary structure of gelatin are generally studied using FTIR spectroscopy and the amide I band occurring between 1600 and 1700 cm^{-1} wavenumber is the most crucial to analyse proteins secondary structure using infrared spectroscopy [63]. Amide-I denotes C=O stretching vibration hydrogen bonding coupled with COO, coupled to contributions from the CN stretch, CCN deformation and in-plane NH bending mode [64]. Hydrogen bonding and the conformation of protein structure determines its exact location [65]. Absorption peak at 1633 cm^{-1} is the characteristic of the coiled structure of gelatin [66] and this is in the agreement with our observation of the amide-I peak in the range of 1631–1635 cm^{-1}.

FTIR spectra of UA2, UA4, UA6 and UAC have been depicted in Figure 2 and peak position of different bands has been presented in Table 4. With slight differences, the major peaks were detected in the amide regions. These spectra were in accordance with those reported by [63]. Amide I bands for UA2, UA4, UA6 and UAC were observed at the wavenumbers of 1632, 1632, 1636 and

1632 cm^{-1}, respectively. The amide A amplitudes for all the samples were high and similar. The higher wavenumber along with high amplitude of UA6 showed that the inter-molecular crosslinks were opened thermally resulting in higher loss of molecular order [30] indicating that longer duration of ultrasound treatment along with actinidin pretreatment had caused increased thermal uncoupling of inter-molecular crosslink. Tu et al. [15] also reported higher amide I band for gelatin extracted by ultrasound treatment than that extracted by waterbath method.

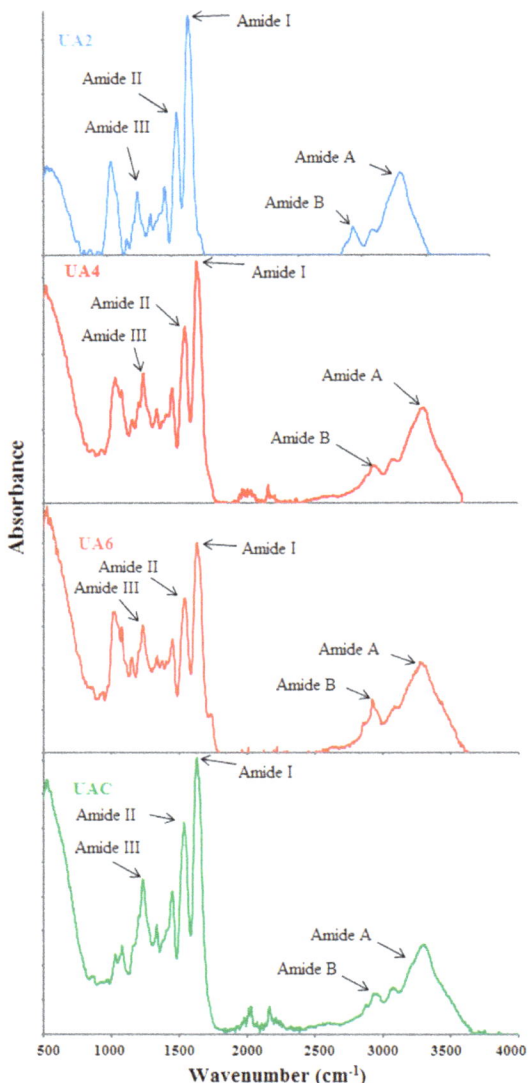

Figure 2. FTIR spectra of gelatin extracted using ultrasound for the time duration of 2, 4 and 6 h (UA2, UA4 and UA6, respectively) with actinidin pretreatment. UAC: control gelatin extracted using ultrasound without enzyme pretreatment.

Table 4. FTIR spectra peak position of gelatin samples extracted for the time duration of 2, 4 and 6 h (UA2, UA4 and UA6, respectively) using ultrasound in conjugation with actinidin pretreatment at level 20 units/g of hide. UAC: control gelatin extracted using ultrasound (U) without enzyme pretreatment.

Band	Peak Wavenumber (cm^{-1})			
	UA2	UA4	UA6	UAC
Amide I	1632	1632	1636	1632
Amide II	1547	1543	1539	1539
Amide III	1238	1238	1242	1234
Amide A	3302	3310	3279	3291
Amide B	2928	2924	2936	2936

Amide II also show alteration in the gelatin secondary structure [67] but specifically it reflects more about the degree of gelatin hydration than its structure [29]. Amide II vibrational modes indicate an out-of-phase combination of CN stretch and in-plane NH deformation modes of the peptide group [30]. Dry collagen had the amide II band in the infrared spectrum range of 1530–1540 cm^{-1} and often had minor bands at lower frequencies [15]. The shifting of amide II to lower wavenumber with lower amplitude suggested H-bond formation with adjacent chains by NH groups [60]. The characteristic absorption bands of UA2, UA4, UA6 and UAC gelatin in amide-II region were shifted to lower wavenumber as the time duration for ultrasound irradiation increased and were observed at the wavenumbers of 1547, 1543, 1539 and 1539 cm^{-1}, respectively indicative of higher NH group involvement in hydrogen bonding particularly in UA6 since its showed lower wavenumber and amplitude. Although the amplitudes of UB2, UA4 and UA6 were lower compared to UAC, UA2 and UA4 displayed amide II at higher wavenumber than UAC.

Amide III spectra for UA2, UA4, UA6 and UAC gelatin were detected at wavenumber of 1238, 1238, 1242 and 1234 cm^{-1}. Amide III absorption spectra represents a complex vibrational mode having components due to C-N stretching and N-H in plane bending arising due to amide linkages as well as significant absorptions arising from the wagging vibrations of CH$_2$ groups from the glycine back bone and proline side chains and this is generally seen in the region of 1200–1400 cm^{-1} [68]. Amide III band displayed in the range of 1233–1234 cm^{-1} suggested triple helical structure loss resulting from disordered gelatin molecules [37]. In addition, lower amplitude exhibited by amide III indicated loss of triple helix structure into random coiled structure resulting from denaturation of collagen into gelatin due to disruption in natural α helix structure of protein chains [63]. The occurrence of amide III of UAC near the 1234 cm^{-1} and its relatively lower amplitude compared to other treatment samples indicated loss of triple helical structure in UAC. Additionally, some more peaks for all the samples were observed at lower than amide III regions because of stretching vibrations of C-O group present in the smaller peptides [68].

Amide A band arising from NH-stretching coupled with hydrogen which is detected in the range of wavenumber of 3400–3440 cm^{-1} for gelatin samples [63] and involvement of N-H group of a peptide in hydrogen bonding shifts this band to lower wavenumber of around 3300 cm^{-1} [15]. The amide A band of the triple-helix biopolymer shifted to lower frequencies because of hydrogen bond formation by the N-H group of a peptide [28]. For samples UA2, UA4, UA6 and UAC, amide A appeared at 3302, 3310, 3279 and 3291 cm^{-1}, respectively. The lower wavenumber of UA6 and UAC compared to UA2 and UA4 indicated higher hydrogen bond formation with the participation of N-H group in α chains. Shifting to lower wavenumber as well as high amplitude of amide A suggested gelatin degradation [28]. The lowest wavenumber along with high amplitude of UA6 implied degraded gelatin. Although UAC displayed amide A at lower wavenumber but its amplitude was lowest amongst the samples. The concurrent effect of actinidin and ultrasound might have brought this difference between the UA6 and UAC.

Asymmetric stretching vibration of =C-H and NH$_3{}^+$ is represented by amide B bands [30]. The amide B for UA2, UA4, UA6 and UAC were discovered at 2928, 2924, 2936 and 2936 cm^{-1}, respectively. Lower wavenumber of UA2 and UA4 compared to UA6 and UAC suggested higher interaction of -NH$_3$ group between peptide chains in UA2 and UA4 [28,30]. Thus, it can be concluded that the secondary structures and functional groups were affected by the ultrasound duration and enzyme pretreatment.

2.10. Microstructure of Gelatin

Microstructure of gelatin is associated with the gelatin physical properties. UA2 displayed less dense, sheet-like structure having particles of bigger size compared to other samples. As the duration of ultrasonic treatment increased, the gelatins structure became denser, inter-connected and disorganized with increasing smaller particles size with increasing voids (Figure 3). Density of the structure increased with the duration of ultrasonic treatment. The particles size of UA6 was smaller and more disorganized than UAC. This might be due to proteolytic degradation by actinidin. Partial unfolding of protein took place under high-power ultrasound whereby, functional groups (such as hydrophobic groups) were exposed and this led to immediate interaction with each other resulting in protein aggregation and network formation [69]. Taking into account the gel strength of the different samples, it seemed that rather than voids, it was the density, large particles size and absence of sheet structure that had more assertive positive effects on the physical properties of gelatin. The result indicated that actinidin pretreatment with ultrasonication resulted in change in the gelatin microstructure.

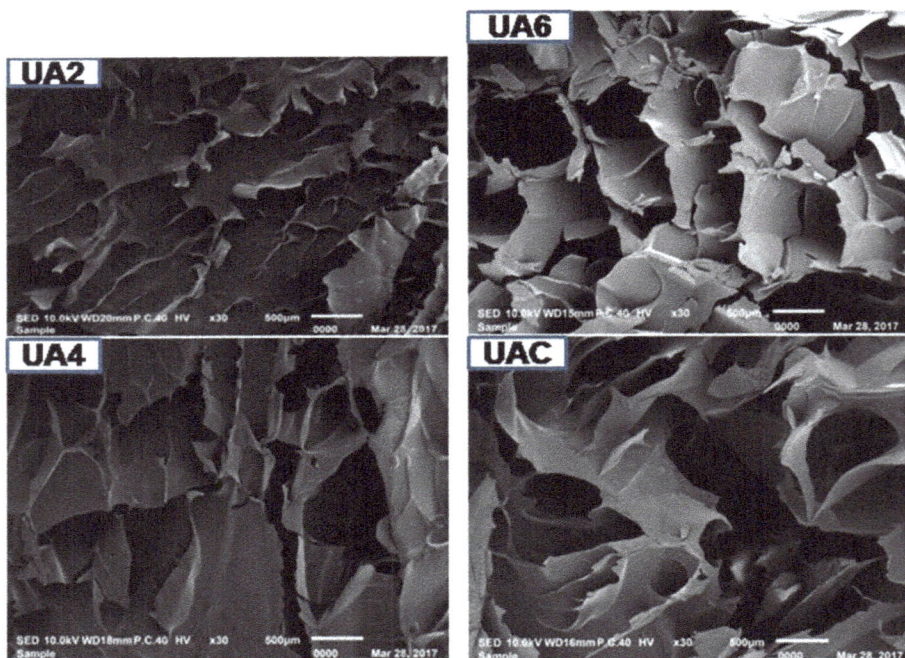

Figure 3. SEM images of gelatin extracted using ultrasound for the time duration of 2, 4 and 6 h (UA2, UA4 and UA6, respectively) with actinidin pretreatment. UAC: control gelatin extracted using ultrasound without enzyme pretreatment.

3. Materials and Methods

3.1. Chemicals

Acrylamide, sodium dodecyl sulphate (SDS), *N,N,N′,N′*-tetramethyl ethylene diamine (TEMED), coomassie brilliant blue R-250, 2-mercaptoethanolwere purchased from Merck, Darmstadt, Germany. Other chemicals and reagents used were of analytical grade. Enzyme actinidin (>30 casein unit/g) obtained from kiwi fruit (*Actinidadeliciosa*) was kindly gifted by Ingredient Resources Pty Ltd., Warriewood, NSW, Australia. Reagents and amino acid standards were purchased from Waters Corporation, Milford, MA, USA and hydroxyproline standard supplement was procured from Agilent Technologies, Santa Clara, CA, USA. The internal standard (*S*)-(+)-2-Aminobutyric Acid (AABA) was purchased from TCI (Tokyo Chemicals Industry Co., Ltd., Chuo-ku, Japan).

3.2. Preparation of Hide

Hide from three- to four-year-old female Brahma cross was procured from a local commercial ruminant abattoir located in Shah Alam, Selangor, Malaysia and transported in ice and stored at −20 °C. The subcutaneous fat was removed by scrapping. The hide was washed thoroughly and stored at −20 °C until further gelatin extraction. It was thawed overnight at 4 °C before being used.

3.3. Ultrasound Assisted Extraction of Gelatin from Bovine Hide in Conjugation with Enzyme Actinidin

3.3.1. Removal of Non-Collagenous Proteins

The non-collagenous materials were removed by treating the hide with 0.1 M NaOH (*w/v*) solution at a hide/solution ratio of 1:5 (*w/v*) stirred at room temperature (25 ± 1 °C) for 6 h, and solution was changed at every 2 h interval. Thereafter, the hairs on the hide were removed by scrapping with scalpel and cut into 1 cm × 2 cm size. The hide was rinsed thoroughly with distilled water until neutral pH wash water was obtained.

3.3.2. Ultrasound Assisted Gelatin Extraction in Conjugation with Enzyme Actinidin

The hide was soaked in 1% HCl for 20 h with discontinuous stirring at a ratio of 1:10 (*w/v*) at room temperature for swelling. The samples were washed thoroughly with distilled water until neutral wash water was obtained. From our previous unpublished results, the level of 20 unit/g of actinidin was found to improve the extraction and quality characteristics of gelatin. Therefore, the swollen hides were incubated with enzymes actinidin for 48 h at the level of 20 unit per g of wet hide at their optimum temperature and pH (20 °C and 7.5, respectively) as indicated by the manufacturers.

The swollen hide samples were kept in the optimum pH solution at hide to solution ratio of 1:3 (*w/v*) and the enzymes were added. The mixture was kept in the orbital shaker incubator (LM-510RD, Yihder, Xinbei, Taiwan) at 20 °C and stirred for 48 h. Thereafter, the mixture was kept in water bath at 90 °C for 15 min to terminate the enzyme activity. Gelatin was extracted at 60 °C for the time duration of 2, 4 or 6 h in ultrasonic bath (SK8210HP, Kudos, Shanghai, China) using 53 kHz frequency and ultrasonic power of 500 W. The mixture was filtered using cheese cloth and then centrifuged (Beckman Coulter Avanti J-26 XPI, Brea, CA, USA) at 12,800× *g* for 20 min. The supernatant was dried using freeze drier (Labconco FreeZone[18], Kansas City, MO, USA) and the dry matter obtained, referred to as "gelatin powder", was stored at 4 °C for further analysis. Control gelatin was extracted using ultrasound treatment at 60 °C for 6 h without the above mentioned enzymatic treatment. The extraction was performed in triplicate.

3.4. Analyses of Gelatin

3.4.1. Yield

The yield of the gelatin was calculated on the wet weight basis of the hide as reported previously [34–36,70].

$$\text{Yield (\%)} = \frac{\text{Weight of the freeze dried gelatin (g)} \times 100}{\text{Wet weight of the hide (g)}}$$

3.4.2. Determination of Colour

ColorFlexHunterLab (Hunter Associates Laboratory Inc., Reston, VA, USA) was used to measure the colour of the gelatin samples. Three colour co-ordinates, namely L^* (lightness), a^* (redness/greenness) and b^* (yellowness/blueness) were used [71]. The sample was filled in a 64 mm glass sample cup with three readings in the same place and triplicate determinations were taken per sample.

3.4.3. Determination of pH

The BSI 757 of British Standard Institute method was used [72]. One percent (w/v) of gelatin solution (0.2 g in 20 mL distilled water) was prepared and it was cooled to room temperature of about 25 °C. The pH meter (Mettler Toledo, AG 8603, Schwerzenbach, Switzerland) was standardized with pH 4.0 and 7.0 buffers and pH determination was carried out in triplicates.

3.4.4. Determination of Amino Acid Composition

Procedure of Awad et al. [73] was used with a slight modification to determine the amino acid (AA) content of the gelatin samples using high performance liquid chromatography (HPLC, Milford, MA, USA).Shortly, 5 mL of 6 M HCl was used to hydrolyse 0.1 to 0.2 g of sample at 110 °C for 24 h. Upon cooling, 4 mL of internal standard (L-amino-N-butyric acid; AABA) was added to the hydrolysate and aliquot was paper and syringe filtered. Ten microlitres of the filtered sample was mixed with 70 μL of borate buffer and 20 μL of ACCQ reagent (Waters Corporation, Milford, MA, USA). A mixture of amino acid standard H (Waters Corporation, MA, USA) and the AABA internal standard (TCI, Chuo-ku, Japan) was spiked with hydroxyproline (Agilent Technologies, CA, USA). The resulting solution was used for derivatization as the working standard. The method was followed to determine the concentration of all AA except methionine, cysteine and tryptophan. Then, an AA column (AccQ Tag 3.9 150 mm; Waters Corporation, MA, USA) was used for peaks separation. Peaks were detected by a fluorescent detector (2475; Waters Corporation, MA, USA). Triplicate determinations were performed and data corresponds to mean values. Standard deviations in all cases lower than 2%.

3.4.5. Electrophoretic Analysis

The molecular weight distributions of the extracted gelatins were determined by SDS-PAGE [74]. Dry gelatin (10 mg) was dissolved in distilled water (1 mL) at 60 °C to create a 10 mg/mL solution. The sample solution was mixed in a 1:2 (v/v) ratio with a 5-fold-concentrated loading buffer (3.55 mL deionized water, 1.25 mL 0.5 M Tris-HCl, pH 6.8, 2.5 mL glycerol, 2.0 mL 10% (w/v) SDS, 0.2 mL 0.5% (w/v) bromophenol blue) containing β-mercaptoethanol (50 μL β-mercaptoethanol+ 950 μL sample buffer prior to use). The mixed solution was heated in boiling water (95 °C) for 5 min before loading onto 4% stacking gels and 7.5% resolving gels. Gel electrophoresis (Mini-PROTEAN Tetra System, Bio-Rad Laboratories, Irvine, CA, USA) was run at a constant current of 15 mA/gel for 15 min; followed by 25 mA/gel until the bromophenol blue dye reached at the bottom of the gel. Following electrophoresis, the gel was stained with 0.1% (w/v) coomassie blue R-250 in 15% (v/v) methanol and 5% (v/v) acetic acid for 2 h and destained with 30% (v/v) methanol and 10% (v/v) acetic acid until the zones on the blue background were clear. Prestained protein ladder (BLUeye, GeneDireX, Keelung

City, Taiwan) was used to estimate the molecular weight distributions of the gelatins. The gel was scanned with a GS-800 Calibrated Densitometer (Bio-Rad Laboratories, CA, USA) gel imaging system.

3.4.6. Determination of Turbidity

Method of Cho et al. [75] was modified slightly to determine the turbidity of the gelatin solutions. Gelatin sample (0.025 g) was dissolved in distilled water (5 mL) at 60 °C to make 0.5% (*w/w*) solution. Absorbance was measured at 660 nm by spectrophotometer (Shimadzu UV Spectrophotometer, Model UV-1800, Kyoto, Japan).

3.4.7. Determination of Gel Strength

Method of Fernandez-Díaz [76] was slightly modified to determine the gel strength of the extracted gelatin. Gelatin (2.0 g) was dissolved in 30 mL of distilled water at 60 °C using 50 mL-beaker (Schott Duran, Mainz, Germany) to get the final concentration of 6.67% (*w/v*). The solution was stirred until gelatin was solubilized completely, and kept at 7 °C for 16–18 h for gel maturation. Bloom strength was measured using Model TA-XT2*i* Texture Analyzer (Stable Micro Systems, Surrey, UK) using a load cell of 5 kN equipped with a 1.27 cm diameter flat-faced cylindrical Teflon plunger (P/0.5R). The dimensions of the sample were3.8 cm in diameter and 2.7 cm in height. The maximum force (in grams) was recorded when the probe penetrated a distance of 4mm inside the sample. The speed of the plunger was 0.5 mm/s. All determinations are means of three measurements.

3.4.8. Determination of Viscosity

Gelatin solution of 6.67% was prepared by dissolving 1.34 g of gelatin in 20 mL of distilled water and heated to 60 °C. RheolabQC (Anton Paar, Graz, Austria) viscometer was used to measure the viscosity of the samples. The measurement was performed in triplicate.

3.4.9. Fourier Transform Infrared (FTIR) Spectroscopy

FTIR spectra were obtained using spectrometer (Perkin Elmer Ltd., Model: Spectrum 100, Tempe, AZ, USA) equipped with a deuterated triglycine sulphate (DTGS) detector. The attenuated total reflectance (ATR) accessory was mounted into the sample compartment. Diamond internal reflection crystal had a 45° angle of incidence to the IR beam. Resolution of 4 cm^{-1} was used to acquire the spectra and 4000–500 cm^{-1} (mid-IR region) was chosen as measurement range at room temperature. Automatic signals were collected in 16 scans and were normalized against a background spectrum recorded from the clean, empty cell at 25 °C.

3.4.10. Microstructure Analysis of Gelatin

Scanning electron microscope (SEM) (JEOL JSM-IT100 InTouchScope, Tokyo, Japan) was used to elucidate the microstructures of gelatin. Dried gelatin samples having a thickness of 2–3 mm were mounted on a bronze stub and sputter-coated with gold (BAL-TEC SCD 005 sputter coater, Schalksmühle, Germany). An acceleration voltage of 10 kV was used to observe the specimen at 30×.

3.5. Statistical Analysis

All statistical analyses were carried out using GLM procedure of Statistical Analysis System package (SAS) Version 9.4 software (Statistical Analysis System, SAS Institute Inc., Cary, NC, USA) and statistical significance was set at $p < 0.05$. Significant differences between means were evaluated by Duncan's Multiple Range Test.

4. Conclusions

Ultrasonication (53 kHz and 500 W) for 6 h at 60 °C can significantly increase the gelatin recovery in conjugation with enzyme actinidin pretreatment. The obtained gelatin showed higher gel strength and viscosity. SDS-PAGE analysis showed progressive degradation of protein chains as the time duration of ultrasound treatment increased. UA2 samples revealed the presence of β, α1 and α2 chains but there was complete absence of β and α2 chains in UA6 and very faint presence of β and α2 chains in UAC. Both UAC and UA6 showed the presence of α1 chain. Amino acids content of the extracted gelatin increased as the time duration of ultrasonic treatment increased. FTIR spectra demonstrated greater loss of molecular order in UA6 and its degradation which might be due to thermal uncoupling of inter-molecular crosslink resulting from longer duration of ultrasound treatment and actinidin pretreatment. SEM images indicated increasing time of ultrasound extraction caused protein aggregation and network formation in the gelatins resulting in increased density and decreased structural integrity.

Acknowledgments: The research work was supported by Putra Grant vide letter No. UPM/700-2/1/GP-IPS/2015/9467000. The first author acknowledgesIndian Council of Agricultural Research, New Delhi, Indiafor providing ICAR-International Fellowship vide letter number F. No. 29-1/2009-EQR/Edn (pt. III) and Department of Agricultural Research & Education, Ministry of Agriculture, Government of India for granting study leave to him (letter number F. No. 7-46/2014-IC II). A special thanks to Director, ICAR-Central Institute of Post-Harvest Engineering and Technology, Ludhiana, Punjab, India for relieving the first author to pursue Ph.D. study. The authors are obliged to Ingredient Resources Pty Ltd., NSW, Australia for gifting actinidin.

Author Contributions: Tanbir Ahmad did the experiments, analysed the data and wrote the article. Amin Ismail and Khalilah A. Khalil helped in designing experiments and contributed in analytical reagents/materials. Siti A. Ahmad facilitated and supervised the research work. Teik K. Leo, Elmutaz A. Awad and Jurhamid C. Imlan helped in some of the analyses of the gelatin samples. Awis Q. Sazili conceived the idea, supervised, secured the grant and facilitated the research work.

Conflicts of Interest: The authors declare no conflict of interest.

References

1. Giménez, B.; Gómez-Guillén, M.; Montero, P. The role of salt washing of fish skins in chemical and rheological properties of gelatin extracted. *Food Hydrocoll.* **2005**, *19*, 951–957. [CrossRef]
2. Gómez-Guillén, M.C.; Giménez, B.; López-Caballero, M.E.; Montero, M.P. Functional and bioactive properties of collagen and gelatin from alternative sources: A review. *Food Hydrocoll.* **2011**, *25*, 1813–1827. [CrossRef]
3. Stainsby, G. Gelatin gels. In *Collagen as a Food. Advances in Meat Research*; Pearson, A.M., Dutson, T.R., Bailey, A.J., Eds.; Van Nostrand Reinhold Company, Inc.: New York, NY, USA, 1987; pp. 209–222.
4. Djabourov, M.; Lechaire, J.-P.; Gaill, F. Structure and rheology of gelatin and collagen gels. *Biorheology* **1993**, *30*, 191–205. [CrossRef] [PubMed]
5. Johnston-Banks, F.A. Gelatine. In *Food Gels*; Harris, P., Ed.; Elsevier Applied Sciences Publishers: New York, NY, USA, 1990; pp. 233–289.
6. Bailey, A. Round table session 1–Structure of collagen. *Adv. Meat Res.* **1985**, *4*, 131–143.
7. Asghar, A.; Henrickson, R.L. Chemical, biochemical, functional, and nutritional characteristics of collagen in food systems. In *Advances in Food Research*; Chischester, C.O., Mark, E.M., Stewart, G.F., Eds.; Academic Press: London, UK, 1982; pp. 231–372.
8. Shishov, A.; Bulatov, A.; Locatelli, M.; Carradori, S.; Andruch, V. Application of deep eutectic solvents in analytical chemistry. A review. *Microchem. J.* **2017**, *135*, 33–38. [CrossRef]
9. Kabir, A.; Locatelli, M.; Ulusoy, H.I. Recent Trends in Microextraction Techniques Employed in Analytical and Bioanalytical Sample Preparation. *Separations* **2017**, *4*, 36. [CrossRef]
10. Diuzheva, A.; Carradori, S.; Andruch, V.; Locatelli, M.; De Luca, E.; Tiecco, M.; Tiecco, M.; Germani, R.; Menghini, L.; Nocentini, A.; et al. Use of innovative (micro) extraction techniques to characterise *Harpagophytumprocumbens* root and its commercial food supplements. *Phytochem. Anal.* **2017**. [CrossRef] [PubMed]
11. Vilkhu, K.; Mawson, R.; Simons, L.; Bates, D. Applications and opportunities for ultrasound assisted extraction in the food industry—A review. *Innov. Food Sci. Emerg. Technol.* **2008**, *9*, 161–169. [CrossRef]
12. Jia, J.; Ma, H.; Zhao, W.; Wang, Z.; Tian, W.; Luo, L.; He, R. The use of ultrasound for enzymatic preparation of ACE-inhibitory peptides from wheat germ protein. *Food Chem.* **2010**, *119*, 336–342. [CrossRef]
13. Li, D.; Mu, C.; Cai, S.; Lin, W. Ultrasonic irradiation in the enzymatic extraction of collagen. *Ultrason. Sonochem.* **2009**, *16*, 605–609. [CrossRef] [PubMed]

14. Kim, H.K.; Kim, Y.H.; Kim, Y.J.; Park, H.J.; Lee, N.H. Effects of ultrasonic treatment on collagen extraction from skins of the sea bass *Lateolabrax japonicus*. *Fish Sci.* **2012**, *78*, 485–490. [CrossRef]

15. Tu, Z.-C.; Huang, T.; Wang, H.; Sha, X.-M.; Shi, Y.; Huang, X.-Q.; Man, Z.-Z.; Li, D.-J. Physico-chemical properties of gelatin from bighead carp (*Hypophthalmichthys nobilis*) scales by ultrasound-assisted extraction. *J. Food Sci. Technol.* **2015**, *52*, 2166. [CrossRef] [PubMed]

16. Galea, C.A.; Dalrymple, B.P.; Kuypers, R.; Blakeley, R. Modification of the substrate specificity of porcine pepsin for the enzymatic production of bovine skin gelatin. *Protein Sci.* **2000**, *9*, 1947–1959. [CrossRef] [PubMed]

17. Nalinanon, S.; Benjakul, S.; Visessanguan, W.; Kishimura, H. Improvement of gelatin extraction from bigeye snapper skin using pepsin-aided process in combination with protease inhibitor. *Food Hydrocoll.* **2008**, *22*, 615–622. [CrossRef]

18. Chomarat, N.; Robert, L.; Seris, J.L.; Kern, P. Comparative efficiency of pepsin and proctase for the preparation of bovine skin gelatin. *Enzym. Microb. Technol.* **1994**, *16*, 756–760. [CrossRef]

19. Pitpreecha, S.; Damrongsakkul, S. Hydrolysis of raw hide using proteolytic enzyme extracted from papaya latex. *Korean J. Chem. Eng.* **2006**, *23*, 972–976. [CrossRef]

20. Damrongsakkul, S.; Ratanathammapan, K.; Komolpis, K.; Tanthapanichakoon, W. Enzymatic hydrolysis of rawhide using papain and neutrase. *Ind. Eng. Chem. Res.* **2008**, *14*, 202–206. [CrossRef]

21. Badii, F.; Howell, N.K. Fish gelatin: Structure, gelling properties and interaction with egg albumen proteins. *Food Hydrocoll.* **2006**, *20*, 630–640. [CrossRef]

22. Gómez-Guillén, M.C.; Turnay, J.; Fernández-Díaz, M.D.; Ulmo, N.; Lizarbe, M.A.; Montero, P. Structural and physical properties of gelatin extracted from different marine species: A comparative study. *Food Hydrocoll.* **2002**, *16*, 25–34. [CrossRef]

23. Muyonga, J.H.; Cole, C.G.B.; Duodu, K.G. Extraction and physico-chemical characterisation of Nile perch (*Lates niloticus*) skin and bone gelatin. *Food Hydrocoll.* **2004**, *18*, 581–592. [CrossRef]

24. Zhang, F.; Xu, S.; Wang, Z. Pre-treatment optimization and properties of gelatin from freshwater fish scales. *Food Bioprod. Process.* **2011**, *89*, 185–193. [CrossRef]

25. Ahmad, T.; Ismail, A.; Ahmad, S.A.; Khalil, K.A.; Kumar, Y.; Adeyemi, K.D.; Sazili, A.Q. Recent advances on the role of process variables affecting gelatin yield and characteristics with special reference to enzymatic extraction: A review. *Food Hydrocoll.* **2017**, *63*, 85–96. [CrossRef]

26. Ha, M.; Bekhit, A.E.-D.A.; Carne, A.; Hopkins, D.L. Characterisation of commercial papain, bromelain, actinidin and zingibain protease preparations and their activities toward meat proteins. *Food Chem.* **2012**, *134*, 95–105. [CrossRef]

27. Arnesen, J.A.; Gildberg, A. Extraction and characterisation of gelatine from Atlantic salmon (*Salmo salar*) skin. *Bioresour. Technol.* **2007**, *98*, 53–57. [CrossRef] [PubMed]

28. Nagarajan, M.; Benjakul, S.; Prodpran, T.; Songtipya, P.; Kishimura, H. Characteristics and functional properties of gelatin from splendid squid (*Loligo formosana*) skin as affected by extraction temperatures. *Food Hydrocoll.* **2012**, *29*, 389–397. [CrossRef]

29. Benjakul, S.; Oungbho, K.; Visessanguan, W.; Thiansilakul, Y.; Roytrakul, S. Characteristics of gelatin from the skins of bigeye snapper, Priacanthustayenus and Priacanthusmacracanthus. *Food Chem.* **2009**, *116*, 445–451. [CrossRef]

30. Ahmad, M.; Benjakul, S. Characteristics of gelatin from the skin of unicorn leatherjacket (*Aluterus monoceros*) as influenced by acid pretreatment and extraction time. *Food Hydrocoll.* **2011**, *25*, 381–388. [CrossRef]

31. Wang, J.; Sun, B.; Cao, Y.; Tian, Y.; Li, X. Optimisation of ultrasound-assisted extraction of phenolic compounds from wheat bran. *Food Chem.* **2008**, *106*, 804–810. [CrossRef]

32. Rostagno, M.A.; Palma, M.; Barroso, C.G. Ultrasound-assisted extraction of soy isoflavones. *J. Chromatogr. A* **2003**, *1012*, 119–128. [CrossRef]

33. Chemat, F.; Khan, M.K. Applications of ultrasound in food technology: Processing, preservation and extraction. *Ultrason. Sonochem.* **2011**, *18*, 813–835. [CrossRef] [PubMed]

34. Balti, R.; Jridi, M.; Sila, A.; Souissi, N.; Nedjar-Arroume, N.; Guillochon, D.; Nasri, M. Extraction and functional properties of gelatin from the skin of cuttlefish (*Sepia officinalis*) using smooth hound crude acid protease-aided process. *Food Hydrocoll.* **2011**, *25*, 943–950. [CrossRef]

35. Lassoued, I.; Jridi, M.; Nasri, R.; Dammak, A.; Hajji, M.; Nasri, M.; Barkia, A. Characteristics and functional properties of gelatin from thornback ray skin obtained by pepsin-aided process in comparison with commercial halal bovine gelatin. *Food Hydrocoll.* **2014**, *41*, 309–318. [CrossRef]

36. Bougatef, A.; Balti, R.; Sila, A.; Nasri, R.; Graiaa, G.; Nasri, M. Recovery and physicochemical properties of smooth hound (*Mustelus mustelus*) skin gelatin. *LWT-Food Sci. Technol.* **2012**, *48*, 248–254. [CrossRef]

37. Sinthusamran, S.; Benjakul, S.; Kishimura, H. Characteristics and gel properties of gelatin from skin of seabass *(Lates calcarifer)* as influenced by extraction conditions. *Food Chem.* **2014**, *152*, 276–284. [CrossRef] [PubMed]

38. Mohtar, N.F.; Perera, C.; Quek, S.-Y. Optimisation of *gelatine extraction* from hoki (Macruronusnovaezelandiae) skins and measurement of gel strength and SDS–PAGE. *Food Chem.* **2010**, *122*, 307–313. [CrossRef]

39. Park, J.-H.; Choe, J.-H.; Kim, H.-W.; Hwang, K.-E.; Song, D.-H.; Yeo, E.-J.; Kim, H.-Y.; Choi, Y.-S.; Lee, S.-H.; Kim, C.-J. Effects of various extraction methods on quality characteristics of duck feet gelatin. *Korean J. Food Sci. Anim. Resour.* **2013**, *33*, 162–169. [CrossRef]

40. Gómez-Guillén, M.C.; Pérez-Mateos, M.; Gómez-Estaca, J.; López-Caballero, E.; Giménez, B.; Montero, P. Fish gelatin: A renewable material for developing active biodegradable films. *Trends Food Sci. Technol.* **2009**, *20*, 3–16. [CrossRef]

41. Arnesen, J.A.; Gildberg, A. Preparation and characterisation of gelatine from the skin of harp seal (*Phoca groendlandica*). *Bioresour. Technol.* **2002**, *82*, 191–194. [CrossRef]

42. Norland, R.E. Fish gelatin. In *Advances in Fisheries Technology and Biotechnology for Increased Profitability*; Voight, M.N., Botta, J.K., Eds.; Technomic Publishing Co.: Lancaster, UK, 1990; pp. 325–333.

43. Li, K.; Ma, H.; Li, S.; Zhang, C.; Dai, C. Effect of ultrasound on alkali extraction protein from rice dreg flour. *J. Food Process Eng.* **2017**, *40*. [CrossRef]

44. Chandrapala, J.; Zisu, B.; Kentish, S.; Ashokkumar, M. The effects of high-intensity ultrasound on the structural and functional properties of α-Lactalbumin, β-Lactoglobulin and their mixtures. *Food Res. Int.* **2012**, *48*, 940–943. [CrossRef]

45. Ward, A.G.; Courts, A. *Science and technology of gelatin*; Academic Press: London, UK, 1977.

46. Zhou, P.; Regenstein, J.M. Determination of total protein content in gelatin solutions with the Lowry or Biuret assay. *J. Food Sci.* **2006**, *71*, C474–C479. [CrossRef]

47. Kasankala, L.M.; Xue, Y.; Weilong, Y.; Hong, S.D.; He, Q. Optimization of gelatine extraction from grass carp (*Catenopharyngodon idella*) fish skin by response surface methodology. *Bioresour. Technol.* **2007**, *98*, 3338–3343. [CrossRef] [PubMed]

48. Ledward, D.A. Gelation of gelatin. In *Functional Properties of Food Macromolecules*; Mitchell, J.R., Ledward, D.A., Eds.; Elsevier Applied Science Publishers: London, UK, 1986; pp. 171–201.

49. Mizuno, K.; Hayashi, T.; Bächinger, H.P. Hydroxylation-induced Stabilization of the Collagen Triple Helix Further characterization of peptides with 4 (r)-hydroxyproline in the XAA position. *J. Biol. Chem.* **2003**, *278*, 32373–32379. [CrossRef] [PubMed]

50. Krise, K.M. The Effects of Microviscosity, Bound Water and Protein Mobility on the Radiolysis and Sonolysis of Hen Egg White. Ph.D Thesis, the Pennsylvania State University, State College, PA, USA, 2011. Publication Number: 3483787.

51. Hu, H.; Wu, J.; Li-Chan, E.C.Y.; Zhu, L.; Zhang, F.; Xu, X.; Fan, G.; Wang, L.; Huang, X.; Pan, S. Effects of ultrasound on structural and physical properties of soy protein isolate (SPI) dispersions. *Food Hydrocoll.* **2013**, *30*, 647–655. [CrossRef]

52. Karki, B.; Lamsal, B.P.; Jung, S.; van Leeuwen, J.H.; Pometto, A.L.; Grewell, D.; Khanal, S.K. Enhancing protein and sugar release from defatted soy flakes using ultrasound technology. *J. Food Eng.* **2010**, *96*, 270–278. [CrossRef]

53. Gülseren, İ.; Güzey, D.; Bruce, B.D.; Weiss, J. Structural and functional changes in ultrasonicated bovine serum albumin solutions. *Ultrason. Sonochem.* **2007**, *14*, 173–183. [CrossRef] [PubMed]

54. O'Sullivan, J.; Arellano, M.; Pichot, R.; Norton, I. The effect of ultrasound treatment on the structural, physical and emulsifying properties of dairy proteins. *Food Hydrocoll.* **2014**, *42*, 386–396. [CrossRef]

55. Yanjun, S.; Jianhang, C.; Shuwen, Z.; Hongjuan, L.; Jing, L.; Lu, L.; Uluko, H.; Yanling, S.; Wenming, C.; Wupeng, G. Effect of power ultrasound pre-treatment on the physical and functional properties of reconstituted milk protein concentrate. *J. Food Eng.* **2014**, *124*, 11–18. [CrossRef]

56. Jambrak, A.R.; Mason, T.J.; Lelas, V.; Paniwnyk, L.; Herceg, Z. Effect of ultrasound treatment on particle size and molecular weight of whey proteins. *J. Food Eng.* **2014**, *121*, 15–23. [CrossRef]

57. Jambrak, A.R.; Mason, T.J.; Lelas, V.; Krešić, G. Ultrasonic effect on physicochemical and functional properties of α-lactalbumin. *LWT-Food Sci. Technol.* **2010**, *43*, 254–262. [CrossRef]

58. Montero, P.; Fernández-Díaz, M.D.; Gómez-Guillén, M.C. Characterization of gelatin gels induced by high pressure. *Food Hydrocoll.* **2002**, *16*, 197–205. [CrossRef]

59. Jongjareonrak, A.; Benjakul, S.; Visessanguan, W.; Tanaka, M. Skin gelatin from bigeye snapper and brownstripe red snapper: Chemical compositions and effect of microbial transglutaminase on gel properties. *Food Hydrocoll.* **2006**, *20*, 1216–1222. [CrossRef]

60. Ahmad, M.; Benjakul, S.; Ovissipour, M.; Prodpran, T. Indigenous proteases in the skin of unicorn leatherjacket (*Alutherus monoceros*) and their influence on characteristic and functional properties of gelatin. *Food Chem.* **2011**, *127*, 508–515. [CrossRef] [PubMed]

61. Jamilah, B.; Tan, K.W.; UmiHartina, M.R.; Azizah, A. Gelatins from three cultured freshwater fish skins obtained by liming process. *Food Hydrocoll.* **2011**, *25*, 1256–1260. [CrossRef]

62. Gudmundsson, M.; Hafsteinsson, H. Gelatin from cod skins as affected by chemical treatments. *J. Food Sci.* **1997**, *62*, 37–39. [CrossRef]

63. Muyonga, J.H.; Cole, C.G.B.; Duodu, K.G. Fourier transform infrared (FTIR) spectroscopic study of acid soluble collagen and gelatin from skins and bones of young and adult Nile perch (*Lates niloticus*). *Food Chem.* **2004**, *86*, 325–332. [CrossRef]

64. Bandekar, J. Amide modes and protein conformation. *Biochim. Biophys. Acta (BBA)-Protein Struct. Mol. Enzymol.* **1992**, *1120*, 123–143. [CrossRef]

65. Uriarte-Montoya, M.H.; Santacruz-Ortega, H.; Cinco-Moroyoqui, F.J.; Rouzaud-Sández, O.; Plascencia-Jatomea, M.; Ezquerra-Brauer, J.M. Giant squid skin gelatin: Chemical composition and biophysical characterization. *Food Res. Int.* **2011**, *44*, 3243–3249. [CrossRef]

66. Yakimets, I.; Wellner, N.; Smith, A.C.; Wilson, R.H.; Farhat, I.; Mitchell, J. Mechanical properties with respect to water content of gelatin films in glassy state. *Polymer* **2005**, *46*, 12577–12585. [CrossRef]

67. Barth, A. Infrared spectroscopy of proteins. *Biochim. Biophys. Acta (BBA)-Bioenerg.* **2007**, *1767*, 1073–1101. [CrossRef] [PubMed]

68. Jackson, M.; Watson, P.H.; Halliday, W.C.; Mantsch, H.H. Beware of connective tissue proteins: Assignment and implications of collagen absorptions in infrared spectra of human tissues. *Biochim. Biophys. Acta (BBA)-Mol. Basis Dis.* **1995**, *1270*, 1–6. [CrossRef]

69. Jiang, L.; Wang, J.; Li, Y.; Wang, Z.; Liang, J.; Wang, R.; Chen, Y.; Ma, W.; Qi, B.; Zhang, M. Effects of ultrasound on the structure and physical properties of black bean protein isolates. *Food Res. Int.* **2014**, *62*, 595–601. [CrossRef]

70. Ktari, N.; Bkhairia, I.; Jridi, M.; Hamza, I.; Riadh, B.S.; Nasri, M. Digestive acid protease from zebra blenny (*Salaria basilisca*): Characteristics and application in gelatin extraction. *Food Res. Int.* **2014**, *57*, 218–224. [CrossRef]

71. Jamilah, B.; Harvinder, K. Properties of gelatins from skins of fish—Black tilapia (*Oreochromis mossambicus*) and red tilapia (*Oreochromis nilotica*). *Food Chem.* **2002**, *77*, 81–84. [CrossRef]

72. Eastoe, J.; Leach, A. Chemical constitution of gelatin (From mammals, chicken tendon, calf skin, pig skin). In *The Science and Technology of Gelatin*; Academic Press Inc.: London, UK, 1977; pp. 73–105.

73. Awad, E.A.; Zulkifli, I.; Farjam, A.S.; Chwen, L.T. Amino acids fortification of low-protein diet for broilers under tropical climate. 2. Nonessential amino acids and increasing essential amino acids. *Ital. J. Anim. Sci.* **2014**, *13*, 631–636. [CrossRef]

74. Laemmli, U.K. Cleavage of structural proteins during the assembly of the head of bacteriophage T4. *Nature* **1970**, *227*, 680–685. [CrossRef] [PubMed]

75. Cho, S.M.; Kwak, K.S.; Park, D.C.; Gu, Y.S.; Ji, C.I.; Jang, D.H.; Lee, Y.B.; Kim, S.B.I. Processing optimization and functional properties of gelatin from shark (*Isurus oxyrinchus*) cartilage. *Food Hydrocoll.* **2004**, *18*, 573–579. [CrossRef]

76. Fernandez-Díaz, M.D.; Montero, P.; Gomez-Guillen, M.C. Gel properties of collagens from skins of cod (Gadus morhua) and hake (Merluccius merluccius) and their modification by the coenhancers magnesium sulphate, glycerol and transglutaminase. *Food Chem.* **2001**, *74*, 161–167. [CrossRef]

Sample Availability: Extracted gelatin samples are available from the authors.

Article

Modification and Characterization of Fe₃O₄ Nanoparticles for Use in Adsorption of Alkaloids

Linyan Yang [1,2], Jing Tian [1], Jiali Meng [1], Ruili Zhao [1], Cun Li [1,*], Jifei Ma [1,*] and Tianming Jin [1,*]

[1] College of Animal Science and Veterinary Medicine, Tianjin Agricultural University, Tianjin 300384, China; y_linyan@163.com (L.Y.); 15822858982@163.com (J.T.); mjl19930123@163.com (J.M.); zhaoruili1109@126.com (R.Z.)
[2] Guangxi Key Laboratory for the Chemistry and Molecular Engineering of Medicinal Resources, Chemical and Pharmaceutical College of Guangxi Normal University, Guilin 541004, China
*   Correspondence: hhlicun@163.com (C.L.); hbmjfts@126.com (J.M.); JTMSCI@163.com (T.J.); Tel.: +86-22-2378-1303 (T.J.); Fax: +86-22-2378-1297 (T.J.)

Received: 27 January 2018; Accepted: 27 February 2018; Published: 2 March 2018

Abstract: Magnetite (Fe₃O₄) is a ferromagnetic iron oxide of both Fe(II) and Fe(III), prepared by FeCl₂ and FeCl₃. XRD was used for the confirmation of Fe₃O₄. Via the modification of Tetraethyl orthosilicate (TEOS), (3-Aminopropyl)trimethoxysilane (APTMS), and Alginate (AA), Fe₃O₄@SiO₂, Fe₃O₄@SiO₂-NH₂, and Fe₃O₄@SiO₂-NH₂-AA nanoparticles could be obtained, and IR and SEM were used for the characterizations. Alkaloid adsorption experiments exhibited that, as for Palmatine and Berberine, the most adsorption could be obtained at pH 8 when the adsorption time was 6 min. The adsorption percentage of Palmatine was 22.2%, and the adsorption percentage of Berberine was 23.6% at pH 8. Considering the effect of adsorption time on liquid phase system, the adsorption conditions of 8 min has been chosen when pH 7 was used. The adsorption percentage of Palmatine was 8.67%, and the adsorption percentage of Berberine was 7.25%. Considering the above conditions, pH 8 and the adsorption time of 8min could be chosen for further uses.

Keywords: Fe₃O₄; modification; alginate; alkaloid

1. Introduction

Although there are many pure phases of iron oxide in nature, the most popular magnetic nanoparticles (MNPs) are the nanoscale zero-valent iron (nZVI), Fe₃O₄ and γ-Fe₂O₃. Magnetite (Fe₃O₄) is a ferromagnetic black color iron oxide of both Fe(II) and Fe(III), which has been the most extensively studied [1]. In 2001, Asher reported co-precipitation method using oleic acid as the surface modification agent to obtain Fe₃O₄ nanoparticles (2–15 nm) [2]. NaOH and diethylene glycol could also be used as the catalyst and reducing agent to fabricate Fe₃O₄ nanoparticles of 80–180 nm in size [3–5]. However, Fe₃O₄ nanoparticles could easily aggregate due to the nanoscale effect and magnetic gravitational effect. It is an effective method of preventing the aggregate of these nanoparticles to wrap the surface of Fe₃O₄ nanoparticles. Fe₃O₄@SiO₂ composite nanoparticles have the desirable properties of magnetic nanoparticles while also benefiting from the SiO₂ shell, such as good hydrophilicity, stability, and biocompatibility [6–8]. In 2016, Tang reported that (3-aminopropyl)-triethoxysilane (APTES) was used as surface modification reagents to get Fe₃O₄@SiO₂-NH₂, which could be used for selective removal of Zn(II) ions from wastewater [9]. While Fe₃O₄@SiO₂-NH₂ nanoparticles could also be modified to obtain mercaptoamine-functionalised silica-coated magnetic nanoparticles for the removal of mercury and lead ions from wastewater [10]. As for the removal of ions, arsenate removal could be achieved by calcium alginate-encapsulated magnetic sorbent, which was prepared by physical method [11]. Superparamagnetic sodium alginate-coated Fe₃O₄ nanoparticles (Alg-Fe₃O₄) were used for removal of malachite green (MG) from aqueous solutions using batch adsorption technique, and the

Alg-Fe$_3$O$_4$ nanoparticles were synthesized using in situ coprecipitation of FeCl$_2$ and FeCl$_3$ in alkaline solution in the presence of sodium alginate [12]. While multifunctional alginate microspheres could also be used for biosensing, drug delivery, and magnetic resonance imaging [13]. To obtain the good biocompatibility, Fe$_3$O$_4$ nanoparticles need to be modified. Fe$_3$O$_4$@SiO$_2$ composite nanoparticles have the desirable properties of good hydrophilicity. (3-Aminopropyl)trimethoxysilane (APTMS) was used as surface modification reagents to get Fe$_3$O$_4$@SiO$_2$-NH$_2$nanoparticles. While calcium alginate-encapsulated magnetic sorbent could be prepared by physical method. Superparamagnetic sodium alginate-coated Fe$_3$O$_4$ nanoparticles (Alg-Fe$_3$O$_4$) could also be synthesized using in situ coprecipitation of FeCl$_2$ and FeCl$_3$ in alkaline solution in the presence of sodium alginate. Covalent modification methods via alginate have been rarely seen. In order to investigate the effects of the covalent alginate-modified method, alkaloid adsorption experiments were designed to study the properties of alginate-modified Fe$_3$O$_4$@SiO$_2$-NH$_2$ nanoparticles.

2. Experimental Section

2.1. Materials and Physical Measurements

(3-Aminopropyl)trimethoxysilane (APTMS), *N*-Hydroxysuccinimide (NHS) and 1-(3-Dimethylaminopropyl)-3-ethylcarbodiimide hydrochloride (EDC) were purchased from Shanghai source Biological Technology Co., Ltd. (Shanghai, China). Alginate (AA) was purchased from Solarbio Life Science (Beijing Solarbio Biological Technology Co., Ltd., Beijing, China). All commercially available chemicals and solvents were of reagent grade and used without further purification. X-ray powder diffraction (XRD) intensities were measured on a Rigaku D/max-IIIA diffractometer (Cu-Kα, λ = 1.54056 Å). Changes in morphology and size could be characterized by Scanning Electronic Microscopy (SEM) (KAI MEIKE CHEMICAL Co., Ltd., Liaocheng, China).

XPS spectra were recorded using a Kratos Axis Ultra DLD spectrometer (KAI MEIKE CHEMICAL Co., Ltd.) employing a monochromated Al-Kα X-ray source (hv = 1486.6 eV). The vacuum in the main chamber was kept above 3×10^{-6} Pa during XPS data acquisitions. General survey scans (binding energy range: 0–1200 eV; pass energy: 160 eV) and high-resolution spectra (pass energy: 40 eV) in the regions of N1s were recorded. Binding energies were referenced to the C1s binding energy at 284.60 eV.

The adsorption data were obtained by RP-HPLC (Reversed phase high performance liquid chromatography). The HPLC system was from Agilent Technologies 1260 Infinity (Agilent Technologies, SantaClara, CA, USA), and was equipped with a quaternary pump and UV-Vis detector (Agilent Technologies). The chromatographic separation was carried out on an ACE Super C18 column (250 × 4.6 mm i.d., 5 µm, FLM, Guangzhou, China). Mobile phase consisted of 50% solution (*v*/*v*) of acetonitrile in water (0.1% H$_3$PO$_4$ and 0.1% SDS). The flow rate was 1 mL/min and the column temperature was set to 40 °C. The effluent was monitored at 265 nm and the injection volume was 20 µL.

2.2. Preparation and Modification of Fe$_3$O$_4$ Nanoparticles

Magnetite nanoparticles were prepared and modified with TEOS, APTMS, and AA to get Fe$_3$O$_4$@SiO$_2$, Fe$_3$O$_4$@SiO$_2$-NH$_2$, and Fe$_3$O$_4$@SiO$_2$-NH$_2$-AA nanoparticles, respectively (Figure 1).

2.2.1. Preparation of Fe$_3$O$_4$ Nanoparticles

Briefly, 7.5 mL of 0.12 M FeCl$_2$ and 7.5 mL of 0.2 M FeCl$_3$ solutions were mixed in a 100-mL flask. The whole reaction system was completed under nitrogen protection. After the magnetic stirring was uniform, the reaction system was heated to 55 °C, which maintained for 15 min. 7.2 mL of 3 M NaOH solution was then added to the reaction system. The reaction system was kept at 55 °C for 40 min. Then the reaction system was stirred at 90 °C for 30 min and cooled to room temperature. The black precipitate was collected by magnetic decantation and washed with deionized water repeatedly until the washings were neutral. The obtained black precipitate was then dried over vacuum at 40 °C overnight, which could be used for XRD measurement [14,15].

2.2.2. Preparation of Fe₃O₄@SiO₂

Fe_3O_4 (10 mg) was acidized by HCl (0.1 mol/L) under the 100 W of ultrasound for 20 min. The supernatant was discarded after adsorption by the magnet. The residue was washed with ultrapure water for twice, and resuspended in ethanol/ultrapure water (20 mL:5 mL). $NH_3·H_2O$ (250 µL) was added to the samples of Fe_3O_4, and the mixture was reacted for 20 min under the 100 W of ultrasound. TEOS (32 µL) was added into the samples. And then the samples were oscillated at 37 °C and 140 r/min for 6 h, followed by adsorption by the magnet. The supernatant was discarded, and the residue was washed with ethanol for twice to yield $Fe_3O_4@SiO_2$, which was resuspended in ethanol (4 mL) [16].

2.2.3. Preparation of Fe₃O₄@SiO₂-NH₂

APTMS (50 µL) was dropwise added to the samples of $Fe_3O_4@SiO_2$ obtained previously, and the mixture was reacted for 24 h. After rinsing with ethanol for twice, the samples named as $Fe_3O_4@SiO_2-NH_2$ were vacuum-dried at 80 °C overnight [17].

2.2.4. Preparation of Fe₃O₄@SiO₂-NH₂-AA

An AA solution (5 mg/mL in MES buffer, pH 6.0) was mixed with *N,N*-dimethylformamide (DMF; 3:1, *v/v*). Then the AA solution (3.75 mg/mL) was converted to *N*-hydroxysuccinimide esters by sequential reaction with EDC (36.3 mg/mL in MES buffer, pH 6.0) for 15 min and NHS (10.95 mg/mL in MES buffer, pH 6.0) for 60 min. The solution was finally introduced to the freshly $Fe_3O_4@SiO_2-NH_2$ nanoparticles and reacted overnight at room temperature. After washing by ethanol, the samples of $Fe_3O_4@SiO_2-NH_2-AA$ could be obtained by vacuum-dried process [18].

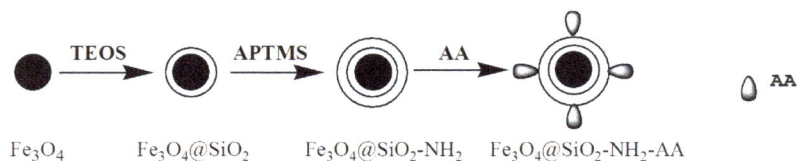

Figure 1. The diagram of surface modification stages.

2.3. Alkaloid Adsorption Test

2.3.1. Preparation of Calibration Standards

100 µg/mL standard solutions in methanol of Palmatine and Berberine were obtained from Solarbio (Beijing, China), and then further diluted in pattern of 1:2 to produce the working solutions with a series of concentrations. The concentration range of calibration standards for Palmatine were 50 µg/mL, 25 µg/mL, 12.5 µg/mL, 6.25 µg/mL, 3.125 µg/mL, 1.5625 µg/mL, 0.78125 µg/mL, while the concentration range of calibration standards for Berberine were 25 µg/mL, 12.5 µg/mL, 6.25 µg/mL, 3.125 µg/mL, 1.5625 µg/mL, 0.78125 µg/mL.

2.3.2. Influence from pH

Approximate 8 mL of mixed standard stock solution (0.5 µg/mL, in methanol, pH 5, 6, 7, 8, 9), 10 mg of $Fe_3O_4@SiO_2-NH_2-AA$ nanoparticles was ultrasonic shocked for 6 min, and then the supernatant and magnetic nanoparticles were obtained by magnetic separation. The magnetic nanoparticles were washed by deionized water (1 mL × 2). The supernatant and detergent were combined. 1.5 mL of the mixture was dried by nitrogen blower at 80 °C. The residue was redissolved in 400 μL of methanol, which was filtered (0.22 μm) for subsequent HPLC analysis.

2.3.3. Influence from Adsorption Time

Approximate 8 mL of mixed standard stock solution (0.5 μg/mL, in methanol), 10 mg of $Fe_3O_4@SiO_2-NH_2-AA$ nanoparticles was ultrasonic shocked for a certain time (2 min, 4 min, 6 min, 8 min, 10 min), and then the supernatant and magnetic particles were obtained by magnetic separation. The magnetic nanoparticles were washed by deionized water (1 mL × 2). The supernatant and detergent were combined. The mixture was dried by nitrogen blower at 80 °C. The residue was redissolved in 400 μL of methanol, which was filtered (0.22 μm) for subsequent HPLC analysis [19–21].

3. Results and Discussion

3.1. XRD Analysis of Fe_3O_4 Nanoparticles

The XRD pattern of Fe_3O_4 nanoparticles is shown in the Figure 2. The peaks at 2θ values of 30.1°, 35.4°, 43.1°, 53.4°, 56.9° and 62.5° are indexed as the diffractions of (220), (311), (222), (422), (511) and (440) respectively, which resembles the standard diffraction spectrum of Fe_3O_4 (JCPDSPDF#19-0629) with respect to its reflection peaks positions [5].

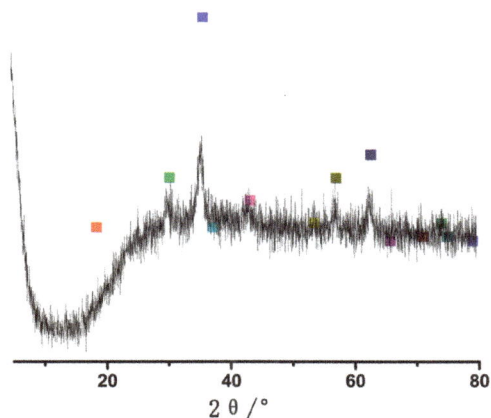

Figure 2. XRD pattern of Fe_3O_4 nanoparticles. (Color squares are the standard diffraction spectrum of Fe_3O_4).

3.2. FTIR Spectra Analysis of Nanoparticles

The $Fe_3O_4@SiO_2-NH_2$ and $Fe_3O_4@SiO_2-NH_2-AA$ nanoparticles were obtained after the surface modification steps. It is apparent that the IR spectra contains not only the peaks in spectra of Fe_3O_4 nanoparticles (Fe-O, 567 cm^{-1}) [15]. 1560 cm^{-1} (C-N vibration) reflected that APTMS was successfully modified onto $Fe_3O_4@SiO_2$nanoparticles [22]. A strong IR peak appears at 1648 cm^{-1}, corresponding to the strong bending vibration of the amide I group, which showed that the modification was successful and $Fe_3O_4@SiO_2-NH_2$nanoparticles were indeed coated with AA (Figure 3) [17,23,24].

3.3. XPS Analysis of Nanoparticles

Figure 4a shows the low-resolution XPS survey spectra of Fe_3O_4, $Fe_3O_4@SiO_2$, $Fe_3O_4@SiO_2-NH_2$ and $Fe_3O_4@SiO_2-NH_2-FA$ samples, all of which are semiquantitative. The low-resolution XPS survey spectra (Figure 4a) of $Fe_3O_4@SiO_2-NH_2$ have peaks of N1s, which showed that APTMS have been modified successfully. High-resolution C1s XPS spectra of the $Fe_3O_4@SiO_2-NH_2$ samples have peaks at 284.603 eV (C-H/C-C) and 285.459 eV (C-O/C-N) (Figure 4b). High-resolution C1s XPS spectra of the $Fe_3O_4@SiO_2-NH_2-AA$ samples have peaks at 284.605 eV (C-H/C-C), 285.891 eV (C-O/C-N), and 287.916 eV (O-C=O/O=C-NH) (Figure 4c), which showed that amide reaction was successful [25].

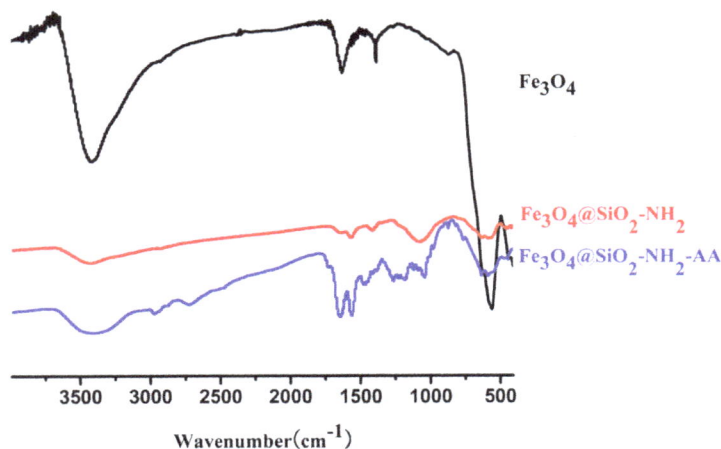

Figure 3. FTIR spectra of: as-prepared Fe_3O_4 nanoparticles (black); $Fe_3O_4@SiO_2-NH_2$ (red); $Fe_3O_4@SiO_2-NH_2-AA$ (blue).

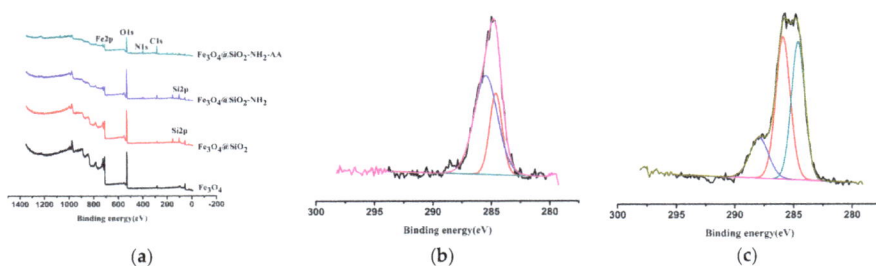

Figure 4. (**a**) XPS wide scan spectra of Fe_3O_4, $Fe_3O_4@SiO_2$, $Fe_3O_4@SiO_2-NH_2$, $Fe_3O_4@SiO_2-NH_2-AA$ nanoparticles; (**b**) High-resolution XPS C1s spectra of $Fe_3O_4@SiO_2-NH_2$; (**c**) High-resolution XPS C1s spectra of $Fe_3O_4@SiO_2-NH_2-AA$.

3.4. SEM Analysis of Nanoparticles

Figure 5a–c show SEM images of Fe_3O_4, $Fe_3O_4@SiO_2-NH_2$, and $Fe_3O_4@SiO_2-NH_2-AA$ nanoparticles. Small particle size of Fe_3O_4 particles is obvious, while a good dispersion effect could be achieved by $Fe_3O_4@SiO_2-NH_2$ nanoparticles. As for $Fe_3O_4@SiO_2-NH_2-AA$ nanoparticles, no good dispersion could be achieved, while better morphology could be achieved, which showed that AA was successfully modified onto $Fe_3O_4@SiO_2-NH_2$ nanoparticles [22]. Almost all particle size of Fe_3O_4 particlesis below 100 nm, as for $Fe_3O_4@SiO_2-NH_2$ nanoparticles and $Fe_3O_4@SiO_2-NH_2-AA$ nanoparticles, particle size is becoming larger and larger, which could also prove that the modification is successful.

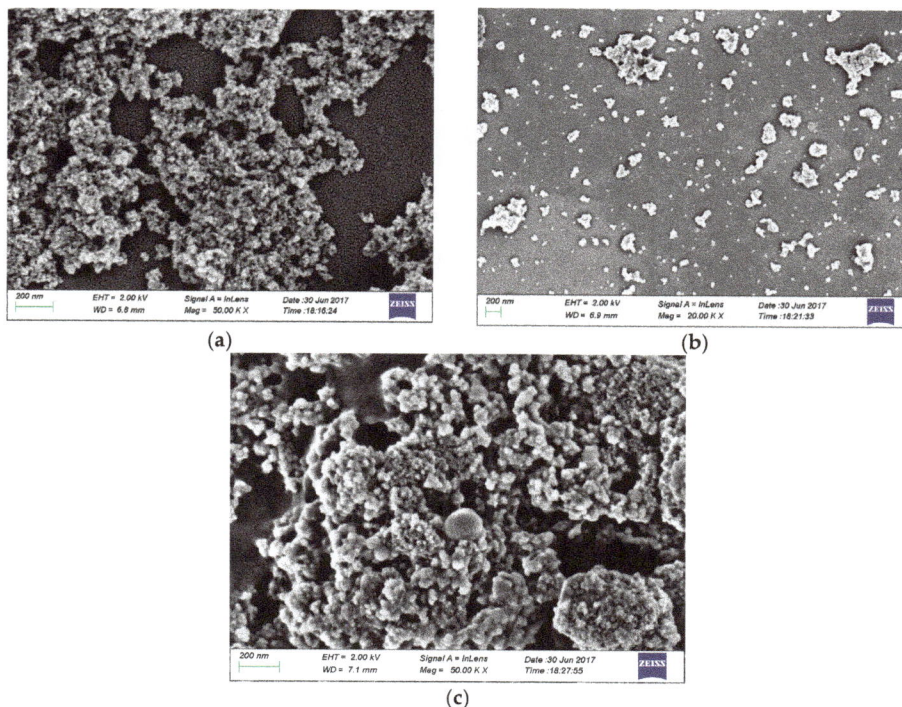

Figure 5. The SEM images of nanoparticles: (a) Fe_3O_4, (b) $Fe_3O_4@SiO_2\text{-}NH_2$, (c) $Fe_3O_4@SiO_2\text{-}NH_2\text{-}AA$.

3.5. Analysis of Alkaloid Adsorption Test

Electrostatic interactions between alkaloids and charged surfaces, therefore, often play a major role in the adsorption behavior of alkaloids. Therefore, Palmatine and Berberine were selected for alkaloid adsorption assay in the current study.

Figure 6a is the chromatogram associated with the concentrations of the standard curve, which belongs to Palmatine. Figure 6b is the chromatogram associated with the concentrations of the standard curve, which belongs to Berberine. The Equation process is as follows:

$$\frac{CV_2}{V_1} \times V_0 = m \tag{1}$$

$$Ap = \frac{C_0V - m}{C_0V} \tag{2}$$

$V_0 = 10$ mL, $V_1 = 1.5$ mL, $V_2 = 0.4$ mL, m is the capacity of alkaloid in the supernatant and detergent, C is the concentration of the supernatant and detergent, which could be obtained by the standard curve.

$C_0 = 0.5$ µg/mL, $V = 8$ mL, Ap is the adsorption percentage of alkaloid.

From Table S1, as for Palmatine and Berberine, the most adsorption could be obtained at pH 8. Considering the effect of alkaline on liquid phase system, the adsorption conditions of pH 8 has been chosen. The adsorption percentage of Palmatine was 22.2%, and the adsorption percentage of Berberine was 23.6%. At pH 8, the carboxylic acid of $Fe_3O_4@SiO_2\text{-}NH_2\text{-}AA$ nanoparticles was converted to a negatively-charged carboxylate ion. Therefore, quaternary ammonium alkaloids were significantly adsorbed onto the carboxylic acid-rich surface, possibly due to electrostatic interactions. The results from this study seem to fit well with a previous report on the study of the charge interaction of alkaloids and polyelectrolyte films.

From Table S2, as for Berberine, the most adsorption could be obtained at 8 min. While the most adsorption could be obtained at 10 min for Palmatine. Considering the effect of adsorption time on liquid phase system, the adsorption conditions of 8 min has been chosen. The adsorption percentage of Palmatine was 8.67%, and the adsorption percentage of Berberine was 7.25%.

The effect of pH was greater than that of adsorption time. Considering the above conditions, pH 8 and the adsorption time of 8 min could be chosen for further uses.

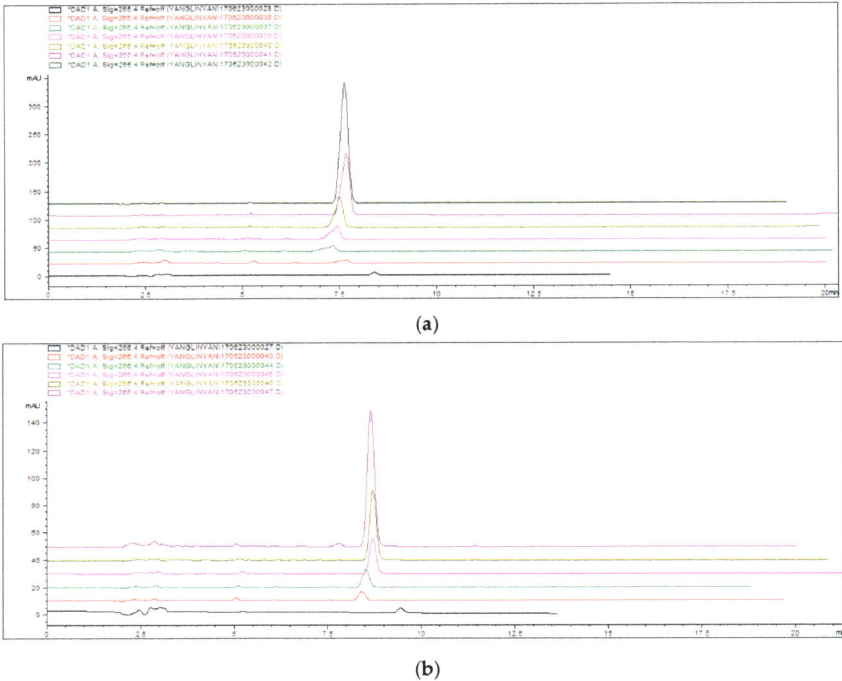

(a)

(b)

Figure 6. *Cont.*

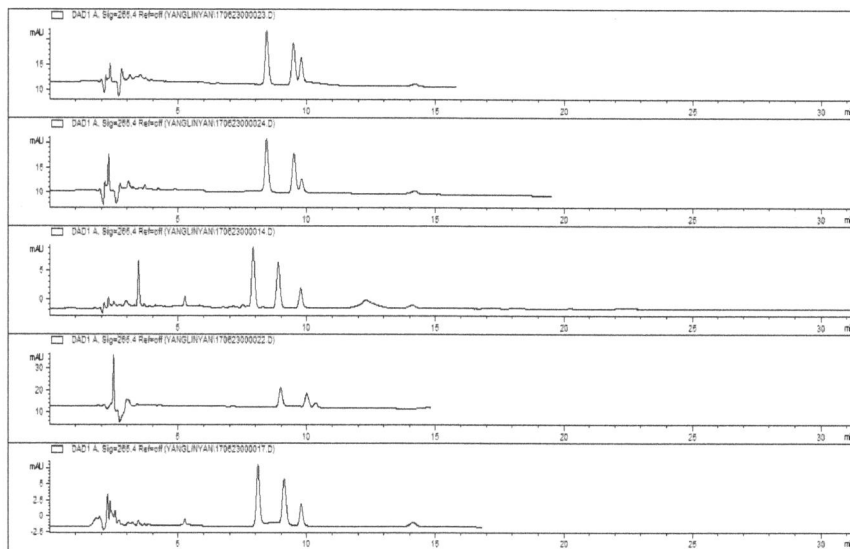

(c)

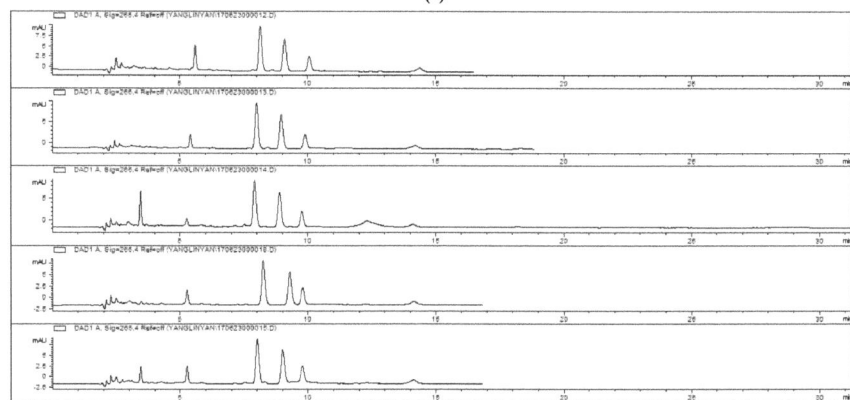

(d)

Figure 6. (**a**) Concentration gradient chromatogram for Palmatine. (Standrad curve: y = 5.41437 + 61.51865x, R = 0.99958, linear range: 0.78125–50 μg/mL); (**b**) Concentration gradient chromatogram for Berberine. (Standard curve: y = −3.38806 + 53.63054x, R = 0.99899, linear range: 0.78125–25 μg/mL); (**c**) HPLC charomatograms of the supernatant after adsorption. Conditions: pH adjustment was as follows: 5, 6, 7, 8, 9; adsorption time was 6 min; (**d**) HPLC charomatograms of the supernatant after adsorption. Conditions: adsorption time adjustment was as follows: 2 min, 4 min, 6 min, 8 min, 10 min, while pH 7 was used.

4. Conclusions

In conclusion, magnetite (Fe_3O_4) could be prepared by $FeCl_2$ and $FeCl_3$, which is a ferromagnetic black color iron oxide of both Fe(II) and Fe(III). XRD was used for the determination of Fe_3O_4 nanoparticles. The peaks at 2θ values of 30.1°, 35.4°, 43.1°, 53.4°, 56.9° and 62.5° resemble the standard diffraction spectrum of Fe_3O_4 (JCPDSPDF#19-0629) with respect to its reflection peaks positions. Fe_3O_4 could be used for modification at the subsequent trials. Fe_3O_4@SiO_2 nanoparticles were successfully obtained by TEOS. Fe_3O_4@SiO_2-NH_2 nanoparticles were prepared by APTMS, while Fe_3O_4@SiO_2-NH_2-AA nanoparticles were obtained by activated AA via amidation reaction. IR, XPS and SEM analysis were used for the characterizations of Fe_3O_4@SiO_2-NH_2 and Fe_3O_4@SiO_2-NH_2-AA nanoparticles. Alkaloid adsorption experiments implied that Fe_3O_4@SiO_2-NH_2-AA nanoparticles as a absorbent could be used for the adsorption of the alkaloids. At pH 8, the carboxylic acid of Fe_3O_4@SiO_2-NH_2-AA nanoparticleswas converted to a negatively-charged carboxylate ion. Therefore, quaternary ammonium alkaloids were significantly adsorbed onto the carboxylic acid-rich surface, possibly due to electrostatic interactions. As for Palmatine and Berberine, the most adsorption could be obtained at pH 8 when the adsorption time was 6 min. The adsorption percentage of Palmatine was 22.2%, while the adsorption percentage of Berberine was 23.6% at pH 8. As for the effect of adsorption time on liquid phase system, the adsorption conditions of 8 min has been chosen when pH 7 was used. Considering the above conditions, pH 8 and the adsorption time of 8 min could be chosen for further uses. This work demonstrates the potential of AA modification in a Fe_3O_4-based alkaloid adsorption study. In further experiments, when the amidation reaction is performed, residual carboxyl groups from AA on the modified Fe_3O_4@SiO_2-NH_2-AA nanoparticles may be used for bio-molecule immobilization.

Supplementary Materials: The supplementary materials are available online.

Acknowledgments: This work was supported by the Research Project of Tianjin Education Commission (2017KJ190), the Open Topic of Guangxi Key Laboratory for the Chemistry and Molecular Engineering of Medicinal Resources (CMEMR2016-B12), the National Natural Science Foundation of China (No. 31572492, No. 31072109, No. 31372482), the Innovative and Entrepreneurial Training Plan for Tianjin College Students (201710061035), the Veterinary Biotechnology Scientific Research Innovation Team of Tianjin, China (Grant No. TD12-5019), the General Fund of Application Foundation & Advanced Technology Program of Tianjin (14JCYBJC30000).

Author Contributions: Linyan Yang conceived and designed the experiments; Jing Tian and Jiali Meng performed the experiments; Ruili Zhao, Cun Li, and Jifei Ma analyzed the data; Tianming Jin contributed reagents/materials/analysis tools, revised and finalized the paper; Linyan Yang wrote the paper.

Conflicts of Interest: The authors declare no conflict of interest.

References

1. Mohammed, L.; Gomaa, H.G.; Ragab, D.; Zhu, J. Magnetic nanoparticles for environmental and biomedical applications: A review. *Particuology* **2017**, *30*, 1–4. [CrossRef]
2. Xu, X.; Friedman, G.; Humfeld, K.D.; Majetich, S.A.; Asher, S.A. Synthesis and utilization of monodisperse superparamagnetic colloidal particles for magnetically controllable photonic crystals. *Chem. Mater.* **2002**, *14*, 1249–1256. [CrossRef]
3. Ge, J.; Hu, Y.; Biasini, M.; Beyermann, W.P.; Yin, Y. Superparamagnetic magnetite colloidal nanocrystal clusters. *Angew. Chem. Int. Ed.* **2007**, *46*, 4342–4345. [CrossRef] [PubMed]
4. Ge, J.; Hu, Y.; Yin, Y. Highly tunable superparamagnetic colloidal photonic crystals. *Angew. Chem. Int. Ed.* **2007**, *46*, 7428–7431. [CrossRef] [PubMed]
5. Wang, W.; Zheng, L.L.; Lu, F.H.; Hong, R.J.; Chen, M.Z.Q.; Zhuang, L. Facile synthesis and characterization of magnetochromatic Fe_3O_4 nanoparticles. *AIP Adv.* **2017**, *7*, 056317. [CrossRef]
6. Ahmed, S.A.; Soliman, E.M. Silica coated magnetic particles using microwave synthesis for removal of dyes from natural water samples: Synthesis, characterization, equilibrium, isotherm and kinetics studies. *Appl. Surf. Sci.* **2013**, *284*, 23–32. [CrossRef]
7. Yan, H.; Zhang, J.C.; You, C.X.; Song, Z.W.; Yu, B.W.; Shen, Y. Surface modification of Fe_3O_4 nanoparticles and their magnetic properties. *Int. J. Miner. Metall. Mater.* **2009**, *16*, 226–229. [CrossRef]
8. Zhang, L.; Shao, H.P.; Zheng, H.; Lin, T.; Guo, Z.M. Synthesis and characterization of Fe_3O_4@SiO_2 magnetic compositenanoparticles by a one-pot process. *Int. J. Miner. Metall. Mater.* **2016**, *23*, 1112–1118. [CrossRef]
9. Bao, S.G.; Tang, L.H.; Li, K.; Ning, P.; Peng, J.H.; Guo, H.B.; Zhu, T.T.; Liu, Y. Highly selective removal of Zn(II) ion from hot-dip galvanizing pickling waste with amino-functionalized Fe_3O_4@SiO_2 magnetic nano-adsorbent. *J. Colloid Interface Sci.* **2016**, *462*, 235–242. [CrossRef] [PubMed]

10. Bao, S.G.; Li, K.; Ning, P.; Peng, J.H.; Jin, X.; Tang, L.H. Highly selective removal of mercury and lead ions from wastewater by mercaptoamine-functionalised silica-coated magnetic nano-adsorbents: Behaviours and mechanisms. *Appl. Surf. Sci.* **2017**, *393*, 457–466. [CrossRef]

11. Lim, S.F.; Zheng, Y.M.; Zou, S.W.; Chen, J.P. Uptake of arsenate by an alginate-encapsulated magnetic sorbent: Process performance and characterization of adsorption chemistry. *J. Colloid Interface Sci.* **2009**, *333*, 33–39. [CrossRef] [PubMed]

12. Mohammadi, A.; Daemi, H.; Barikani, M. Fast removal of malachite green dye using novel superparamagnetic sodium alginate-coated Fe_3O_4 nanoparticles. *Int. J. Biol. Macromol.* **2014**, *69*, 447–455. [CrossRef] [PubMed]

13. Joshi, A.; Solanki, S.; Chaudhari, R.; Bahadur, D.; Aslam, M.; Srivastava, R. Multifunctional alginate microspheres for biosensing, drug delivery and magnetic resonance imaging. *Acta Biomater.* **2011**, *7*, 3955–3963. [CrossRef] [PubMed]

14. Wang, J.Y.; Ren, L.; Wang, X.Q.; Wang, Q.; Wan, Z.F.; Li, L.; Liu, W.M.; Wang, X.M.; Li, M.L.; Tong, D.W.; et al. Superparamagnetic microsphere-assisted fluoroimmunoassay for rapid assessment of acute myocardial infarction. *Biosens. Bioelectr.* **2009**, *24*, 3097–3102. [CrossRef] [PubMed]

15. Rezayan, A.H.; Mousavi, M.; Kheirjou, S.; Amoabediny, G.; Ardestani, M.S.; Mohammadnejad, J. Monodisperse magnetite (Fe_3O_4) nanoparticles modified with water soluble polymers for the diagnosis of breast cancer by MRI method. *J. Magn. Magn. Mater.* **2016**, *420*, 210–217. [CrossRef]

16. Biradar, A.V.; Patil, V.S.; Chandra, P.; Doke, D.S.; Asefa, T. A trifunctional mesoporous silica-based, highly active catalyst for one-pot, three-step cascade reactions. *Chem. Commun.* **2015**, *51*, 8496–8499. [CrossRef] [PubMed]

17. Guo, L.; Ding, W.; Meng, F. Fabrication and in vitro evaluation of folate-modified iron ferrite nanoparticles with high doxorubicin loading for receptors-magnetic-guided drug delivery. *Nano Brief Rep. Rev.* **2014**, *9*, 1450021–1450028. [CrossRef]

18. Cai, Y.; Yuan, F.; Wang, X.; Sun, Z.; Chen, Y.; Liu, Z.; Wang, X.; Yang, S.; Wang, S. Synthesis of core-shell structured Fe_3O_4@carboxymethylcellulose magnetic composite for highly efficient removal of Eu(III). *Cellulose* **2017**, *24*, 175–190. [CrossRef]

19. Liu, L.; Wang, Z.B.; Song, Y.; Yang, J.; Wu, L.J.; Yang, B.Y.; Wang, Q.H.; Wang, L.Q.; Wang, R.X.; Yang, C.J. Simultaneous determination of eight alkaloids in rat plasma by UHPLC-MS/MS after oral administration of Coptisdeltoidea C.Y. Cheng et Hsiao and Coptischinensis Franch. *Molecules* **2016**, *21*, 913–915. [CrossRef] [PubMed]

20. Wang, Q.; Long, Y.; Yao, L.; Ye, M.; Xu, L. C_{18}-COOH silica: preparation, characterization and its application in purification of quaternary ammonium alkaloids from Coptischinensis. *Phytochem. Anal.* **2017**, *28*, 332–343. [CrossRef] [PubMed]

21. Ye, L.H.; Liu, X.D.; Chang, Y.X.; An, M.R.; Wang, S.L.; Xu, J.J.; Peng, L.Q. Analysis of isoquinoline alkaloids using chitosan-assisted liquid-solid extraction followed by microemulsion liquid chromatography employing a sub-2-micron particles stationary phase. *Electrophoresis* **2016**, *37*, 3118–3125. [CrossRef] [PubMed]

22. Jiang, W.; Wu, J.; Shen, Y.; Tian, R.; Zhou, S.; Jiang, W. Synthesis and characterization of doxorubicin loaded pH-sensitive magnetic core-shell nanocomposites for targeted drug delivery applications. *Nano Brief Rep. Rev.* **2016**, *11*, 1650127–1650140. [CrossRef]

23. Jung, Y.C.; Muramatsu, H.; Fujisawa, K.; Kim, J.H.; Hayashi, T.; Kim, Y.A.; Endo, M.; Terrones, M.; Dresselhaus, M.S. Optically and biologically active mussel protein-coated double-walled carbon nanotubes. *Small* **2011**, *7*, 3292–3297. [CrossRef] [PubMed]

24. Andhariya, N.; Upadhyay, R.; Mehta, R.; Chudasama, B. Folic acid conjugated magnetic drug delivery system for controlled release of doxorubicin. *J. Nanopart. Res.* **2013**, *15*, 1–12. [CrossRef]

25. Chen, X.L.; Zhao, T.T.; Zou, J.L. A novel mimetic peroxidase catalyst by using magnetite-containing silica nanoparticles as carriers. *Microchim. Acta* **2009**, *164*, 93–99. [CrossRef]

Sample Availability: Samples of the compounds are not available from the authors.

Article

Isolation and Purification of Two Isoflavones from *Hericium erinaceum* Mycelium by High-Speed Counter-Current Chromatography

Jinzhe He [1], Peng Fan [1], Simin Feng [2,*], Ping Shao [2] and Peilong Sun [1,*]

[1] Department of Food Science and Engineering, Zhejiang University of Technology, Hangzhou 310014, Zhejiang, China; hejzgd@163.com (J.H.); iamfanpeng@163.com (P.F.)
[2] Ocean College, Zhejiang University of Technology, Hangzhou 310014, Zhejiang, China; pingshao325@zjut.edu.cn
* Correspondence: fengsimin@zjut.edu.cn (S.F.); sunpeil388@163.com (P.S.); Tel.: +86-571-8832-0951(S.F.); +86-571-8832-0388 (P.S.)

Received: 28 January 2018; Accepted: 24 February 2018; Published: 2 March 2018

Abstract: High-speed counter-current chromatography (HSCCC) was used to separate and purify two isoflavones for the first time from *Hericium erinaceum* (*H. erinaceum*) mycelium using a two-phase solvent system composed of chloroform-dichloromethane-methanol-water (4:2:3:2, $v/v/v/v$). These two isoflavones were identified as genistein (4′,5,7-trihydroxyisoflavone, $C_{15}H_{10}O_5$) and daidzein (4′,7-dihydroxyisoflavone, $C_{15}H_{10}O_4$), using infrared spectroscopy (IR), electro-spary ionisation mass (ESI-MS), [1]H-nuclear magnetic resonance (NMR) and ^{13}C-NMR spectra. About 23 mg genistein with 95.7% purity and 18 mg daidzein with 97.3% purity were isolated from 150 mg ethanolic extract of *H. erinaceum* mycelium. The results demonstrated that HSCCC was a feasible method to separate and purify genistein and daidzein from *H. erinaceum* mycelium.

Keywords: *Hericium erinaceuns* mycelium; high-speed counter-current chromatography (HSCCC); genistein; daidzein

1. Introduction

Hericium erinaceum (*H. erinaceum*), commonly called monkey head mushroom or lion's mane mushroom, is a wood-rotting fungi that belong to the *Hericiaceae* family [1,2]. It is a traditional edible mushroom widely used as herbal medicines in East Asian countries [3]. *H. erinaceum* was reported to have many bioactivities, including anti-oxidant, immune regulatory, anti-aging, anti-microbial, anti-inflammatory, and anti-cancer activities [4–8].

Many phytochemicals, including polysaccharides, pyrones, terpenoids, phenols are present in the mycelium and fruiting bodies of *H. erinaceum* [9–11]. Flavonoids are found in many plant species and exhibit many bioactivities, including anti-oxidant, anti-inflammatory and anti-cancer effects [12,13]. It was reported that the methanol extract of *H. erinaceum* had strong antioxidant activities as it contained flavonoids and phenolic compounds [14]. However, specific flavonoid spices in *H. erinaceum* have still not been identified. In view of these beneficial properties, study on the separation and purification of flavonoids from *H. erinaceum* is necessary.

Many studies have been focused on the separation and purification of flavonoids from natural plants [15,16]. Most separation and purification methods are based on thin-layer chromatography and other chromatographic techniques based on a solid stationary phase, such as semi-preparative and preparative HPLC [17]. These methods were generally restricted to several disadvantages, including low-yielding, time-consuming, complex processing, high-cost, poor reproducibility, and irreversible adsorption [18,19]. High-speed counter-current chromatography (HSCCC) is an advanced technique based on liquid-liquid partitioning and has played an important role in the preparation of phytochemicals in the last decade [20]. HSCCC provides advantages by eliminating the solid support, which may cause adsorption or degradation of target compounds [18,21]. Therefore, it has been widely used for separation and purification of flavonoids, alkaloids, terpenes, polyphenols, and other natural products [22–26]. To the best of our knowledge, there was no report about isolating and purifying of isoflavones from *H. erinaceum* mycelium by HSCCC.

In this study, we discussed the development of the HSCCC method for the separation and purification of pure isoflavones from *H. erinaceum* mycelium. The two-phase solvent system composed of chloroform-dichloromethane-methanol-water was established. In addition, chemical structures of two purified isoflavones were further identified by infrared spectroscopy (IR), electro-spary ionisation mass (ESI-MS), ^{1}H-nuclear magnetic resonance (NMR) and ^{13}C-NMR spectra. As a result, two isoflavones, genistein (4′,5,7-trihydroxyisoflavone) and daidzein (4′,7-dihydroxyisoflavone), were isolated from *H. erinaceum* mycelium for the first time by HSCCC.

2. Results and Discussion

2.1. HPLC Analysis

Crude extract from *H. erinaceum* mycelium, named as HEM-E-E, was analyzed by HPLC. As showed in Figure 1, compounds **1** and **2** were two main components in HEM-E-E. Compounds **1** and **2** were set as target compounds in further HSCCC separation of HEM-E-E.

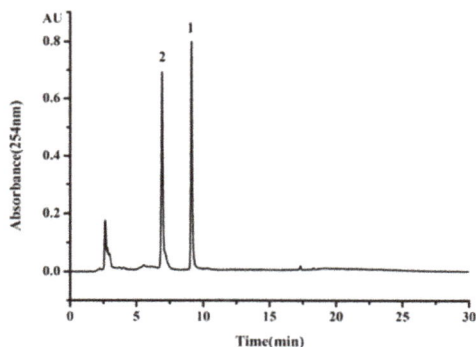

Figure 1. HPLC chromatograms of *H. erinaceum* mycelium crude extract (HEM-E-E). Peak number 1 and 2 refer to compounds **1** and **2**.

2.2. Selection of Two-Phase Solvent System

Choosing a suitable two-phase solvent system is a very important step in HSCCC experiment. The partition coefficient (K) and retention of the stationary phase are key factors for HSCCC separation [20]. The K value usually reflects the distribution between two mutually equilibrated solvent phases. A small K value elutes the solute close to the solvent front with lower resolution. A large K value tends to give better resolution but broader, more dilute peaks [20,27]. The retention of the stationary phase is accomplished by a combination of coiled column configuration and the planetary motion of the column holder. Therefore, the successful separation of HSCCC mainly depends on the selection of the two-phase solvent system. For the target compounds, suitable K values for HSCCC are $0.5 \leq K \leq 2.0$ and the separation factor (α) between two components should be greater than 1.5 [20,28]. To select a suitable solvent system, several solvent systems were tested in this study. The K and α values in different solvent systems were listed in Table 1. At first, the n-hexane-ethyl acetate/methanol water (HEMWat) system was tested, which can be used for analyses over a wide range of polarity [29,30]. As shown in Table 1, the K values of compounds **1** and **2** in HEMWat (1:1:1:1, 3:2:3:2, 4:5:4:5, *v/v/v/v*) were lower than 0.5, which meant compounds **1** and **2** were mainly distributed in the lower phase. This phenomenon indicated that the system's polarity was too high, compared to the target compounds. We tried *n*-hexane-methanol-water and ethyl acetate-methanol-water in different composition, but the K values were still lower than 0.5 (data not shown). Then the chloroform/methanol/water (ChMWat) system, which is extremely useful for separations of various natural products with moderate hydrophobicity, was tested [29,30]. When the solvent system was changed to ChMWat (4:4.5:2.5, 5:4:2, 5:5:2, $v/v/v$), KU/L values of compounds **1** and **2** increased. The results showed that the solvent systems were suitable for the separation of compound **2**. However, the K values for compound **1** were too large, which caused a long time for elution and low resolution. Based on the above data, systems of chloroform-dichloromethane-methanol-water (4:1.5:2:2, 4:2:3:2, 4:4.5:2:5, $v/v/v/v$) were tested. The K values (0.71~0.87 for compound **1** and 0.56~0.91 for compound **2**) were suitable for separation of compounds **1** and **2**. These three solvent systems were selected for further research.

Table 1. The partition coefficient K, separation factor α and settling time of target components in different solvent systems.

No	Solvent System	Ratio (v/v)	K Values [a]		α [b]	Settling Time
			Compound 1	Compound 2		
1	*n*-hexane-ethyl acetate-methanol-water	1:1:1:1	-	0.24	/	/
2	*n*-hexane-ethyl acetate-methanol-water	3:2:3:2	-	0.27	/	/
3	*n*-hexane-ethyl acetate-methanol-water	4:5:4:5	-	0.25	/	/
4	chloroform-methanol-water	4:4.5:2.5	2.87	0.46	6.23	/
5	chloroform-methanol-water	5:4:2	4.36	1.26	3.46	/
6	chloroform-methanol-water	5:5:2	3.45	0.83	4.15	/
7	chloroform-dichloromethane-methanol-water	4:1.5:2:2	0.84	0.91	1.08	19 s
8	chloroform-dichloromethane-methanol-water	4:2:3:2	0.87	0.56	1.55	14 s
9	chloroform-dichloromethane-methanol-water	4:4.5:2:5	0.71	0.63	1.12	24 s

[a] K values expressed as: A_U/A_L, where A_U and A_L are the peak of target compound in the upper and lower phase respectively. [b] α expressed as: separation factor between two target compounds K_1/K_2 or K_2/K_1. "-" stand for that the K value was too small.

In order to improve the retention of the stable phase, the settling time of the solvent system should be less than 20 s [20]. The settling time of three solvent systems was 24, 14 and 19 s, respectively (Table1). The shorter settling time means a higher retention of the stationary phase. In this study, the solvent system chloroform-dichloromethane-methanol-water (4:2:3:2, $v/v/v/v$) with a settling time of 14 s was selected for further HSCCC separation. Several flavonoids were reported to be separated from the seeds of *Vernonia anthelmintica Willd* by HSCCC using a two-step operation. The two solvent systems were chloroform–dichloromethane–methanol–water (2:2:3:2, $v/v/v/v$) and 1,2 dichloroethane–methanol–acetonitrile–water (4:1.1:0.25:2, $v/v/v/v$) [31], respectively. In this study, we used chloroform-dichloromethane-methanol-water (4:2:3:2, $v/v/v/v$) and separated two isoflavones from *H. erinaceum* mycelium using one-step operation.

2.3. HSCCC Separation

We expect that the suitable two-phase solvent system, chromatographic parameters including flow rate, rotary speed and column temperature may also affect the separation of HSCCC [32]. In this study, the upper and lower phase of chloroform-dichloromethane-methanol-water (4:2:3:2, $v/v/v/v$) system was used as the stationary phase and mobile phase, respectively. The effects of flow rate on separation of the target compounds by HSCCC were shown in Table 2. Both separation time and stationary phase retention decreased with the increase of the mobile phase flow rate. Low flow rate (1 mL/min) could increase to the retention of the stationary phase (67.8%), but it also extended the separation time (320 min). High flow rate (3 mL/min) could decrease the separation time (200 min). However, it decreased the retention of stationary phase (54.3%) and the purity of the target compounds. The retention levels of the stationary phase for a given flow-rate of the mobile phase will greatly contribute to the application of HSCCC. The flow rate is a key parameter that influences the chromatographic behavior after all other conditions are set [33]. Based on the separation time, the stationary phase retention and the purity of the target compounds, 2 mL/min was selected as the optimized flow rate. In this condition, the purity of the target compounds was high, the retention of the stationary phase was 65% and the separation time was 250 min. Our results are consistent with other research that shows that HSCCC is an effective method to isolate isoflavones or flavonoids from raw material [34,35].

Table 2. Separation time, stationary phase retention and purities of the two target compounds by high-speed counter-current chromatography (HSCCC) as affected by flow rate.

Flow-Rate (mL/min)	Separation-Time (min)	Retention (%)	Purity (%)	
			Compound 1	Compound 2
1	320	67.8	94.2	96.5
2	250	64.5	95.7	97.3
3	200	54.3	92.4	93.5

The rotary speed could also affect the separation time and the stationary phase retention. Low rotary speed reduces the volume of the stationary phase that is retained in the column, which leads to low chromatographic resolution and purity of targeted compounds. However, the high rotary speed might produce excessive sample band broadening due to the violent pulsation in the column [36]. The optimized HSCCC condition for separation of HEM-E-E was 900 rpm (rotary speed) and 20 °C (column temperature).

Under the optimized HSCCC conditions, an appropriate retention percentage of the stationary phase was 64.5%, and the purified target compounds **1** and **2** were obtained (Figure 2). About 23 mg compound **1** and 18 mg compound **2** were yield from 150 mg HEM-E-E. As shown in Figure 3, the purity of compounds **1** and **2** were 95.7% and 97.3%, respectively. It was reported that four isoflavones including daidzein and genistein was separated by HSCCC under a linear gradient elution, using a solvent system composed of n-hexane-ethyl acetate-1-butanol-methanol-water [37]. In another research, daidzein and genistein was separated from *Trifolium pratense L.* by HSCCC using the solvent system of *n*-hexane-ethyl acetate-ethanol-water [38]. In this manuscript, we developed a new solvent system (chloroform-dichloromethane-methanol-water 4:2:3:2 ($v/v/v/v$)) for the separation of genistein and daidzein. It offered higher yields of genistein and daidzein compared to previews research.

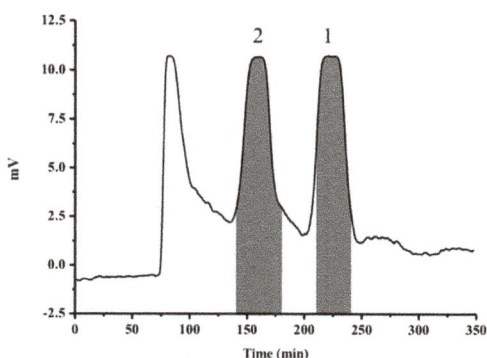

Figure 2. HSCCC chromatogram of HEM-E-E under the optimized condition. The upper phase of chloroform-dichloromethane-methanol-water (4:2:3:2, $v/v/v/v$) system was used as the stationary phase and lower phase of these solvent system was used as mobile phase. HSCCC condition was as follows: flow rate 2.0 mL/min, column temperature 20 °C, sample loading 10 mL, sample content 150 mg/10 mL, detection wavelength = 254 nm, rotary speed= 900 rpm. Peak number 1 and 2 refer to compounds **1** and **2**.

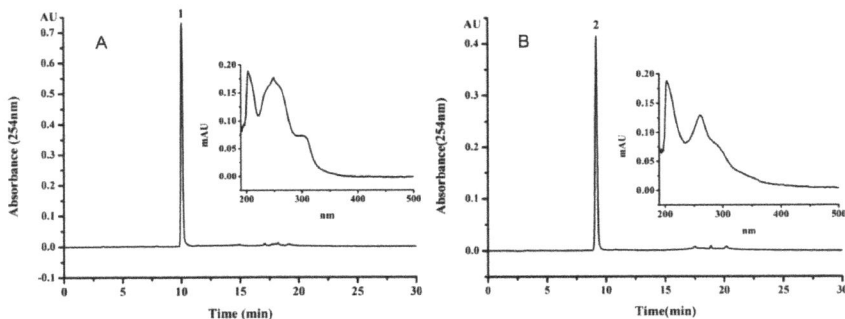

Figure 3. HPLC chromatograms of compound **1** (**A**) and compound **2** (**B**) and UV wavelength scanning of compounds **1** and **2** (inside). Peak number 1 and 2 refer to compounds **1** and **2**.

2.4. Identification of Chemical Structure

The chemical structures of the compounds **1** and **2** were analyzed by IR, MS, UV and NMR chromatography.

The structural data of the compound **1** are listed as follows: The ESI-MS showed the pseudo molecular ion [M + H]$^+$ peak at m/z 271.2, corresponding to molecular formula of $C_{15}H_{10}O_5$. The IR absorption bands at 3409.97 cm^{-1} indicated the presence of O-H; the absorption bands at 1615.64, 1650.44 cm^{-1} indicated the presence of C=O; the absorption bands at 1519.11cm^{-1} indicated the presence of aromatic ring; the absorption bands at 1274.23–1043.52 cm^{-1} indicated the presence of C-O. The UV spectrum showed conjugated groups by presenting maximum absorptions at 209 and 254 nm.

The structural data of the compound **2** are listed as follows: The ESI-MS showed the pseudo molecular ion [M + H]$^+$ peak at m/z 255.2, corresponding to molecular formula of $C_{15}H_{10}O_4$. The IR absorption bands at 3219.56 cm^{-1} indicated the presence of O-H; the absorption bands at 1632.21, 1606.62 cm^{-1} indicated the presence of C=O; the absorption bands at 1460.88 cm^{-1} indicated the presence of aromatic ring; the absorption bands at 844.02 cm^{-1} indicated the presence of =C-H on the benzene ring; the absorption bands at 1239.76 and 1193.19 cm^{-1} indicated the presence of C-O. The UV spectrum showed conjugated groups by presenting maximum absorptions at 204, 240 and 299 nm.

Their spectroscopic data for ^{1}H-NMR (400MHz) and ^{13}C-NMR (100MHz) were summarized in Table 3.

Table 3. ^{1}H (400 MHz) and ^{13}C-NMR (100 MHz) spectroscopic data of genistein and daidzein. [a,b]

Pos.	Genistein		Daidzein	
	δ_C	δ_H	δ_C	δ_H
2	154.77	8.30 (1H, s, H-2)	154.68	8.13 (1H, s, H-2)
3	124.71		125.95	
4	182.23		178.18	
5	163.84		128.52	8.06 (1H, d, *J* = 8.83Hz, H-5)
6	100.10	6.21 (1H, s, H-6)	116.52	6.94 (1H, dd, *J* = 2.24, 8.83Hz, H-6)
7	165.92		164.60	
8	94.27	6.33 (1H, s, H-8)	103.23	6.86 (1H, d, H-8)
9	159.69		159.80	
10	106.28		118.20	
1′	123.29		124.29	
2′	131.38	7.36 (1H, d, *J* = 8.48Hz, H-1′)	131.42	7.37 (1H, d, H-1′)
3′	116.25	6.84 (1H, d, *J* = 8.50Hz, H-2′)	116.22	6.84 (1H, d, H-2′)
4′	158.81		158.69	
5′	116.25	6.84 (1H, d, *J* = 8.50Hz, H-3′)	116.22	6.84 (1H, d, H-3′)
6′	131.38	7.36 (1H, d, *J* = 8.48 Hz, H-4′)	131.42	7.37 (1H, d, H-4′)

[a] Chemical shifts in ppm, coupling constants in 400 Hz. [b] Genistein and Daidzein were measured in CH$_3$DO.

The molecular formulas of compounds **1** and **2** were established as $C_{15}H_{10}O_5$ and $C_{15}H_{10}O_4$, respectively. Based on the IR and ^{1}H and ^{13}C-NMR spectra data, compounds **1** and **2** were identified as genistein (4′,5,7-trihydroxyisoflavone) and daidzein (4′,7-dihydroxyisoflavone), respectively. Our results were consistent with the NMR spectra data of genistein and daidzein in other studies [38–40]. Their chemical structures were shown in Figure 4. Genistein and daidzein were two isoflavones isolated from *H. erinaceum* mycelium.

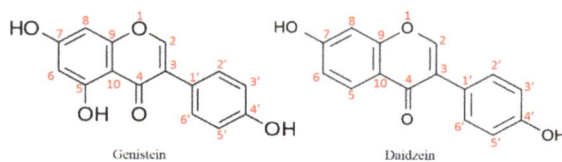

Figure 4. Chemical structures of genistein and daidzein.

3. Experimental Section

3.1. HSCCC Apparatus

HSCCC was performed on a TEB 300A (Tauto Biotechnique Company, Shanghai, China) high-speed counter-current chromatography apparatus. The apparatus consisted of three preparative coils connected in series (the inner diameter of tube, 1.5 mm; total volume, 280 mL) and a 10 mL sample loop. The revolution radius was 5 cm, and the β value was varied from 0.5 at the internal terminal to 0.8 at the external terminal (β = r/R, where r is the distance from the coil to the holder shaft, and R is the distance between the holder axis and the central axis of the centrifuge) [41]. The rotation speed was ranged from 0 to 1000 rpm. The system was equipped with a model TBP-50A constant-flow pump (Tauto Bioteh, Shanghai, China), a model UV-500 detector (XUYUKJ Instruments, Hangzhou, China) operating at 254 nm, and a model N2000 workstation (Zhejiang University, Hangzhou, China). DC-2010 constant temperature-circulating implement (Hanagzhou Dawei Instrument, Hangzhou, China) was used to adjust the experimental temperature.

3.2. Reagents and Materials

All organic solvents for HSCCC (analytical grade) were purchased from Shanghai Lingfeng Chemical Reagent Co. Ltd. (Shanghai, China). Acetonitrile used for HPLC analysis (chromatographic grade) were purchased from Tianjin Shiled Excellence Technology Co., Ltd. (Tianjin, China). Water was commercial ultrapure water. *H. erinaceus* (Bull.: Fr.) Pers. mycelium powder was purchased from Beijing Fuerkang Biotechnology research institute (Beijing, China) in August 2016. The scientific name was identified by one of the authors (Peilong Sun). The voucher specimen (ZJUT13000) was deposited at the Herbarium College of Pharmacy in Zhejiang University of Technology.

3.3. Preparation of H. erinaceum Mycelium Extracts

H. erinaceum mycelium powder (1000 g) was extracted three times with 95% ethanol (4L) under reflux for 4 h and at 80 °C using a stir bar. After removing ethanol by vacuum distillation at 55 °C, 76 g obtained ethanol extract was suspended in water (100 mL), and was extracted by petroleum ether and ethyl acetate in sequence. Ethyl acetate fraction was vacuum distilled, and 25 g crude extract was yield. The crude extract was named as HEM-E-E and stored at 4 °C for the HSCCC separation.

3.4. Selection of Two-Phase Solvent

Two-phase solvent systems were selected on the base of the partition coefficient value (K) of the two target compounds. Three solvent systems: (1) *n*-hexane-ethyl acetate-methanol-water systems (*v/v/v/v*, 1:1:1:1 or 3:2:3:2 or 4:5:4:5), (2) chloroform-methanol-water systems (*v/v/v/v*, 4:4.5:2.5 or

5:4:2 or 5:5:2) and (3) chloroform-dichloromethane-methanol-water systems ($v/v/v/v$, 4:1.5:2:2 or 4:2:3:2 or 4:4.5:2:5) were investigated. The K values were determined by HPLC as follows: HEMP-E-E (5 mg) was added to the equilibration of two-phase solvent system, followed by vigorous shaking for 1 min. After two phases were completely separated, 1 mL of each phase was evaporated to dryness, dissolved in 1 mL of acetonitrile and the K values were determined by HPLC analysis. The peak areas of the upper phase and the lower phase were recorded as A_U and A_L, respectively. K value was obtained by the equation: $K = A_U/A_L$. The HPLC conditions were described in Section 3.7.

3.5. Preparation of the Two-Phase Solvent System and Sample Solution

The selected two-phase solvent system composed of chloroform-dichloromethane-methanol-water ($v/v/v/v$, 4:2:3:2) was used for HSCCC separation. The solvent system in a separatory funnel was violently shaken for thorough mixing. After equilibration, the two phases were separated and degassed by sonication for 15 min before used. The lower and upper phase were used as the mobile and stationary phase, respectively. About 150 mg of HEM-E-E were dissolved in the solvent mixture containing 5 mL the lower phase and 5 mL upper phase.

3.6. HSCCC Separation

Head-tail elution was performed for the separation of HEMP-E-E. The coiled column was first entirely filled with the upper phase of the solvent system. Then the apparatus was rotated at a speed of 900 rpm, and the lower phase was pumped into the column at a flow rate of 2 mL/min. When the hydrodynamic equilibrium was achieved, as indicated by a clear mobile phase eluting at the tail outlet. About 10 mL of HEMP-E-E solution was injected into the separation column through the injection valve. The separation temperature was controlled at 20 °C. The effluent was continuously monitored at 254 nm, and collected using a fraction collector set at 5 min for each tube. Each fraction was collected according to the chromatogram and evaporated under vacuum. Three HSCCC fractions were obtained, the first fraction contained some impurities (confirmed by HPLC analysis, data not shown), were no longer investigated. The other two fractions are set as two target compounds and named as compounds **1** and **2**, respectively.

3.7. HPLC Analysis of HEM-E-E and Its HSCCC Fractions

The analytical HPLC equipment was a Waters 1525 system consisting of a Waters 1525 Binary pump, a Waters 2487 UV-vis Photodiode array detector, a Waters 2707 injection valve with a 20 μL loop, and a Waters HPLC workstation (Waters, Milford, MA, USA). The column applied in this work was a XTerra MS C18 column (250 mm × 4.6 mm, 5 μm, Waters, USA). The system run with a gradient program at 1 mL/min, and two solvents acetonitrile (A) and water (B) with the following gradient combinations: 0–10 min 30% B; 10–20 min, 30–60% B; 20–25 min, 60–90% B; 25–30 min, 90–60 % B; The eluent was monitored at 254 nm, and the purity was calculated by the target analytic peak area divided by the total peak area (unitary area method).

3.8. Identification of HSCCC Peak Fractions

The UV-vis spectra were recorded by a UV-1900 spectrophotometer (Puxi, Beijing, China) using a 1 cm path length cell with absorption wavelength at 254 nm.

The IR spectrum was recorded in KBr disc and the spectrum was scanned from 400 to 4000 cm^{-1} with a 6700 Nicolet Fourier transform-infrared spectrophotometer (Madison, WI, USA).

ESI-MS was performed by Waters SQD2 mass spectrometer (Waters, Milford, MA, USA), operating in positive mode. The MS conditions were as follows: capillary voltage 3 kV and temperature maintained at 300 °C, cone voltage 40 V, Mass-scan range were measured from m/z 50 to 1000, source temperature 120 °C, the gas flow rate for cone and desolvation (N_2) were 500 mL/min. Mass data in this manner provided for the collection of information of intact precursor ions.

Molecules **2018**, *23*, 560

In addition, two purified compounds were concentrated to dryness under reduced pressure and lyophilized, followed by dissolution in deuterated methanol (CH$_3$DO) for NMR analysis. The ^{1}H and ^{13}C spectra were obtained on a Bruker Avance III 400 MHz NMR spectrometer (Bruker Biospin Co., Billerica, MA Rheinstetten, Germany). ^{1}H-nuclear magnetic resonance (NMR) and ^{13}C-NMR spectra were obtained at the center of analysis, Shanghai Microspectrcum Chemical Technology Service Co. Ltd. (Shanghai, China).

4. Conclusions

In this study, the upper and low phase of chloroform-dichloromethane-methanol-water at a volume ratio of 4:2:3:2 (*v/v/v/v*) was selected as the mobile and stationary phase, and the separation condition of *H. erinaceum* mycelium crude extract (named HEM-E-E) were selected as follow: flow rate 2.0 mL/min, rotary speed 900 rpm, column temperature 20 °C. Under the optimized HSCCC conditions, 23 mg compound **1** with the purity of 95.7% and 18 mg compound **2** with the purity of 97.5 % were isolated from 150 mg HEM-E-E. These two compounds were confirmed as genistein (4′,5,7-Trihydroxyisoflavone) and daidzein (4′,7-Dihydroxyisoflavone). To the best of our knowledge, this is the first report in which two isoflavones, genistein and daidzein, are isolated and discovered from *H. erinaceum* mycelium. The results also demonstrated that HSCCC method is a powerful tool for the quick and efficient separation and purification of bioactive compounds from natural products.

Acknowledgments: This work was supported by National high technology research and development program (863 plan, PR China) (2014AA 022205), and by the National Natural Science Foundation of China (No. 31671813).

Author Contributions: Jinzhe He and Peilong Sun conceived and designed the experiments; Peng Fan performed the experiments; Simin Feng and Ping Shao analyzed the data; Peng Fan contributed reagents/materials/analysis tools; Simin Feng wrote the paper.

Conflicts of Interest: The author declares no conflict of interest.

References

1. Chang, C.H.; Chen, Y.; Yew, X.X.; Chen, H.X.; Kim, J.X.; Chang, C.C.; Peng, C.C.; Peng, R.Y. Improvement of Erinacine A Productivity in *Hericium erinaceus* Mycelia and Its Neuroprotective Bioactivity against the Glutamate-Insulted Apoptosis. *LWT Food Sci. Technol.* **2016**, *19*, 616–626. [CrossRef]
2. Jiang, S.; Wang, S.; Sun, Y.; Zhang, Q. Medicinal properties of *Hericium erinaceus* and its potential to formulate novel mushroom-based pharmaceuticals. *Appl. Microbiol. Biotechnol.* **2014**, *98*, 7661–7670. [CrossRef] [PubMed]
3. Li, W.; Zhou, W.; Cha, J.Y.; Kwon, S.U.; Baek, K.H.; Shim, S.H.; Lee, Y.M.; Kim, Y.H. Sterols from *Hericium erinaceum* and their inhibition of TNF-α and NO production in lipopolysaccharide-induced RAW 264.7 cells. *Phytochemistry* **2015**, *115*, 231–238. [CrossRef] [PubMed]
4. Mizuno, T. Yamabushitake, *Hericium erinaceum*: Bioactive substances and medicinal utilization. *Food Rev. Int.* **1995**, *11*, 173–178. [CrossRef]
5. Okamoto, K.; Sakai, T.; Shimada, A.; Shirai, R.; Sakamoto, H.; Yoshida, S.; Ojima, F.; Ishiguro, Y.; Kawagishi, H. Antimicrobial chlorinated orcinol derivatives from mycelia of *Hericium erinaceum*. *Phytochemistry* **1993**, *34*, 1445–1446. [CrossRef]
6. Mori, K.; Obara, Y.; Hirota, M.; Azumi, Y.; Kinugasa, S.; Inatomi, S.; Nakahata, N. Nerve growth factor-inducing activity of *Hericium erinaceus* in 1321N1 human astrocytoma cells. *Biol. Pharm. Bull.* **2008**, *31*, 1727. [CrossRef] [PubMed]
7. Wong, K.H.; Vikineswary, S.; Abdullah, N.; Naidu, M.; Keynes, R. Activity of Aqueous Extracts of Lion's Mane Mushroom *Hericium erinaceus* (Bull.: Fr.) Pers. (Aphyllophoromycetideae) on the Neural Cell Line NG108-15. *Int. J. Med. Mushrooms* **2007**, *9*, 57–65. [CrossRef]
8. Chaiyasut, C.; Sivamaruthi, B.S. Anti-hyperglycemic property of *Hericium erinaceus*—A mini review. *Asian Pac. J. Trop. Biomed.* **2017**, *7*, 1036–1040. [CrossRef]

9. Lu, C.C.; Huang, W.S.; Lee, K.F.; Lee, K.C.; Hsieh, M.C.; Huang, C.Y.; Lee, L.Y.; Lee, B.O.; Teng, C.C.; Shen, C.H. Inhibitory effect of Erinacines A on the growth of DLD-1 colorectal cancer cells is induced by generation of reactive oxygen species and activation of p70S6K and p21. *J. Funct. Foods* **2016**, *21*, 474–484. [CrossRef]

10. Kawagishi, H.; Ando, M.; Shinba, K.; Sakamoto, H.; Yoshida, S.; Ojima, F.; Ishiguro, Y.; Ukai, N.; Furukawa, S. Chromans, hericenones F, G and H from the mushroom *Hericium erinaceum*. *Phytochemistry* **1992**, *32*, 175–178. [CrossRef]

11. Kawagishi, H.; Shimada, A.; Hosokawa, S.; Mori, H.; Sakamoto, H.; Ishiguro, Y.; Sakemi, S.; Bordner, J.; Kojima, N.; Furukawa, S. Erinacines E, F, and G, stimulators of nerve growth factor (NGF)-synthesis, from the mycelia of *Hericium erinaceum*. *Tetrahedron Lett.* **1994**, *35*, 1569–1572. [CrossRef]

12. Qin, L.; Chen, H. Enhancement of flavonoids extraction from fig leaf using steam explosion. *Ind. Crops Prod.* **2015**, *69*, 1–6. [CrossRef]

13. Raffa, D.; Maggio, B.; Raimondi, M.V.; Plescia, F.; Daidone, G. Recent discoveries of anticancer flavonoids. *Eur. J. Med. Chem.* **2017**, *142*, 213–228. [CrossRef] [PubMed]

14. Li, H.; Park, S.; Moon, B.; Yoo, Y.B.; Lee, Y.W.; Lee, C. Targeted phenolic analysis in *Hericium erinaceum* and its antioxidant activities. *Food Sci. Biotechnol.* **2012**, *21*, 881–888. [CrossRef]

15. Xi, J.; Yan, L. Optimization of pressure-enhanced solid-liquid extraction of flavonoids from Flos Sophorae and evaluation of their antioxidant activity. *Sep. Purif. Technol.* **2017**, *175*, 170–176. [CrossRef]

16. Huang, P.; Zhang, Q.; Pan, H.; Luan, L.; Liu, X.; Wu, Y. Optimization of integrated extraction-adsorption process for the extraction and purification of total flavonoids from *Scutellariae barbatae* herba. *Sep. Purif. Technol.* **2017**, *175*, 203–212. [CrossRef]

17. Xiao, X.H.; Yuan, Z.Q.; Li, G.K. Preparation of phytosterols and phytol from edible marine algae by microwave-assisted extraction and high-speed counter-current chromatography. *Sep. Purif. Technol.* **2013**, *104*, 284–289. [CrossRef]

18. Luo, L.; Yan, C.; Zhang, S.; Li, L.; Li, Y.; Zhou, P.; Sun, B. Preparative separation of grape skin polyphenols by high-speed counter-current chromatography. *Food Chem.* **2016**, *212*, 712–721. [CrossRef] [PubMed]

19. Ma, C.; Wang, J.; Chu, H.; Zhang, X.; Wang, Z.; Wang, H.; Li, G. Purification and characterization of aporphine alkaloids from leaves of *Nelumbo nucifera* Gaertn and their effects on glucose consumption in 3T3-L1 adipocytes. *Int. J. Mol. Sci.* **2014**, *15*, 3481. [CrossRef] [PubMed]

20. Ito, Y. Golden rules and pitfalls in selecting optimum conditions for high-speed counter-current chromatography. *J. Chromatogr. A* **2005**, *1065*, 145–168. [CrossRef] [PubMed]

21. Chen, B.; Peng, Y.; Wang, X.; Li, Z.; Sun, Y. Preparative Separation and Purification of Four Glycosides from Gentianae radix by High-Speed Counter-Current Chromatography and Comparison of Their Anti-NO Production Effects. *Molecules* **2017**, *22*, 2. [CrossRef] [PubMed]

22. Ma, C.; Hu, L.; Fu, Q.; Gu, X.; Tao, G.; Wang, H. Separation of four flavonoids from *Rhodiola rosea* by on-line combination of sample preparation and counter-current chromatography. *J. Chromatogr. A* **2013**, *1306*, 12–19. [CrossRef] [PubMed]

23. Tang, Q.; Yang, C.; Ye, W.; Liu, J.; Zhao, S. Preparative isolation and purification of bioactive constituents from *Aconitum coreanum* by high-speed counter-current chromatography coupled with evaporative light scattering detection. *J. Chromatogr. A* **2007**, *1144*, 203–207. [CrossRef] [PubMed]

24. Rts, F.; Welendorf, R.M.; Nigro, E.N.; Frighetto, N.; Siani, A.C. Isolation of ursolic acid from apple peels by high speed counter-current chromatography. *Food Chem.* **2008**, *106*, 767–771.

25. Gutzeit, D.; Winterhalter, P.; Jerz, G. Application of preparative high-speed counter-current chromatography/ electrospray ionization mass spectrometry for a fast screening and fractionation of polyphenols. *J. Chromatogr. A* **2007**, *1172*, 40–46. [CrossRef] [PubMed]

26. Yao, S.; Liu, R.; Huang, X.; Kong, L. Preparative isolation and purification of chemical constituents from the root of *Adenophora tetraphlla* by high-speed counter-current chromatography with evaporative light scattering detection. *J. Chromatogr. A* **2007**, *1139*, 254–262. [CrossRef] [PubMed]

27. Zhu, Y.; Liu, Y.; Zhan, Y.; Liu, L.; Xu, Y.; Xu, T.; Liu, T. Preparative isolation and purification of five flavonoid glycosides and one benzophenone galloyl glycoside from *Psidium guajava* by high-speed counter-current chromatography (HSCCC). *Molecules* **2013**, *18*, 15648–15661. [CrossRef] [PubMed]

28. Zhang, S.; Li, L.; Cui, Y.; Luo, L.; Li, Y.; Zhou, P.; Sun, B. Preparative high-speed counter-current chromatography separation of grape seed proanthocyanidins according to degree of polymerization. *Food Chem.* **2017**, *219*, 399–407. [CrossRef] [PubMed]

29. Friesen, J.B.; Pauli, G.F. G.U.E.S.S.—A generally useful estimate of solvent systems in CCC. *J. Liq. Chromatogr. Relat. Technol.* **2005**, *28*, 2777–2806. [CrossRef]

30. Oka, F.; Oka, H.; Ito, Y. Systematic Search for Suitable 2-Phase Solvent Systems for High-Speed Countercurrent Chromatography. *J. Chromatogr. A* **1991**, *538*, 99–108. [CrossRef]

31. Tian, G.L.; Zhang, U.; Zhang, T.Y.; Yang, F.Q.; Ito, Y. Separation of flavonoids from the seeds of Vernonia anthelmintica Willd by high-speed counter-current chromatography. *J. Chromatogr. A* **2004**, *1049*, 219–222. [CrossRef]

32. Peng, J.; Jiang, Y.; Fan, G.; Chen, B.; Zhang, Q.; Chai, Y.; Wu, Y. Optimization suitable conditions for preparative isolation and separation of curculigoside and curculigoside B from Curculigo orchioides by high-speed counter-current chromatography. *Sep. Purif. Technol.* **2006**, *52*, 22–28. [CrossRef]

33. Du, Q.; Wu, C.; Qian, G.; Wu, P.; Ito, Y. Relationship between the flow-rate of the mobile phase and retention of the stationary phase in counter-current chromatography. *J. Chromatogr. A* **1999**, *835*, 231–235. [CrossRef]

34. Ma, X.F.; Tu, P.F.; Chen, Y.J.; Zhang, T.Y.; Wei, Y.; Ito, Y. Preparative isolation and purification of two isoflavones from Astragalus membranaceus Bge. var. mongholicus (Bge.) Hsiao by high-speed counter-current chromatography. *J. Chromatogr. A* **2003**, *992*, 193–197. [CrossRef]

35. Liu, F.; Han, S.; Ni, Y.Y. Isolation and purification of four flavanones from peel of Citrus changshanensis. *J. Food Process. Pres.* **2017**, *41*. [CrossRef]

36. Deng, S.; Deng, Z.; Fan, Y.; Peng, Y.; Li, J.; Xiong, D.; Liu, R. Isolation and purification of three flavonoid glycosides from the leaves of Nelumbo nucifera (Lotus) by high-speed counter-current chromatography. *J. Chromatogr. B* **2009**, *877*, 2487–2492. [CrossRef] [PubMed]

37. Shinomiya, K.; Kabasawa, Y.; Nakazawa, H.; Ito, Y. Countercurrent chromatographic separation of soybean isoflavones by two different types of coil planet centrifuges with various two-phase solvent systems. *J. Liq. Chromatogr. Relat. Technol.* **2003**, *26*, 3497–3509. [CrossRef]

38. Wang, Y.Q.; Tang, Y.; Liu, C.M.; Shi, C.; Zhang, Y.C. Determination and isolation of potential alpha-glucosidase and xanthine oxidase inhibitors from *Trifolium pratense* L. by ultrafiltration liquid chromatography and high-speed countercurrent chromatography. *Med. Chem. Res.* **2016**, *25*, 1020–1029. [CrossRef]

39. Kinjo, J.E.; Furusawa, J.I.; Baba, J.; Takeshita, T.; Yamasaki, M.; Nohara, T. Studies on the Constituents of Pueraria-Lobata. 3. Isoflavonoids and Related-Compounds in the Roots and the Voluble Stems. *Chem. Pharm. Bull.* **1987**, *35*, 4846–4850. [CrossRef]

40. Singh, H.; Singh, S.; Srivastava, A.; Tandon, P.; Bharti, P.; Kumar, S.; Dev, K.; Maurya, R. Study of hydrogen-bonding, vibrational dynamics and structure activity relationship of genistein using spectroscopic techniques coupled with DFT. *J. Mol. Struct.* **2017**, *1130*, 929–939. [CrossRef]

41. Zhao, C.X.; Xu, Y.L.; He, C.H. Axial dispersion coefficient in high-speed counter-current chromatography. *J. Chromatogr. A* **2009**, *1216*, 4841–4846. [CrossRef] [PubMed]

Sample Availability: Samples of the *H. erinaceus* (Bull.: Fr.) Pers. mycelium powder, compounds **1** and **2** are available from the authors.

Article

Fast Determination of Yttrium and Rare Earth Elements in Seawater by Inductively Coupled Plasma-Mass Spectrometry after Online Flow Injection Pretreatment

Zuhao Zhu and Airong Zheng *

College of Ocean and Earth Sciences, Xiamen University, Xiamen 361102, China; zhz@stu.xmu.edu.cn
* Correspondence: arzheng@xmu.edu.cn; Tel.: +86-189-5928-8323

Received: 29 November 2017; Accepted: 26 January 2018; Published: 23 February 2018

Abstract: A method for daily monitoring of yttrium and rare earth elements (YREEs) in seawater using a cheap flow injection system online coupled to inductively coupled plasma-mass spectrometry is reported. Toyopearl AF Chelate 650M® resin permits separation and concentration of YREEs using a simple external calibration. A running cycle consumed 6 mL sample and took 5.3 min, providing a throughput of 11 samples per hour. Linear ranges were up to 200 ng kg^{-1} except Tm (100 ng kg^{-1}). The precision of the method was <6% (RSDs, $n = 5$), and recoveries ranged from 93% to 106%. Limits of detection (LODs) were in the range 0.002 ng kg^{-1} (Tm) to 0.078 ng kg^{-1} (Ce). Good agreement between YREEs concentrations in CASS-4 and SLEW-3 obtained in this work and results from other studies was observed. The proposed method was applied to the determination of YREEs in seawater from the Jiulong River Estuary and the Taiwan Strait.

Keywords: rare earth elements; flow injection; inductively coupled plasma-mass spectrometry; seawater

1. Introduction

Due to their similar chemical properties, yttrium and fourteen rare earth elements (La, Ce, Pr, Nd, Sm, Eu, Gd, Tb, Dy, Ho, Er, Tm, Yb, Lu), collectively named YREEs, have always been studied together. Rare earth elements have a narrow range of relative atomic weights, ranging from 138.91 (La) to 173.04 (Lu), which results in extremely coherent chemical properties [1,2]. YREEs in seawater are drawing increasing attention due to the following aspects: (1) YREEs can be used as powerful tracers in marine biogeochemistry (ocean circulation, scavenging processes, trace metal cycles, etc.) and the redox-sensitive nature of Ce and Eu make them valuable for oxidation-reduction reactions [1–6]; (2) as YREEs are analogues of the radioactive actinides, understanding the geochemical cycling of YREEs can provide clues on the behavior of actinides (Am and Cm), which is important for monitoring the migration of actinides in radioactive waste repositories [7,8]; and (3) YREEs are widely used in industry (superconductor, functional materials, etc.), medical diagnostics (MRI, magnetic resonance imaging) and agriculture (as fertilizer) [9–14]. Anomalous concentrations of Gd, La and Sm discovered in estuarine and coastal seawaters reveals that the risks of YREEs' release into the ocean through runoff and sewage are increasing, threatening marine ecosystems and altering the distribution patterns of YREEs in estuaries and oceans [2,9–14]. Therefore, it is of great interest to monitor YREEs in seawater.

However, quickly and accurately determining the concentration of YREEs in seawater is a challenge because of their low concentrations and high salt matrix. The emergence of inductively coupled plasma-mass spectrometry (ICP-MS) has made trace and multi-element analysis more readily available and has been commonly used for the determination of YREEs. Nevertheless, little tolerance to total dissolved solids (<0.1%) makes direct (or after dilution) determinations of YREEs in seawater by ICP-MS inadvisable, since large amounts of salts will not only cause clogging of nebulizer, torch and cones, leading to signal drift, but also introduce severe polyatomic interferences. Thus YREEs must be separated and concentrated before detection. As pretreatment procedures, liquid-liquid extraction [5,15–17], co-precipitation [18–21], and solid-phase extraction [22–27] techniques have been utilized. Among these procedures, solid-phase extraction using chelating absorbents is becoming popular due to its wide selectivity of YREEs, low risk of contamination, freedom from toxic reagents and its simplicity for interfacing with ICP-MS to permit online determination when using flow injection (FI) as sample preconcentration systems. Compared to offline determination (batch method), online approaches using FI-ICP-MS has the advantages of sensitive response, high sample throughput, small volumes consumption of sample and reagents, little risk of contamination and labor savings [9,12,28–36]. The commercially available FIAS-400 (Perkin-Elmer, Waltham, MA, USA) [29,31–33] and SeaFAST (Elemental Scientific, Omaha, NE, USA) [36] FI system has been used in many studies devoted to the development of online determination of YREEs in seawater when coupled with ICP-MS. However, the relative high expense for a typical ICP-MS lab has restricted their wide application. Amongst the various absorbents used for FI-ICP-MS, Toyopearl AF Chelate 650M® resin, featuring iminodiacetate functional groups, can sequester all YREEs and offers stability (no shrinkage of the resin under both strong acid and high salt environments), while remaining inexpensive and commercially available [31,37–39]. Although Willie and Sturgeon [31] applied this resin for the online determination of YREEs in seawater using FI-ICP-TOF-MS, the relatively high LODs and the poor sample throughput (5 h^{-1}) were not attractive for routine measurements.

In this study, based on previous work [40,41], Toyopearl AF Chelate 650M® resin was used to separate YREEs from seawater and a fast FI-ICP-MS method for the online determination of YREEs in seawater was established. The FI system was readily set-up and automated to ensure ease of operation, as well as much cheaper than the FIAS-400 and SeaFAST. Experimental parameters were investigated and optimized to minimize sample consumption and improve the LODs. Finally, the developed method was used for the determination of YREEs in estuarine and coastal seawaters.

2. Results and Discussion

2.1. Effects of Sample Loading Rate and Time

The sample loading rate and time determine the volume of sample analyzed. To optimize the sample loading rate, 7.5 mL seawater (salinity = 33) was processed through the minicolumn using flow rates ranging from 1.5 to 4.0 mL min^{-1}. For all YREEs, it was found that peak areas decreased with increasing loading rate. When the loading rate was 2.0 mL min^{-1}, response from Ce dropped the most, to about 89% (Supplementary Information Table S1) of the peak area achieved at 1.5 mL min^{-1} (generally the rate of sample passing a column packed with Toyopearl AF Chelate 650M® resin by gravity is 1.5 mL min^{-1}, under which the YREEs can be 100% retained). Taking the analysis time and retention efficiency into consideration, 2.0 mL min^{-1} was subsequently used as the loading rate.

Theoretically, the peak area response should increase linearly with loading time. In this study, loading times from 120 s to 540 s were examined, with results showing that the relative coefficient (R^2) between the peak areas and loading times were all >0.9934 for all YREEs over the entire time range (Supplementary Information Table S2), indicating that the capacity of the minicolumn would not be overloaded even by processing 18 mL of high salinity seawater. In order to shorten the running time and improve throughput, 180 s was selected as the sample loading time, i.e., 6 mL sample was consumed. Loading time can be extended if the YREEs concentrations are significantly lower.

2.2. Influence of Interferences and Effect of Rinsing Conditions

Although ICP-MS detect YREEs may suffer from polyatomic interference, proper selection of target isotopes may help minimize such effects. However, interferences from lower REE oxides such as $^{143}Nd^{16}O^+$ on ^{159}Tb, $^{147}Sm^{16}O^+$ on ^{163}Dy, $^{149}Sm^{16}O^+$ on ^{165}Ho, $^{150}Nd^{16}O^+$ and $^{150}Sm^{16}O^+$ on ^{166}Er,

$^{153}Eu^{16}O^+$ on ^{169}Tm, $^{159}Tb^{16}O^+$ on ^{175}Lu are inevitable. Nevertheless, the oxide level was minimized during the tuning step by optimizing the ICP-MS parameters to ensure $^{140}Ce^{16}O/^{140}Ce$ <2%, plus the concentration variances of YREEs are generally within one order of magnitude, thus corrections for interference caused by light YREEs oxides were not considered, and polyatomic interferences such as $^{131}Ru^{16}O^+$ on ^{147}Sm, $^{140}Ce^{35}Cl^+$ on ^{175}Lu, and $^{135}Ba^{16}O^+$ on ^{151}Eu can be overcome using collision gas (He) mode [26,31,33].

However, physical interference induced by the salt matrix on instrument should also be resolved, since large amounts of salts (e.g., Na^+, K^+, Ca^{2+}, Cl^-) may deposit on the torch and the cones of the ICP-MS to cause considerable signal depression and drift in YREEs response. After sample loading, a mixture of buffer solution and ultrapure water was passed through the minicolumn to remove the residual salts. The flow rate was the same as the loading rate (2.0 mL min^{-1}) to reduce flow pulses in the minicolumn and the rinse time ranged from 30 to 70 s was studied. To minimize the rinse time, a test sample of salinity 33 was processed and response of Na, Mg, Cl, Ba and the YREEs were monitored. The relative peak areas (%) of each elements using rinsing time 40–70 s to 30 s are shown in Figure 1. With a rinsing time of 60 s, peak areas of Na, Mg and Cl decreased to 39%, 36% and 35%, respectively. While no significant decrease in peak area was observed with a further 10 s rinsing. The peak areas of Ba were nearly the same as the procedural blank, demonstrating that Ba in the seawater was not retained by the minicolumn, such that its interferences were negligible. All YREEs were approximately 100% recovered irrespective of the rinse time range. Consequently, 60 s was determined to be the optimal rinse time.

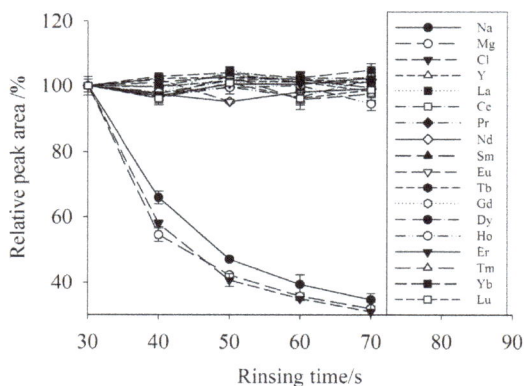

Figure 1. Effect of rinse time on Na, Mg, Cl and YREEs peak areas; test sample salinity = 33. The relative peak areas were normalized to that obtained using a rinsing time of 30 s.

2.3. Effects of Eluting Condition

To ensure the retained YREEs were totally eluted, the concentration of eluent (HNO$_3$) and elution rate and time were optimized. Results for La, Gd and Yb are shown in Figures 2 and 3 as examples. The concentrations of HNO$_3$ used ranged from 0.5 to 2.0 mol L^{-1}. The results showed that there were no significant differences between the peak shapes for the various HNO$_3$ concentrations (Figure 2) with the result that 0.8 mol L^{-1} HNO$_3$ was selected as the eluent since 0.5 mol L^{-1} HNO$_3$ is not strong enough to elute all sequestered trace metals which will stay on the column and compete with YREEs from the next sample to chelate with the resin. While more concentrated HNO$_3$ than 0.8 mol L^{-1} was not considered to protect ICP-MS, and our study showed that when used 0.8 mol L^{-1} HNO$_3$ as the eluting acid, the retention efficiency of the column would not decrease after 400 runs. As for elution rate, 1.0 mL min^{-1} was selected to obtain the fastest elution (Figure 3), since a higher flow rate may lead to instability of the plasma. With eluting use 0.8 mol L^{-1} HNO$_3$ under a flow rate of 1.0 mL min^{-1}, 50 s was required to elute all YREEs.

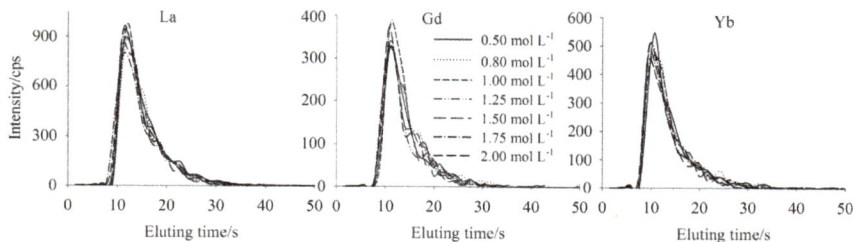

Figure 2. Elution profiles for La, Gd and Yb when eluted with different concentrations of HNO$_3$; test sample salinity = 33.

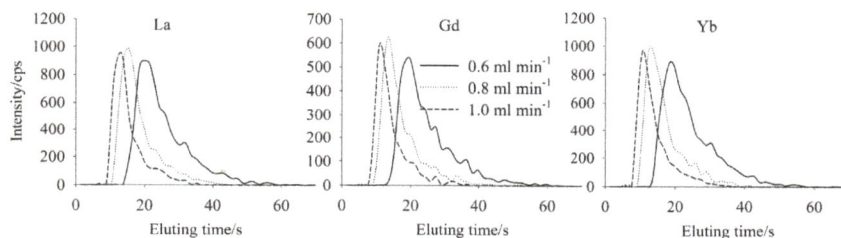

Figure 3. Elution profiles for Y when eluted with 0.8 mol L^{-1} HNO$_3$ at different flow rates; test sample salinity = 33.

2.4. Calibration and Effect of Salinity

When analyzing high matrix samples, the method of standard addition is often used for quantification to compensate for matrix effects. However, this methodology is tedious and labor intensive. Research by Willie and Sturgeon [31] concluded that the recoveries of YREEs from seawater using Toyopearl AF Chelate 650M® resin were independent of the salinity, and many studies have utilized a simple external calibration based on dilute HNO$_3$ for quantification. To confirm this conclusion and expand the simple calibration to seawater samples having a wide range of salinities (i.e., estuarine waters), four calibration curves (0–10 ng kg^{-1}) based on four different matrices (0.02 mol L^{-1} HNO$_3$, estuarine water sample with salinities of 2, 15 and 33) were prepared and analyzed using FI-ICP-MS. All four samples were using standard additions methodology (briefly, YREEs working standards were added to separate aliquots of a sample, and then the standard-containing samples plus the original sample were analyzed using FI-ICP-MS). The four calibration curves for Y are displayed in Figure 4 as an example.

The slopes of the four curves for each YREEs were tested for significant difference by SPSS (IBM SPSS Statistics 19.0, New York, NY, USA) using covariance analysis and all achieved a σ score (probability value) >0.05 (confidence level), indicating that the four slopes were not statistically different, demonstrating that the retention and elution properties of Toyopearl AF Chelate 650M® were not influenced by salinity. YREEs concentrations were thus calculated using external calibration curves comprising a 0.02 mol L^{-1} HNO$_3$ as the matrix.

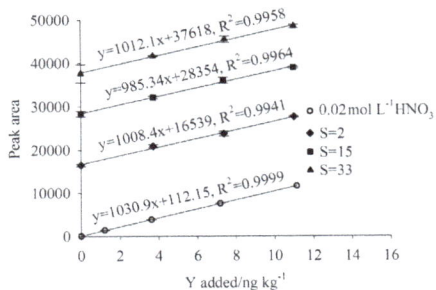

Figure 4. Standard additions calibration curves for Y based on samples having different matrices.

2.5. Analytical Figures of Merit

Using the optimized conditions, about 5.3 min was required for processing a 6 mL sample, resulting in a sample throughput of 11 h^{-1}. The linear range of the method was examined using a ten-point external calibration curve with concentrations from 0 to 200 ng kg^{-1}. The R^2 of the 7 ranges (0–5, 0–7, 0–15, 0–25, 0–50, 0–100 and 0–200 ng kg^{-1}) were all larger than 0.9917 (except for Tm, the R^2 of which was 0.9479 for the 0–200 ng kg^{-1} range). Covariance analysis was used to investigate the significant difference between the slopes of the 7 curves. Results showed the σ scores for all YREEs >0.05 (excepted Tm in 0–200 ng kg^{-1} range), indicating no significant differences between the 7 slopes for each YREEs. Therefore, the developed FI-ICP-MS procedure was capable of accurate measurements of YREEs concentration up to 200 ng kg^{-1} (100 ng kg^{-1} for Tm), suitable for the determination of YREEs in almost all seawater samples.

The precision and accuracy of the method were also examined using samples having differing salinities (salinity = 2, 15 and 33). The repeatability and reproducibility of the method was evaluated via repetitive inter-day (n = 5) and separate-day (n = 4) measurements of three samples, with inter-day RSDs of 0.3–6% and separate-day RSDs of 2–8%. Spike recovery testing was conducted and recoveries of 93–106% were obtained, confirming the accuracy of the method. Also we tested the performance of the method by analysis some high salinity aged seawater sample (salinity = 35–40, to simulate open ocean water with low level dissolved organic matter), and better RSDs of 0.47–4.81% and recovery of 94.5–104% were obtained, likely due to the aged seawater contained much less dissolved organic matter, which may compete with YREEs to absorb on the resin or compete with the resin to absorb the YREEs, than the estuarine and coastal seawater. These results suggested that the developed method provides satisfactory analytical results for seawater with wide range salinity.

Although Certified Reference Materials for YREEs in seawater (like newly released NASS-7 and some GEOTRACES intercalibration reference materials) were not commercially available at the time of this study, compiled results of multiple reports on YREE concentrations in CASS-4 and SLEW-3 can provide valuable reference data for the validation of the methodology developed in this study [35]. Such reference data, plus that obtained in this study are summarized in Table 1.

The data show good agreement with other studies (except the Y, Sm, Gd, Ho and Tm in CASS-4, whose RSD of this study value and the reference compiled were >5%), indicating the proposed method can provide reliable YREEs concentrations in estuarine and coastal seawater. As to the durability of the minicolumn, no decrease of YREEs retention efficiency was detected after 400 runs.

The procedural blank was obtained by processing 0.02 mol L^{-1} HNO$_3$ (pH ~1.6). The LODs of this FI-ICP-MS system were calculated based on 3 s (standard deviation) of 11 procedure blanks. The results are shown in Table 2. The blank and LODs are sufficiently low to permit determination of YREEs in estuarine and coastal seawaters. However if the open ocean seawater was subject to analysis, the HNO$_3$ and the buffer should be further purified to reduce the procedure blank. For the HNO$_3$, a second or even more times of distillation can be adopted, while for the buffer an additional minicolumn packed with Toyopearl AF Chelate 650M® resin placed on the buffer line can be implemented.

Table 1. YREEs concentrations in CASS-4 (Salinity = 30.7) and SLEW-3 (Salinity = 15) from references and this study.

Elements	CASS-4 (ng kg^{-1})			SLEW-3 (ng kg^{-1})		
	Reference Compiled [a]	This Study [b]	RSD [c] (%)	Reference Compiled [d]	This Study [b]	RSD [c] (%)
Y	20.93 ± 0.40	18.89 ± 0.12	7.25	40.55 ± 2.05	38.10 ± 2.39	4.41
La	9.37 ± 0.38	9.96 ± 0.15	4.33	7.80 ± 0.13	8.22 ± 0.25	3.71
Ce	4.69 ± 0.92	4.90 ± 0.07	3.13	7.08 ± 0.68	7.19 ± 0.45	1.06
Pr	1.33 ± 0.06	1.37 ± 0.01	2.10	1.68 ± 0.05	1.64 ± 0.03	1.78
Nd	5.39 ± 0.47	5.49 ± 0.04	1.30	8.18 ± 0.35	7.97 ± 0.19	1.82
Sm	5.55 ± 0.17	6.00 ± 0.21	5.45	7.10 ± 0.15	7.38 ± 0.21	2.69
Eu	0.23 ± 0.03	0.23 ± 0.02	0.00	0.54 ± 0.08	0.55 ± 0.02	1.72
Gd	1.29 ± 0.1	1.46 ± 0.04	8.74	3.09 ± 0.01	3.20 ± 0.06	2.36
Tb	0.20 ± 0.03	0.20 ± 0.01	0.00	0.45 ± 0	0.43 ± 0.02	2.94
Dy	1.41 ± 0.08	1.42 ± 0.05	0.38	3.37 ± 0.02	3.33 ± 0.08	0.95
Ho	0.38 ± 0.05	0.35 ± 0.02	6.32	0.91 ± 0	0.91 ± 0.07	0.13
Er	1.20 ± 0.1	1.27 ± 0.08	4.15	2.71 ± 0.01	2.78 ± 0.05	1.72
Tm	0.23 ± 0.07	0.20 ± 0.02	10.75	0.37 ± 0	0.35 ± 0.01	3.93
Yb	1.21 ± 0.14	1.16 ± 0.01	3.29	1.95 ± 0.14	1.85 ± 0.06	3.85
Lu	0.20 ± 0.03	0.19 ± 0.01	4.56	0.31 ± 0.03	0.30 ± 0.01	3.51

[a] Mean ± 1 standard deviation, *n* = 6, based on results in references [16,19–21,25,27]; [b] Mean ± 1 standard deviation, *n* = 5; [c] Relative standard deviation of this study from the data of reference compiled; [d] Mean ± 1 standard deviation, *n* = 3, based on results in references [16,19].

Table 2. Procedural blanks and LODs.

Elements	Blank [a] (ng kg^{-1})	LODs (ng kg^{-1})
Y	0.126 ± 0.023	0.034
La	0.172 ± 0.05	0.045
Ce	0.61 ± 0.112	0.078
Pr	0.038 ± 0.023	0.019
Nd	0.124 ± 0.038	0.048
Sm	0.046 ± 0.021	0.027
Eu	0.007 ± 0.004	0.009
Gd	0.05 ± 0.01	0.022
Tb	0.007 ± 0.003	0.003
Dy	0.03 ± 0.007	0.021
Ho	0.006 ± 0.003	0.003
Er	0.02 ± 0.007	0.012
Tm	0.003 ± 0.003	0.002
Yb	0.009 ± 0.007	0.005
Lu	0.002 ± 0.002	0.002

[a] Mean ± 1 standard deviation, *n* = 5.

2.6. Comparison with Other FI-ICP-MS Systems

Figures of merit of the developed FI-ICP-MS procedures are summarized in Table 3. Compared with other FI systems, this arrangement is easy to construct and short sample duration to ensure high sample throughput (11 h^{-1}), and provided excellent LODs based on only 6 mL sample consumption, moreover the total cost of this FI system is only about 20% of the cost of FIAS-400 (let alone the much more expensive seaFAST), making the daily and large scale determination of YREEs in seawater samples relatively inexpensive in a regular ICM-MS lab. Though report of Benkhedda et al. [30]

indicated shorter duration (4 min) and less sample consumption than this study, but an ICP-TOF-MS was required and the separation unit based on a knotted reactor was not as easily prepared as a minicolumn used here; results of Wang et al. [34] had the shortest duration (2.8 min) amongst all arrangements, while much more sample was needed and the fast loading rate may risk decreased retention efficiency and pump tubing aging. Advantages of this work were especially obviously when compared with thetudy of Willie and Sturgeon [31], in which the same Toyopearl AF Chelate 650M® absorbent was used, while the sensitivity provided by the ICP-TOF-MS was much lower than the quadrupole ICP-MS used in our work.

2.7. Applications

The established method was used for the determination of YREEs in seawater collected from the Jiulong River Estuary and the Taiwan Strait. Results are presented in Supplementary Information Tables S3 and S4. Post-Archean Australian Shales (PAAS) [42] normalized YREEs distributions patterns of the Jiulong River Estuary and the Taiwan Strait are plotted in Figure 5.

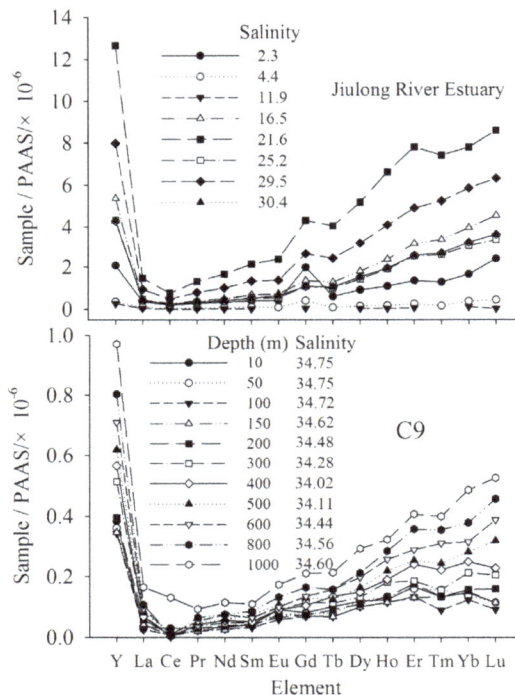

Figure 5. PAAS normalized YREEs patterns in surface water of Jiulong River Estuary and in station C9 (22°07′13″ N, 118°24′41″ E) of Taiwan Strait.

Relatively flat YREEs patterns were observed in samples having salinities of 4.4 and 11.9 from the Jiulong River Estuary, while all YREEs patterns show obviously negative Ce anomaly and positive Gd anomaly (except salinity 11.9). The latter could be attributed to anthropogenic Gd discharge (e.g., MRI contrast reagent). In the seawater of Taiwan Strait, slightly negative Ce anomalies were obtained from the YREEs patterns (except the surface and bottom water), which are commonly observed in the world oceans [6]. The geochemistry of YREEs in the Jiulong River Estuary and the Taiwan Strait will be further studied in future work.

Table 3. Comparison with other FI-ICP-MS methodologies for the determination of YREEs in seawater.

Loading Rate (mL min⁻¹)	Eluting Rate (mL min⁻¹)	Eluting Time (s)	HNO₃ (mol L⁻¹)	Duration (min)	Absorbent	Sample (mL)	LODs (ng kg⁻¹)	Reference
5	1.2	80	1.0	~5.5	-ᵃ	10	0.06–0.27	[8]
12	1.5	300	2.0	~12	Amberlite XAD-7 + 8HQ	100	0.002–0.016ᵇ	[9]
2.0	1.0	30	0.1	>30	APARᶜ	60	0.001–0.013	[12]
2	1.0	100	0.8ᵈ	~9.5	I-8-HQᵉ	3	0.06–0.6	[29]
4.4	0.8	90	0.4	~4	PMBPᶠ	2.2	0.003–0.04	[30]
5	1.5	30	1.5	12	Toyopearl AF Chelate 650M®	50	0.02–0.29	[31]
3.2	1.7	61	1.0	7	Muromac A-1	6.4	0.04–0.251	[32]
3.2	2.0	60	1.4	7	MAF-8HQ ᵍ	6.4	0.11–0.30	[33]
7.4	0.5	35	0.9	2.8	M-PTFE ʰ	14.8	0.001–0.02	[34]
5	0.5	120	2	6	Nobias chelate PB1M	10	0.005–0.09	[35]
1.0	0.3	5	1.5ⁱ	15	-	7	0.001–0.036	[36]
2.0	1.0	60	0.8	~5.3	Toyopearl AF Chelate 650M®	6	0.002–0.078	This study

ᵃ precipitation reagent; ᵇ only Eu, Tb, Ho, Tm, Lu detected; ᶜ alkyl phosphinic acid resin; ᵈ 2 mol L⁻¹ HCl + 0.8 mol L⁻¹ HNO₃; ᵉ 8-hydroxyquinoline; ᶠ 1-phenyl-3-methyl-4-benzoylpyrazol-5-one; ᵍ 8-quinoline-immoblized fluorinated metal alkoxide glass; ʰ polytetrafluoroethylene; ⁱ 1.5 mol L⁻¹ HNO₃ + 0.4% HAc.

3. Materials and Methods

3.1. Reagents and Samples

All solutions were prepared with ultrapure water (18.20 MΩ cm, Millipore, Darmstadt, Germany). Trace metal free nitric acid was obtained by purifying nitric acid (Merck, Darmstadt, Germany) using a sub-boiling distillation system. Standard stock solutions of YREEs (1000 ppm) were obtained from the National Institute of Metrology (Beijing, China). Working standards were prepared via serial dilutions of the stock solution with 0.02 mol L^{-1} purified HNO_3 (equal to acidified sample pH ~1.6). Ammonium acetate (NH_4Ac) buffer solution was prepared by mixing 30 mL aqueous ammonia (Sinopharm Chemical Reagent Co., Nanjing, China) and 20 mL glacial acetate acid (HAc, Sinopharm Chemical Reagent Co., Nanjing, China) and diluting to 1 L using ultrapure water; the pH was subsequently adjusted to 5.5 ± 0.2 with HAc or NH_4OH [31]. The buffer solution was further purified to remove potential YREEs by passing it through a column packed with Toyopearl AF Chelate 650M® resin. Two Certified Reference Materials (SLEW-3 and CASS-4) were purchased from the National Research Council Canada (Ottawa, Canada). Estuarine samples (salinity of 2 and 15) collected from the Jiulong River Estuary and coastal seawater (salinity = 33) collected from the South China Sea were used to optimize the method, All samples were acidified to pH ~1.6 using purified HNO_3 after filtered using 0.45 μm polycarbonate membranes. Trace metal clean procedures were used for the water sample collection.

All reagents and samples were stored in fluorinated ethylene propylene, low-density polyethylene or polypropylene acid washed bottles (Nalgene, Rochester, NY, USA). The cleaning procedure for all labware is detailed in Wen et al. [43].

3.2. Instrumentation

An Agilent 7700× ICP-MS (Agilent, Tokyo, Japan) operating in time-resolved-analysis mode was used for the measurement of YREEs. The ICP-MS was equipped with an octopole reaction/collision system which was employed to help overcome oxide and polyatomic interferences. The operating conditions were daily optimized with a 1 μg L^{-1} tuning solution (Co, Y, In, Tl, Ce) in the eluting acid at a flow rate equal to the elution rate. The typical operating parameters are summarized in Table 4.

Table 4. Typical ICP-MS operating conditions.

Rf Power	1500 W
Plasma gas	15.0 L min^{-1}
Auxiliary gas	1.0 L min^{-1}
Carrier gas	0.85 L min^{-1}
Collision gas (He)	4.1 mL min^{-1}
Integration time	0.1 s per isotope
Sampling depth	8 mm
Target isotopes	^{89}Y ^{139}La ^{140}Ce ^{141}Pr ^{143}Nd ^{147}Sm ^{151}Eu ^{157}Gd ^{159}Tb ^{163}Dy ^{165}Ho ^{166}Er ^{169}Tm ^{174}Yb ^{175}Lu

3.3. FI System and FI-ICP-MS Analysis Procedure

The construction of the FI system used in this study is shown in Figure 6. Apart from the metal free minicolumn assembly (MC-2CNME, Global FIA, Fox Island, WA, USA) with a tapered inner chamber (2 cm long with 27 μL internal volume) packed with Toyopearl AF Chelate 650M® (particle size: 40–90 μm; Tosoh Bioscience GmbH, Griesheim, Germany) resin and the T joint (i.d. 0.75 mm, VICI, Houston, TX, USA), all other parts of the FI system were the same as those used in our previous study [40,41]. The FI system was controlled using a computer running LabVIEW program (National Instruments, Austin, TX, USA). The schematic of FI-ICP-MS procedure is given in Figure 6 and the optimized FI program is summarized in Table 5.

Table 5. Typical flow injection program and valve position description.

Step	Duration/s	Pump 1/mL min^{-1}	Pump 2/mL min^{-1}	8-Position Valve	6-Way Valve
Conditioning	20	0.5	1.5	1	A
Loading	180	0.5	2.0	1	A
Rinsing	60	0.5	2.0	1	A
Eluting	50	1.0	0	2	B
		Return to conditioning			

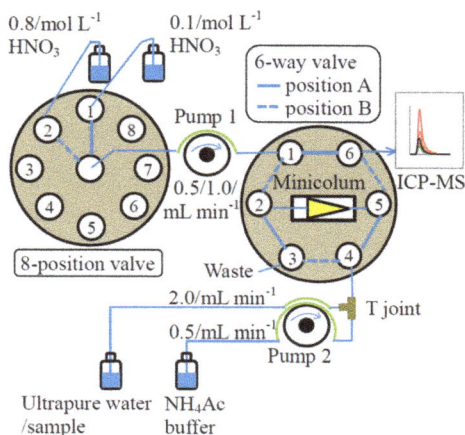

Figure 6. Schematic of the FI system and the FI-ICP-MS procedure.

A run cycle comprises four steps: step 1, conditioning, buffer solution and ultrapure water are mixed at the T joint and then passed through the minicolumn; step 2, loading, sample tube is placed into the sample bottle, sample (pH ~1.6) is online buffered to pH 5.5 ± 0.2 before entering the minicolumn and YREEs are retained on the column while the matrix salts pass to waste; step 3, rinsing, sample tube is transferred to the ultrapure water bottle and the mixture of buffer solution and ultrapure water is passed through the minicolumn to remove residual salts; step 4, 8-position valve is switched from position 1 to position 2 and 6-way valve is switched from position A to position B, 0.8 mol L^{-1} HNO$_3$ is pumped through the minicolumn in the reverse direction to elute the sequestered YREEs to the ICP-MS, and the data acquisition by the ICP-MS is manually activated at the same time. The elution profiles are recorded and the peak areas are integrated using the Agilent MassHunter workstation (Agilent, Santa Clara, CA, USA). The integration range was determined based on the comparison of YREEs signal intensity between the sample and the baseline (0–5 cps). The concentrations of the YREEs are determined using both standard addition as well as external standard calibration (see details in Section 2.4. Calibration and effect of salinity).

4. Conclusions

An automated FI system coupled online with ICP-MS to determine YREEs in seawater was developed. The components of the FI system in this work are all commercially available and the FI system is easy and cheap to assemble. With low LODs (0.002–0.078 ng kg^{-1}), the method only needs 6 mL of sample and achieves accurate and fast sample analysis (11 h^{-1}), making the regular monitoring of YREEs in seawater affordable. The analytical results of YREEs in CASS-4 and SLEW-3 confirmed that the proposed method can provide reliable results. The proposed method has been successfully applied to the determination of YREEs in seawater from the Jiulong River Estuary and Taiwan Strait,

Molecules **2018**, *23*, 489

and the procedure blank can be further reduced to meet the requirement of measurement of open ocean seawater by the further purification of HNO_3 and the buffer. The developed FI system can also be used as a preconcentration manifold for the offline detection (batch method) of not only YREEs but also other transition metals (Fe, Mn, Cu, Zn, etc.), and both the online and offline methods will be used in future work for trace metal detection in seawater.

Supplementary Materials: The following are available online, Table S1: Percentage of REEs peak area (%) of different loading rates to peak area of 1.5 mL min^{-1}. Table S2: Relative coefficients (R^2) between REEs peak areas and loading times. Table S3: REEs concentrations measured by the presented method from samples collected in the Jiulong River Estuary (water samples were collected in April 2015). Table S4: REEs concentrations measured by the presented method from samples collected in Taiwan Strait (seawaters were collected at the station C9, 22°07′13″ N, 118°24′41″ E, in April 2014).

Acknowledgments: This work was financially supported by the project "Special Fund for Marine Scientific Research in the Public Interest—A Nearshore of Micro/Nano Particles Biological Ecological Effect Assessment Techniques and Demonstration Application" from State Oceanic Administration of China (No. 201505034). Ralph Sturgeon (National Research Council Canada) is thanked for polishing the English and giving valuable comments, which improved this manuscript.

Author Contributions: Zuhao Zhu and Airong Zheng conceived and designed the experiments. Zuhao Zhu performed the experiment and analyzed the data. Zuhao Zhu and Airong Zheng wrote the paper.

Conflicts of Interest: The authors declare no conflict of interest.

References

1. Elderfield, H.; Greaves, M.J. The rare earth elements in seawater. *Nature* **1982**, *296*, 214–219. [CrossRef]
2. Hatje, V.; Bruland, K.W.; Flegal, A.R. Determination of rare earth elements after pre-concentration using NOBIAS-chelate PA-1® resin: Method development and application in the San Francisco Bay plume. *Mar. Chem.* **2014**, *160*, 34–41. [CrossRef]
3. Byrne, R.H.; Kim, K.-H. Rare earth element scavenging in seawater. *Geochim. Cosmochim. Acta* **1990**, *54*, 2645–2656. [CrossRef]
4. German, C.R.; Masuzawa, T.; Greaves, M.J.; Elderfield, H.; Edmond, J.M. Dissolved rare earth elements in the Southern Ocean: Cerium oxidation and the influence of hydrography. *Geochim. Cosmochim. Acta* **1995**, *59*, 1551–1558. [CrossRef]
5. Nozaki, Y.; Alibo, D.S. Importance of vertical geochemical processes in controlling the oceanic processes of dissolved rare earth elements in the northeastern Indian Ocean. *Earth Planet. Sci. Lett.* **2003**, *205*, 155–172. [CrossRef]
6. Zheng, X.-Y.; Plancherel, Y.; Saito, M.A.; Scott, P.M.; Henderson, G.M. Rare earth elements (REEs) in the tropical South Atlantic and quantitative deconvolution of their non-conservative behavior. *Geochim. Cosmochim. Acta* **2016**, *177*, 217–237. [CrossRef]
7. Barretto, P.M.C.; Fujimori, K. Natural analogue studies: Geology and mineralogy of Morro do Ferro, Brazil. *Chem. Geol.* **1986**, *55*, 297–312. [CrossRef]
8. Yan, X.-P.; Kerrich, R.; Hendry, M.J. Flow injection on-line group preconcentration and separation of (ultra) trace rare earth elements in environmental and geological samples by precipitation using a knotted reactor as a filterless collector for inductively coupled plasma mass spectrometric determination. *J. Anal. At. Spectrom.* **1999**, *14*, 215–221.
9. Vicente, O.; Padro, A.; Martinez, L.; Olsina, R.; Marchevsky, E. Determination of some rare earth elements in seawater by inductively coupled plasma mass spectrometry using flow injection preconcentration. *Spectrochim. Acta B* **1998**, *53*, 1281–1287. [CrossRef]
10. Bau, M.; Dulski, P. Anthropogenic origin of positive gadolinium anomalies in river waters. *Earth Planet. Sci. Lett.* **1996**, *143*, 245–255. [CrossRef]
11. Elbaz-Poulichet, F.; Seidel, J.-L.; Othoniel, C. Occurrence of an anthropogenic gadolinium anomaly in river and coastal waters of Southern France. *Water Res.* **2002**, *36*, 1102–1105. [CrossRef]
12. Fu, Q.; Yang, L.; Wang, Q. On-line preconcentration with a novel alkyl phosphinic acid extraction resin coupled with inductively coupled plasma mass spectrometry for determination of trace rare earth elements in seawater. *Talanta* **2007**, *72*, 1248–1254. [CrossRef] [PubMed]

13. Shellock, F.G.; Spinazzi, A. MRI Safety Update 2008: Part 1, MRI contrast agents and nephrogenic systemic fibrosis. *Am. J. Roentgenol.* **2008**, *191*, 1129–1139. [CrossRef] [PubMed]

14. Kulaksız, S.; Bau, M. Anthropogenic dissolved and colloid/nanoparticle-bound samarium, lanthanum and gadolinium in the Rhine River and the impending destruction of the natural rare earth element distribution in rivers. *Earth Planet. Sci. Lett.* **2013**, *362*, 43–50. [CrossRef]

15. Shabani, M.B.; Akagi, T.; Shimizu, H.; Masuda, A. Determination of trace lanthanides and yttrium in seawater by inductively coupled plasma mass spectrometry after preconcentration with solvent extraction and back-extraction. *Anal. Chem.* **1990**, *62*, 2709–2714. [CrossRef]

16. Lawrence, M.G.; Kamber, B.S. Rare earth element concentrations in the natural water reference materials (NRCC) NASS-5, CASS-4 and SLEW-3. *Geostand. Geoanal. Res.* **2007**, *31*, 95–103. [CrossRef]

17. Labrecque, C.; Lariviere, D. Quantification of rare earth elements using cloud point extraction with diglycolamide and ICP-MS for environmental analysis. *Anal. Methods* **2014**, *6*, 9291–9298. [CrossRef]

18. Greaves, M.J.; Elderfield, H.; Klinkhammer, G.P. Determination of the rare earth elements in natural waters by isotope-dilution mass spectrometry. *Anal. Chim. Acta* **1989**, *218*, 265–280. [CrossRef]

19. Bayon, G.; Birot, D.; Bollinger, C.; Barrat, J.A. Multi-element determination of trace elements in natural water reference materials by ICP-SFMS after Tm addition and iron co-precipitation. *Geostand. Geoanal. Res.* **2010**, *35*, 145–153. [CrossRef]

20. Freslon, N.; Bayon, G.; Birot, D.; Bollinger, C.; Barrat, J.A. Determination of rare earth elements and other trace elements (Y, Mn, Co, Cr) in seawater using Tm addition and Mg(OH)$_2$ co-precipitation. *Talanta* **2011**, *85*, 582–587. [CrossRef] [PubMed]

21. Zheng, X.-Y.; Yang, J.; Henderson, G.M. A robust procedure for high-precision determination of rare earth element concentrations in seawater. *Geostand. Geoanal. Res.* **2014**, *39*, 277–292. [CrossRef]

22. Zhang, T.; Shan, X.; Liu, R.; Tang, H.; Zhang, S. Preconcentration of rare earth elements in seawater with poly(acrylaminophosphonic dithiocarbamate) chelating fiber prior to determination by inductively coupled plasma mass spectrometry. *Anal. Chem.* **1998**, *70*, 3964–3968. [CrossRef]

23. Wen, B.; Shan, X.; Xu, S. Preconcentration of ultratrace rare earth elements in seawater with 8-hydroxyquinoline immobilized polyacrylonitrile hollow fiber membrane for determination by inductively coupled plasma mass spectrometry. *Analyst* **1999**, *124*, 621–626. [CrossRef]

24. Haley, B.A.; Klinkhammer, G.P. Complete separation of rare earth elements from small volume seawater samples by automated ion chromatography: Method development and application to benthic flux. *Mar. Chem.* **2003**, *82*, 197–220. [CrossRef]

25. Zhu, Y.; Itoh, A.; Fujimori, E.; Umemura, T.; Haraguchi, H. Determination of rare earth elements in seawater by ICP-MS after preconcentration with a chelating resin-packed minicolumn. *J. Alloys Compd.* **2006**, *408–412*, 985–988. [CrossRef]

26. Karadas, C.; Kara, D.; Fisher, A. Determination of rare earth elements in seawater by inductively coupled plasma mass spectrometry with off-line column preconcentration using 2,6-diacetylpyridine functionalized Amberlite XAD-4. *Anal. Chim. Acta* **2011**, *689*, 184–189. [CrossRef] [PubMed]

27. Zereen, F.; Yilmaz, V.; Arslan, Z. Solid phase extraction of rare earth elements in seawater and estuarine water with 4-(2-thiazolylazo) resorcinol immobilized Chromosorb 106 for determination by inductively coupled plasma mass spectrometry. *Microchem. J.* **2013**, *110*, 178–184. [CrossRef] [PubMed]

28. Rodriguez, J.A.; Hernandez, P.; Salazar, V.; Castrillejo, Y.; Barrado, E. Amperometric biosensor for oxalate determination in urine using sequential injection analysis. *Molecules* **2012**, *17*, 8859–8871. [CrossRef] [PubMed]

29. Halicz, L.; Gavrieli, I.; Dorfman, E. On-line method for inductively coupled plasma mass spectrometric determination of rare earth elements in highly saline brines. *J. Anal. At. Spectrom.* **1996**, *11*, 811–814. [CrossRef]

30. Benkhedda, K.; Infante, H.G.; Ivanova, E.; Adams, F.C. Determination of sub-parts-per-trillion levels of rare earth elements in natural waters by inductively coupled plasma time-of-flight mass spectrometry after flow injection on-line sorption preconcentration in a knotted reactor. *J. Anal. At. Spectrom.* **2001**, *16*, 995–1001. [CrossRef]

31. Willie, S.N.; Sturgeon, R.E. Determination of transition and rare earth elements in seawater by flow injection inductively coupled plasma time-of-flight mass spectrometry. *Spectrochim. Acta B* **2001**, *56*, 1707–1716. [CrossRef]

32. Hirata, S.; Kajiya, T.; Aihara, M.; Honda, K.; Shikino, O. Determination of rare earth elements in seawater by on-line column preconcentration inductively coupled plasma mass spectrometry. *Talanta* **2002**, *58*, 1185–1194. [CrossRef]

33. Kajiya, T.; Aihara, M.; Hirata, S. Determination of rare earth elements in seawater by inductively coupled plasma mass spectrometry with on-line column pre-concentration using 8-quinolinole-immobilized fluorinated metal alkoxide glass. *Spectrochim. Acta B* **2004**, *59*, 543–550. [CrossRef]

34. Wang, Z.-H.; Yan, X.-P.; Wang, Z.-P.; Zhang, Z.-P.; Liu, L.-W. Flow injection on-line solid phase extraction coupled with inductively coupled plasma mass spectrometry for determination of (ultra) trace rare earth elements using maleic acid grafted polytetrafluoroethylene fibers as sorbent. *J. Am. Soc. Mass Spectrom.* **2006**, *17*, 1258–1264. [CrossRef] [PubMed]

35. Zhu, Y.; Umemura, T.; Haraguchi, H.; Inagaki, K.; Chiba, K. Determination of REEs in seawater by ICP-MS after on-line preconcentration using a syringe-driven chelating column. *Talanta* **2009**, *78*, 891–895. [CrossRef] [PubMed]

36. Hathorne, E.C.; Haley, B.; Stichel, T.; Grasse, P.; Zieringer, M.; Frank, M. Online preconcentration ICP-MS analysis of rare earth elements in seawater. *Geochem. Geophys. Geosyst.* **2012**, *13*. [CrossRef]

37. Arslan, Z.; Paulson, A.J. Solid phase extraction for analysis of biogenic carbonates by electrothermal vaporization inductively coupled plasma mass spectrometry (ETV-ICP-MS): An investigation of rare earth element signatures in otolith microchemistry. *Anal. Chim. Acta* **2003**, *476*, 1–13. [CrossRef]

38. Shelley, R.U.; Zachhuber, B.; Sedwick, P.N.; Worsfold, P.J.; Lohan, M.C. Determination of total dissolved cobalt in UV-irradiated seawater using flow injection with chemiluminescence detection. *Limnol. Oceanogr. Methods* **2010**, *8*, 352–362. [CrossRef]

39. Sturgeon, R.E.; Willie, S.N.; Yang, L.; Greenberg, P.; Spatz, R.O.; Chen, Z.; Scriver, C.; Clancy, V.; Lam, J.W.H.; Thorrold, S. Certification of a fish otolith reference material in support of quality assurance for trace element analysis. *J. Anal. At. Spectrom.* **2005**, *20*, 1067–1071. [CrossRef]

40. Zhu, Z.; Zheng, A. Determination of rhenium in seawater from the Jiulong River Estuary and Taiwan Strait, China by automated flow injection inductively coupled plasma-mass spectrometry. *Anal. Lett.* **2017**, *50*, 1422–1434. [CrossRef]

41. Zhu, Z.; Zheng, A. Automated flow injection coupled with ICP-MS for the online determination of trace silver in seawater. *Spectroscopy* **2017**, *32*, 50–59.

42. Pourmand, A.; Dauphas, N.; Ireland, T.J. A novel extraction chromatography and MC-ICP-MS technique for rapid analysis of REE, Sc and Y: Revising CI-chondrite and Post-Archean Australian Shale (PAAS) abundances. *Chem. Geol.* **2012**, *291*, 38–54. [CrossRef]

43. Wen, L.-S.; Stordal, M.C.; Tang, D.; Gill, G.A.; Santschi, P.H. An ultraclean cross-flow ultrafiltration technique for the study of trace metal phase speciation in seawater. *Mar. Chem.* **1996**, *55*, 129–152. [CrossRef]

Sample Availability: Samples of the those seawater samples from Jiulong River Estuary and Taiwan Strait are available from the authors, the other samples and reagents are not available from the authors.

molecules

MDPI

Article

Qualitative and Quantitative Phytochemical Analysis of Different Extracts from *Thymus algeriensis* Aerial Parts

Nassima Boutaoui [1], Lahcene Zaiter [1], Fadila Benayache [1], Samir Benayache [1], Simone Carradori [2,*], Stefania Cesa [3], Anna Maria Giusti [4], Cristina Campestre [2], Luigi Menghini [2], Denise Innosa [5] and Marcello Locatelli [2]

[1] Unité de recherche Valorisation des Ressources Naturelles, Molécules Bioactives et Analyses Physicochimiques et Biologiques, Université Frères Mentouri, Constantine 1, Route d'Aïn El Bey, 25000 Constantine, Algérie; boutaoui.nassima@gmail.com (N.B.); lahcene.zaiter@yahoo.fr (L.Z.); fbenayache@yahoo.fr (F.B.); sbenayache@yahoo.com (S.B.)

[2] Department of Pharmacy, University "G. d'Annunzio" of Chieti-Pescara, Via dei Vestini 31, 66100 Chieti, Italy; cristina.campestre@unich.it (C.C.); luigi.menghini@unich.it (L.M.); marcello.locatelli@unich.it (M.L.)

[3] Dipartimento di Chimica e Tecnologie del Farmaco, Sapienza Università di Roma, P.le Aldo Moro 5, 00185 Rome, Italy; stefania.cesa@uniroma1.it

[4] Dipartimento di Medicina Sperimentale, Sapienza Università di Roma, P.le Aldo Moro 5, 00185 Rome, Italy; annamaria.giusti@uniroma1.it

[5] Facoltà di Bioscienze e tecnologie agro-alimentari e ambientali, Università di Teramo, Via Renato Balzarini 1, 64100 Teramo, Italy; dinnosa@unite.it

* Correspondence: simone.carradori@unich.it; Tel./Fax: +39-0871-3554-583

Received: 15 January 2018; Accepted: 17 February 2018; Published: 20 February 2018

Abstract: This study was performed to evaluate the metabolite recovery from different extraction methods applied to *Thymus algeriensis* aerial parts. A high-performance liquid chromatographic method using photodiode array detector with gradient elution has been developed and validated for the simultaneous estimation of different phenolic compounds in the extracts and in their corresponding purified fractions. The experimental results show that microwave-assisted aqueous extraction for 15 min at 100 °C gave the most phenolics-enriched extract, reducing extraction time without degradation effects on bioactives. Sixteen compounds were identified in this extract, 11 phenolic compounds and five flavonoids, all known for their biological activities. Color analysis and determination of chlorophylls and carotenoids implemented the knowledge of the chemical profile of this plant.

Keywords: color analysis; pigments; MAE; HPLC-PDA; SFE; *Thymus algeriensis*

1. Introduction

The genus *Thymus* belongs to the Lamiaceae family, which comprises about 400 genera. They are mainly herbaceous, perennials, small shrubs occurring within the Mediterranean region, which is the center of the entire genus, and are also characteristic in Asia, Southern Europe and North Africa [1]. Historically, the aerial parts of *Thymus* species, rich in volatile constituents, have been commonly used as herbal teas, condiments and spices. In addition, they have shown many ethnomedicinal properties such as tonic, carminative, digestive, antispasmodic, antimicrobial, antioxidant, antiviral, and anti-inflammatory activities [2]. *Thymus* leaves extracts, despite their frequent use as spice and infusions, are used in traditional medicine as astringent, expectorant, antiseptic, anti-rheumatic, diuretic, analgesic and cicatrizing agents. Thyme can also be used as a phytotherapy agent in veterinary (antispasmodic, antiseptic and digestive); it is applied as feed additives and for treating diseases of pets and farm animals [3,4].

Thymus algeriensis Boiss. & Reut. (*Thymus hirtus* Willd. ssp. *algeriensis*) is the most widespread North African species. It is endemic in Libya, Tunisia, Algeria and Morocco. *T. algeriensis* is largely used, fresh or dried, mainly as a culinary herb. On the other hand, the species is used in traditional medicine in respiratory disorders, against illnesses of the digestive tube and as anti-abortive [5–8]. The chemical composition of its essential oil has been previously studied with an exclusive attention to the volatile components, although results of its biological activity are limited [9–12].

Due to the growing interest in the characterization of endemic plants in pharmaceutical, cosmetic and food industry and as a part of our continuing investigation seeking new ways to enhance the recovery of bioactive substances, different extracts from *Thymus algeriensis* aerial parts were obtained by classical maceration, microwave-assisted extraction and supercritical fluid extraction, and compared by means of their phenolic content by a validated quantitative and analytical high performance liquid chromatography (HPLC-PDA) method. To the best of our knowledge this is the first attempt to characterize this plant species for the presence of these compounds which are well known to exert modulatory effects on biological systems.

To approach this issue, we decided to study the ability of three different extraction techniques on the recovery of the most abundant phenolic secondary metabolites according to an HPLC-PDA method validated in our laboratory [13]. After each extraction, some parameters were optimized in order to improve the phenolic pattern profile in terms of recovery and amount of each constituent. Hydroalcoholic extraction was then performed and compared with supercritical fluid (SFE) and microwave-assisted (MAE) extractions in terms of yield and recovery. In particular, the application of microwaves for heating the solvents and plant tissues in extraction process is known to increase the kinetic of extraction, to reduce extraction time and solvent waste, to promote higher extraction rate and to save costs compared to classical methods [14].

Moreover, SFE furnishes some operational advantages since it works with supercritical solvents with different physicochemical properties such as graduable density, relatively high diffusivity and low viscosity, thereby providing enhanced transport properties and faster extraction rates by means of an easy diffusion through solid materials [15].

To gain an efficient and adequate metabolite recovery, a crucial control and successive optimization of each parameter is necessary. Initially, in our experiments (MAE and SFE) we aimed at limiting thermal degradation phenomena setting the temperature at 40 °C. Then, we modified pressure for SFE, in order to modify the density of the supercritical fluid, and temperature for MAE, to study the impact on phenolic recovery.

Successively, in order to improve the chemical composition knowledge of this plant, the aerial parts were further characterized by color analysis using a device-independent color space (CIELAB parameters) as defined by the "Commission Internationale de l'Eclairage" and specific pigments (carotenoids and chlorophylls) pattern. This comprehensive phytochemical profile could be used to better corroborate the traditional use of this plant.

2. Experimental Section

2.1. Materials

Chemical standards: gallic acid, catechin, caffeic acid, chlorogenic acid, 4-hydroxy-benzoic acid, vanillic acid, epicatechin, syringic acid, 3-hydroxy-benzoic acid, isovanillin, *p*-coumaric acid, rutin, sinapinic acid, *t*-ferulic acid, naringin, 2,3-dimethoxy-benzoic acid, benzoic acid, *o*-coumaric acid, quercetin, *t*-cinnamic acid, naringenin, carvacrol, harpagoside (all purity >98%) were purchased by Sigma-Aldrich (Milan, Italy). Methanol, chloroform, ethyl acetate and *n*-butanol (HPLC-grade), acetic acid (99%), acetonitrile (HPLC-grade) were obtained from Carlo Erba Reagenti (Milan, Italy). Double-distilled water was obtained using a Millipore Milli-Q Plus water treatment system (Millipore Bedford Corp., Bedford, MA, USA). All extractions were monitored by thin layer chromatography (TLC) performed on 0.2 mm thick silica gel plates (60 F254 Merck) and the spots were detected under an ultraviolet (UV) lamp (at 254 and 365 nm). Column chromatography was carried out using Sigma-Aldrich silica gel (high purity grade, pore size 60 Å, 200–425 mesh particle size).

2.2. Plant Material

Samples of full bloom plants were collected from wild population in M'Sila region (Coordinates: 35°42′N 4°33′E), Algeria, in May 2016 and identified by Professor Mohamed Kaabeche (Biology Department, University of Setif 1, Algeria). A voucher specimen has been deposited in the Herbarium of the VARENBIOMOL research unit, University Frères Mentouri Constantine 1. Aerial parts were manually separated, dried at controlled temperature (40 ± 1 °C) in the dark until constant weight. Then plant material was powdered to a uniform granulometry and stored in the dark at –20 °C, in vacuum bags, until extractions and further phytochemical analyses.

3. Extraction Procedures

3.1. Hydroalcoholic Extraction and Fractionation

The air-dried aerial parts (leaves and flowers, 2.0 kg) of *T. algeriensis* were powdered (slight grinding at controlled temperature, up to 35 °C) and macerated at room temperature with EtOH–H_2O 70:30, (*v/v*) (15 L) for 24 h, four times with fresh solvent. After a filtration step, the extracts were combined, concentrated under reduced pressure, diluted in H_2O (800 mL) under magnetic stirring and maintained at 4 °C overnight to precipitate chlorophylls. After a second filtration step, the resulting solution was extracted with solvents with increasing polarities (chloroform, ethyl acetate and *n*-butanol). Each extract was dried with anhydrous Na_2SO_4, filtered over Chromafil® PET 20/25 (0.2 μm pore size, Machery-Nagel AG, Oensingen, Switzerland) into brown glass vials and concentrated under vacuum (up to 35 °C) to yield the following extracts: $CHCl_3$ (7.42 g), EtOAc (4.19 g), *n*-BuOH (33.15 g).

The chloroform extract was further fractionated by column chromatography (on silica gel; cyclohexane/diethyl ether, step gradients) to yield 31 fractions (F1–F31), combined according to their TLC profiles. The ethyl acetate extract was also further fractionated by column chromatography (on silica gel; $CHCl_3$/MeOH, step gradients) to yield 23 fractions (F1–F23), combined according to their TLC profiles. The extracts/subfractions were collected in a vial at room temperature, the extraction solvent was dried under a gentle N_2 flow at room temperature and the residue stored at −20 °C until chromatographic analysis (Figure 1).

Figure 1. Schematic flowchart performed on *Thymus algeriensis* aerial parts.

3.2. Supercritical Fluid Extraction (SFE)

The SFE extractor consists of a CO_2 delivery pump (PU-2080-CO_2, Jasco, Tokyo, Japan), a thermostatic chamber with a 50 mL extraction column, an UV-Vis detector equipped with high-pressure cell (875-UV, Jasco) and an automatic back pressure regulator (BP-2080 plus, Jasco). A sample of triturated aerial parts (20 g) was packed in a 50 mL SFE extraction bulk. The plant material was exposed to a dynamic extraction at 40 °C for 60 min with a CO_2 flow-rate of 3 mL/min. Two different pressure values (10 and 30 MPa) were applied in order to modulate supercritical fluid density as also reported in the literature regarding *Thymus*-related species [16]. After filtration over Chromafil® PET 20/25 (0.2 µm pore size, Machery-Nagel AG, Oensingen, Switzerland) into brown glass vials, the extracts (yield 0.73% for 10 MPa and 0.65% for 30 MPa) were collected at room temperature and stored at −20 °C until chromatographic analysis.

3.3. Microwave-Assisted Extraction (MAE)

MAE was performed using an automatic Biotage Initiator™ 2.0 (Uppsala, Sweden) characterized by 2.45 GHz high-frequency microwaves and power range 0–300 W. The internal vial temperature was strictly controlled by an infrared (IR) sensor probe. Ground samples were added with water (20:1 *v:w*, liquid-to-solid ratio). Then, the suspension was transferred in a 10 mL sealed vessel suitable for an automatic single-mode microwave reactor. MAE was carried out heating by microwave irradiation at 40, 60, 80, 100 or 120 °C for 5, 10 or 15 min and then cooling with pressurized air. After filtration over Chromafil® PET 20/25 (0.2 µm pore size, Machery-Nagel AG, Oensingen, Switzerland) into brown glass vials, the extraction solvent was dried under a gentle N_2 flow at room temperature. The dried mixtures (yields between 8.3–10.1%) were stored at −20 °C until further chromatographic analysis [17].

3.4. HPLC Analysis

HPLC-PDA phenolic pattern was evaluated by the validated method reported in the literature [18], using an HPLC Waters liquid chromatography (model 600 solvent pump, 2996 PDA) and a Phenomenex prodigy ODS(3) 100A 250 mm × 4.6 mm, 5 µm as column. Mobile phase was directly *on-line* degassed by using a Biotech 4CH DEGASI Compact (Onsala, Sweden). Empower v.2 Software (Waters Spa, Milford, MA, USA) was used to collect and analyze data. All extracts were weighted, dissolved in mobile phase and then 20 µL were directly injected into HPLC-PDA system. For over range samples, 1:10 dilution factor was applied. Data are reported as mean ± standard deviation of three independent measurements. The identification of individual compounds was carefully performed on the basis of their retention time (verified also by UV-Vis spectra) by comparison with those of pure standard compounds, without difficulties in peak tracking when multiple substances co-elute (see supplementary Materials, section S1).

4. Color Analysis

CIELAB parameters (L*, a*, b*, C^*_{ab} and h_{ab}), as defined by the "Commission Internationale de l'Eclairage", were determined on the powdered aerial parts of *T. algeriensis* using a colorimeter X-Rite SP-62 (X-Rite Europe GmbH, Regensdorf, Switzerland), equipped with a D65 illuminant and an observer angle of 10°. Color description was based on three parameters: L* that defines the lightness and varies between 0 (absolute black) and 100 (absolute white), a* that measures the greenness (−a*) or the redness (+a*) and b* that measures the blueness (−b*) and the yellowness (+b*). C^*_{ab} (chroma, saturation) expresses a measure of color intensity and h_{ab} (hue, color angle) is the attribute of appearance by which a color is identified according to its resemblance to red, yellow, green, or blue, or a combination of two of these attributes in sequence. Cylindrical coordinates C^*_{ab} and h_{ab} are calculated from the parameters a* and b* using the equations $C^*_{ab} = (a^{*2} + b^{*2})^{\frac{1}{2}}$ and $h_{ab} = \tan^{-1}(b^*/a^*)$ [19]. Three different powdered *T. algeriensis* aerial parts samples were analyzed. The results are expressed as the mean value ± standard deviation (SD).

5. Carotenoids and Chlorophylls Analysis

The total carotenoids and chlorophylls a and b analysis in *Thymus* sample was performed according to Solovchenko with some modifications [20]. The sample was homogenized with mortar and pestle in 5 mL of chloroform-methanol (2:1, *v/v*) containing 0.01% butylated hydroxytoluene (BHT) to inhibit peroxidation process. Moreover, homogenization was carried out with 50 mg of MgO to prevent chlorophyll pheophytinization. The homogenate was passed through a paper filter and after, distilled water was added to the amount of 0.2 of the extract volume. Finally, the mixture was centrifuged in glass tube test for 18 min at 3000 g at 10 °C to complete separation of chloroform fraction from methanol/aqueous one. The chloroform phase (lower phase) contained the hydrophobic molecules (chlorophylls, carotenoids, lipids, etc.), while the methanol-water phase (upper phase) contained the hydrophilic molecules. Absorption spectrum of the chloroform phase was recorded with a Beckman Coulter DU 800 instrument in the range of 350–800 nm with a spectral resolution of 0.5 nm at a temperature of 20 °C. Both chlorophylls and the total carotenoid contents were determined using absorption coefficient according to Wellburn (1994) [21]. Equations to determine the concentrations of chlorophyll a (C_a) and b (C_b), as well as total carotenoid (C_{tot}) contents are reported below:

$C_a = 11.47\ A_{665.6} - 2A_{647.6}$;

$C_b = 21.85\ A_{647.6} - 4.53\ A_{665.6}$;

$C_{tot} = (1000\ A_{480} - 1.33C_a - 23.93\ C_b)/202$.

The data are reported as means of three replications and expressed as µg/mg DW (dry weight) ± SD (standard deviation).

6. Results and Discussion

6.1. Hydroalcoholic Extracts and Subfractions

The aim of this analysis was to carry out a qualitative and quantitative study of the different extracts or subfractions of *T. algeriensis* aerial parts, and also to compare the extractive procedures in order to develop extraction methods with better yields (Table 1). The use of different solvents with increasing polarity led to a preliminary but metabolite-oriented purification. Successive subfractioning of CHCl$_3$ and EtOAc extracts highlighted better the presence of specific secondary metabolites based on the results obtained by means of our validated high performance liquid chromatography-photodiode array detector (HPLC-PDA) procedure.

Ethyl acetate extract was the richest of phenolic constituents reaching *p*-coumaric acid and benzoic acid the highest concentration in some isolated fractions (40.62 µg/g and 5.71 µg/g, respectively), and catechin having the highest concentration in a successive isolated one (6.23 µg/g).

Finally, eleven compounds were identified in chloroform extract, among which naringenin and benzoic acid resulted with the highest concentrations in the fractions F16 and F24 (8.97 µg/g and 10.92 µg/g, respectively), and only epicatechin was identified in the fraction F30 with the concentration of 6.78 µg/g.

Nine compounds were identified in the *n*-butanol fraction, including three flavonoids among which epicatechin had the highest concentration (48.03 µg/g), while of the six phenolic acids present in this fraction, *o*-coumaric acid resulted in the highest concentration (9.83 µg/g).

Table 1. Phenolic profile of main subfractions of *Thymus algeriensis* aerial part extracts obtained by maceration.

Identified Compound	Content (µg/g ± SD, DW)						
	F16 (CHCl$_3$)	F24 (CHCl$_3$)	F30 (CHCl$_3$)	F13 (EtOAc)	F22 (EtOAc)	F27 (EtOAc)	*n*-BuOH
Catechin	1.12 ± 0.01					6.23 ± 0.05	
4-Hydroxy-benzoic acid					16.31 ± 0.91	3.61 ± 0.30	0.66 ± 0.02
Vanillic acid	5.17 ± 0.11	0.23 ± 0.01			0.22 ± 0.01		
Epicatechin			6.78 ± 0.12	0.55 ± 0.01			48.03 ± 2.98

Table 1. *Cont.*

Identified Compound	Content (µg/g ± SD, DW)						
	F16 (CHCl₃)	F24 (CHCl₃)	F30 (CHCl₃)	F13 (EtOAc)	F22 (EtOAc)	F27 (EtOAc)	n-BuOH
Syringic acid							1.93 ± 0.11
p-Coumaric acid				1.26 ± 0.81	40.62 ± 3.01	1.63 ± 0.88	1.70 ± 0.58
Rutin	0.57 ± 0.02						4.52 ± 0.41
t-Ferulic acid	0.20 ± 0.01	0.23 ± 0.01			1.46 ± 0.13	0.84 ± 0.01	0.55 ± 0.01
Naringin		0.16 ± 0.01		4.02 ± 0.39	0.46 ± 0.01		
2,3-Dimethoxy-benzoic acid	6.51 ± 0.59				7.51 ± 0.47		3.52 ± 0.20
Benzoic acid		10.92 ± 1.21		5.71 ± 0.47			
o-Coumaric acid						1.03 ± 0.09	9.83 ± 0.87
Naringenin	8.97 ± 0.74	0.90 ± 0.03					0.47 ± 0.01
Carvacrol	0.43 ± 0.01						
Total	22.97 ± 1.01	12.44 ± 1.03	6.78 ± 0.12	11.54 ± 0.99	66.58 ± 2.70	13.34 ± 1.12	71.21 ± 2.40

DW: dry weight.

6.2. Supercritical Fluid Extraction (SFE)

The aim of this experiment was to study the influence of the operating pressure parameter on the kinetics of the supercritical extraction process, in order to optimize the operative conditions and to compare the performance with the other extraction methods, using the secondary metabolites profile as discriminant marker.

We carried out two separate plant sample extractions at a fixed temperature of 40 °C, operating with different pressure values (10 MPa and 30 MPa) in order to find the leading coordinates towards the best extraction yield. We found that the increasing pressure did not improve either yield or the phenolic recovery, with only slight effects on the vanillic acid amount. According to the quality and the quantity of metabolites, the more interesting results were obtained under the following operating conditions: 40 °C for the temperature and 30 MPa for the pressure (Table 2). Collectively, this method afforded low amounts for phenolic compounds from *T. algeriensis* aerial parts.

Table 2. Phenolic profile of *T. algeriensis* aerial part extracts obtained by supercritical fluid extraction obtained at two discrete pressure values.

Identified Compound	Content (µg/g ± SD, DW)	
	40 °C, 10 MPa	40 °C, 30 MPa
Gallic acid	0.10 ± 0.01	0.10 ± 0.01
Catechin	0.05 ± 0.01	0.05 ± 0.01
Vanillic acid		0.18 ± 0.02
Epicatechin	0.15 ± 0.02	0.15 ± 0.01
Isovanillin	1.49 ± 0.09	1.48 ± 0.06
p-Coumaric acid	0.17 ± 0.02	0.14 ± 0.01
Naringin	0.06 ± 0.01	0.06 ± 0.01
Harpagoside	0.10 ± 0.01	0.10 ± 0.01
Total	2.12 ± 0.10	2.26 ± 0.10

DW: dry weight.

6.3. Microwave-Assisted Extraction (MAE)

Design and optimization of the microwave-assisted extraction of this plant were performed keeping in mind the impact of different parameters (temperature, time, and solvent volume) on the recovery of main metabolites profile used as discriminant marker.

The choice of the solvent, to afford the best extraction yield, is one of the most important steps for the development of an extraction method. In order to obtain good results, a preliminary microwave-assisted extraction based on water as a polar solvent (solvent volume: 20 mL; plant material: 1 g; extraction time: 10 min) was applied and the best extraction temperature ranging from 40 to 120 °C was evaluated by means of the results provided by HPLC method (see supplementary Materials, section S2). Once selected the best temperature condition (100 °C), to obtain the most diversified extract in phenolics, we tried extraction times of 5 and 15 min, obtaining the best extraction yield with an extraction time of 15 min in the same solvent (water). Finally, we have applied these conditions to select the best extraction solvent. The aqueous extract resulted richer in plant secondary phenolic metabolites than the EtOH/H$_2$O (70/30, v/v) extract, obtained in the same conditions (Table 3).

The richest phenolic pattern was obtained in the previously selected conditions (temperature: 100 °C, time: 15 min) using water as the solvent. Several compounds were identified in this extract, whose composition is reported in Table 3, which contained a complex mixture of plant secondary metabolites belonging to the chemical classes of phenolic acids and flavonoids, both known for their pharmacological activities.

The phenolic acids were identified as gallic acid, chlorogenic acid, vanillic acid, syringic acid, 3-hydroxybenzoic acid, *p*-coumaric acid, sinapinic acid, *t*-ferulic acid, benzoic acid, *o*-coumaric acid. Benzoic acid displayed the highest concentration (4145.75 µg/g) followed by epicatechin (246.752 µg/g), chlorogenic acid (1745.98 µg/g), syringic acid (615.20 µg/g), naringin (376.60 µg/g), catechin (359.80 µg/g), and *o*-coumaric acid (341.55 µg/g). Flavonoids and phenolic compounds are the most important groups of secondary metabolites and bioactive compounds in plants [22,23]. Two important trends could be also extrapolated from the impact of the increasing temperature on the recovery: first, some compounds (chlorogenic acid, *t*-ferulic acid, quercetin, isovanillin, epicatechin, syringic acid, catechin and *p*-coumaric acid) reached the highest amount until 100 °C and then their concentrations tended to diminish at higher temperatures. Secondly, other compounds (vanillic acid, gallic acid, sinapinic acid, naringin, 2,3-dimethoxybenzoic acid, and *o*-coumaric acid) had their maximum recovery at lower temperature and extraction time.

Table 3. High performance liquid chromatography-photodiode array detector (HPLC-PDA) analysis of the phenolic profile of *Thymus algeriensis* microwave-assisted extracts.

Identified Compound	Content (µg/g ± SD, DW)							
	40 °C, 10 min, water	60 °C, 10 min, water	80 °C, 10 min, water	100 °C, 10 min, water	120 °C, 10 min, water	100 °C, 5 min, water	100 °C, 15 min, water	100 °C, 15 min, EtOH/H$_2$O 50:50
Gallic acid	36.51 ± 0.23	72.84 ± 0.36	24.57 ± 0.21	28.05 ± 0.19	20.31 ± 0.20	9.70 ± 0.09	37.97 ± 0.25	
Catechin				46.69 ± 0.33	238.91 ± 1.66	45.23 ± 0.18	359.80 ± 1.98	
Chlorogenic acid	118.34 ± 1.01	1203.98 ± 5.29	1090.00 ± 4.77	1766.64 ± 5.13	930.04 ± 3.99	1570.40 ± 4.33	1745.98 ± 5.65	
4-Hydroxybenzoic acid				18.38 ± 0.20				37.38 ± 0.66
Vanillic acid		37.56 ± 0.71	50.17 ± 0.98	9.73 ± 0.12		176.74 ± 1.02	23.92 ± 0.66	
Epicatechin			813.21 ± 2.29	1675.51 ± 3.95	821.40 ± 3.02	200.62 ± 1.15	2462.75 ± 2.00	66.11 ± 0.99
Syringic acid			17.86 ± 0.99	52.22 ± 1.39	315.18 ± 3.00		747.63 ± 3.23	615.20 ± 4.03
3-Hydroxybenzoic acid		243.31 ± 2.21	57.94 ± 0.98	212.38 ± 1.75	70.62 ± 0.67	75.06 ± 0.60	166.73 ± 1.02	19.82 ± 0.51
Isovanillin		223.75 ± 1.97	248.12 ± 1.88	271.11 ± 2.04	59.05 ± 0.77	68.21 ± 0.67	40.42 ± 0.78	200.37 ± 1.09
p-Coumaric acid	15.78 ± 0.13	14.07 ± 0.29	21.65 ± 0.28	40.00 ± 0.32	23.05 ± 0.27	18.82 ± 0.23	106.99 ± 0.77	62.02 ± 0.19
Rutin	39.70 ± 0.23	375.42 ± 1.01	37.63 ± 0.65	89.45 ± 0.69		16.22 ± 0.09	196.89 ± 1.00	55.32 ± 0.55
Sinapinic acid	4.96 ± 0.13	31.58 ± 0.60	5.37 ± 0.17	5.53 ± 0.17	5.26 ± 0.18	29.64 ± 0.60	46.20 ± 0.63	8.48 ± 0.10
t-Ferulic acid		19.89 ± 0.99	14.22 ± 0.50	27.56 ± 0.52	21.72 ± 0.54	1869.72 ± 2.01	140.64 ± 0.73	
Naringin	89.63 ± 0.98	385.92 ± 2.12	179.46 ± 2.03	29.64 ± 0.23	31.31 ± 0.24	132.37 ± 1.00	376.60 ± 2.77	1611.91 ± 2.49
2,3-dimethoxybenzoic acid	520.37 ± 2.02	1212.75 ± 5.84	222.84 ± 1.95					507.77 ± 2.39
Benzoic acid	1455.82 ± 3.39	4896.51 ± 5.77	2425.66 ± 4.23	2393.82 ± 4.78	2697.97 ± 5.00	3889.74 ± 4.77	4157.75 ± 4.67	
o-Coumaric acid	227.74 ± 1.04	348.86 ± 2.20	37.66 ± 0.45	23.51 ± 0.29	30.27 ± 0.28	812.41 ± 2.00	341.55 ± 1.17	163.21 ± 0.99
Quercetin			63.04 ± 0.20	74.71 ± 0.23	59.52 ± 0.20	68.60 ± 0.28	180.72 ± 0.77	
Total	2508.85 ± 5.36	9084.30 ± 8.12	5362.13 ± 4.99	7009.51 ± 8.54	5009.41 ± 4.12	9731.14 ± 7.98	11000.12 ± 9.96	2732.38 ± 5.88

DW: dry weight.

The experimental results show that aqueous extraction, assisted by microwave at temperature 100 °C and for a time of 15 min, produced a phenolic-enriched extract of *Thymus algeriensis* aerial parts.

6.4. Color Analysis

Dry samples of *Thymus algeriensis* aerial parts were blended in a mixer and further homogenized in a mortar. The samples of blended powdered dry leaves of *Thymus* showed a nonhomogeneous color in which three different shades could be evidenced, a pale beige, a pale pink and a pale green. Conversely, the mortar homogenized samples displayed a more homogeneous light brown color characterized by two only pale nuances, a pale green and an undefined pale shade among beige, pink and orange. The tristimulus colorimetry was employed in this study to evaluate the color properties: the colorimetric data and the mean value, completed by the standard deviation, of three different samples, are reported in Table 4.

It is not possible to give an interpretation for comparison of this CIELAB colorimetric analyses, because only few data are available in literature about colorimetric studies on aerial parts in general, and no data are disposable for this plant in particular. Moreover, as known, the colorimetric parameters are deeply influenced by several different factors [24,25], among which storage temperature, humidity, light exposition, besides the standard conditions applied, according to the Commission Internationale de l'Éclairage (CIE). In this work a D65 illuminant, sunlight simulating, with an observer angle of 10°, was used. Reflectance curve is reported in Figure 2. Only a weak yellow parameter (light positive b*) can be shown, in line with the phenolic profile denoted by the HPLC analysis.

Table 4. Color analysis using a device-independent color space (CIELAB parameters), as defined by the "Commission Internationale de l'Eclairage", data of *Thymus algeriensis* aerial parts.

CIELAB Parameters	Mean Value	SD
L*	51.25	1.91
a*	1.09	0.36
b*	5.53	0.69
C^*_{ab}	5.64	0.72
h_{ab}	78.90	3.28

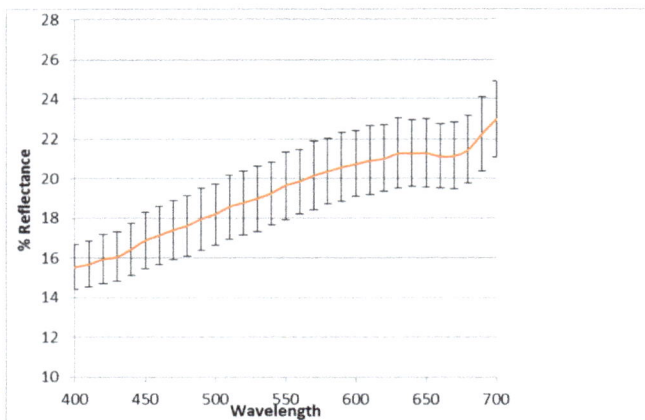

Figure 2. Reflectance analysis of *Thymus algeriensis* aerial parts.

6.5. Pigments Determination (Total Carotenoids, Chlorophyll a and Chlorophyll b)

T. algeriensis aerial parts plant displayed significant differences in analyzed pigments concentration. It is possible to notice that chlorophyll a content (0.0529 ± 0.0072 µg/mg) is higher compared with the other pigments (chlorophyll b = 0.0452 ± 0.0069 µg/mg and carotenoids 0.0165 ± 0.007 µg/mg) ($p < 0.002$) (Figure 3), but chlorophyll b was present in good proportion, as expected in shade plants. Generally, the concentration of chlorophyll a (Chl a) is two to three times the concentration of chlorophyll b (Chl b) due to the importance of chlorophyll a as the primary pigment in photosynthesis. However, a greater proportion of Chl b improved light-collection ability of the leaf in the region of far-red light. In this plant, the value of the Chl a/b ratio was 1.17, indicating a high chlorophyll a content. Chl a/b ratio can be a useful indicator of nitrogen partitioning within a leaf and this ratio is predicted to respond to light and nitrogen availability. In particular, Chl a/b ratio should increase with increasing irradiance at a given

Molecules **2018**, *23*, 463

nitrogen availability [26]. Regarding total carotenoid concentration, our finding showed an amount about three times lower than chlorophyll a (0.0165 ± 0.007 µg/mg and 0.0529 ± 0.0072 µg/mg, respectively) (Figure 3). High carotenoid contents are revealed under high insolation level, because in this case they act as protectors from photoinhibition [26]. The quali-quantitative evaluation of these ubiquitous bulk substances is of crucial interest not only due to their role in the protection against oxidative stress [27], but also for their interference with biological assays generating false positive and negative results [28].

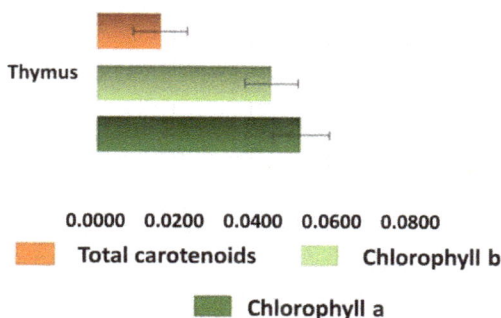

Figure 3. Pigments concentration in *T. algeriensis* aerial parts (µg/mg ± SD).

7. Conclusions

The study of endemic plants could support their ethnobotanical use by means of a wide comprehension of the qualitative and quantitative phytochemical profile. After the evaluation of color parameters and pigment content of the aerial parts of *T. algeriensis*, we developed an HPLC method with diode array detection for the quantitative and qualitative estimation of phenolic compounds obtained by different extraction methods. Three extraction methods have been carried out in this work for the purpose of developing extractions under the best conditions. The results of the present study revealed important data regarding the phenolic composition of *Thymus algeriensis* aerial parts up to now; several phenolic compounds known for their pharmacological properties were identified and quantified in different extracts of this plant. MAE was shown to be the best-performing procedure. The presence of a significant amount of respective bioactive components in this plant and the variation of quantity based on the polarity of the solvent used for the extraction process ensured its unequivocal recommendation for use in the pharmaceutical and nutraceutical sector.

Supplementary Materials: The following are available online, Section S1: Chemical standards resolution in HPLC-PDA method and gradient elution profile, Section S2: HPLC-PDA chromatograms obtained for MAE optimization.

Acknowledgments: This work was supported by University "G. d'Annunzio" of Chieti–Pescara, Chieti, Italy.

Author Contributions: F.B. and S.C. conceived and designed the experiments; S.B. collected the material and prepared it for the analysis, N.B., L.M., and L.Z. performed the extractions; M.L. and D.I. performed the HPLC analyses. S.Ce. carried out the color analysis, A.M.G. performed the pigment analyses. C.C. analyzed the data and contributed reagents and materials; M.L. and S.C. wrote the paper.

Conflicts of Interest: The authors declare no conflict of interest.

References

1. Figueiredo, A.C.; Barroso, J.G.; Pedro, L.G. Volatiles from *Thymbra* and *Thymus* species of the western Mediterranean basin, Portugal and Macaronesia. *Nat. Prod. Commun.* **2010**, *5*, 1465–1476. [PubMed]
2. Nabavi, S.M.; Marchese, A.; Izadi, M.; Curti, V.; Daglia, M.; Nabavi, S.F. Plants belonging to the genus *Thymus* as antibacterial agents: from farm to pharmacy. *Food Chem.* **2015**, *173*, 339–347. [CrossRef] [PubMed]

3. Maksimović, Z.; Stojanović, D.; Sostaric, I.; Dajić, Z.; Ristić, M. Composition and radical-scavenging activity of *Thymus glabrescens* Willd. (Lamiaceae) essential oil. *J. Sci. Food Agric.* **2008**, *88*, 2036–2041. [CrossRef]

4. Jarić, S.; Mitrović, M.; Pavlović, P. Review of ethnobotanical, phytochemical, and pharmacological study of *Thymus serpyllum* L. *Evid. Based Complement. Alternat. Med.* **2015**, *2015*, 101978. [CrossRef] [PubMed]

5. Guesmi, F.; Ben, F.M.; Mejri, M.; Landoulsi, A. In-vitro assessment of antioxidant and antimicrobial activities of methanol extracts and essential oil of *Thymus hirtus* sp. algeriensis. *Lipids Health Dis.* **2014**, *13*, 114–125.

6. Guesmi, F.; Ali, M.B.; Barkaoui, T.; Tahri, W.; Mejri, M.; Ben-Attia, M.; Bellamine, H.; Landoulsi, A. Effects of *Thymus hirtus* sp. *algeriensis* Boiss. et Reut. (Lamiaceae) essential oil on healing gastric ulcers according to sex. *Lipids Health Dis.* **2014**, *13*, 138–150. [PubMed]

7. Espina, L.; García-Gonzalo, D.; Laglaoui, A.; Mackey, B.M.; Pagán, R. Synergistic combinations of high hydrostatic pressure and essential oils or their constituents and their use in preservation of fruit juices. *Int. J. Food Microbiol.* **2013**, *161*, 23–30. [CrossRef] [PubMed]

8. Guesmi, F.; Beghalem, H.; Tyagi, A.K.; Ali, M.B.; Mouhoub, R.B.; Bellamine, H.; Landoulsi, A. Prevention of H_2O_2 induced oxidative damage of rat testis by *Thymus algeriensis*. *Biomed. Environ. Sci.* **2016**, *29*, 275–285. [PubMed]

9. Giordani, R.; Hadef, Y.; Kaloustina, J. Compositions and antifungal activities of essential oils of some Algerian aromatic plants. *Fitoterapia* **2008**, *79*, 199–203. [CrossRef] [PubMed]

10. Hazzit, M.; Baaliouamer, A.; Verissimo, A.R.; Faleiro, M.L.; Miguel, M.G. Chemical composition and biological activities of Algerian *Thymus* oils. *Food Chem.* **2009**, *116*, 714–721. [CrossRef]

11. Chemat, S.; Cherfouh, R.; Meklati, B.Y.; Belanteur, K. Composition and microbial activity of thyme (*Thymus algeriensis genuinus*) essential oil. *J. Essent. Oil Res.* **2012**, *24*, 5–11. [CrossRef]

12. Nikolic, M.; Glamoclija, J.; Ferreira, I.C.F.R.; Calhelha, R.C.; Fernandes, A.; Markovic, T.; Marković, D.; Giweli, A.; Soković, M. Chemical composition, antimicrobial, antioxidant and antitumor activity of *Thymus serpyllum* L., *Thymus algeriensis* Boiss. and Reut and *Thymus vulgaris* L. essential oils. *Ind. Crops Prod.* **2014**, *52*, 183–190. [CrossRef]

13. Diuzheva, A.; Carradori, S.; Andruch, V.; Locatelli, M.; De Luca, E.; Tiecco, M.; Germani, R.; Menghini, L.; Nocentini, A.; Gratteri, P.; et al. Use of innovative (micro)extraction techniques to characterize *Harpagophytum procumbens* root and its commercial food supplements. *Phytochem. Anal.* **2018**. [CrossRef]

14. Delazar, A.; Nahar, L.; Hamedeyazdan, S.; Sarker, S.D. Microwave-assisted extraction in natural products isolation. *Methods Mol. Biol.* **2012**, *864*, 89–115. [PubMed]

15. da Silva, R.P.F.F.; Rocha-Santos, T.A.P.; Duarte, A.C. Supercritical fluid extraction of bioactive compounds. *Trends Analyt. Chem.* **2016**, *76*, 40–51. [CrossRef]

16. Petrović, N.V.; Petrović, S.S.; Džamić, A.M.; Ćirić, A.D.; Ristić, M.S.; Milovanović, S.L.; Petrović, S.D. Chemical composition, antioxidant and antimicrobial activity of *Thymus praecox* supercritical extracts. *J. Supercrit. Fluids* **2016**, *110*, 117–125. [CrossRef]

17. Mollica, A.; Locatelli, M.; Macedonio, G.; Carradori, S.; Sobolev, A.P.; De Salvador, R.F.; Monti, S.M.; Buonanno, M.; Zengin, G.; Angeli, A.; et al. Microwave-assisted extraction, HPLC analysis, and inhibitory effects on carbonic anhydrase I, II, VA, and VII isoforms of 14 blueberry Italian cultivars. *J. Enzyme Inhib. Med. Chem.* **2016**, *31*, 1–6. [CrossRef] [PubMed]

18. Zengin, G.; Menghini, L.; Malatesta, L.; De Luca, E.; Bellagamba, G.; Uysal, S.; Aktumsek, A.; Locatelli, M. Comparative study of biological activities and chemical fingerprint of two wild Turkish species: *Asphodeline anatolica* and *Potentilla speciosa*. *J. Enzyme Inhib. Med. Chem.* **2016**, *31*, 203–208. [CrossRef] [PubMed]

19. Clydesdale, F.M.; Ahmed, E.M. Colorimetry-methodology and applications. *CRC Crit. Rev. Food Sci. Nutr.* **1978**, *10*, 243–301. [CrossRef] [PubMed]

20. Solovchenko, A.E.; Chivkunova, O.B.; Merzlyak, M.N.; Reshetnikova, I.V. A spectrophotometric analysis of pigments in apples. *Russian J. Plant. Physiol.* **2001**, *48*, 693–700. [CrossRef]

21. Wellburn, A. The spectral determination of chlorophylls a and b, as well as total carotenoids, using various solvents with spectrophotometers of different resolution. *J. Plant Physiol.* **1994**, *144*, 307–313. [CrossRef]

22. Williamson, G. The role of polyphenols in modern nutrition. *Nutr. Bull.* **2017**, *42*, 226–235. [CrossRef] [PubMed]

23. Cushnie, T.P.; Lamb, A.J. Antimicrobial activity of flavonoids. *Int. J. Antimicrob. Agents.* **2005**, *26*, 343–356. [CrossRef] [PubMed]

24. Cesa, S.; Carradori, S.; Bellagamba, G.; Locatelli, M.; Casadei, M.A.; Masci, A.; Paolicelli, P. Evaluation of processing effects on anthocyanin content and colour modifications of blueberry (*Vaccinium* spp.) extracts: Comparison between HPLC-DAD and CIELAB analyses. *Food Chem.* **2017**, *232*, 114–123. [CrossRef] [PubMed]
25. Cesa, S.; Casadei, M.A.; Cerreto, F.; Paolicelli, P. Infant milk formulas: effect of storage conditions on the stability of powdered products towards autoxidation. *Foods* **2015**, *4*, 487–500. [CrossRef] [PubMed]
26. Ivanov, L.A.; Ivanova, L.A.; Ronzhina, D.A.; Yudina, P.K. Changes in the chlorophyll and carotenoid contents in the leaves of steppe plants along a latitudinal gradient in South *Ural. Russian J. Plant Physiol.* **2013**, *60*, 812–820. [CrossRef]
27. Park, J.-Y.; Park, C.-M.; Kim, J.-J.; Noh, K.-H.; Cho, C.-W.; Song, Y.-S. The protective effect of chlorophyll a against oxidative stress and inflammatory processes in LPS-stimulated macrophages. *Food Sci. Biotechnol.* **2007**, *16*, 205–211.
28. Atanasov, A.G.; Waltenberger, B.; Pferschy-Wenzig, E.M.; Linder, T.; Wawrosch, C.; Uhrin, P.; Temml, V.; Wang, L.; Schwaiger, S.; Heiss, E.H.; et al. Discovery and resupply of pharmacologically active plant-derived natural products: A review. *Biotechnol. Adv.* **2015**, *33*, 1582–1614. [CrossRef] [PubMed]

Sample Availability: Samples of the compounds are available from the authors of the Unité de recherche Valorisation des Ressources Naturelles, Molécules Bioactives et Analyses Physicochimiques et Biologiques. Université Frères Mentouri, Constantine 1, Route d'Aïn El Bey, 25000 Constantine, Algérie.

molecules

MDPI

Article

A Comparative Assessment of Biological Effects and Chemical Profile of Italian *Asphodeline lutea* Extracts

Dora Melucci [1], Marcello Locatelli [2,3,*], Clinio Locatelli [1], Alessandro Zappi [1],
Francesco De Laurentiis [1], Simone Carradori [2], Cristina Campestre [2], Lidia Leporini [2],
Gokhan Zengin [4], Carene Marie Nancy Picot [5], Luigi Menghini [2] and
Mohamad Fawzi Mahomoodally [5]

[1] Department of Chemistry "G. Ciamician", University of Bologna, 40126 Bologna, Italy;
dora.melucci@unibo.it (D.M.); clinio.locatelli@unibo.it (C.L.); alessandro.zappi4@unibo.it (A.Z.);
francesco.de5@studio.unibo.it (F.D.L.)
[2] Department of Pharmacy, University "G. D'Annunzio" of Chieti-Pescara, 66100 Chieti, Italy;
simone.carradori@unich.it (S.C.); cristina.campestre@unich.it (C.C.); l.leporini@unich.it (L.L.);
luigi.menghini@unich.it (L.M.)
[3] Interuniversity Consortium of Structural and Systems Biology, 00136 Rome, Italy
[4] Department of Biology, Selcuk University, Campus, 42250, 42130 Konya, Turkey;
gokhanzengin@selcuk.edu.tr
[5] Department of Health Sciences, University of Mauritius, Réduit 80837, Mauritius;
picotcarene@yahoo.com (C.M.N.P.); f.mahomoodally@uom.ac.mu (M.F.M.)
* Correspondence: m.locatelli@unich.it; Tel.:+39-08713554590

Received: 16 January 2018; Accepted: 17 February 2018; Published: 19 February 2018

Abstract: The present study aims to highlight the therapeutic potential of *Asphodeline lutea* (AL), a wild edible plant of the Mediterranean diet. Roots, aerial parts, and flowers of AL at two different phenological stages were collected from three locations in Italy. The inhibitory activities of extracts on strategic enzymes linked to human diseases were assessed. The antioxidant properties were evaluated in vitro, using six standard bioassays. The phenolic and anthraquinone profiles were also established using HPLC-PDA. Zinc, cadmium, lead, and copper contents were also determined. All the samples inhibited acetylcholinesterase (from 1.51 to 2.20 mg GALAEs/g extract), tyrosinase (from 7.50 to 25.3 mg KAEs/g extract), and α-amylase (from 0.37 to 0.51 mmol ACAEs/g extract). Aloe-emodin and physcion were present in all parts, while rhein was not detected. The phenolic profile and the heavy metals composition of specimens gathered from three different regions of Italy were different. It can be argued that samples collected near the street can contain higher concentrations of heavy metals. The experimental data confirm that the *A. lutea* species could be considered as a potential source of bioactive metabolites, and its consumption could play a positive and safe role in human health maintenance.

Keywords: *Asphodeline lutea*; HPLC-PDA; heavy metals; tyrosinase; diabetes; neurodegenerative disease

1. Introduction

Asphodeline lutea (AL) Reichenb (synonym: *Asphodelus luteus* L., family Xanthorrhoeaceae), also known as King's Spear or Yellow Asphodel, is a perennial landscaping plant native to South-eastern Europe, North Africa and Turkey, characterized by a single stem with semi hollow leaves and yellow-orange flowers [1,2]. The stems and leaves are traditionally consumed in the Mediterranean region as an edible plant due to their nutritional protein quality of [3,4].

The chemotherapeutic value of Bulgarian and Turkish AL root extracts has been evidenced only in recent years, revealing the presence of different therapeutically useful compounds. The anti-microbial and anti-mutagenic activities of methanol root extracts of *AL* have been reported [5], while the hepatoprotective and antioxidant capacity of the ethanol root extracts of *AL* both in vivo and in vitro has also been evidenced in an animal model of CCl_4-injured liver [6]. Moreover, the methanol and chloroform extracts from *AL* roots caused a marked inhibition of multidrug resistance in mouse tumour cells transfected with the human *MDR1* gene [7], whereas methanol, acetone and aqueous extracts of different *Asphodeline* spp. parts were reported to moderately inhibit elastase, collagenase and hyaluronidase enzymes at 100 µg/mL [8]. Other studies have reported the use of extracts of *A. lutea* among local populations for skin diseases and haemorrhoids [9,10].

The methanol root extracts of *AL* of Bulgarian origin were found to be rich in caffeic acid, catechin and epicatechin [11]. Anthraquinones (1,5,8-trihydroxy-3-methylanthraquinone, 1-hydroxy-8-methoxy-3-methylanthraquinone, chrysophanol, 1,1′,8,8′,10-pentahydroxy-3,3′-dimethyl-10,7′-bianthracene-9,9′,10′-trione) [12], naphthalenes and naphthoquinones [13] were also previously isolated from *AL*. The antioxidant activity of *AL* chloroform extracts in lard and sunflower oil was attributed to 2-acetyl-1-hydroxy-8-methoxy-3-methylnaphthalene [14]. However, little is known about the chemical variability and the potential therapeutic ability of *AL* of Italian origin.

Based on these considerations, we aimed to evaluate the antioxidant activities, the enzyme (acetylcholinesterase, butyrylcholinesterase, tyrosinase, α-amylase, and α-glucosidase) inhibitory potential of extracts from different anatomical regions of *AL*, collected in diverse sites in the Italian Central Apennines, at different phenological stages, as well as the determination of anthraquinones, phenolics and heavy metal profiles.

2. Results

After extraction, each sample was fully characterized to establish a comprehensive chemical fingerprint of total phenolic and flavonoid content (Table 1), specific content of phenolics (Table 2), and anthraquinones (Table 3) and heavy metals bioaccumulation (Table 4). Then, the *AL* extracts were tested in order to assess their pharmacological properties such as antioxidant, metal chelating and enzyme inhibition.

Table 1. Total phenolic and flavonoid content of different parts of *A. lutea* collected from three different locations in Italy.

Location	Stage/Parts	Total Phenolic Content (mg GAE/g Extract) *	Total Flavonoid Content (mg RE/g Extract) *
Perugia	PF-R	12.5 ± 0.2 [a]	5.4 ± 0.7 [a]
	PF-AP	24.7 ± 0.9 [a]	19.8 ± 0.3 [a]
	F-R	17.7 ± 0.4 [a]	4.8 ± 0.1 [a]
	F-AP	27.7 ± 0.6 [b]	22.1 ± 0.2 [b]
	F-Fl	19.4 ± 0.6 [b]	11.4 ± 0.1 [b]
Novele	PF-R	12.4 ± 0.7 [a]	3.7 ± 0.2 [b]
	PF-AP	23.8 ± 0.3 [a]	14.8 ± 0.5 [b]
	F-R	12.7 ± 0.2 [b]	4.6 ± 0.1 [a]
	F-AP	23.9 ± 0.3 [c]	17.3 ± 0.1 [c]
	F-Fl	17.5 ± 0.7 [c]	11.0 ± 0.4 [b]
Pescosansonesco	PF-R	9.8 ± 0.2 [b]	2.9 ± 0.2 [c]
	PF-AP	24.0 ± 0.8 [a]	19.3 ± 0.2 [a]
	F-R	10.7 ± 0.4 [c]	3.2 ± 0.1 [b]
	F-AP	38.2 ± 0.8 [a]	28.0 ± 0.3 [a]
	F-Fl	24.7 ± 0.5 [a]	13.5 ± 0.3 [a]

* Values expressed are means ± S.D. of three simultaneous measurements. GAEs, gallic acid equivalents; REs, rutin equivalents. PF: preflowering plant, F: flowering plant, R: roots, AP: aerial parts, Fl: flowers. Data marked with different letters within the same column indicate statistically significant differences in the same stages/parts for each location ($p < 0.05$).

Table 2. Phenolic profile of different parts of *A. lutea* collected from three different locations in Italy *.

Phenolic Components	Perugia					Novele					Pescosansonesco				
	PF-R	PF-AP	F-R	F-AP	F-Fl	PF-R	PF-AP	F-R	F-AP	F-Fl	PF-R	PF-AP	F-R	F-AP	F-Fl
Gallic acid	nd	nd	0.9 ± 0.1	nd	0.52 ± 0.05	nd	nd	1.1 ± 0.1	0.31 ± 0.05	1.1 ± 0.8	1.96 ± 0.04	nd	1.09 ± 0.02	1.1 ± 0.6	nd
Catechin	nd	0.84 ± 0.03	0.54 ± 0.05	nd	nd	nd	nd	nd	0.43 ± 0.07	nd	1.8 ± 0.1	1.1 ± 0.03	0.51 ± 0.01	nd	0.41 ± 0.02
p-OH benzoic acid	nd	nd	nd	nd	nd	0.57 ± 0.04	nd	nd	nd	nd	0.42 ± 0.05	nd	nd	nd	nd
Epicatechin	nd	nd	0.35 ± 0.03	nd	nd	nd	nd	nd	nd	nd	nd	nd	nd	nd	nd
3-OH benzoic acid	0.36 ± 0.04	nd	0.40 ± 0.05	nd	nd	nd	nd	nd	nd	nd	nd	1.9 ± 0.5	nd	nd	nd
p-Coumaric acid	nd	nd	nd	nd	nd	2.2 ± 0.2	0.28 ± 0.02	0.32 ± 0.05	nd	nd	0.24 ± 0.03	0.46 ± 0.03	0.31 ± 0.07	nd	0.42 ± 0.02
Rutin	nd	1.6 ± 0.9	nd	1.1 ± 0.1	9.0 ± 0.1	3.0 ± 0.3	nd	0.67 ± 0.08	nd	5.6 ± 0.4	10.0 ± 0.9	8.9 ± 0.7	nd	nd	0.9 ± 0.1
Naringin	nd	nd	nd	nd	nd	nd	nd	nd	nd	0.24 ± 0.03	nd	1.5 ± 0.2	0.28 ± 0.02	nd	nd
2,3-diMeOBA	nd	1.2 ± 0.6	0.33 ± 0.09	nd	nd	0.56 ± 0.04	nd	nd	0.37 ± 0.02	nd	nd	53.0 ± 5.0	nd	nd	nd
Benzoic acid	nd	nd	nd	0.50 ± 0.05	2.4 ± 0.2	2.6 ± 0.2	nd	0.26 ± 0.09	0.35 ± 0.01	nd	0.33 ± 0.02	18.0 ± 3.0	nd	0.4 ± 0.1	0.87 ± 0.09
Quercetin	0.31 ± 0.07	2.3 ± 0.3	nd	0.52 ± 0.05	2.3 ± 0.24	nd	0.31 ± 0.05	nd	2.7 ± 0.4	3.0 ± 1.0	12.0 ± 1.0	0.7 ± 0.1	0.63 ± 0.04	0.3 ± 0.8	5.0 ± 1.0
Naringenin	nd	nd	nd	nd	0.63 ± 0.03	nd	nd	nd	nd	nd	nd	0.79 ± 0.08	nd	nd	nd
Total (µg/mg)	0.67	6.01	2.54	2.12	14.88	8.94	0.59	2.31	4.20	10.29	27.32	86.15	2.81	1.89	7.66

* Values expressed are means ± S.D. of three simultaneous measurements. nd: not detected, PF: preflowering plant, F: flowering plant, R: roots, AP: aerial parts, Fl: flowers. 2,3-diMeOBA = 2,3-dimethoxy- benzoic acid. Chlorogenic acid, *trans*-cinnamic acid, *o*-coumaric acid, sinapinic acid, *trans*-ferulic acid, syringic acid, vanillic acid, and 3-OH-4-MeO-benzaldehyde were not detected in any sample.

Table 3. Anthraquinone profiles of different parts of *A. lutea* collected from three different locations in Italy *.

Location	Stage/Parts	Aloe-emodin	Emodin	Chrysophanol	Physcion	Total (µg/mg)
Perugia	PF-R	3.10 ± 0.30	nd	0.51 ± 0.06	15.0 ± 2.0	18.3
	PF-AP	0.82 ± 0.08	nd	1.60 ± 0.70	1.70 ± 0.90	4.12
	F-R	0.72 ± 0.07	nd	2.00 ± 1.00	0.77 ± 0.09	3.47
	F-AP	0.81 ± 0.06	nd	nd	5.00 ± 1.00	6.20
	F-Fl	2.70 ± 0.30	nd	0.69 ± 0.04	2.40 ± 0.50	5.81
Novele	PF-R	2.00 ± 1.00	nd	nd	18.0 ± 3.0	19.5
	PF-AP	0.85 ± 0.07	nd	1.51 ± 0.9	5.00 ± 1.00	7.57
	F-R	0.65 ± 0.09	nd	1.10 ± 0.50	1.00 ± 0.40	2.72
	F-AP	0.65 ± 0.08	nd	nd	3.00 ± 1.00	4.07
	F-Fl	0.71 ± 0.05	nd	0.39 ± 0.04	0.84 ± 0.08	1.94
Pescosansonesco	PF-R	1.00 ± 0.20	0.82 ± 0.09	2.00 ± 1.00	6.00 ± 1.00	10.6
	PF-AP	1.00 ± 0.30	nd	nd	4.00 ± 1.00	5.39
	F-R	1.10 ± 0.20	nd	2.00 ± 1.00	5.00 ± 1.00	5.81
	F-AP	2.00 ± 1.00	nd	0.44 ± 0.04	4.87 ± 2.00	7.41
	F-Fl	2.00 ± 1.00	nd	0.60 ± 0.07	8.00 ± 1.00	11.0

* Values expressed are means ± S.D. of three simultaneous measurements. nd: not detected, PF: preflowering plant, F: flowering plant, R: roots, AP: aerial parts, Fl: flowers. Rhein was not detected in any sample.

Table 4. Heavy metals (Zn, Cd, Pb and Cu) content of different parts of *A. lutea* collected from three different locations in Italy *.

Location	Stage/Parts	Zn (ppm)	Cd (ppm)	Pb (ppm)	Cu (ppm)
Perugia	PF-R	1100 ± 300	<LoD	<LoD	1100 ± 350
	PF-AP	<LoD	<LoD	<LoD	<LoD
	F-R	500 ± 100	90 ± 30	<LoD	<LoD
	F-AP	53 ± 5	<LoD	<LoD	35 ± 6
	F-Fl	300 ± 50	<LoD	<LoD	<LoD
Novele	PF-R	460 ± 90	700 ± 200	<LoD	<LoD
	PF-AP	<LoD	<LoD	<LoD	<LoD
	F-R	140 ± 40	<LoD	180 ± 40	<LoD
	F-AP	250 ± 20	50 ± 10	<LoD	<LoD
	F-Fl	190 ± 50	130 ± 30	70 ± 30	<LoD

* Values expressed are means ± S.D.; <LoD = below limit of detection; PF: preflowering plant, F: flowering plant, R: roots, AP: aerial parts, Fl: flowers. For Pescosansonesco samples (PF-R, PF-AP, F-R, F-AP, F-Fl) all values are <LoD.

From the data in Table 1 it is possible to state that both phenolic and flavonoid contents reached the maximum in the flowering stage with respect to the preflowering one. Aerial parts were the richest in these metabolites (AP > Fl > R) disregarding the sampling site. For the phenolic content, the following order was observed: aerial parts of *AL* collected in Pescosansonesco > Perugia > Novele. A similar pattern was observed for the flavonoid content. The lowest amount of phenolics was recorded in the roots of the preflowering plant. The phenolic content in whole plant increases evidently during blooming, and this aspect is mainly related to the flowers development, while the amount of phenolics in roots and aerial parts were constant. The contribution of the flavonoid fractions, which represent more than fifty percent of the floral phenolic content, is particularly evident. The distribution of phenolics mainly in the aerial parts is consistent with the physiological function of such class of metabolites, and could support their nutraceutical value when used as edible parts. These data showed some differences with total phenolic content (13.02 mg GAE/g DW) and total flavonoid content (7.63 mg RE/g DW) found in the roots of *A. lutea* from Syrian origin [15].

The detailed phenolic profiles of the tested extracts of *AL* are summarized in Table 2. The aerial part of preflowering *AL* (86.15 µg/mg) collected in Pescosansonesco contained the highest amount of phenolic compounds, with a high quantity of 2,3-dimethoxybenzoic acid. Benzoic acid and quercetin were present in the aerial parts of flowering *AL* collected in all the three locations. Aerial parts of preflowering *AL* (0.588 µg/mg) collected in Novele contained the lowest amount of phenolic compounds (rutin 0.28 ± 0.04 µg/mg and quercetin 0.31 ± 0.05 µg/mg). Chlorogenic acid, vanillic acid, syringic acid, 3-hydroxy-4-methoxybenzaldehyde, sinapic acid, *trans*-ferulic acid, *o*-coumaric acid, and *trans*-cinnamic acid were not detected in any sample. Naringenin was found only in aerial parts and only in two samples (Fl from Perugia and AP from Novele at the preflowering stage). On the other hand, epicatechin was detected only in roots from plants collected in Perugia in full bloom. Naringin, naringenin, benzoic acid and its derivatives (2,3-dimethoxybenzoic acid, 3-hydroxybenzoic acid, 4-hydroxybenzoic acid), gallic acid, catechin and epicatechin, *p*-coumaric acid and rutin were identified and quantified in this species for the first time. This information could enhance the knowledge about the nutritional and medical value of this plant.

The quantification of anthraquinones in the extracts of *AL* collected from different spontaneous growing sites in Italy is presented in Table 3. Aloe-emodin and physcion were present in all the extracts, whereas rhein was not detected. Emodin was present only in the roots collected before flowering in the samples from Pescosansonesco. The roots proved to be the plant part with the highest anthraquinone content. This quantity was higher in samples of non-flowering plants. The data are consistent with the general rules indicating the preferred harvest time for hypogeous organs is during the vegetative stage.

Table 4 reports the extrapolated concentrations of Zn(II), Cd(II), Pb(II), and Cu(II) in all the extracts. The *AL* samples collected in Pescosansonesco showed contents of all metals lower than the limit of detection (LoD) or too low to be detectable by the polarograph (whose LoD is usually on the order of 10 ppb). The Perugia samples showed the presence of Zn(II) in almost all parts of the plant, both before and after flowering, and a low amount of Cd(II) and Cu(II) in post-flowering parts of the plant. The samples collected in Novele, instead, showed a significant content of Zn(II), Cd(II) and Cu(II), mostly in post-flowering aerial parts and flowers.

This higher concentration of metals in the plants of Novele can be explained by the fact that their sampling sites were closer to vehicular traffic. Thus, it is conceivable that the *AL* collected in the proximity of main roads can contain high amounts of heavy metals due to car traffic and anthropogenic factors, but it could be important to consider the possibility of potential correlation of pollutants, as external stressors, that can act as modulator of plant secondary metabolism.

Table 5 summarizes the radical scavenging, reducing, antioxidant, and metal chelating properties of the plants parts of *AL* collected in Perugia, Novele, and Pescosansonesco. The following order was observed for DPPH radical scavenging of samples collected from the three locations: aerial parts of flowering *A. lutea* > aerial parts of preflowering *A. lutea* > flowers of flowering *A. lutea* > roots of flowering *A. lutea* > roots of preflowering *A. lutea*. Similar promising results were also reported by other authors regarding *A. lutea* species from different origin [15,16]. The aerial parts of flowering

A. lutea collected from Perugia and Pescosansonesco showed potent reducing capacity against FRAP (84 and 128 mg TEs/g extract, respectively) and CUPRAC (108 and 160 mg TEs/g extract, respectively). The roots of flowering *A. lutea* collected in Perugia showed the highest antioxidant activity. Conversely, aerial parts of *A. lutea* harvested in Novele showed the most potent metal chelation ability (Table 5).

Table 5. Antioxidant evaluation and metal chelating activity of different parts of *A. lutea* collected from three different locations in Italy *.

Location	Stage/Parts	DPPH Scavenging **	ABTS Scavenging **	FRAP **	CUPRAC **	Phosphomolybdenum Assay ‡	Metal Chelating Activity †
	PF-R	39.8 ± 0.5 [a]	56.0 ± 2.0 [a]	37.0 ± 1.0 [b]	57.5 ± 1.5 [a]	0.91 ± 0.05 [a]	12.8 ± 0.7 [a]
	PF-AP	68.0 ± 1.0 [b]	74.0 ± 3.0 [c]	64.0 ± 2.0 [b]	86.0 ± 2.0 [b]	0.87 ± 0.07 [b]	16.0 ± 2.0 [b]
Perugia	F-R	43.8 ± 0.1 [a]	75.0 ± 4.0 [a]	52.0 ± 4.0 [a]	71.0 ± 0.9 [a]	1.03 ± 0.07 [a]	8.5 ± 0.1 [a]
	F-AP	86.5 ± 1.5 [b]	102 ± 3 [b]	84.0 ± 2.0 [b]	108 ± 1 [b]	0.93 ± 0.09 [b]	18.0 ± 2.0 [a]
	F-Fl	52.0 ± 1.0 [b]	63.0 ± 2.0 [b]	50.0 ± 1.0 [b]	65.0 ± 1.0 [b]	0.73 ± 0.04 [b]	15.2 ± 0.6 [a]
	PF-R	30.4 ± 0.3 [b]	57.0 ± 4.0 [a]	42.0 ± 2.0 [a]	52.0 ± 2.0 [b]	0.71 ± 0.04 [c]	7.3 ± 0.1 [b]
	PF-AP	58.6 ± 0.5 [c]	121 ± 3 [a]	65.0 ± 1.0 [b]	82.0 ± 3.0 [b]	0.99 ± 0.06 [a]	19.7 ± 0.4 [a]
Novele	F-R	32.0 ± 1.0 [b]	56.0 ± 2.0 [b]	42.8 ± 0.9 [b]	60.0 ± 2.0 [b]	0.92 ± 0.04 [c]	5.8 ± 0.1 [c]
	F- AP	75.5 ± 0.3 [c]	97.0 ± 1.0 [c]	68.0 ± 1.0 [c]	86.7 ± 0.6 [c]	0.79 ± 0.01 [c]	13.1 ± 1.5 [c]
	F-Fl	44.0 ± 1.0 [c]	54.0 ± 2.0 [c]	45.5 ± 0.4 [c]	59.0 ± 1.0 [c]	0.77 ± 0.05 [b]	12.9 ± 0.3 [b]
	PF-R	20.5 ± 0.3 [c]	36.0 ± 1.0 [b]	29.0 ± 1.0 [c]	40.4 ± 0.3 [c]	0.80 ± 0.04 [b]	6.5 ± 0.7 [c]
	PF-AP	70.9 ± 0.4 [a]	85.0 ± 1.0 [b]	75.0 ± 1.0 [a]	92.2 ± 0.6 [a]	1.00 ± 0.10 [a]	15.4 ± 1.5 [b]
Pescosansonesco	F-R	26.0 ± 0.8 [c]	53.6 ± 0.8 [c]	38.0 ± 1.0 [c]	46.0 ± 1.0 [c]	0.99 ± 0.07 [b]	6.9 ± 0.9 [b]
	F-AP	90.4 ± 0.4 [a]	180 ± 1 [a]	128 ± 4 [a]	160 ± 1 [a]	1.43 ± 0.01 [a]	16.3 ± 0.1 [b]
	F-Fl	64.0 ± 1.0 [a]	89.0 ± 1.0 [a]	62.0 ± 1.0 [a]	86.0 ± 1.0 [a]	0.96 ± 0.06 [a]	14.5 ± 0.7 [a]

* Values expressed are means ± S.D. of three simultaneous measurements. ** mg TE/g extract (TE = Trolox equivalents); ‡ mmol TE/g extract; † (mg EDTAE/g extract) (EDTAE = EDTA equivalents). Data marked with different letters within the same column indicate statistically significant differences in the same stages/parts for each location ($p < 0.05$).

Table 6 presents the inhibitory activity of extracts from different parts of the plant (aerial part, flowers, and roots) of *AL* collected from three separate accessions from Central Italy. All the studied samples inhibited AChE, with values ranging from 1.51 to 2.20 mg GALAEs/g extract. Conversely, other authors found a very low AChE inhibition for both leaves and bulb extracts of *A. lutea* from Palestinian flora [16]. Similarly, the studied plant samples inhibited tyrosinase (ranging from 7.50 to 25.30 mg KAEs/g extract) and α-amylase (ranging from 0.37 to 0.51 mmol ACAEs/g extract). The aerial parts of preflowering *AL* collected in Perugia and those of flowering *AL* collected in Pescosansonesco showed no activity against BChE. The best α-glucosidase inhibitory effects (44.2 mmol ACAEs/g extract) was observed in the roots of flowering *AL* collected in Novele.

Table 6. Enzyme inhibitory effects of different parts of *A. lutea* collected from three different locations in Italy *.

Location	Stage/Parts	AChE Inhibition **	BChE Inhibition **	Tyrosinase Inhibition †	α-Amylase Inhibition ‡	α-Glucosidase Inhibition ‡
	PF-R	1.74 ± 0.07 [c]	2.03 ± 0.06 [a]	12.0 ± 2.0 [a]	0.39 ± 0.01 [a]	na
	PF-AP	1.59 ± 0.01 [c]	na	7.5 ± 0.3 [c]	0.45 ± 0.01 [a]	14.0 ± 0.1 [b]
Perugia	F-R	2.20 ± 0.40 [a]	1.90 ± 0.20 [b]	14.0 ± 0.2 [a]	0.45 ± 0.03 [a]	2.1 ± 0.4 [b]
	F-AP	1.77 ± 0.09 [b]	0.31 ± 0.06 [b]	7.0 ± 2.0 [c]	0.41 ± 0.04 [c]	na
	F-Fl	1.65 ± 0.05 [a]	0.77 ± 0.03 [b]	23.0 ± 2.0 [a]	0.45 ± 0.01 [b]	32.1 ± 0.2 [b]
	PF-R	1.88 ± 0.08 [b]	2.02 ± 0.04 [a]	12.0 ± 1.0 [a]	0.39 ± 0.01 [a]	15.8 ± 0.3 [b]
	PF-AP	1.82 ± 0.04 [a]	1.10 ± 0.10 [b]	15.0 ± 2.0 [b]	0.41 ± 0.01 [b]	0.25 ± 0.02 [c]
Novele	F-R	1.92 ± 0.04 [b]	2.05 ± 0.02 [a]	21.0 ± 1.0 [a]	0.39 ± 0.01 [b]	44.2 ± 0.4 [a]
	F-AP	1.77 ± 0.06 [b]	1.06 ± 0.01 [a]	21.0 ± 2.0 [b]	0.43 ± 0.01 [b]	2.01 ± 0.01 [b]
	F-Fl	1.71 ± 0.04 [a]	1.37 ± 0.07 [a]	12.0 ± 2.0 [c]	0.45 ± 0.01 [b]	42.7 ± 0.3 [a]
	PF-R	1.92 ± 0.02 [a]	1.83 ± 0.05 [b]	12.0 ± 1.0 [a]	0.37 ± 0.01 [b]	19.3 ± 0.2 [a]
	PF-AP	1.67 ± 0.05 [b]	1.36 ± 0.04 [a]	20.0 ± 2.0 [a]	0.46 ± 0.03 [a]	37.4 ± 0.9 [a]
Pescosansonesco	F-R	1.88 ± 0.02 [c]	1.60 ± 0.20 [c]	14.0 ± 2.0 [b]	0.38 ± 0.02 [b]	na
	F-AP	2.10 ± 0.60 [a]	na	25.3 ± 0.5 [a]	0.51 ± 0.01 [a]	34.1 ± 0.4 [a]
	F-Fl	1.51 ± 0.08 [b]	0.58 ± 0.07 [c]	18.8 ± 0.8 [b]	0.48 ± 0.01 [a]	na

* Values expressed are means ± S.D. of three simultaneous measurements. ** mg GALAE/g extract (GALAE = galantamine equivalents); † mg KAE/g extract (KAE = kojic acid equivalents); ‡ mmol ACAE/g extract (ACAE = acarbose equivalents); na: not active. PF: preflowering plant, F: flowering plant, R: roots, AP: aerial parts, Fl: flowers. Data marked with different letters within the same column indicate statistically significant differences in the same stages/parts for each location ($p < 0.05$).

3. Discussion

Chronic pathologies, such as type II diabetes and neurodegenerative diseases such as Alzheimer's disease (AD) are pandemics affecting every segment of the population. Existing therapies alleviate the trauma caused by these pathologies, but harbour several side effects. The concern of the scientific community is to find novel therapeutic agents with a minimum of side effects, to manage type II diabetes and AD. Plants possess a large array of phenolic compounds endowed with numerous therapeutic properties.

Neurodegenerative disorders encompass more than 600 diseases involving progressive and irreversible deterioration of the nervous system, subsequently resulting in neuronal cell death [17]. The most prevalent form of neurodegenerative disorder afflicting the global population is AD. AD is believed to occur because of the accumulation of amyloid-beta protein and the alteration of tau proteins in the brain, and an apparent oxidative stress [18]. A hypothesis has been suggested that AD is related to type II diabetes, though the association is complex and not fully understood. Notwithstanding, conditions are interlinked by inflammatory response, insulin resistance, glycogen synthase kinase 3β signalling mechanism, insulin growth factor signalling, oxidative stress, neurofibrillary tangle formation, acetylcholinesterase activity regulation, and amyloid beta formation. Thus, the inhibition of acetylcholinesterase, butyrylcholinesterase, tyrosinase, α-amylase, and α-glucosidase has been considered to mitigate the deleterious effects of type II diabetes and AD or as adjunctive treatment modalities [19]. Additionally, scientific interest is gradually shifting towards enzymes which have not yet been considered by the pharmaceutical industries. The inhibition of such enzymes, so far considered as non-pharma target, can be of potential relevance and can proved to be a promising strategy for the management of these debilitating complications [20].

AChE has attracted much attention for the management of this devastating irreversible disorder, owing to its ability to hydrolyse acetylcholine, a neurotransmitter [21]. The activity of BChE was found to increase in areas of the brain most affected by AD [22]. Furthermore, in the late stage of Alzheimer's disease, the level of AChE was found to drop to 55–67% of normal value, while BChE activity increases up to 120% [23]. Finding cholinesterase inhibitors having different specificities might help to allay the trauma associated to the different stages of AD.

In the present study, the aerials parts of flowering *AL* collected from Pescosansonesco showed potent inhibition of AChE, but no activity against BChE. Wang and co-workers previously reported that aloe-emodin was a potent inhibitor of AChE [24]. Interestingly, aerials parts of flowering *AL* from Pescosansonesco contained significant amount of aloe-emodin and physcion. *A. anatolica*, another species of the genus *Asphodeline*, was reported to contained physcion and also inhibited AChE [19]. The roots of *AL* at the preflowering stage collected at Perugia and Novele demonstrated potent BChE inhibitory activity. These plant samples contained high level of physcion. This anthraquinone was also previously reported to exert moderate to strong inhibition on tyrosinase [25]. Tyrosinase is a multifunctional copper-containing metalloenzyme required for the production of melanin pigment in humans [26]. Melanin plays a vital role by shielding ionizing radiation and absorbing free radicals [27]. Additionally, depletion of neuromelanin, which is structurally related to melanin, present in the substantia nigra, was associated to Parkinson's disease [28]. The inhibition of tyrosinase might prevent the aggravation of neurodegenerative disorders, such as Parkinson's disease. It was observed that mostly the aerial parts of *AL* possessed pronounced tyrosinase inhibitory potential. The aerial parts contained the highest amount of phenolic compounds. *AL* was used traditionally for the management of skin ailments. This virtue might be ascribed to its tyrosinase inhibitory potential, as reported in the present study.

Chrysophanol was identified in all the investigated samples. This anthraquinone was shown to inhibit mammalian intestinal α-glucosidase activity [29], thus delaying glucose surge. Most of the studied plant samples were potent inhibitors of α-glucosidase, showing higher acarbose equivalent values, compared to α-amylase inhibition. Indeed, mild α-amylase inhibition versus marked α-glucosidase inhibition was requested, since a pronounced activity of the former enzyme advocated the event of gastrointestinal problems [30]. Specifically, the roots of flowering *AL* collected in Novele showed high α-glucosidase inhibitory activity and low inhibition against α-amylase. This sample contained the highest amount of gallic acid, previously reported to possess α-amylase and α-glucosidase inhibitory activity [31], while *p*-coumaric acid was shown to be inactive [32]. Among the detected antraquinones, aloe-emodin, crysophanol and physcion can contribute to α-glucosidase inhibition [29,33]. The antioxidant potential of different parts of *AL* collected from different locations was also evaluated. Indeed, oxidative stress has been associated to the onset and/or worsening of various pathologies, including type II diabetes and AD. FRAP, CUPRAC and phosphomolybdenum assay are methods suitable to measure the reducing abilities of the samples. FRAP is based on the reduction of Fe^{3+}–TPTZ (2,4,6-tri(2-pyridyl)-*s*-triazine) complex, to produce coloured Fe^{2+}–TPTZ [24], Cu(II)–Nc (neocuproine) is reduced to coloured Cu(I)–Nc in the CUPRAC assay, while the phosphomolybdenum assay is based on the reduction of molybdenum (VI) to molybdenum (V) producing a green phosphomolybdenum (V) complex [34]. Aerial parts of *AL* collected in Pescosansonesco at flowering stage showed the highest reducing potential against the three reducing methods used. Indeed, this sample contained the highest amount of phenolics and flavonoids, known to possess strong reducing potential.

From the reported results is important to highlight that the activities obtained from phosphomolybdenum assay for the Italian *AL* roots (from 0.71 to 1.03 mmol TE/g extract) were lower than the corresponding *Asphodeline* spp. from Turkey [35–37] (from 1.18 to 2.94 mmol TE/g extract). No differences were observed for metal chelating activity assay.

Our CUPRAC and FRAP results are in accordance with the data reported in literature [35–37], even if is important to highlight that slightly lower values were observed for the Novele PF-R root samples. On the other hand, ABTS and DPPH assays measure the ability of the plant samples to scavenge free radicals. DPPH is a protonated free radical, which is reduced to a stable diamagnetic molecule [38]. The data collected in the present investigation demonstrated that aerial parts of flowering *AL* were the most potent scavengers of DPPH radical compared to the other extracts, irrespective of the sampling site. The herein reported DPPH and ABTS results are in accordance with the data reported in literature for root samples [35–37], even if the sample Perugia F-R shows on DPPH test the higher value (43.8 mg TE/g extract) respect to the other *Asphodeline* spp.

Additional comparisons could be done considering another edible plant, as just reported in literature [37,39]. Also in this case for *Asphodeline* root samples very similar biological activities respect to phosphomolybdenum, CUPRAC, FRAP, DPPH and ABTS assays were observed, and it is further confirmed that *Asphodeline* spp. had lower activity compared to *Potentilla* spp. [39–41].

The increased interest in the use of ABTS radical arises from its ability to act in both organic and aqueous conditions and its stability in a wide range of pH [42]. The aerial parts of flowering *AL* collected in Perugia and Pescosansonesco showed strong ABTS radical quenching abilities compared to the other plant parts as also reported by Karandeniz et al. [43]. Interestingly, the phenolic and flavonoid contents were the highest in the same samples. Previous studies have appraised the potent scavenging capacities of phenolics and flavonoids on ABTS [44].

Transition metals, especially Cu and Fe, act as catalysts in the production of reactive oxygen species, which react with other molecules resulting in oxidative stress and cell damage/death [45]. In vitro techniques used to assess the ability of phytochemicals to chelate these potentially toxic transition metals offer the scope for the development of nutraceuticals. In the present study, the different plants parts of *AL* showed variable degree of metal chelating potential. The preflowering aerial parts of *AL* collected in Novele showed the highest metal chelating ability. This sample contained rutin and quercetin, previously reported to exert metal chelating properties [46,47]. Comparing the chemical composition and activities of different parts of the same plant, as well as the parts from plants in different phenological stages, an evident quantitative and qualitative variability was recorded. Ecological, climatic, and genetic factors are probably involved in the variation of secondary metabolite profiles observed from samples of *AL* of different geographical origin [48,49].

4. Conclusions

The present study could be considered as the first comparative investigation on chemical composition (phenolics, anthraquinones and heavy metals) and biological activities of extracts from different parts of *A. lutea* collected from three different wild populations in Central Italy. Owing to the key role of oxidative stress and heavy metals in the onset/progression of large number of diseases, the antioxidant and chelating potential of the plant was also assessed. The phenolic and anthraquinone profile of *A. lutea* extracts evidenced qualitative and quantitative differences that, at least in part, are coherent with the presence of anthraquinones mainly in roots and phenolics in the aerial parts. The dynamics of production and accumulation of selected active metabolites during the floral induction phase are still not clear, apart from the quantity of phenolics in the aerial part that resulted strongly increased. Similarly, their enzyme inhibitory and antioxidant potential also varied. However, the potent enzyme inhibitory activity and antioxidant properties observed are worth further scientific consideration. The experimental data confirm that the *A. lutea* species could be rationally considered as a potential source of bioactive metabolites, and its consumption could play a positive role in human health maintenance.

5. Materials and Methods

5.1. Plant Materials

Plant material was collected from wild populations in three different locations in Central Italy, Perugia (43°06′39.2′′N 12°20′53.6′′E, 350 m a.s.l., in the surroundings of a residential building, probably subspontaneous), Novele (42°45′50.6′′N 13°20′35.0′′E, 550 m a.s.l., AP, close to the main road) and Pescosansonesco (42°14′31.4′′N 13°52′29.8′′E, PE, 530 m a.s.l., on a calcareous cliff). From each site a representative sampling (at least ten plants) was done, taking care to avoid causing damage to the wild population. Sampling was performed on March and May 2016, in order to collect plants during the vegetative phase (PF—pre-flowering stage) and in full bloom (F—flowering). The botanical identity was confirmed by a senior taxonomist (Prof. F. Tammaro) and voucher specimens are conserved in the Herbarium of the Department of Pharmacy, "G. d'Annunzio" University of Chieti-Pescara (Italy). Each plant was manually separated in root (R), aerial parts (AP, consisting in stem and leaves) and flowers (Fl) and air-dried in an oven at 40 ± 1 °C, until achievement of constant weight. Then plant material was ground using a mixer grinder to a fine powder, passing through a 40 mesh sieve to obtain a uniform granulometry and stored in a vacuum box in the dark at 4 °C until use. Methanol extracts were obtained by maceration with 250 mL of organic solvent at room temperature (25 ± 1 °C) overnight, as previously reported [50].

5.2. Chemicals

All the phenolic chemical standards (gallic acid, catechin, chlorogenic acid, *p*-hydroxybenzoic acid, vanillic acid, epicatechin, syringic acid, 3-hydroxybenzoic acid, 3-hydroxy-4-methoxy-benzaldehyde, *p*-coumaric acid, rutin, sinapic acid, *trans*-ferulic acid, naringin, 2,3-dimethoxy-benzoic acid, benzoic acid, *o*-coumaric acid, quercetin, *trans*-cinnamic acid, naringenin) (purity > 98%) were purchased from Sigma Aldrich (Milan, Italy). Methanol (HPLC-grade), formic acid (99%), nitric acid (65%, supra-pure metal grade) and sulfuric acid (98%, ultrapure grade) were obtained from Carlo Erba Reagents (Milan, Italy). Double-distilled water was obtained using a Millipore Milli-Q Plus water treatment system (Millipore Bedford Corp., Bedford, MA, USA). Standard metal solutions of Cd(II), Cu(II), Pb(II) and Zn(II) (1000 mg L^{-1}, suprapure grade), hydrochloric acid (30%) and sodium acetate (anhydrous, analytical grade) were purchased by Merck (Darmstad, Germany). Anthraquinone chemical standards (all >99%) were purchased from Extrasynthese (Genay, France).

5.3. Total Bioactive Components (Phenolics and Flavonoids)

Total phenolic content was determined using the Folin-Ciocalteu colorimetric method [51], and expressed as gallic acid equivalents (GAEs/g extract). Total flavonoid content was determined by previously reported method [52] and expressed as rutin equivalents (REs/g extract).

5.4. HPLC Analyses for Phenolics and Anthraquinones

High-performance liquid chromatography with diode array detection (HPLC-PDA) was employed. Phenolic as well as anthraquinone patterns were evaluated by validated methods reported in the literature [35,53], using an HPLC Waters liquid chromatography (model 600 solvent pump, 2996 DAD, Waters S.p.A., Milford, MA, USA). Mobile phase was directly degassed *on-line* using a Biotech 4CH DEGASI Compact (Onsala, Sweden). Empower v.2 Software (Waters S.p.A., Milford, MA, USA) was used to collect and analyse data.

For the quantitative analyses on investigated compounds (both anthraquinones and phenolics), the HPLC-PDA methods were validated using external calibration (for the identification of the analytes retention times, and UV/Vis spectra) with pure chemical standards at different concentration levels. Precision and trueness were validated using fortified samples (with pure chemical standard working solutions) at three different concentration levels and the results fulfil international guideline references [35,53]. The instrument configurations, using also a column oven for the reproducibility of the analytes retention times, and the validated methods allow to the correct identification and quantification of the investigated compounds.

5.5. Heavy Metals Determination

Cd(II), Cu(II), Pb(II) and Zn(II) were determined by differential pulse adsorptive stripping voltammetry (DPASV). An AMEL Mod. 433 Multipolarograph (AMEL Instrumentation s.r.l., Milan, Italy), equipped with a hanging mercury drop electrode (HDME) as working electrode, Ag | AgCl | KCl$_{satd.}$ as reference electrode and a platinum-wire auxiliary electrode [54], was employed. The supporting electrolyte was an acetic acid/sodium acetate buffer (1 M, pH = 4.65).

The relevant instrumental parameters were the following. Initial potential: E_i = −1.300 V; deposition potential: E_d = −1.300 V; final potential: E_f = 0.100 V; electrodeposition time: t_d = 120 s; delay time before the potential sweep: t_r = 5 s; potential scan rate: dE/dt = 20 mV/s; stirring rate: r = 600 r.p.m.; the potential values were referred to Ag | AgCl | KCl$_{satd.}$.

Each sample was previously digested by acid attack [55]: 0.25 g of plant sample was put in a 100-mL Pyrex digestion tube, and 20 mL of a mixture 1:1 of nitric acid and sulfuric acid was added. The tube, connected to a Vigreux column condenser, was then heated at 180 °C for 120 min. For the analytical measures, 15 mL of buffer solution and 300 μL of the digested solution were put into the voltammetric cell, and then 3 standard additions of 300 μL of a solution containing all the four metals at 10 mg L^{-1} were carried on. For each addition, two replicates were performed. The blank was obtained by mixing 15 mL of buffer solution and 300 μL of a digested solution containing only the mixture 1:1 of nitric acid and sulfuric acid. Finally, for each sample a standard addition line for each metal was computed, using peak areas as dependent variable and the metal content was extrapolated. The limit of detection (LoD) was calculated using the three-sigma approach, as in previous works [56].

5.6. Antioxidant Activity

The DPPH (1,1-diphenyl-2-picrylhydrazyl) radical and ABTS (2,2′-azino-bis(3-ethyl- benzothiazoline)-6-sulphonic acid) radical cation scavenging activities were determined, and the results were expressed as Trolox equivalents (TEs/g extract). The reducing power of the extracts was measured according to the reported method, using cupric ion reducing antioxidant power (CUPRAC) and ferric ion reducing antioxidant power (FRAP), and results were expressed as Trolox equivalents (TEs/g extract). Total antioxidant capacity was determined using phosphomolybdenum method. Metal chelating activity of the extracts against ferrous ions was also determined, and the results were expressed as EDTA equivalents (EDTAE/g extract). All these antioxidant procedures were performed as previously documented [57].

5.7. Enzyme Inhibitory Activities

Enzyme inhibitory activities (acetylcholinesterase (AChE), butyrylcholinesterase (BChE), tyrosinase, α-amylase and α-glucosidase) of the extracts were determined using the already published methodology [58]. Results for enzyme inhibitory activities were expressed as standard compound equivalents (galantamine for AChE and BChE, kojic acid for tyrosinase, acarbose for both α-amylase and α-glucosidase).

5.8. Statistical Analysis

All the assays were carried out in triplicate. The results are expressed as mean values and standard deviation (SD). The differences between some stages/parts in each location were calculated using one-way analysis of variance (ANOVA) followed by Tukey's honest significant difference post hoc test with $\alpha = 0.05$. This treatment was carried out using SPSS v. 14.0 program (SPSS Institute Inc., Cary, NC, USA).

Acknowledgments: This work was supported by grants from the Italian Ministry of University (FAR 2016). The research is part of the project "Filiera delle piante officinali in Abruzzo".

Author Contributions: M.L. and D.M. conceived and designed the experiments; L.M. and L.L. collected the material and prepared it for the analysis, S.C. performed the extractions; M.L. performed the HPLC analyses. C.L., A.Z. and F.D.L. carried out the heavy metal analysis, G.Z. and M.F.M. performed the biological assays. C.C. and C.M.N.P. analyzed the data and contributed reagents and materials; M.L. and S.C. wrote the paper.

Conflicts of Interest: The authors declare no conflict of interest.

References

1. Lazarova, I.; Gevrenova, R. *Asphodeline lutea* (L.) Rchb.: A review of its botany, phytochemistry and ethnopharmacology. *Pharmacia* **2013**, *60*, 21–25.
2. AlRawashdeh, I.M.I. Comparison diversity of *Asphodeline lutea* plant species among six locations at Alshoubak and Alnaqab ecosystems in Jordan. *ARPN J. Agric. Biol. Sci.* **2016**, *11*, 160–164.
3. Guarrera, P.M.; Savo, V. Wild food plants used in traditional vegetable mixtures in Italy. *J. Ethnopharmacol.* **2016**, *185*, 202–234. [CrossRef] [PubMed]
4. Zengin, G.; Aktumsek, A.; Guler, G.-O.; Cakmak, Y.-S.; Girón-Calle, J.; Alaiz, M.; Vioque, J. Nutritional quality of protein in the leaves of eleven *Asphodeline* species (*Liliaceae*) from Turkey. *Food Chem.* **2012**, *135*, 1360–1364. [CrossRef] [PubMed]
5. Uysal, A.; Lazarova, I.; Zengin, G.; Gunes, E.; Aktumsek, A.; Gevrenova, R. New Perspectives on *Asphodeline lutea* from Bulgaria and Turkey: Anti-mutagenic, Anti-microbial and Anti-methicillin Resistant *Staphylococcus aureus* (MRSA) Activity. *Br. J. Pharm. Res.* **2016**, *10*, 1–10. [CrossRef]
6. Lazarova, I.; Simeonova, R.; Vitcheva, V.; Kondeva-Burdina, M.; Gevrenova, R.; Zheleva-Dimitrova, D.; Zengin, G.; Danchev, N.D. Hepatoprotective and antioxidant potential of *Asphodeline lutea* (L.) Rchb. roots extract in experimental models in vitro/in vivo. *Biomed. Pharmacoth.* **2016**, *83*, 70–78. [CrossRef] [PubMed]
7. Ivanova, A.; Todorova-Nikolova, G.; Lazarova, I.; Engi, H.; Molnár, J. Modulation of multidrug resistance by selected edible plants-*Asphodeline lutea* and *Allium ursinum*. *CR Acad. Bulgar. Sci.* **2010**, *63*, 221–224.
8. Ilhan, M.; Zengin, G.; Küpeli Akkol, E.; Aktümsek, A.; Süntar, I. The importance of Asphodeline species on enzyme inhibition: anti-elastase, anti-hyaluronidase and anti-collagenase potential. *Turk. J. Pharm. Sci.* **2016**, *13*, 323–327.
9. Ali-Shtayeh, M.S.; Yaniv, Z.; Mahajna, J. Ethnobotanical survey in the Palestinian area: a classification of the healing potential of medicinal plants. *J. Ethnopharmacol.* **2000**, *73*, 221–232. [CrossRef]
10. Kargioglu, M.; Cenkci, S.; Serteser, A.; Evliyaoglu, N.; Konuk, M.; Kök, M.S.; Bagci, Y. An ethnobotanical survey of Inner-West Anatolia, Turkey. *Hum. Ecol.* **2008**, *36*, 763–777. [CrossRef]

11. Lazarova, I.; Zengin, G.; Aktumsek, A.; Gevrenova, R.; Ceylan, R.; Uysal, S. HPLC–DAD analysis of phenolic compounds and antioxidant properties of *Asphodeline lutea* roots from Bulgaria and Turkey. *Ind. Crops Prod.* **2014**, *61*, 438–441. [CrossRef]

12. Todorova, G.; Lazarova, I.; Mikhova, B.; Kostova, I. Anthraquinone, naphthalene, and naphthoquinone components of *Asphodeline lutea*. *Chem. Nat. Compd.* **2010**, *46*, 322–323. [CrossRef]

13. Ivanova, A.; Todorova-Nikolova, G.; Platikanov, S.; Antonova, D.; Kostova, I. Comparative GC/MS study of volatiles from different parts of *Asphodeline lutea* Rchb. *CR Acad. Bulgar. Sci.* **2008**, *61*, 727–730.

14. Lazarova, I.; Marinova, E.; Todorova-Nikolova, G.; Kostova, I. Antioxidant properties of *Asphodeline lutea* of Bulgarian origin. *Riv. Ital. Sostanze Grasse* **2009**, *86*, 181–188.

15. Adawia, K. Comparison of the total phenol, flavonoid contents and antioxidant activity of methanolic root extracts of *Asphodelus microcarpus* and *Asphodeline lutea* growing in Syria. *Int. J. Pharmacogn. Phytochem. Res.* **2017**, *9*, 159–164. [CrossRef]

16. Ali-Shtayeh, M.S.; Jamous, R.M.; Zaitoun, S.Y.A.; Qasem, I.B. In-vitro screening of acetylcholinesterase inhibitory activity of extracts from Palestinian indigenous flora in relation to the treatment of Alzheimer's disease. *Funct. Foods Health Dis.* **2014**, *4*, 381–400.

17. Bose, S.; Cho, J. Targeting chaperones, heat shock factor-1, and unfolded protein response: Promising therapeutic approaches for neurodegenerative disorders. *Ageing Res. Rev.* **2016**, *35*, 155–175. [CrossRef] [PubMed]

18. Song, M.K.; Bischoff, D.S.; Song, A.M.; Uyemura, K.; Yamaguchi, D.T. Metabolic relationship between diabetes and Alzheimer's disease affected by Cyclo(His-Pro) plus zinc treatment. *BBA Clin.* **2017**, *7*, 41–54. [CrossRef] [PubMed]

19. Kandimalla, R.; Thirumala, V.; Reddy, P.H. Is Alzheimer's disease a Type 3 Diabetes? A critical appraisal. *Biochim. Biophys. Acta* **2016**, *1863*, 1078–1089. [CrossRef] [PubMed]

20. Singh, H.; Singh, J.V.; Gupta, M.K.; Singh, P.; Sharma, S.; Nepali, K.; Bedi, P.M.S. Benzoflavones as cholesterol esterase inhibitors: Synthesis, biological evaluation and docking studies. *Bioorg. Med. Chem. Lett.* **2017**, *27*, 850–854. [CrossRef] [PubMed]

21. Luo, L.; Li, Y.; Qiang, X.; Cao, Z.; Xu, R.; Yang, X.; Xiao, G.; Song, Q.; Tan, Z.; Deng, Y. Multifunctional thioxanthone derivatives with acetylcholinesterase, monoamine oxidases and β-amyloid aggregation inhibitory activities as potential agents against Alzheimer's disease. *Bioorg. Med. Chem.* **2017**, *25*, 1997–2009. [CrossRef] [PubMed]

22. Hanafy, D.M.; Prenzler, P.D.; Burrows, G.E.; Ryan, D.; Nielsen, S.; El Sawi, S.A.; El Alfy, T.S.; Abdelrahman, E.H.; Obied, H.K. Biophenols of mints: Antioxidant, acetylcholinesterase, butyrylcholinesterase and histone deacetylase inhibition activities targeting Alzheimer's disease treatment. *J. Funct. Foods* **2017**, *33*, 345–362. [CrossRef]

23. Li, J.; Cesari, M.; Liu, F.; Dong, B.; Vellas, B. Effects of diabetes mellitus on cognitive decline in patients with Alzheimer disease: a systematic review. *Can. J. Diabetes* **2017**, *41*, 114–119. [CrossRef] [PubMed]

24. Wang, Y.; Pan, W.-L.; Liang, W.-C.; Law, W.-K.; Tsz-Ming, I.D.; Ng, T.-B.; Miu-Yee, W.M.; Chi-Cheong, W.D. Acetylshikonin, a novel AChE inhibitor, inhibits apoptosis via upregulation of heme oxygenase-1 expression in SH-SY5Y cells. *Evid. Based Complement. Alternat. Med.* **2013**, *2013*, 937370. [CrossRef] [PubMed]

25. Leu, Y.L.; Hwang, T.L.; Hu, J.W.; Fang, J.Y. Anthraquinones from *Polygonum cuspidatum* as tyrosinase inhibitors for dermal use. *Phytother. Res.* **2008**, *22*, 552–556. [CrossRef] [PubMed]

26. Gou, L.; Lee, J.; Hao, H.; Park, Y.-D.; Zhan, Y.; Lü, Z.-R. The effect of oxaloacetic acid on tyrosinase activity and structure: Integration of inhibition kinetics with docking simulation. *Int. J. Biol. Macromol.* **2017**, *101*, 59–66. [CrossRef] [PubMed]

27. Cespedes, C.L.; Balbontin, C.; Avila, J.G.; Dominguez, M.; Alarcon, J.; Paz, C.; Burgos, V.; Ortiz, L.; Peñaloza-Castro, I.; Seigler, D.S. Inhibition on cholinesterase and tyrosinase by alkaloids and phenolics from *Aristotelia chilensis* leaves. *Food Chem. Toxicol.* **2017**, *109 Pt 2*, 984–995. [CrossRef] [PubMed]

28. Mann, D.M.; Yates, P.O. Possible role of neuromelanin in the pathogenesis of Parkinson's disease. *Mech. Ageing Dev.* **1983**, *21*, 193–203. [CrossRef]

29. Jung, H.A.; Ali, M.Y.; Choi, J.S. Promising inhibitory effects of anthraquinones, naphthopyrone, and naphthalene glycosides, from *Cassia obtusifolia* on α-glucosidase and human protein tyrosine phosphatases 1B. *Molecules* **2017**, *22*, 28. [CrossRef] [PubMed]

30. Figueiredo-González, M.; Grosso, C.; Valentão, P.; Andrade, P.B. α-Glucosidase and α-amylase inhibitors from *Myrcia* spp.: A stronger alternative to acarbose? *J. Pharm. Biomed. Anal.* **2016**, *118*, 322–327. [CrossRef] [PubMed]

31. Oboh, G.; Ogunsuyi, O.B.; Ogunbadejo, M.D.; Adefegha, S.A. Influence of gallic acid on α-amylase and α-glucosidase inhibitory properties of acarbose. *J. Food Drug Anal.* **2016**, *24*, 627–634. [CrossRef] [PubMed]

32. Torres-Naranjo, M.; Suárez, A.; Gilardoni, G.; Cartuche, L.; Flores, P.; Morocho, V. Chemical constituents of *Muehlenbeckia tamnifolia* (Kunth) Meisn (*Polygonaceae*) and its in vitro α-amilase and α-glucosidase inhibitory activities. *Molecules* **2016**, *21*, 1461. [CrossRef] [PubMed]

33. Arvindekar, A.; More, T.; Payghan, P.V.; Laddha, K.; Ghoshal, N.; Arvindekar, A. Evaluation of anti-diabetic and alpha glucosidase inhibitory action of anthraquinones from *Rheum emodi*. *Food Funct.* **2015**, *6*, 2693–2700. [CrossRef] [PubMed]

34. Rajurkar, N.S.; Hande, S. Estimation of phytochemical content and antioxidant activity of some selected traditional indian medicinal plants. *Indian J. Pharm. Sci.* **2011**, *73*, 146–153. [CrossRef] [PubMed]

35. Zengin, G.; Locatelli, M.; Ceylan, R.; Aktumsek, A. Anthraquinone profile, antioxidant and enzyme inhibitory effect of root extracts of eight *Asphodeline* taxa from Turkey: Can *Asphodeline* roots be considered as a new source of natural compounds? *J. Enzyme Inhib. Med. Chem.* **2016**, *31*, 754–759. [CrossRef] [PubMed]

36. Locatelli, M.; Zengin, G.; Uysal, A.; Carradori, S.; De Luca, E.; Bellagamba, G.; Aktumsek, A.; Lazarova, I. Multicomponent pattern and biological activities of seven Asphodeline taxa: Potential sources of natural-functional ingredients for bioactive formulations. *J. Enzyme Inhib. Med. Chem.* **2016**, *32*, 60–67. [CrossRef] [PubMed]

37. Zengin, G.; Menghini, L.; Malatesta, L.; De Luca, E.; Bellagamba, G.; Uysal, S.; Aktumsek, A.; Locatelli, M. Comparative study of biological activities and multicomponent pattern of two wild Turkish species: Asphodeline anatolica and Potentilla speciosa. *J. Enzyme Inhib. Med. Chem.* **2016**, *31*, 203–208. [CrossRef] [PubMed]

38. Chaouche, T.M.; Haddouchi, F.; Ksouri, R.; Atik-Bekkara, F. Evaluation of antioxidant activity of hydromethanolic extracts of some medicinal species from South Algeria. *J. Chin. Med. Assoc.* **2014**, *77*, 302–307. [CrossRef] [PubMed]

39. Grochowski, D.M.; Uysal, S.; Aktumsek, A.; Granica, S.; Zengin, G.; Ceylan, R.; Locatelli, M.; Tomczyk, M. In vitro enzyme inhibitory properties, antioxidant activities, and phytochemical profile of Potentilla thuringiaca. *Phytochem. Lett.* **2017**, *20*, 365–372. [CrossRef]

40. Uysal, S.; Zengin, G.; Locatelli, M.; Bahadori, M.B.; Mocan, A.; Bellagamba, G.; De luca, E.; Mollica, A.; Aktumsek, A. Cytotoxic and enzyme inhibitory potential of two Potentilla species (P. speciosa L. and P. reptans Willd.) and their chemical composition. *Front. Pharmacol. Sect. Ethnopharmacol.* **2017**, *8*, 290. [CrossRef] [PubMed]

41. Paduch, R.; Wiater, A.; Locatelli, M.; Pleszczyńska, M.; Tomczyk, M. Aqueous Extracts of Selected Potentilla Species Modulate Biological Activity of Human Normal Colon Cells. *Curr. Drug Targets* **2015**, *16*, 1495–1502. [CrossRef] [PubMed]

42. Amaeze, O.; Ayoola, G.; Sofidiya, M.; Adepoju-Bello, A.; Adegoke, A.; Coker, H. Evaluation of antioxidant activity of *Tetracarpidium conophorum* (Müll. Arg) Hutch & Dalziel leaves. *Oxid. Med. Cell. Longev.* **2011**, *2011*, 1–8.

43. Karadeniz, A.; Cinbilgel, I.; Gün, S.S.; Cetin, A. Antioxidant activity of some Turkish medicinal plants. *Nat. Prod. Res.* **2015**, *29*, 2308–2312. [CrossRef] [PubMed]

44. Formagio, A.S.N.; Volobuff, C.R.F.; Santiago, M.; Cardoso, C.A.L.; Vieira, M.D.C.; Valdevina Pereira, Z. Evaluation of antioxidant activity, total flavonoids, tannins and phenolic compounds in *Psychotria* leaf extracts. *Antioxidants* **2014**, *3*, 745–757. [CrossRef] [PubMed]

45. Aboul-Ela, H.M.; Saad, A.A.; El-Sikaily, A.M.; Zaghloul, T.I. Oxidative stress and DNA damage in relation to transition metals overload in Abu-Qir Bay, Egypt. *J. Genet. Eng. Biotechnol.* **2011**, *9*, 51–58. [CrossRef]

46. Da Silva, E.R.; do Carmo Maquiaveli, C.; Magalhães, P.P. The leishmanicidal flavonols quercetin and quercitrin target *Leishmania* (*Leishmania*) *amazonensis* arginase. *Exp. Parasitol.* **2012**, *130*, 183–188. [CrossRef] [PubMed]

47. Ganeshpurkar, A.; Saluja, A.K. The pharmacological potential of rutin. *Saudi Pharm. J.* **2017**, *25*, 149–164. [CrossRef] [PubMed]

48. Menghini, L.; Leporini, L.; Pintore, G.; Chessa, M.; Tirillini, B. Essential oil content and composition of three sage varieties grown in Central Italy. *J. Med. Plants Res.* **2013**, *7*, 480–489.

49. Nescatelli, R.; Carradori, S.; Marini, F.; Caponigro, V.; Bucci, R.; De Monte, C.; Mollica, A.; Mannina, L.; Ceruso, M.; Supuran, C.T.; et al. Geographical characterization by MAE-HPLC and NIR methodologies and carbonic anhydrase inhibition of Saffron components. *Food Chem.* **2017**, *221*, 855–863. [CrossRef] [PubMed]

50. Zengin, G. A study on in vitro enzyme inhibitory properties of *Asphodeline anatolica*: New sources of natural inhibitors for public health problems. *Ind. Crops Prod.* **2016**, *83*, 39–43. [CrossRef]

51. Slinkard, K.; Singleton, V.L. Total phenol analysis: Automation and comparison with manual methods. *Am. J. Enol. Viticult.* **1977**, *28*, 49–55.

52. Berk, S.; Tepe, B.; Arslan, S.; Sarikurkcu, C. Screening of the antioxidant, antimicrobial and DNA damage protection potentials of the aqueous extract of *Asplenium ceterach* DC. *Afr. J. Biotechnol.* **2011**, *10*, 8902–8908.

53. Locatelli, M.; Genovese, S.; Carlucci, G.; Kremer, D.; Randic, M.; Epifano, F. Development and application of high-performance liquid chromatography for the study of two new oxyprenylated anthraquinones produced by *Rhamnus* species. *J. Chromatogr. A* **2012**, *1225*, 113–120. [CrossRef] [PubMed]

54. Melucci, D.; Locatelli, M.; Locatelli, C. Trace level voltammetric determination of heavy metals and total mercury in tea matrices (*Camellia sinensis*). *Food Chem. Toxicol.* **2013**, *62*, 901–907. [CrossRef] [PubMed]

55. Hoenig, M.; Baeten, H.; Vanhentenrijk, S.; Vassileva, E.; Quevauviller, P. Critical discussion on the need for an efficient mineralization procedure for the analysis of plant material by atomic spectrometric methods. *Anal. Chim. Acta* **1998**, *358*, 85–94. [CrossRef]

56. Locatelli, C.; Melucci, D. Voltammetric determination of metals as food contaminants: An excellent alternative to spectroscopic measurements. Application to meals, vegetables, mussels, clams and fishes. In *Voltammetry: Theory, Types and Applications*; Saito, Y., Kikuchi, T., Eds.; Nova Science Publishers: Hauppauge, NY, USA, 2014; pp. 225–248.

57. Rinaldi, F.; Hanieh, P.N.; Longhi, C.; Carradori, S.; Secci, D.; Zengin, G.; Ammendolia, M.G.; Mattia, E.; Del Favero, E.; Marianecci, C.; et al. Neem oil nanoemulsions: Characterisation and antioxidant activity. *J. Enzyme Inhib. Med. Chem.* **2017**, *32*, 1265–1273. [CrossRef] [PubMed]

58. Zengin, G.; Uysal, A.; Gunes, E.; Aktumsek, A. Survey of phytochemical composition and biological effects of three extracts from a wild plant (*Cotoneaster nummularia* Fisch. et Mey.): A potential source for functional food ingredients and drug formulations. *PLoS ONE* **2014**, *9*, e113527. [CrossRef] [PubMed]

molecules

MDPI

Article

A Novel Flavonoid Glucoside from the Fruits of *Lycium ruthenicun*

Jing-Jing Qi [1,2], Yong-Ming Yan [3], Li-Zhi Cheng [3], Bao-Hua Liu [3,*], Fu-Ying Qin [2,3] and Yong-Xian Cheng [1,2,3,4,*]

1   School of Pharmacy, Yunnan University of Traditional Chinese Medicine, Kunming 650500, China;
    qijing023@163.com
2   State Key Laboratory of Phytochemistry and Plant Resources in West China, Kunming Institute of Botany,
    Chinese Academy of Sciences, Kunming 650201, China; qinfuying@mail.kib.ac.cn
3   Guangdong Key Laboratory for Genome Stability & Disease Prevention, School of Pharmaceutical Sciences,
    School of Medicine, Shenzhen University Health Science Center, Shenzhen 518060, China;
    yanym@szu.edu.cn (Y.-M.Y.); 13424039397@163.com (L.-Z.C.)
4   School of Pharmacy, Henan University of Chinese Medicine, Zhengzhou 450008, China
*   Correspondence: ppliew@szu.edu.cn (B.-H.L.); yxcheng@szu.edu.cn (Y.-X.C.);
    Tel./Fax: +86-0755-8617-2799 (Y.-X.C.)

Received: 3 January 2018; Accepted: 28 January 2018; Published: 3 February 2018

Abstract: A novel flavonoid glucoside, ruthenicunoid A (**1**), together with eight known substances, were isolated from the fruits of *Lycium ruthenicun* Murr. Their structures were elucidated by extensive spectroscopic data and chemical methods. Especially, the absolute configuration of glucose residue in **1** was assigned by acid hydrolysis followed by derivatization and GC analysis. Biological evaluation towards Sirtuin 1 (SIRT1) found that compounds **1** and **2** exhibit inhibitory activity against SIRT1 in a concentration-dependent manner, indicating its potential on SIRT1-associated disorders.

Keywords: *Lycium ruthenicun*; flavonoid; ruthenicunoid A; SIRT1

1. Introduction

Lycium ruthenicun Murr. is found in the northwest regions of China. Its fruit is edible and has been used as a remedy for the treatment of hypertension, ureteral stones, tinea and furuncle, and gingvial bleeding [1–3]. The fruits of *L. ruthenicun* contains a variety of bioactive ingredients, in particular, polyphenols such as anthocyanins, which have antioxidant effects and are beneficial for the prevention and treatment of cardiovascular diseases are rich in the fruits [4,5]. A literature search found that the major research in the past focused on the extraction methods and measurement of the total anthocyanins [6–8]; no comprehensive study has been conducted to explore the chemical constituents of *L. ruthenicun*. This attracted our attention. In the course of continuous study, a new flavonoid glucoside, ruthenicunoid A, and eight known compounds were isolated and identified. All the compounds were tested for their biological activity on SIRT1, a nicotinamide adenosine dinucleotide (NAD)-dependent deacetylase. Our efforts will be described below.

2. Results and Discussion

2.1. Structure Elucidation of the Compounds

The EtOH extract of *L. ruthenicun* was suspended in water and partitioned with EtOAc. The EtOAc soluble part was submitted to a combination of chromatography to afford compounds **1–9** (Figure 1).

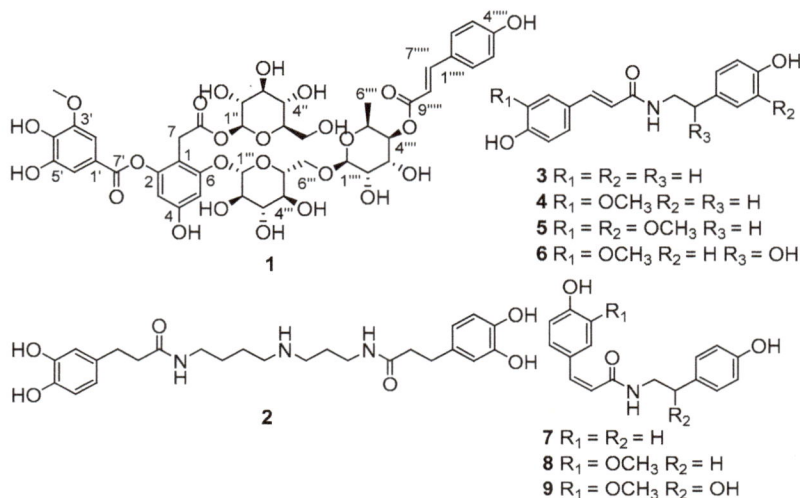

Figure 1. Chemical structures of compounds **1–9**.

Compound **1**, obtained as a brownish auburn gum, has the molecular formula $C_{43}H_{50}O_{25}$ (19 degrees of unsaturation) based on analysis of its HRESIMS at *m/z* 989.2546 [M + Na]$^+$ (calcd. for $C_{43}H_{50}O_{25}Na$, 989.2539). The ^1H NMR spectrum of **1** (Table 1) shows an AABB coupling system characteristic of a group of protons at δ_H 7.48 (2H, d, *J* = 8.5 Hz, H-2$''''''$, 6$''''''$) and 6.81 (2H, d, *J* = 8.5 Hz, H-3$''''''$, 5$''''''$), four aromatic protons at δ_H 6.42 (1H, d, *J* = 1.8 Hz, H-3), δ_H 6.67 (1H, d, *J* = 1.8 Hz, H-5), δ_H 7.30 (1H, d, *J* = 1.8 Hz, H-2$'$), and δ_H 7.35 (1H, d, *J* = 1.8 Hz, H-6$'$), suggesting the presence of two 1,2,3,5-tetrasubstituted benzene rings. In addition, one methoxy group at δ_H 3.88 (3H, s, 3$'$-OCH$_3$) and two olefinic protons respectively at δ_H 7.63 (1H, d, *J* = 15.9 Hz, H-7$''''''$) and δ_H 6.37 (1H, d, *J* = 16.0 Hz, H-8$''''''$) were observed. The ^{13}C NMR and DEPT spectra of **1** (Table 1) show 43 carbon signals attributed to two methyl (one oxygenated), three sp^3 methylene, twenty-five methine (ten olefinic and fifteen aliphatic), and thirteen quaternary carbons (three carbonyls, ten sp^2 including seven oxygenated). Inspection of these NMR data found that the partial signals resemble those of malvone [9,10], differing in that 5$'$-OMe in malvone was replaced by 5$'$-OH in **1**. The HMBC correlation (Figure 2) of OCH$_3$/C-3$'$ and ROESY correlation of OCH$_3$/H-2$'$ (Figure 2), in consideration of the chemical shifts of C-4$'$ (δ_C 141.6), C-5$'$ (δ_C 146.5), secured the presence of 3-methoxy,4,5-dihydroxyl substituted pattern. Further HMBC correlations of H-1$''$/C-8, H-1$'''$/C-6, H-7/C-1, C-2, C-6, in consideration of chemical shifts of C-2, C-4, and C-6 indicated the position of two glucose residues. HMBC correlations of H-2$'$, H-6$'$/C-7$'$ and the significant upfield shift of C-2 (δ_C 152.1) secured an ester carbonyl attached to C-2 instead of C-4. Apart from the red part, the remaining signals (blue part) are in accordance with those of 4-*p*-cumaroyl-α-rhamnosyl-(1 → 6)-β-glucose [11]. The observation of the above-mentioned AABB coupling system, a transformed double bond ($J_{H-7'''''',H-8''''''}$ = 15.9 Hz), and two sugar moieties in the middle field supported our conclusion. Additional HMBC cross peaks of H-1$''''$/C-6$'''$, H-4$''''$/C-9$''''''$ further indicated the linkage pattern in the blue part of **1**. The red and blue parts were connected via C-6-O-C-1$'''$ supported by the HMBC correlation of H-1$'''$/C-6 and the

ROESY correlation of H-5/H-1‴. Thus, the planar structure of **1** was deduced. For the configuration of the sugar moieties, acid hydrolysis of **1** followed by TLC comparison and GC analysis allowed the assignment of D-glucose and L-rhamnose. In detail, the L-cysteine methyl ester hydrochloride derivatives of the hydrolysis product of **1**, D-, L-glucose and L-rhamnose were prepared and subjected to GC analysis. The retention time for that of **1** is 17.698 min and 21.290 min, close to that of L-rhamnose (17.847 min) and D-glucose (21.276 min) rather than L-glucose (21.768 min), clarifying the type of sugar and its configuration. It should be noted that D-rhamnose or D,L-rhamnose in this study was not readily available, so that the derivative of D-rhamnose couldn't be prepared and analyzed by GC. However, it is possible to differentiate L- from D-form of rhamnose by comparing the consistency of retention time between the derivative of L-rhamnose and that of the mixture of L-rhamnose with **1**. In this way, we found that the retention time for L-cysteine methyl ester hydrochloride derivative of L-rhamnose is identical with that of co-injection of the mixture (16.827 min for the latter) by GC/MS analysis, securing the type of rhamnose and its configuration accordingly. Taken together, the structure of **1** was identified and named as ruthenicunoid A.

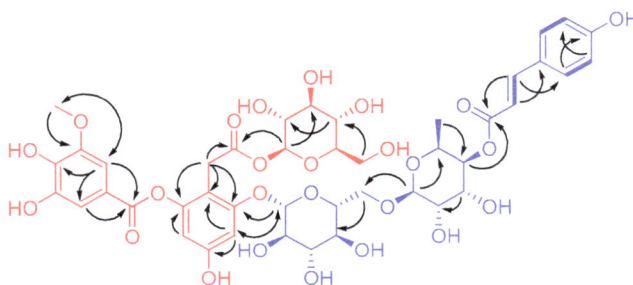

Figure 2. ^{1}H-^{1}H COSY (—) and key HMBC (⌒) and ROESY (⌒) correlations of **1**.

Table 1. ^{1}H (600 MHz) and ^{13}C NMR (150 MHz) data of **1** (δ in ppm, *J* in Hz, methanol-d_4).

			1		
No.	δ_H	δ_C	No.	δ_H	δ_C
1		109.2	1‴	4.87, brs	103.2
2		152.1	2‴	3.51, m	74.8
3	6.42, d, 1.8	105.2	3‴	3.47, m	77.8
4		159.1	4‴	3.32, overlap	71.0
5	6.67, d, 1.8	102.4	5‴	3.32, overlap	77.7
6		158.6	6‴	3.96, m	67.9
7	3.73, m	30.4		3.62, m	
	3.66, m		1⁗	4.76, brs	102.2
8		172.2	2⁗	3.43, m	78.2
1′		120.1	3⁗	3.86, m	70.4
2′	7.30, d, 1.8	106.9	4⁗	5.00, m	75.3
3′		149.3	5⁗	3.79, m	67.9
4′		141.6	6⁗	1.04, d, 6.2	17.8
5′		146.5	1″‴		127.2
6′	7.35, d, 1.8	112.8	2″‴	7.48, d, 8.5	131.3
7′		166.5	3″‴	6.81, d, 8.5	116.8
1″	5.45, d, 8.2	96.0	4″‴		161.2
2″	3.89, m	72.1	5″‴	6.81, d, 8.5	116.8
3″	3.30, m	73.8	6″‴	7.48, d, 8.5	131.3
4″	3.42, m	71.2	7″‴	7.63, d, 15.9	146.9
5″	3.50, m	77.7	8″‴	6.37, d, 15.9	115.2
6″	3.92, m	62.5	9″‴		169.1
	3.74, m		-OCH$_3$	3.88, s	56.9

By analysis of the NMR spectroscopic data and comparison with the literature, the known compounds were respectively identified as N^1,N^{10}-bis(dihydrocaffeoyl)spermidine (**2**) [12], *N-trans*-coumaroyltyramine (**3**) [13], *N-trans*-feruloyltyramine (**4**) [14], *N-trans*-feruloyl 3'-O-methyldopamine (**5**) [15], *N-trans*-feruloyloctopamine (**6**) [14], *N-cis*-coumaroyltyramine (**7**) [16], *N-cis*-feruloyltyramine (**8**) [14], and *N-cis*-feruloyloctopamine (**9**) [14].

2.2. Biological Evaluation

SIRT1 is a nicotinamide adenosine dinucleotide (NAD)-dependent deacetylase which regulates a wide range of cellular functions and is implicated in many diseases such as aging, cancer and so on [17–20]. So far, several SIRT1 activators and inhibitors such as nicotinamide (IC$_{50}$ value less than 50 μM), salermide (IC$_{50}$ value = 76.2 μM), and cambinol (IC$_{50}$ value = 56 μM) were documented [21]. With this assay in hand and considering the title species is used for aging, compounds **1–9** were thus tested for their inhibitory activity against SIRT1. The results showed that compounds **1** and **2** are active towards SIRT1 (Figure 3) with **2** to be more potent than **1**, comparable to that of nicotinamide at the concentration of 100 μM, whereas compounds **3–9** are not active (data not shown). The finding of **2** as a SIRT1 inhibitory substance indicated that such type of amide or aliphatic amine might be of important structure class for antiaging drug design.

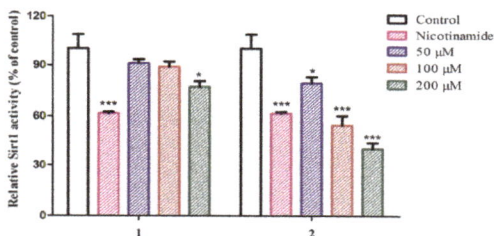

Figure 3. SIRT1 activation of compounds **1** and **2**. SIRT1 enzyme activity was measured using the SIRT1 Fluorometric Drug Discovery Kit. Statistical analysis was performed using one-way analysis of the variance (ANOVA) followed by Bonferroni's multiple comparison tests. All error bars are S.E.M. * $p < 0.05$, *** $p < 0.001$ versus control ($n = 3$).

3. Experimental Section

3.1. General Procedures

Optical rotations were recorded on a Horiba SEPA-300 polarimeter. UV spectrum was recorded on a Shimadzu UV-2401PC spectrometer (Shimadzu Corporation, Tokyo, Japan). GC analysis was performed using an Agilent 6890N gas chromatography instrument (Agilent Technologies, Santa Clara, CA, USA). GC/MS analysis was performed using an Agilent 7890B GC System (Agilent Technologies, Santa Clara, CA, USA) and a Asilent 5977 MSD inrun (Agilent Technologies, Santa Clara, CA, USA). NMR spectra were recorded on a Bruker AV-400 (Bruker, Karlsruhe, Germany) or an AV-600 spectrometer (Bruker, Karlsruhe, Germany), with TMS as an internal standard. ESIMS, and HRESIMS were measured on an Agilent G6230TOF MS spectrometer (Agilent Technologies, Santa Clara, CA, USA). C-18 silica gel (40–60 μm; Daiso Co., Tokyo, Japan), MCI gel CHP 20P (75–150 μm, Mitsubishi Chemical Industries, Tokyo, Japan) and Sephadex LH-20 (Amersham Pharmacia, Uppsala, Sweden) were used for column chromatography. Semi-preparative HPLC was carried out using an Agilent 1200 liquid chromatograph with a YMC-Pack ODS-A column (250 mm × 10 mm, i.d., 5 μm) and Thermo Hypersil GOLD-C$_{18}$ column (250 mm × 21.2 mm, i.d., 5 μm).

3.2. Plant Material

The fruits of *L. ruthenicum* were collected from the market of herbal medicine in Yunnan province, People's Republic of China, in September 2016. The material was identified by Mr. Bin Qiu at Yunnan Institute of Materia Medica, and a voucher specimen (CHYX-0605) is deposited at the State Key Laboratory of Phytochemistry and Plant Resources in West China, Kunming Institute of Botany, Chinese Academy of Sciences, People's Republic of China.

3.3. Extraction and Isolation

The fruits of *L. ruthenicum* (5 kg) were powdered and soaked by 80% aqueous EtOH (3 × 25 L × 24 h) to give a crude extract, which was suspended in water followed by extraction with EtOAc to afford an EtOAc soluble extract (85 g). The EtOAc extract was divided into six parts (Fr.1–Fr.6) by using a MCI gel CHP 20P column eluted with gradient aqueous MeOH (20–100%). Fr.2 (3.5 g) was purified by Sephadex LH-20 (MeOH) followed by semipreparative HPLC (MeOH/H_2O, 27:73, containing 0.05% formic acid) to afford compound **2** (78.4 mg, t_R = 9.8 min). Fr.4 (10.1 g) was separated by Sephadex LH-20 (MeOH) to yield six fractions (Fr.4.1–Fr.4.6). Fr.4.3 (2.1 g) was separated by RP-18 column (MeOH/H_2O, 30–100%) to get three fractions (Fr.4.3.1–Fr.4.3.3). Fr.4.3.3 (490 mg) was separated by Sephadex LH-20 (MeOH) to yield four fractions (Fr.4.3.3.1–Fr.4.3.3.4). Among these, Fr.4.3.3.4 (48 mg) was purified by semi-preparative HPLC (MeCN/H_2O, 28:72) to yield compounds **4** (2.1 mg, t_R = 16.1 min) and **5** (2.3 mg, t_R = 21.3 min). Fr.4.4 (1.0 g) was separated by RP-18 column (MeOH/H_2O, 35–100%) to get five fractions (Fr.4.4.1–Fr.4.4.5). Fr.4.4.2 (180 mg) was separated by preparative HPLC (MeOH/H_2O, 10–100%) to get three fractions (Fr.4.4.2.1–Fr.4.4.2.3). Fr.4.4.2.1 (23 mg) was purified by semi-preparative HPLC (MeCN/H_2O, 21:79) to afford compound **1** (4.9 mg, t_R = 15.4 min). Fr.4.4.3 (380 mg) was separated by preparative HPLC (MeOH/H_2O, 10–100%) to get nine fractions (Fr.4.4.3.1–Fr.4.4.3.9). Of which, Fr.4.4.3.3 (56.3 mg) was purified by semipreparative HPLC (MeCN/H_2O, 18:82) to afford compounds **6** (5.4 mg, t_R = 27.9 min) and **9** (1.0 mg, t_R = 30.3 min). Fr.4.4.3.7 (23 mg) was purified by semipreparative HPLC (MeCN/H_2O, 27:73) to yield compound **7** (2.3 mg, t_R = 22.8 min). Fr.4.4.3.8 (44 mg) was purified by semi-preparative HPLC (MeCN/H_2O, 23:77) to afford compounds **3** (7.1 mg, t_R = 27.0 min) and **8** (2.1 mg, t_R = 29.6 min).

3.4. Compound Characterization Data

Ruthenicunoid A (**1**): Brownish auburn gum; $[\alpha]_D^{21}$: −23.5 (*c* 0.49, MeOH). UV (MeOH) λ_{max} (log ε): 203 (4.66), 313 (4.47) nm. ESIMS *m/z*: 989 [M + Na]$^+$. HRESIMS *m/z*: 989.2546 [M + Na]$^+$ (calcd. for $C_{43}H_{50}O_{25}Na$, 989.2539); ^{1}H- and ^{13}C-NMR, see Table 1.

3.5. Acid Hydrolysis and Sugar Analysis

A solution of **1** (1.0 mg) in 1 N HCl was stirred at 70 °C for 5 h. After cooling, the mixtures were extracted with EtOAc. The aqueous layer was neutralized with 1 N NaOH and concentrated in vacuo, which was subsequently dissolved in anhydrous pyridine (2 mL). To these solutions L-cysteine methyl ester hydrochloride (2.0 mg) was added, and the mixtures were stirred at 60 °C for 1 h and concentrated in vacuo at 0 °C. Slow addition of 1-(trimethylsiyl) imidazole to the mixtures was followed by stirring at 60 °C for 2 h. Aliquots (4 μL) of the supernatants were subjected to chiral GC analysis to determine that D-glucose and L-rhamnose unitis are present in **1** [22,23].

3.6. SIRT1 Inhibition

For examination of SIRT1 inhibition of the compounds, each well contained 0.5 U (1 U = 1 pmol/min at 37 °C) of SIRT1 enzyme, 1000 μM of NAD$^+$ (Enzo Life Sciences, Farmingdale, NY, USA), 100 μM of SIRT1 peptide substrate (Enzo Life Sciences) and SIRT1 assay buffer (50 mM Tris-HCl, pH 8.0, 137 mM NaCl, 2.7 mM KCl, 1 mM $MgCl_2$, 1 mg/mL BSA) along with the test compounds at a concentration of 50, 100 and 200 μM, respectively. Nicotinamide, a known inhibitor of

Molecules **2018**, *23*, 325

SIRT1 enzyme was used as a control at a concentration of 100 μM. The plate was incubated at 37 °C for 30 min and the reaction was stopped using Fluor de Lys developer II solution (Enzo Life Sciences) containing 2 mM nicotinamide. The plate was further incubated at 37 °C for another 30 min and the samples were read by a fluorimeter with an excitation wavelength of 360 nm and emission wavelength of 460 nm [24].

4. Conclusions

To conclude, this study led to the isolation of a new flavonoid glucoside and eight known amide derivatives from the edible fruits of *L. ruthenicun*. Biological evaluation found that both **1** and **2** showed inhibitory activity against SIRT1, indicating their roles in SIRT1-associated disorders and suggesting **2** to be a potent structure template worth for further optimization as SIRT1 inhibitors.

Supplementary Materials: The following data are available online.

Acknowledgments: This study was supported by National Key Research and Development Program of China (2017YFA0503900) and National Natural Science Fund for Distinguished Young Scholars (81525026).

Author Contributions: Y.-X.C. conceived and designed the experiments, J.-J.Q. performed the experiments. Y.-M.Y., L.-Z.C.., F.-Y.Q. and B.-H.L. analyzed the data; Y.-X.C. wrote the paper. All authors read and approved the final manuscript.

Conflicts of Interest: The authors declare that there is no conflict of interest.

References

1. Zhao, J.; Xu, F.; Ji, T.F.; Li, J. A new spermidine from the fruits of *Lycium ruthenicum*. *Chem. Nat. Compd.* **2014**, *50*, 880–883. [CrossRef]

2. Rao, A.V.; Snyde, D.M. Raspberries and human health: A review. *J. Agric. Food Chem.* **2010**, *58*, 3871–3883. [CrossRef] [PubMed]

3. Zilic, S.; Serpen, A.; Akillioglu, G.; Gokmen, V.; Vancetovic, J. Phenolic compounds, carotenoids, anthocyanins, and antioxidant capacity of colored maize (*Zea mays* L.) Kernels. *J. Agric. Food Chem.* **2012**, *60*, 1224–1231. [CrossRef] [PubMed]

4. Li, J.; Qu, W.J.; Zhang, S.J.; Lv, H.Y. Study on antioxidant activity of pigment of *Lycium ruthenicum*. *Chin. J. Chin. Mater. Med.* **2006**, *31*, 1179–1183. [CrossRef]

5. Zheng, J.; Ding, C.X.; Wang, L.S.; Li, G.L.; Shi, J.Y.; Li, H.; Wang, H.L.; Suo, Y.R. Anthocyanins composition and antioxidant activity of wild *Lycium ruthenicum* Murr. from Qinghai-Tibet Plateau. *Food Chem.* **2011**, *126*, 859–865. [CrossRef]

6. Li, J.; Qu, W.J.; Lv, H.Y.; Yuan, H. Study on extracting and refining of the pigments from *Lycium ruthenicum* Murr. *Nat. Prod. Res. Dev.* **2006**, *18*, 650–654. [CrossRef]

7. Li, S.Z.; Li, J.; Yang, Z.J.; Yuan, H. Study on separation and purifeation of total flavonoids from *Lycium ruthenicum* Murr. with macroreticular resin. *Food Sci.* **2009**, *30*, 19–24. [CrossRef]

8. Luo, H.; Jin, L.; Gao, S.F.; Chen, H.G. Determination of Anthocyanin in *Lycium ruthenicum* Murr. from different producing areas in Hei River basin by UV. *Chin. Med. J. Res. Pract.* **2015**, *29*, 24–27. [CrossRef]

9. Lopes, P.; Richard, T.; Saucier, C.; Teissedre, P.; Monti, J.; Glories, Y. Anthocyanone A: A quinone methide derivative resulting from malvidin 3-*O*-glucoside degradation. *J. Agric. Food Chem.* **2007**, *55*, 2698–2704. [CrossRef] [PubMed]

10. Kamiya, H.; Yanase, E.; Nakatsuka, S. Novel oxidation products of cyanidin 3-*O*-glucoside with 2,2′-azobis-(2,4-dimethyl)valeronitrile and evaluation of anthocyanin content and its oxidation in black rice. *Food Chem.* **2014**, *155*, 221–226. [CrossRef] [PubMed]

11. Qi, J.J.; Yan, Y.M.; Wang, C.X.; Cheng, Y.X. Compounds from *Lycium ruthenicum*. *Nat. Prod. Res. Dev.* **2017**. Available online: http://kns.cnki.net/kcms/detail/51.1335.Q.20170904.1418.016.html (accessed on 4 September 2017).

12. Yingyongnarongkul, B.; Apiratikul, N.; Aroonrerk, N.; Suksamrarn, A. Synthesis of bis, tris and tetra (dihydrocaffeoyl) polyamine conjugates as antibacterial agents against VRSA. *Arch. Pharm. Res.* **2008**, *31*, 698–704. [CrossRef] [PubMed]

13. Zhao, G.X.; Hui, Y.X.; Rupprecht, J.K.; Mclaughlin, J.L.; Wood, K.V. Additional bioactive compounds and trilobacin, a novel highly cytotoxic acetogenin, from the bark of *Asimina triloba. J. Nat. Prod.* **1992**, *55*, 347–356. [CrossRef] [PubMed]

14. King, R.R.; Calhoun, L.A. Characterization of cross-linked hydroxycinnamic acid amides isolated from potato common scab lesions. *Phytochemistry* **2005**, *66*, 2468–2473. [CrossRef] [PubMed]

15. Cutillo, F.; Dabrosca, B.; Dellagreca, M.; Marino, C.D.; Golino, A.; Previtera, L.; Zarrelli, A. Cinnamic acid amides from *Chenopodium album*: Effects on seeds germination and plant growth. *Phytochemistry* **2003**, *64*, 1381–1387. [CrossRef]

16. Kim, D.K.; Lim, J.P.; Kim, J.W.; Park, H.W.; Eun, J.S. Antitumor and antiinflammatory constituents from *Celtis sinensis. Arch. Pharm. Res.* **2005**, *28*, 39–43. [CrossRef] [PubMed]

17. Ma, L.; Li, Y. SIRT1: Role in cardiovascular biology. *Clin. Chim. Acta* **2015**, *440*, 8–15. [CrossRef] [PubMed]

18. Luo, X.Y.; Qu, S.L.; Tang, Z.H.; Zhang, Y.; Liu, M.H.; Peng, J.; Tang, H.; Yu, K.L.; Zhang, C.; Ren, Z.; et al. SIRT1 in cardiovascular aging. *Clin. Chim. Acta* **2014**, *437*, 106–114. [CrossRef] [PubMed]

19. Clark-Knowles, K.V.; He, X.H.; Jardine, K.; Coulombe, J.; Dewar-Darch, D.; Caron, A.Z.; Gray, D.A.; McBurney, M.W. Reversible modulation of SIRT1 activity in a mouse strain. *PLoS ONE* **2017**. [CrossRef] [PubMed]

20. Sun, S.W.; Buer, B.C.; Marshab, E.N.G.; Kennedy, R.T. A label-free Sirtuin 1 assay based on dropletelectrospray ionization mass spectrometry. *Anal. Methods* **2016**, *8*, 3458–3465. [CrossRef] [PubMed]

21. Kumar, A.; Chauhan, S. How much successful are the medicinal chemists in modulation of SIRT1: A critical review. *Eur. J. Med. Chem.* **2016**, *119*, 45–69. [CrossRef] [PubMed]

22. Shi, Y.N.; Tu, Z.C.; Wang, X.L.; Yan, Y.M.; Fang, P.; Zuo, Z.L.; Hou, B.; Yang, T.H.; Cheng, Y.X. Bioactive compounds from the insect *Aspongopus chinensis. Bioorg. Med. Chem. Lett.* **2014**, *24*, 5164–5169. [CrossRef] [PubMed]

23. Jordan, D.S.; Daubenspeck, J.M.; Dybvig, K. Rhamnose biosynthesis in mycoplasmas requires precursor glycans larger than monosaccharide. *Mol. Microbiol.* **2013**, *89*, 918–928. [CrossRef] [PubMed]

24. Karbasforooshan, H.; Karimi, G. The role of SIRT1 in diabetic cardiomyopathy. *Biomed. Pharmacother.* **2017**, *90*, 386–392. [CrossRef] [PubMed]

Sample Availability: Sample of the compound **1** is available from the authors.

molecules

MDPI

Article

Metabolite Profiling of 14 Wuyi Rock Tea Cultivars Using UPLC-QTOF MS and UPLC-QqQ MS Combined with Chemometrics

Si Chen [1,2], Meihong Li [1,2], Gongyu Zheng [1], Tingting Wang [1], Jun Lin [2], Shanshan Wang [2], Xiaxia Wang [2], Qianlin Chao [3], Shixian Cao [3], Zhenbiao Yang [1,2,4] and Xiaomin Yu [1,2,*]

[1] College of Horticulture, Fujian Agriculture and Forestry University, Fuzhou 350002, China;
   cstc1990@hotmail.com (S.C.); limei123home@gmail.com (M.L.); zgy5403@126.com (G.Z.);
   lalaxiaojie0527@163.com (T.W.); yang@ucr.edu (Z.Y.)
[2] FAFU-UCR Joint Center for Horticultural Biology and Metabolomics, Fujian Provincial Key Laboratory of
   Haixia Applied Plant Systems Biology, Fujian Agriculture and Forestry University, Fuzhou 350002, China;
   realnadal@163.com (J.L.); shanshanwang22@yeah.net (S.W.); wangxiaxia530@126.com (X.W.)
[3] Wuyi Star Tea Industry Co., Ltd., Wuyishan 354300, China; chaoqianlin@wuyistar-tea.com (Q.C.);
   caoshixian@wuyistar-tea.com (S.C.)
[4] Center for Plant Cell Biology, Institute of integrated Genome Biology,
   and Department of Botany and Plant Sciences, University of California, Riverside, CA 92521, USA
* Correspondence: xmyu0616@fafu.edu.cn; Tel.: +86-591-8639-1591

Received: 25 December 2017; Accepted: 20 January 2018; Published: 24 January 2018

Abstract: Wuyi Rock tea, well-recognized for rich flavor and long-lasting fragrance, is a premium subcategory of oolong tea mainly produced in Wuyi Mountain and nearby regions of China. The quality of tea is mainly determined by the chemical constituents in the tea leaves. However, this remains underexplored for Wuyi Rock tea cultivars. In this study, we investigated the leaf metabolite profiles of 14 major Wuyi Rock tea cultivars grown in the same producing region using UPLC-QTOF MS and UPLC-QqQ MS with data processing via principal component analysis and cluster analysis. Relative quantitation of 49 major metabolites including flavan-3-ols, proanthocyanidins, flavonol glycosides, flavone glycosides, flavonone glycosides, phenolic acid derivatives, hydrolysable tannins, alkaloids and amino acids revealed clear variations between tea cultivars. In particular, catechins, kaempferol and quercetin derivatives were key metabolites responsible for cultivar discrimination. Information on the varietal differences in the levels of bioactive/functional metabolites, such as methylated catechins, flavonol glycosides and theanine, offers valuable insights to further explore the nutritional values and sensory qualities of Wuyi Rock tea. It also provides potential markers for tea plant fingerprinting and cultivar identification.

Keywords: Wuyi Rock tea; quality; UPLC-QTOF MS; UPLC-QqQ MS; metabolite profiling; metabolomics; cluster analysis; cultivars

1. Introduction

Oolong tea is a partially-fermented tea manufactured in southeast China, mainly in Fujian and Guangdong. In recent years, the production and consumption of oolong tea has increased greatly worldwide, attributed to its pleasurable aroma and taste favored by consumers [1,2]. As a functional drink, oolong tea exhibits many health-promoting benefits, such as anti-oxidant, anti-cancer, anti-obesity, anti-atherosclerosis, anti-diabetes and anti-allergic activities [1].

Wuyi Rock tea is a distinctive and premium subcategory of oolong tea grown in Wuyi Mountain, which is a UNESCO World Heritage site and considered the birthplace of oolong tea, as well as nearby regions in the north part of Fujian Province. Recognized as the most prestigious oolong tea in China,

Wuyi Rock tea boasts a history of over 1500 years and is renowned for its rich flavor and long-lasting fragrance, so-called 'rock charm and floral fragrance' [3]. Consumer demand for Wuyi Rock tea, both domestic and abroad, is increasing year by year but is often hindered by limited supplies and resulting high market price. Production of high-quality Wuyi Rock tea involves very complicated procedures, including leaf-picking, withering, zuoqing (partial fermentation, which includes alternating rotation and cooling steps), fixation (enzyme inactivation), rolling, roasting, grading and packaging [1] and usually relies on experienced workers [4]. Apart from manufacturing procedures, like the production of other types of tea, the quality of Wuyi Rock tea is also determined by the initial metabolite contents in fresh tea leaves, which depends on both cultivars and environmental factors [5–7]. According to the conventional classification by local people on the basis of the natural environment where tea plants have been grown, Wuyi Rock tea is subdivided into authentic rock tea, half rock tea, riverbank tea and tea grown outside the main production area in descending grade order [3]. This may suggest the geographic location as a key factor influencing the quality of Wuyi Rock tea.

On the other hand, the choice of cultivars to produce Wuyi Rock tea also matters but remains underexplored except for a few comparative studies, which have focused only on a small number of major constituents in processed tea and were not performed under controlled environmental conditions [8,9]. Contributed by unique climate and soil conditions, Wuyi Mountain is home to a large collection of tea germplasms. Historically, some tea cultivars have been used to produce Wuyi Rock tea since ancient times. Primary cultivars are 'Shuixian' and 'Rougui'; the former was registered as a national tea cultivar whereas the latter as a provincial tea cultivar due to their stable quality and higher yields. Other elite clonal cultivars include 'Dahongpao', 'Tieluohan', 'Baijiguan', 'Shuijingui', 'Guazijin' and 'Jinsuoshi', which are among the estimated 216 cultivars listed as Wuyi Rock tea cultivars. Such diverse genetic resources are valuable for producing Wuyi Rock tea. However, for most of these cultivars, research to examine quality-related traits at the genetic and metabolomic level is critical yet insufficient. Therefore, it would be helpful to comprehensively survey the metabolomes of representative tea cultivars, and identify important varietal differences relevant to tea quality.

Non-targeted metabolomics approach based on UPLC-QTOF MS, GC-TOF MS or NMR is a powerful technique capable of detecting a high number of endogenous metabolites simultaneously [10]. It has been widely applied in tea research to study impacts of environmental factors on tea metabolites [11–13], characterize dynamic changes during tea manufacture [14,15] and discover key compounds for tea type discrimination [16,17]. In this study, by combining UPLC-QTOF MS-based non-targeted analysis with UPLC-QqQ MS-based targeted quantifications of catechins, rutin, amino acids and caffeine, we analyzed the metabolite profiles of unprocessed fresh tea leaves of 14 major Wuyi Rock tea cultivars grown in the same environmental conditions subjected to the same cultivation practices. Data processing by principal component analysis (PCA), partial least squared discriminant analysis (PLS-DA) and hierarchical cluster analysis revealed differences as well as commonalities between the leaf phytochemical compositions among cultivars. It offered a comprehensive view for leaf metabolomes of Wuyi Rock tea cultivars in general and provided basis for future characterizations of nutritional values, sensory qualities and biological properties of Wuyi Rock tea.

2. Results and Discussion

2.1. Major Tea Leaf Metabolites Showed both Universal and Cultivar-Dependent Accumulation Patterns

To identify abundant metabolites and assess metabolite differences in 14 Wuyi Rock cultivars, which included 'Dahongpao' (DHP), 'Tieluohan' (TLH), 'Baijiguan' (BJG), 'Shuijingui' (SJG), 'Bantianyao' (BTY), 'Shuixian' (SX), 'Rougui' (RG), 'Beidou' (BD), 'Queshe' (QS), 'Xiaoyemaoxie' (XYMX), 'Jinfenghuang' (JFH), 'Aijiaowulong' (AJWL), 'Guazijin' (GZJ) and 'Jinsuoshi' (JSS) (Figure 1), non-targeted analysis based on UPLC-QTOF MS was performed to profile tea leaves. Forty-nine major metabolites were tentatively assigned based on their accurate masses, MS/MS fragmentation patterns and UV absorbance, in comparison to standard compounds and references (Table 1). Catechins, caffeine and free amino acids have been shown in a large body of literatures to contribute significantly to the taste and flavor quality of tea [6,18–20]. Therefore, absolute quantifications of these compounds, along with rutin, were carried out using UPLC-QqQ MS to enable comparisons with tea cultivars from other studies. The quantification results were shown in Table 2. Relative differences in the metabolites found in each sample were depicted in a heat map, which integrated measurements from both non-targeted and targeted analyses (Figure 2 and Figure S1).

Figure 1. Leaf phenotypes of 14 Wuyi Rock tea cultivars. (**A**) 'Dahongpao' (DHP); (**B**) 'Tieluohan' (TLH); (**C**) 'Baijiguan' (BJG); (**D**) 'Shuijingui' (SJG); (**E**) 'Bantianyao' (BTY); (**F**) 'Shuixian' (SX); (**G**) 'Rougui' (RG); (**H**) 'Beidou' (BD); (**I**) 'Queshe' (QS); (**J**) 'Xiaoyemaoxie' (XYMX); (**K**) 'Jinfenghuang' (JFH); (**L**) 'Aijiaowulong' (AJWL); (**M**) 'Guazijin' (GZJ); (**N**) 'Jinsuoshi' (JSS).

Most compounds were detected in all tea cultivars suggesting the presence of common machinery for secondary metabolism in tea plants (Figure 2). However, sharp variations between cultivars (VIP > 1 and $p < 0.05$) were found for many metabolite classes, such as flavan-3-ols, proanthocyanidins, flavonol glycosides, flavone glycosides, flavonone glycosides, phenolic acid derivatives, hydrolysable tannins, alkaloids and amino acids (Figure 2).

In PCA score plot (Figure 3A), the first and the second principal components explained 36.0% and 17.0% of the variation, respectively. Samples were clustered mainly according to their biological replicates in the PCA score plot, except that cultivars RG and BTY showed close aggregation, indicating inter-cultivar variations in metabolite profiles of Wuyi Rock tea cultivars. In addition, cultivars JFH and SJG were clearly separated from other cultivars along PC1 whereas cultivars QS and BD were separated from others along PC2. As these cultivars were grown in the same geographic region under the same cultivation condition, influences of varying environmental conditions on the chemical make-ups of tea leaves could be minimized. As a result, differences in metabolite compositions were largely attributed to particular genotype traits.

To investigate major differential metabolites, a PCA loading plot was applied (Figure 3B). The major groups that stood out in the plot corresponded to the MS signals of catechins (e.g., (−)-epigallocatechingallate (EGCG), (−)-epicatechingallate (ECG), (−)-epigallocatechin (EGC), (−)-epicatechin (EC), (−)-gallocatechin (GC), and (−)-epigallocatechin 3-(3-*O*-methylgallate) (EGCG3″Me)) and flavonol glycosides (e.g., rutin, quercetin galactosyl rutinoside, quercetin glucosyl rutinoside, kaempferol rutinoside and kaempferol glucosyl rutinoside). This inferred that catechins as well as quercetin and kaempferol derivatives were the most critical parameters for cultivar discrimination.

Table 1. Metabolites putatively identified in 14 tea cultivars by UPLC–QTOFMS.

Compound#	Tentative Assignments	RT (min)	Detected [M − H]⁻ (m/z)	Theoretical [M − H]⁻ (m/z)	Mass Error (ppm)	Formula	MS/MS Fragments	Ref.
Catechins								
1	GC	3.84	305.0668	305.0661	0.51	$C_{15}H_{14}O_7$	219.0660, 179.0348, 167.0347, 139.0397, 125.0242	Authentic standard [b]
2	EGC	4.93	305.0680	305.0661	1.04	$C_{15}H_{14}O_7$	219.0663, 179.0350, 167.0349, 139.0400, 125.0245	Authentic standard [b]
3	C	5.36	289.0718	289.0712	0.20	$C_{15}H_{14}O_6$	245.0817, 203.0710, 125.0242	Authentic standard [b]
4	EC	6.28	289.0722	289.0712	2.13	$C_{15}H_{14}O_6$	245.0820, 203.0711, 123.0450	Authentic standard [b]
5	EGCG	6.35	457.0783	457.0771	2.60	$C_{22}H_{18}O_{11}$	305.0662, 169.0143, 125.0244	Authentic standard [b]
6	8-C-ascorbylepigallocatechin 3-gallate	6.65	631.0938	631.0935	-0.37	$C_{28}H_{24}O_{17}$	479.0821, 316.0218	[21]
7	EGCG3"Me	7.42	471.0933	471.0927	0.16	$C_{23}H_{20}O_{11}$	305.0667, 287.0560, 183.0300, 161.0243	Authentic standard [b]
8	ECG	7.86	441.0827	441.0822	0.51	$C_{22}H_{18}O_{10}$	331.0458, 289.0719, 245.0818, 169.0145, 125.0245	Authentic standard [b]
9	ECG3"Me	8.92	455.0984	455.0978	-0.02	$C_{23}H_{20}O_{10}$	289.0717, 183.0298	Authentic standard [b]
10	epiafzelechin 3-gallate	8.97	425.0880	425.0873	0.41	$C_{22}H_{18}O_9$	273.0765, 255.0661	[21]
Proanthocyanidins								
11	prodelphinidin B	4.11	609.1251	609.1244	0.13	$C_{30}H_{26}O_{14}$	441.0825, 423.0718, 305.0667, 125.0243	[2]
12	EC-GC dimer	4.80	593.1301	593.1295	0.06	$C_{30}H_{26}O_{13}$	423.0714, 305.0659, 289.0717	[21]
13	prodelphinidin B2 (or B4) 3'-O-gallate	5.12	761.1357	761.1354	-0.27	$C_{37}H_{30}O_{18}$	609.1236, 591.1144, 577.1347, 423.0717	[2]
14	procyanidin dimer (B type)	5.68	577.1352	577.1346	0.02	$C_{30}H_{26}O_{12}$	451.1029, 425.0874, 407.0768, 289.0716	[22]
15	EGC-ECG dimer	6.04	745.1409	745.1405	-0.19	$C_{37}H_{30}O_{17}$	593.1298, 423.0714, 407.0768, 177.0191	[23]
16	3-galloylprocyanidin B1/3'-galloylprocyanidin B2	6.78	729.1458	729.1456	-0.36	$C_{37}H_{30}O_{16}$	407.0766, 289.0716	[23]
Flavonol/flavone glycosides								
17	isovitexin glucoside	6.08	595.1655 [a]	595.1663 [a]	-0.40	$C_{27}H_{30}O_{15}$	313.0711	[24]
18	apigenin 6-C-glucoside 8-C-arabinoside	6.91	563.1405	563.1401	-0.22	$C_{26}H_{28}O_{14}$	473.1086, 443.0980, 383.0769, 353.0664	[24]
19	myricetin 3-robinobioside (or 3-neohesperidoside)	6.93	625.1407	625.1405	-0.44	$C_{27}H_{30}O_{17}$	316.0219	[21]
20	myricetin 3-galactoside	7.02	479.0829	479.0826	0.36	$C_{21}H_{20}O_{13}$	316.0223, 315.0141, 271.0245	[25]
21	myricetin 3'-glucoside	7.12	479.0830	479.0826	-0.22	$C_{21}H_{20}O_{13}$	316.0224, 315.0146, 271.0245	[25]
22	quercetin 3-O-galactosyl rutinoside	7.21	771.1990	771.1984	0.03	$C_{33}H_{40}O_{21}$	609.1434, 463.0903, 301.0339, 300.0266	[11]
23	quercetin 3-O-glucosyl rutinoside	7.36	771.1991	771.1984	0.23	$C_{33}H_{40}O_{21}$	609.1458, 301.0348, 300.0272	[11]

Table 1. *Cont.*

Compound#	Tentative Assignments	RT (min)	Detected [M − H]⁻ (m/z)	Theoretical [M − H]⁻ (m/z)	Mass Error (ppm)	Formula	MS/MS Fragments	Ref.
24	camellianin B	7.69	579.1704 [a]	579.1714 [a]	−0.75	$C_{27}H_{30}O_{14}$	433.1129, 313.0709	[26]
25	rutin	7.70	609.1455	609.1456	−0.93	$C_{27}H_{30}O_{16}$	300.0274, 299.0195	Authentic standard [b]
26	kaempferol-3-O-galactosylrutinoside	7.72	755.2040	755.2035	−0.05	$C_{33}H_{40}O_{20}$	533.1294, 285.0398, 284.0319	[2]
27	tricetin	7.90	303.0504 [a]	303.0505 [a]	1.56	$C_{15}H_{10}O_7$	285.0402, 257.0450	[21]
28	kaempferol-3-O-glucosylrutinoside	8.00	755.2042	755.2035	0.19	$C_{33}H_{40}O_{20}$	593.1511, 285.0403, 284.0325	[2]
29	quercetin 3-O-glucoside	8.02	463.0879	463.0877	−0.62	$C_{21}H_{20}O_{12}$	300.0274, 299.0195, 243.0297	[25]
30	kaempferol 3-O-rutinoside	8.43	593.1511	593.1506	−0.21	$C_{27}H_{30}O_{15}$	501.0102, 285.0399, 284.0326	Authentic standard [b]
31	kaempferol galactoside	8.51	447.0929	447.0927	−0.84	$C_{21}H_{20}O_{11}$	285.0387, 284.0317	[2]
32	kaempferol glucoside	8.78	447.0929	447.0927	−0.76	$C_{21}H_{20}O_{11}$	284.0324, 255.0295, 227.0349	Authentic standard [b]
33	capilliposide I isomer 1	9.74	1065.3052 [a]	1065.3087 [a]	−2.68	$C_{48}H_{56}O_{27}$	617.2078, 449.1078, 303.0506	[27]
34	capilliposide II isomer 1	10.19	1049.3113 [a]	1049.3138 [a]	−1.84	$C_{48}H_{56}O_{26}$	741.2036, 595.1495, 287.0553	[27]
35	capilliposide I isomer 2	10.60	1065.3061 [a]	1065.3087 [a]	−1.97	$C_{48}H_{56}O_{27}$	617.2083, 449.1086, 303.0514	[27]
36	capilliposide II isomer 2	10.88	1049.3114 [a]	1049.3138 [a]	−1.78	$C_{48}H_{56}O_{26}$	741.2048, 287.0564	[27]
Flavonone glycosides								
37	eriodictyol 5,3′-di-O-glucoside	6.08	611.1617	611.1612	−0.10	$C_{27}H_{32}O_{16}$	491.1189, 449.1292, 329.0869	[21]
38	naringenin diglucoside	6.16	595.1664	595.1663	−0.78	$C_{27}H_{32}O_{15}$	577.1552, 475.1243, 433.1348, 381.0827, 313.0923	[21]
39	eriodictyol 7-O-glucoside	6.57	449.1086	449.1084	−0.68	$C_{21}H_{22}O_{11}$	329.0657, 197.0455	[21]
Phenolic acids								
40	theogallin	2.90	343.0669	343.0665	0.16	$C_{14}H_{16}O_{10}$	191.0560	Authentic standard [b]
41	3-p-coumaroylquinic acid	5.18	337.0928	337.0923	−0.25	$C_{16}H_{18}O_8$	163.0399	[2]
42	4-p-coumaroylquinic acid	6.15	337.0924	337.0923	−0.36	$C_{16}H_{18}O_8$	191.0542, 173.0454, 163.0398, 119.0500, 111.0441, 93.0343	[2]
43	5-p-coumaroylquinic acid	6.42	337.0925	337.0923	−0.08	$C_{16}H_{18}O_8$	173.0457, 163.0396, 119.0499, 93.0343	[2]
Hydrolysable tannins								
44	monogalloyl glucose	2.44	331.0668	331.0665	−0.80	$C_{13}H_{16}O_{10}$	271.0454, 211.0247, 169.0140, 125.0242	[28]
45	methyl 6-O-galloyl-β-D-glucose	3.67	345.0823	345.0822	−1.09	$C_{14}H_{18}O_{10}$	285.0611, 225.0401, 183.0296	[21]
46	1,4,6-tri-O-galloyl-β-D-glucose	6.64	635.0894	635.0884	1.60	$C_{27}H_{24}O_{18}$	483.0777, 465.0666, 423.0524, 313.0562, 241.0348, 169.0142, 125.0236	[25]
Alkaloids								
47	theobromine	3.80	181.0725 [a]	181.0726 [a]	3.53	$C_7H_8N_4O_2$	138.0671	Authentic standard [b]
48	caffeine	5.60	195.0893 [a]	195.0882 [a]	5.60	$C_8H_{10}N_4O_2$	138.0670	Authentic standard [b]
Amino acids								
49	theanine	1.43	175.1085 [a]	175.1083 [a]	4.72	$C_7H_{14}N_2O_3$	158.0823, 129.1030	Authentic standard [b]

[a] $[M + H]^+$. [b] This letter indicates that identification of the compound was confirmed by the authentic standard.

Table 2. Abundance (mg/g DW) of catechins, rutin, caffeine and amino acids in tea leaves in relation to cultivars.

Compound	DHP	TLH	BJG	SJG	BTY	SX	RG	BD	QS	XYMX	JFH	AJWL	GZJ	JSS
Catechins	210.09 ± 12.55 bc	201.42 ± 5.35 c	155.29 ± 6.42 e	199.58 ± 4.77 cd	225.73 ± 6.42 ab	233.28 ± 4.11 a	195.99 ± 8.44 cd	210.14 ± 16.03 abc	177.60 ± 2.85 de	162.51 ± 5.88 e	218.92 ± 3.72 abc	167.86 ± 4.68 e	201.94 ± 9.50 c	165.06 ± 6.09 e
EGCG	94.67 ± 5.17 d	95.07 ± 2.93 d	66.93 ± 3.97 f	111.71 ± 1.76 bc	97.81 ± 2.50 d	103.73 ± 1.66 cd	81.87 ± 3.43 e	116.21 ± 8.70 b	96.45 ± 2.08 d	69.71 ± 2.25 f	128.08 ± 1.63 a	71.20 ± 2.23 ef	81.60 ± 3.61 e	68.27 ± 3.35 f
ECG	22.83 ± 1.56 c	15.17 ± 0.56 fg	18.13 ± 0.93 de	13.09 ± 0.32 gh	22.83 ± 0.60 c	26.51 ± 0.26 b	14.91 ± 0.67 fg	31.31 ± 2.56 a	27.92 ± 0.29 b	11.12 ± 0.29 h	15.28 ± 0.14 fg	16.77 ± 0.32 ef	20.80 ± 0.89 bc	18.48 ± 0.71 de
EC	11.73 ± 0.70 b	7.60 ± 0.14 d	9.04 ± 0.28 c	5.12 ± 0.14 e	11.65 ± 0.58 b	13.95 ± 0.40 a	9.73 ± 0.68 c	9.15 ± 0.74 c	8.59 ± 0.30 cd	7.55 ± 0.20 d	5.09 ± 0.17 e	11.44 ± 0.21 b	12.43 ± 0.39 b	11.41 ± 0.28 b
EGC	72.83 ± 5.50 abcd	71.60 ± 1.42 bcd	50.99 ± 1.45 g	65.09 ± 2.37 def	76.43 ± 2.91 abc	81.44 ± 2.20 a	80.21 ± 3.28 ab	49.92 ± 3.92 g	42.48 ± 0.65 g	69.68 ± 3.09 cde	60.53 ± 2.29 f	64.80 ± 1.81 def	73.84 ± 3.98 abc	61.44 ± 1.89 ef
EGCG''Me	4.81 ± 0.22 c	6.48 ± 0.36 b	6.89 ± 0.49 b	0.03 ± 0.00 f	12.05 ± 0.45 a	2.62 ± 0.15 e	4.49 ± 0.05 c	0.03 ± 0.00 f	0.05 ± 0.00 f	0.03 ± 0.00 f	6.89 ± 0.39 b	0.13 ± 0.01 f	4.77 ± 0.18 c	3.38 ± 0.18 d
GC	2.48 ± 0.14 d	4.80 ± 0.16 b	2.45 ± 0.05 d	3.84 ± 0.21 c	4.00 ± 0.16 c	3.81 ± 0.05 c	3.81 ± 0.33 c	2.75 ± 0.30 d	1.44 ± 0.00 e	3.81 ± 0.09 c	2.67 ± 0.05 d	2.69 ± 0.12 d	6.75 ± 0.47 a	1.55 ± 0.05 e
C	0.75 ± 0.05 de	0.69 ± 0.12 def	0.85 ± 0.05 cd	0.69 ± 0.05 def	0.96 ± 0.08 c	1.23 ± 0.05 b	0.96 ± 0.08 c	0.77 ± 0.05 cde	0.67 ± 0.05 def	0.61 ± 0.05 ef	0.37 ± 0.05 g	0.83 ± 0.05 cd	1.76 ± 0.08 a	0.53 ± 0.05 fg
Rutin	0.68 ± 0.02 bcd	0.38 ± 0.01 de	0.89 ± 0.12 b	0.40 ± 0.04 de	0.44 ± 0.02 de	0.70 ± 0.07 bcd	0.60 ± 0.04 bcd	0.50 ± 0.07 cde	0.18 ± 0.02 e	0.42 ± 0.01 de	5.40 ± 0.39 a	0.83 ± 0.11 bc	0.85 ± 0.08 bc	0.23 ± 0.02 e
Caffeine	26.00 ± 2.28 b	18.35 ± 0.40 de	21.81 ± 0.44 cd	29.79 ± 1.40 a	23.97 ± 1.09 bc	25.41 ± 1.11 b	20.93 ± 1.38 cd	19.28 ± 1.42 de	22.93 ± 0.62 bc	14.27 ± 1.02 f	13.81 ± 0.58 f	14.16 ± 0.76 f	18.85 ± 1.72 de	16.51 ± 0.36 ef
Amino acids	4.60 ± 0.32 g	6.01 ± 0.36 ef	21.07 ± 0.80 a	15.83 ± 0.72 c	6.68 ± 0.31 ef	6.84 ± 0.06 ef	5.83 ± 0.22 f	3.11 ± 0.15 h	18.24 ± 0.41 b	4.29 ± 0.39 g	4.33 ± 0.17 g	5.77 ± 0.35 f	8.05 ± 0.23 d	7.09 ± 0.19 de
l-Theanine	2.61 ± 0.30 ef	3.31 ± 0.33 de	14.32 ± 0.68 a	12.08 ± 0.77 b	4.35 ± 0.36 cd	3.97 ± 0.05 cd	3.63 ± 0.28 de	0.80 ± 0.08 g	11.63 ± 0.62 b	1.81 ± 0.23 fg	2.08 ± 0.08 f	2.59 ± 0.26 ef	4.91 ± 0.30 c	4.13 ± 0.18 cd
Glu	1.15 ± 0.05 g	1.73 ± 0.10 cde	2.53 ± 0.10 b	2.33 ± 0.11 b	1.38 ± 0.07 fg	1.82 ± 0.07 cd	1.41 ± 0.09 fg	1.22 ± 0.05 g	3.05 ± 0.16 a	1.57 ± 0.08 def	1.45 ± 0.08 efg	1.76 ± 0.15 cd	1.68 ± 0.05 cdef	1.93 ± 0.17 c
Asp	0.22 ± 0.03 e	0.43 ± 0.01 d	1.13 ± 0.05 b	0.56 ± 0.02 c	0.27 ± 0.01 e	0.42 ± 0.03 d	0.22 ± 0.02 e	0.43 ± 0.02 d	1.37 ± 0.06 a	0.41 ± 0.05 d	0.23 ± 0.02 e	0.55 ± 0.01 c	0.55 ± 0.03 c	0.31 ± 0.04 e
Ser	0.22 ± 0.01 h	0.23 ± 0.03 gh	0.39 ± 0.02 c	0.35 ± 0.04 cde	0.29 ± 0.02 efg	0.22 ± 0.02 gh	0.24 ± 0.01 gh	0.28 ± 0.01 efgh	0.58 ± 0.01 a	0.24 ± 0.04 gh	0.24 ± 0.02 fgh	0.37 ± 0.02 def	0.48 ± 0.03 b	0.31 ± 0.02 def
Trp	0.06 ± 0.00 a	0.02 ± 0.00 def	0.05 ± 0.00 b	0.05 ± 0.00 b	0.03 ± 0.00 c	0.05 ± 0.00 b	0.03 ± 0.00 cde	0.02 ± 0.00 fgh	0.03 ± 0.00 cd	0.01 ± 0.00 h	0.01 ± 0.00 gh	0.01 ± 0.00 gh	0.03 ± 0.00 cd	0.02 ± 0.00 efg
Phe	0.05 ± 0.00 bc	0.04 ± 0.00 bcde	0.05 ± 0.00 bc	0.04 ± 0.00 def	0.03 ± 0.00 efg	0.05 ± 0.00 bc	0.04 ± 0.00 bcde	0.05 ± 0.00 b	0.12 ± 0.01 a	0.04 ± 0.00 cdef	0.03 ± 0.00 g	0.03 ± 0.00 fg	0.03 ± 0.00 g	0.04 ± 0.00 bcd

Table 2. *Cont.*

Compound	DHP	TLH	BJG	SJG	BTY	SX	RG	BD	QS	XYMX	JFH	AJWL	GZJ	JSS
Gln	0.05 ± 0.00 g	0.07 ± 0.01 fg	0.18 ± 0.01 c	0.16 ± 0.02 cd	0.10 ± 0.00 ef	0.12 ± 0.01 de	0.07 ± 0.00 fg	0.10 ± 0.01 ef	1.19 ± 0.05 a	0.07 ± 0.01 fg	0.10 ± 0.00 ef	0.27 ± 0.01 b	0.23 ± 0.01 b	0.14 ± 0.00 de
Leu	0.04 ± 0.00 b	0.02 ± 0.00 ef	0.07 ± 0.00 a	0.04 ± 0.00 bc	0.03 ± 0.00 cd	0.02 ± 0.00 efg	0.03 ± 0.00 ef	0.03 ± 0.00 ef	0.03 ± 0.00 de	0.02 ± 0.00 i	0.03 ± 0.00 de	0.02 ± 0.00 hi	0.02 ± 0.00 ghi	0.02 ± 0.00 fgh
Lys	0.04 ± 0.00 b	0.02 ± 0.00 defgh	0.15 ± 0.01 a	0.02 ± 0.00 cd	0.02 ± 0.00 de	0.02 ± 0.00 defgh	0.02 ± 0.00 defg	0.02 ± 0.00 def	0.03 ± 0.00 bc	0.01 ± 0.00 h	0.01 ± 0.00 efgh	0.01 ± 0.00 fgh	0.01 ± 0.00 gh	0.02 ± 0.00 de
Tyr	0.03 ± 0.00 b	0.02 ± 0.00 bc	0.02 ± 0.00 cde	0.01 ± 0.00 hi	0.02 ± 0.00 bcd	0.02 ± 0.00 def	0.02 ± 0.00 bc	0.02 ± 0.00 bcd	0.02 ± 0.00 efg	0.02 ± 0.00 fg	0.03 ± 0.00 a	0.02 ± 0.00 gh	0.01 ± 0.00 i	0.02 ± 0.00 fg
Pro	0.02 ± 0.00 cde	0.02 ± 0.00 f	0.03 ± 0.00 bc	0.03 ± 0.00 b	0.02 ± 0.00 ef	0.03 ± 0.00 bc	0.02 ± 0.00 def	0.03 ± 0.00 bcd	0.03 ± 0.00 a	0.02 ± 0.00 g	0.03 ± 0.00 bc	0.03 ± 0.00 bcd	0.03 ± 0.00 bc	0.02 ± 0.00 def
Thr	0.02 ± 0.00 bcde	0.02 ± 0.00 cde	0.03 ± 0.00 ab	0.03 ± 0.01 abcd	0.03 ± 0.01 abcd	0.02 ± 0.00 e	0.02 ± 0.00 bcde	0.02 ± 0.00 de	0.04 ± 0.01 a	0.03 ± 0.00 abcd	0.02 ± 0.00 cde	0.03 ± 0.00 abc	0.03 ± 0.00 bcde	0.03 ± 0.00 abcd
Val	0.02 ± 0.00 ab	0.01 ± 0.00 fg	0.02 ± 0.00 a	0.02 ± 0.00 cde	0.01 ± 0.00 defg	0.02 ± 0.00 bcd	0.02 ± 0.00 bc	0.02 ± 0.00 ab	0.01 ± 0.00 g	0.01 ± 0.00 efg	0.02 ± 0.00 bcd	0.01 ± 0.00 cdef	0.01 ± 0.00 efg	0.01 ± 0.00 cdef
GABA	0.02 ± 0.00 fg	0.02 ± 0.00 ef	0.11 ± 0.00 a	0.05 ± 0.00 b	0.03 ± 0.00 cd	0.02 ± 0.00 e	0.03 ± 0.00 d	0.01 ± 0.00 fgh	0.01 ± 0.00 i	0.01 ± 0.00 hi	0.01 ± 0.00 gh	0.03 ± 0.00 c	0.01 ± 0.00 hi	0.03 ± 0.00 d
His	0.02 ± 0.00 bc	0.01 ± 0.00 d	0.06 ± 0.01 a	0.01 ± 0.00 cd	0.01 ± 0.00 cd	0.01 ± 0.00 cd	0.01 ± 0.00 d	0.01 ± 0.00 d	0.02 ± 0.00 b	0.01 ± 0.00 d	0.01 ± 0.00 d	0.01 ± 0.00 d	0.01 ± 0.00 h	0.01 ± 0.00 d
Ile	0.01 ± 0.00 de	0.01 ± 0.00 efg	0.03 ± 0.00 a	0.01 ± 0.00 fg	0.02 ± 0.00 cd	0.01 ± 0.00 fg	0.01 ± 0.00 fg	0.01 ± 0.00 fg	0.01 ± 0.00 def	0.01 ± 0.00 gh	0.02 ± 0.00 bc	0.01 ± 0.00 gh	ND	0.02 ± 0.00 b
Arg	0.01 ± 0.00 b	0.02 ± 0.00 b	1.73 ± 0.06 a	0.03 ± 0.01 b	0.01 ± 0.00 b	0.01 ± 0.00 b	0.01 ± 0.00 b	0.02 ± 0.00 b	0.03 ± 0.00 b	ND	0.01 ± 0.00 b	0.01 ± 0.00 b	ND	0.02 ± 0.00 b
Asn	0.01 ± 0.01 b	0.01 ± 0.00 b	0.16 ± 0.02 a	0.02 ± 0.00 b	0.01 ± 0.01 b	0.01 ± 0.01 b	0.01 ± 0.01 b	0.01 ± 0.00 b	0.03 ± 0.01 b	0.01 ± 0.00 b	0.01 ± 0.00 b	0.01 ± 0.01 b	0.01 ± 0.00 b	0.01 ± 0.00 b

Results are expressed as mean ± standard deviation ($n = 3$). Means with different letters in row are significantly different according to Tukey's HSD test ($p < 0.05$). ND = non-detectable.

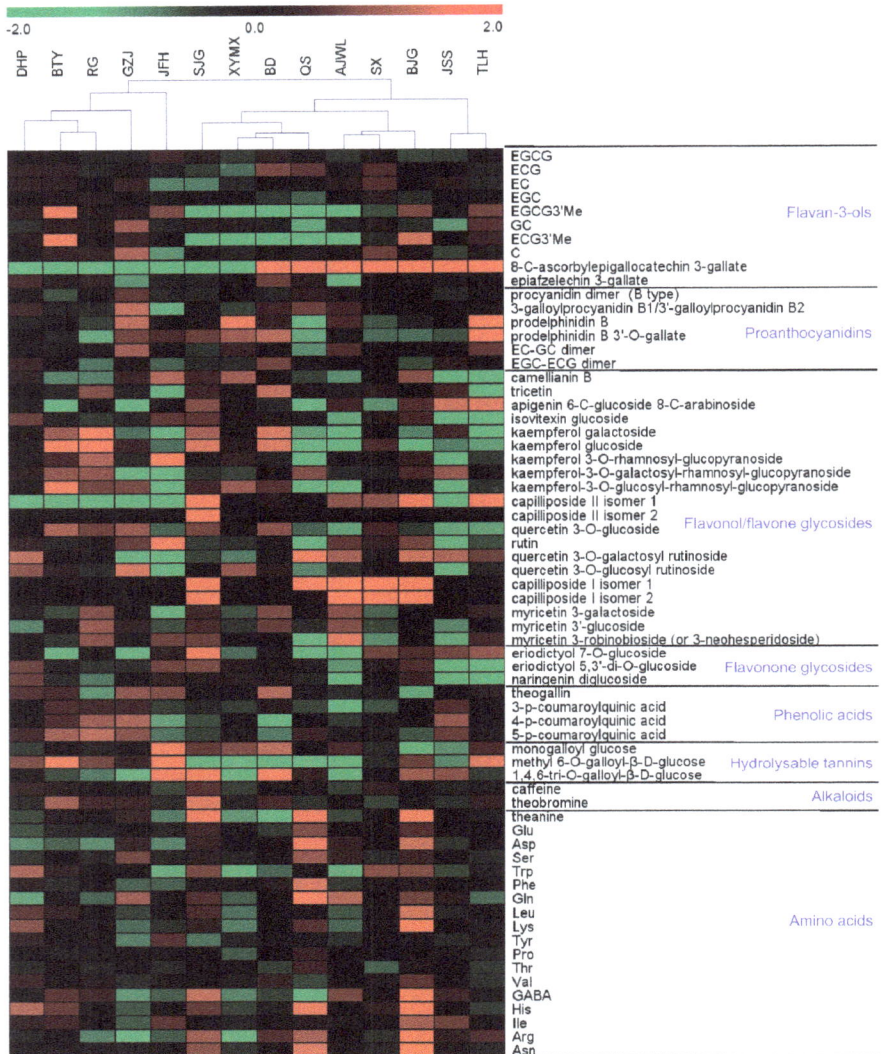

Figure 2. Comparisons of metabolite levels in 14 tea cultivars. The analysis is based on the normalized average signal abundance from three biological replicates for each cultivar. Normalized values are shown on a color scale proportional to the content of each metabolite and are expressed as log2 using the MultiExperiment Viewer software (MeV v4.9.0, J. Craig Venter Institute, La Jolla, CA, USA).

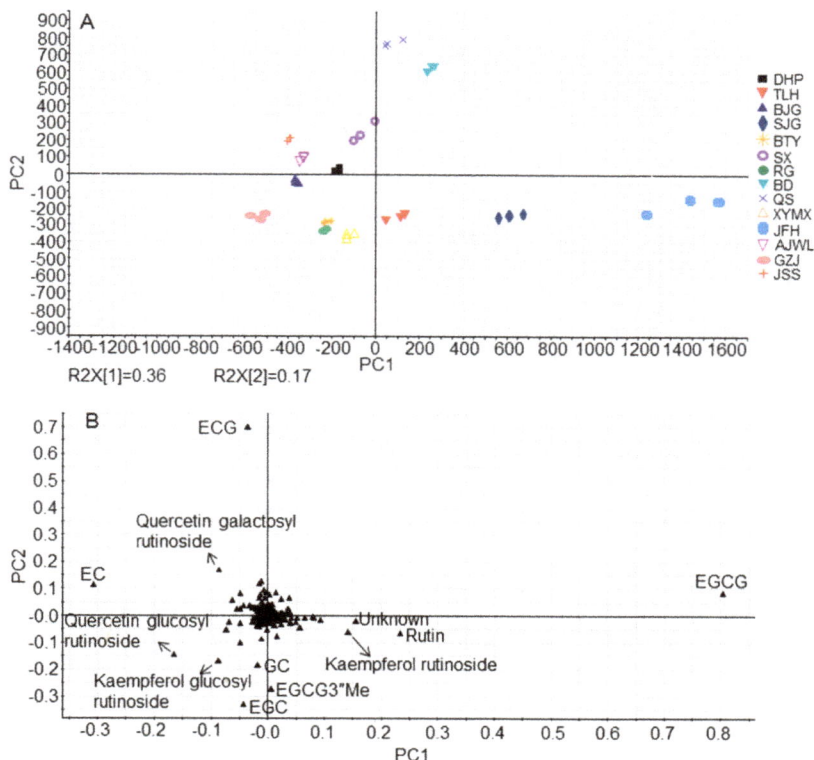

Figure 3. Principal component analysis (PCA) of methanol extracts of tea leaves. (**A**) Score plot of PCA demonstrating differences in metabolite profiles between leaf samples based on 466 filtered single molecular features detected by UPLC-QTOF MS in ESI⁻. The principal components 1 and 2 explained 36.0% and 17.0% of total variance, respectively. For each cultivar, three biological replicates were prepared, where one replicate was a pool of 7–8 tea leaves. R2X, explained variation; (**B**) Loading plot of PCA indicating primary differential metabolites.

2.2. Flavan-3-ols Exhibited Variable Levels in Tea Leaves

A total of 10 flavan-3-ols was tentatively identified by UPLC-QTOF MS, including GC, EGC, (−)-catechin (C), EC, EGCG, 8-C-ascorbylepigallocatechin 3-gallate, EGCG3″Me, ECG, epicatechin 3-(3-O-methylgallate) (ECG3″Me) and epiafzelechin 3-gallate (Table 1). Major flavan-3-ols included EGCG, EGC, EC and ECG, which occurred at descending levels in all cultivars examined (Table 2). In general, phenolic contents of tea cultivars applied to black and oolong tea are higher than that of green tea [7]. EGCG, as the most dominant flavan-3-ol, ranged from 66.93 mg/g in cultivar BJG to 128.08 mg/g in cultivar JFH, whereas GC (1.44–6.75 mg/g) and C (0.37–1.76 mg/g) were only minor components (Table 2).

EGCG3″Me, an O-methylated catechin, was detected in all tea cultivars, albeit in very low abundance in cultivars SJG, BD, QS, XYMX and AJWL (Figure 2 and Table 2). Methylated catechins have attracted much attention because of their stronger anti-allergic activities than catechins, including EGCG [29,30]. Efforts have been made to screen different tea varieties to identify cultivars enriched in EGCG3″Me [6,31]. Lv and coworkers identified four out of 71 Chinese tea cultivars with EGCG3″Me contents higher than 10 mg/g; interestingly, they are all oolong tea cultivars, implying that oolong tea cultivars may be a good source for finding EGCG3″Me-rich tea cultivars [31]. Supporting this notion, we found that cultivar BTY contained the highest content of EGCG3″Me (12.05 mg/g) (Table 2). Moreover, the other three cultivars, TLH, BJG and JFH, also produced medium levels of EGCG3″Me (≥6 mg/g). The distribution of a second *O*-methylated catechin, ECG3″Me, which also exhibited a strong anti-inflammatory activity in vitro [29], resembled that of EGCG3″Me, ranging from barely detectable in cultivars SJG, BD, QS, XYMX and AJWL to being highest in cultivar BTY. Due to the lack of an authentic standard, the absolute quantification of ECG3″Me was impossible. Nevertheless, the level of this compound somewhat demonstrated a positive correlation with the EGCG3″Me level in tea cultivars (Figure 2 and Figure S1). As a result, there is potentially a higher chance of finding tea cultivars rich in ECG3″Me among cultivars rich in EGCG3″Me.

An ascorbic acid-appended EGCG derivative, namely, 8-C-ascorbylepigallocatechin 3-gallate, was another interesting flavan-3-ol present as a minor constituent in seven cultivars, TLH, BJG, SX, BD, QS, AJWL and JSS (Figure 2). Initially isolated from a commercial oolong tea sample, this compound was structurally elucidated through NMR spectroscopy by Hashimoto and coworkers [32]. Subsequent activity tests showed that it demonstrated inhibitory effects against HIV replication in H9 lymphocyte cells [33] and pancreatic lipase [34]. Information on the distribution of 8-C-ascorbylepigallocatechin 3-gallate among tea cultivars is scarce. Nonetheless, considering where this compound was first isolated and its high occurrence in the current study, Wuyi Rock tea cultivars may be a promising source for compound isolation to further explore its therapeutic potential.

2.3. Cultivar JFH Possessed High Contents of Rutin and Kaempferol Rutinoside

Flavonol glycosides (FOGs) are one of most important phenolic compounds in tea besides catechins. Though less abundant than catechins, they confer velvety and astringent tastes to tea infusions at much lower thresholds and hence are key tea taste determinants [35]. Unambiguous FOG assignments are difficult due to the fact that many authentic standards of FOGs in tea are not commercially available. Moreover, FOGs usually contain several positional isomers. Nevertheless, galactosyl flavonols were reported to elute earlier than glucosyl flavonols [15]. Taking account of the differences in chromatographic retention behaviors, in combined with analyses of MS^2 fragmentation patterns and UV absorbance (if available), we tentatively identified 18 FOGs in the current study. These FOGs, most commonly kaempferol and quercetin derivatives, were mainly present in the form of mono-, di-, tri- and tetraglycosides (Table 1). Many FOGs have been previously detected in processed tea products or fresh tea leaves [2,36].

Contents of kaempferol and quercetin glycosides varied widely between tea cultivars (Figure 2). In particular, the rutin (also called quercetin 3-O-rutinoside) content in cultivar JFH (5.40 mg/g) was found to be significantly higher ($p < 0.01$) than in other cultivars (between 0.18–0.89 mg/g) (Figure 4A and Table 2). Apart from rutin, the highest level of kaempferol 3-O-rutinoside was also detected in cultivar JFH (Figure 4B). In contrast, four flavonol triglycosides, kaempferol 3-O-galactosyl rutinoside, kaempferol 3-O-glucosyl rutinoside, quercetin 3-O-galactosyl-rutinoside and quercetin 3-O-glucosyl rutinoside, were barely detectable in this cultivar (Figure 4A,B). Quercetin 3-O-galactosyl rutinoside and quercetin 3-O-glucosyl rutinoside could be synthesized from rutin by glycosyltransferases. Kaempferol 3-O-galactosyl rutinoside and kaempferol 3-O-glucosyl rutinoside may derive from kaempferol 3-O-rutinoside catalyzed by the same type of enzymes. In cultivar JFH, we speculate that enzyme(s) responsible for glycosylation of flavonol diglycosides are either not functional or expressed at very low levels, accounting for the high accumulations of rutin and kaempferol 3-O-rutinoside. Alternatively, genes which are critical for flavonoid metabolism and catabolism in cultivar JFH may be differentially regulated. Comparing gene expressions in the phenylpropanoid pathway among these cultivars would further shed light on flavonoid biosynthesis in tea. A number of pharmacological properties of rutin, such as anti-inflammation, anti-microbial, anti-tumor and anti-asthma, have been well documented [37]. Therefore, finding tea cultivars with high yields of flavonoids such as rutin could be useful in diversifying the utilization of functional components in tea resources.

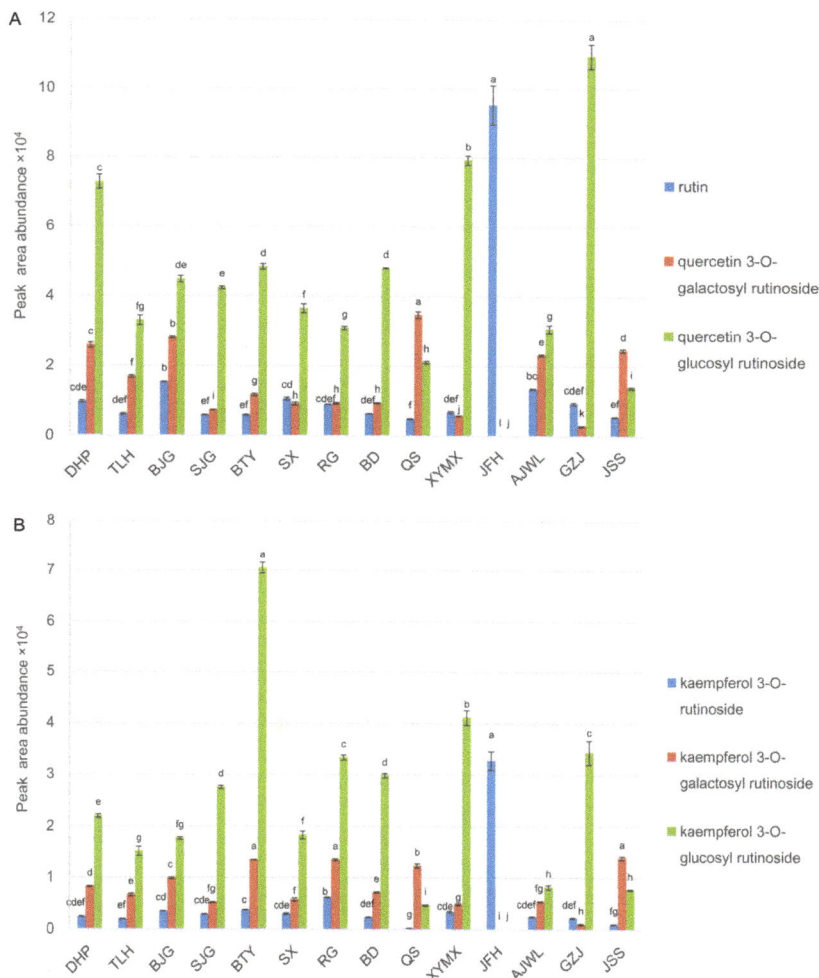

Figure 4. Mean peak area abundance values (±SD) of (**A**) some quercetin glycosides and (**B**) some kaempferol glycosides in leaves of 14 tea cultivars. Different letters on top of the vertical bars of the graph indicate significant differences among the samples, which were determined by Tukey's HSD test at $p < 0.05$.

2.4. Cultivars BJG, SJG and QS Demonstrated High Levels of Amino Acids in Leaves

Amino acid constituents of tea leaves have a large impact on the taste and aroma properties of processed tea [18]. There exists a positive correlation between the quality of tea and amino acid contents [18]. Theanine, glutamate and serine are key components imparting "umami" or "brothy" taste to tea infusions [19,38]. To compare amino acid profiles between leaves of different cultivars, hydrophilic interaction liquid chromatography (HILIC) tandem mass spectrometry was applied since amino acids are typically not well resolved in C18 columns. In total, 18 amino acids in varying concentrations were detected (Table 2). Theanine, a non-proteinogenic amino acid synthesized from glutamate and ethylamine by theanine synthetase, has been shown in many studies as the most abundant free amino acid in tea plants [20,39–41]. As expected, theanine was found to be the most predominant free amino acid in 13 out of 14 cultivars examined, accounting for 42.3–72.3% of total free amino acids in leaves. Other abundant amino acids included glutamate, aspartate and serine (Table 2). In contrast, cultivar BD contained a higher level ($p < 0.01$) of glutamate (1.2 mg/g) than theanine (0.8 mg/g). Moreover, the total amino acid content (3.11 mg/g) in leaves of cultivar BD was lowest. Cultivars BJG, SJG and QS were characterized by high levels of amino acids (Table 2). Not only containing markedly higher ($p < 0.05$) amounts of theanine (14.32, 12.08 and 11.63 mg/g for cultivars BJG, SJG and QS, respectively), they also demonstrated high accumulations of other amino acids, suggesting an overall up regulation of amino acid metabolism. For example, glutamate and aspartate were at high levels in both cultivars. The glutamine level in cultivar QS and the arginine level in cultivar BJG were significantly higher ($p < 0.01$) than other cultivars.

Interestingly, cultivar BJG is a light-sensitive albino tea variety (Figure 1C) [42]. A recent study showed that cultivar BJG exhibited yellow leaf phenotype, and reduced synthesis of chlorophyll and carotenoid under high light intensity but could turn green when transferred to low intensity light [42]. Similar scenarios were described for other albino tea cultivars such as 'Anji Baicha', whose albinism is induced by temperature instead of light [43]. Feng and coworkers reported that theanine and glutamate levels in some temperature-sensitive albino tea cultivars were higher than in normal green cultivars [19]. Similar to temperature-sensitive cultivars, cultivar BJG also contained the highest level of theanine and a relatively higher level of glutamate, second only to cultivar QS (Table 2). One possible explanation is that suppressed chlorophyll biosynthesis leads to elevated levels of glutamate, which provides more substrates for theanine synthesis [43].

2.5. Purine Alkaloids Exhibited Variable Levels in Tea Leaves

Purine alkaloids are naturally found in tea. The biosynthesis of purine alkaloids, mainly caffeine and theobromine, has been extensively investigated in tea plants [41,44,45]. The caffeine is the most abundant purine alkaloid in tea leaves, ranging between 1.5–5% [36]. Among the 14 cultivars, the caffeine content in leaves was found to vary between 1.38% (cultivar JFH) and 2.98% (cultivar SJG) (Table 2), consistent with the previous report [36]. Moreover, cultivar SJG also had markedly a higher ($p < 0.05$) content of theobromine than other cultivars, followed by cultivar BTY (Figure 2). Previous studies suggested that genotypic factors other than environmental factors may have more effects on the caffeine content [46]. Therefore, differences in the abundances of purine alkaloids in the current study may be the result of genetic variations.

3. Materials and Methods

3.1. Chemicals

(−)-Epigallocatechingallate, (−)-epigallocatechin, (−)-catechin, (−)-epicatechingallate, (−)-epicatechin, (−)-gallocatechin, rutin and L-theanine (all with purity≥95%) were obtained from Sigma-Aldrich (St. Louis, MO, USA). (−)-Epigallocatechin3-(3-O-methylgallate) (≥95%) and kaempferol 3-rutinoside (≥98%) were purchased from ChemFaces (Wuhan, China). Caffeine (≥98%) was obtained from Yuanye Biotechnology Inc. (Shanghai, China). Theobromine (≥99%) and kaempferol glucoside (≥98%) were obtained from BioBioPha Co., Ltd. (Kunming, China). Theogallin (≥95%) was kindly provided by Dr. Qingxie Chen of Fujian Agriculture and Forestry University, China. Acetonitrile (MS grade), methanol (HPLC grade) and formic acid (98%) were obtained from Sigma-Aldrich. Deionized water was produced by a Milli-Q water purification system (Millipore, Billerica, MA, USA).

3.2. Tea Samples and Sample Preparation

All tea plants of *Camellia sinensis* (five-year-old) used in the current study were grown at the same tea germplasm garden and under the same cultivation practice, which was managed by the Wuyi

Star Tea Industry Co., Ltd., Wuyishan City, Fujian, China (latitude: 27.71° N, longitude: 118.00° E). The fully expanded second leaves were collected from 14 Wuyi Rock tea cultivars on 9 May 2015. The cultivars included 'Dahongpao', 'Tieluohan', 'Baijiguan', 'Shuijingui', 'Bantianyao', 'Shuixian', 'Rougui', 'Beidou', 'Queshe', 'Xiaoyemaoxie', 'Jinfenghuang', 'Aijiaowulong', 'Guazijin' and 'Jinsuoshi' (Figure 1). For each cultivar, three biological replicates were collected with each replicate gathered from 7–8 individual tea plants. The excised leaf samples were immediately frozen in liquid nitrogen, brought back to lab and stored at −80 °C until analysis.

Extraction of tea leaves were carried out as previously described with some minor modifications [47]. Briefly, frozen tea leaves were individually ground to fine powders using precooled mortars and pestles. Following lyophilization, 30 mg (± 0.5 mg) of ground samples was weighted and 1.2 mL of 70% (v/v) methanol was added for metabolite extraction. Samples were vortexed, sonicated at 25 °C for 20 min and centrifuged (10 min, 12,000 g). Supernatants were diluted 50-fold with 70% (v/v) methanol, filtered through a 0.22 μm PVDF filter (Millipore) and stored at −20 °C until analyzed. Three biological sample replicates were prepared for each cultivar.

3.3. UPLC-QTOF MS-Based Non-Targeted Metabolite Analysis

Aliquots (1 μL) of above extracts were analyzed on a Waters Acquity UPLC system coupled in tandem to a Waters photodiode array (PDA) detector and a SYNAPT G2-Si HDMS QTOF mass spectrometer (Waters, Manchester, UK). Chromatographic separation was performed on a Waters Acquity UPLC HSS T3 column (2.1 × 100 mm, 1.8 μm) at 40 °C with water containing 0.1% formic acid (phase A) and acetonitrile containing 0.1% formic acid (phase B) for chromatographic elution: 0–2 min (99–93% A), 2–13 min (93–60% A), 13–14 min (60–1% A) and 14–17 min (1–1% A). The flow rate was set at 0.3 mL/min.

Samples were run in both positive and negative ionization modes as separate chromatographic runs. Following settings were applied during LC-MS runs: capillary voltage, 2.0 kV (ESI$^+$) and 2.5 kV (ESI$^-$); cone voltage, 40 eV; collision energy, 4 eV; source temperature, 120 °C; desolvation temperature, 500 °C; cone gas flow, 50 L/h; desolvation gas flow, 800 L/h; m/z range, 50–1200 Da. The collision energy ramp for MS^e (continuum mode) was set from 10 to 50 eV. LockSpray (leucine encephalin) reference ions with m/z of 556.2771 (for ESI$^+$) or 554.2615 (for ESI$^-$) were infused during data acquisition for online calibration. Each triplicate tea sample was analyzed once.

Quality control (QC) samples were prepared by mixing 30 mg of one leaf sample to become a combined sample. QC samples were injected throughout the analytical runs (every five samples) to check the instrument performance. The MassLynx software (version 4.1, Waters, Milford, MA, USA) was used to control all instruments and calculate accurate masses.

3.4. UPLC-QqQ MS-Based Targeted Quantification of Catechins, Rutin, Caffeine and Amino Acids

For quantifications of catechins, rutin, caffeine and amino acids, 2 μL of above extracts, with appropriate dilutions within the range of the calibration curve, were injected on a Waters Acquity UPLC system coupled in tandem to a Waters photodiode array (PDA) detector and a XEVO TQ-S MS triple quadrupole mass spectrometer (Waters, Milford, MA, USA).

To detect catechins, rutin and caffeine, chromatographic separation was achieved on a Waters Acquity UPLC BEH C18 column (2.1 × 100 mm, 1.7 μm) at 40 °C with water containing 0.1% formic acid (phase A) and acetonitrile containing 0.1% formic acid (phase B) for chromatographic elution: 0–12 min (95–83% A), 12–13 min (83–0% A) and 13–16.5 min (0–0% A). The flow rate was set at 0.3 mL/min. Mass spectrometry was performed in the ESI$^-$ mode for catechins and rutin, and in the ESI$^+$ mode for caffeine with the following settings: capillary voltage: 3.0 kV (ESI$^+$) and 2.0 kV (ESI$^-$); desolvation temperature: 400 °C; source temperature: 150 °C; cone gas flow: 150 L/h; desolvation gas flow: 800 L/h. Collision energy and cone voltage were optimized for above compounds with multiple reaction monitoring (MRM) for quantification. Calibration curves generated by injecting increasing concentrations of chemical standards were used to determine the absolute concentrations of catechins, rutin and caffeine.

Amino acids were measured in the same manner except that the chromatographic separation was achieved on a Merck SeQuant ZIC-HILIC column (2.1 × 100 mm, 5 μm) at 40 °C with water containing 5 mM ammonium acetate (phase A) and acetonitrile containing 0.1% formic acid (phase B) for chromatographic elution: 0–13 min (5–41% A), 13–15 min (41–60% A) and 15–20 min (60–5% A). The flow rate was set at 0.4 mL/min. Mass spectrometry was performed in the ESI$^+$ mode using the same setting for caffeine. In all cases, the MassLynx software (version 4.1, Waters, Milford, MA, USA) was used for instrument control and data acquisition. Each triplicate tea sample was analyzed once.

3.5. Data Processing and Statistical Analysis

Resulting chromatograms from UPLC-QTOFMS were processed using Progenesis QI software (version 2.1, Nonlinear Dynamics, New Castle upon Tyne, UK) with default settings for peak picking, normalization (normalized to all compounds), signal integration and initial peak assignments. Only chromatograms between elution time 1–14 min were included in the analysis. The final data set contained 1550 molecular features in the ESI$^-$ mode and 992 molecular features in the ESI$^+$ mode. For comparing the abundances of molecular features, the data matrix consisting of mass features (including retention time and accurate mass values) and peak area values was exported from Progenesis QI to Excel. The mean peak area abundance values from three biological replicates of the same cultivar were calculated and differences in metabolite signal abundances were compared across cultivars.

The raw 1550 molecular features detected in the ESI$^-$ mode were filtered to only include single features. After filtering, the remaining 466 molecular features were fed into PCA analysis to observe intrinsic metabolite variance between cultivars; PLS-DA was performed to identify differential metabolites using Progenesis QI extension EZinfo after Pareto scaling. One-way ANOVA was used to measure the significance of metabolites in cultivar discrimination using SPSS (version 13.0, SPSS, Chicago, IL, USA). Significantly different metabolites between samples were selected with variable importance in the projection (VIP) > 1 and a p value < 0.05. Heat map with hierarchical clustering (Pearson's correlation, average linkage) was generated with MultiExperiment Viewer software (version 4.9.0, J. Craig Venter Institute, La Jolla, CA, USA) to visualize accumulation patterns of annotated major metabolites between sample types. Before analysis, the data were log2 transformed and normalized to the median level of individual compounds, combining data from metabolite analyses by UPLC-QTOF MS and UPLC-QqQ MS. The data matrix used for PCA and PLS-DA analyses was listed in Table S1.

3.6. Metabolite Identification

Annotation obtained from Progenesis QI was used as a starting point for manual peak identification. Metabolites were identified by comparing accurate masses, MS/MS fragmentation patterns and isotope patterns with authentic standards, online metabolite databases of Metlin [21], HMDB [48], MassBank [49], ReSpect [22], KNApSAcK [50] and literature references [2,11,27,28,51]. Each mass spectrum was manually inspected to verify whether software-predicted fragments were derived from a single metabolite. UV spectra (Waters, Milford, MA, USA) were used for identification whenever possible.

4. Conclusions

To the best of our knowledge, this is the first exhaustive study of leaf metabolite profiles of different Wuyi Rock tea cultivars. A combined UPLC-QTOF MS and UPLC-QqQ MS approach coupled with multivariate data analysis revealed fundamental varietal differences in primary and secondary metabolism between cultivars. Those differential metabolites mainly include phenolic compounds (e.g., flavan-3-ols, flavonol glycosides and phenolic acids), alkaloids and amino acids. Major catechins as well as quercetin and kaempferol glycosides were determined as critical for cultivar discrimination. The functional compounds found in leaves of Wuyi Rock tea cultivars as well as the knowledge on the cultivar-specific differences provides insights for their potential applications as dietary supplements or nutraceuticals. For instance, cultivar BTY would be an excellent target for anti-allergic study owing to the production of high levels of methylated catechins. Cultivar JFH may serve as a stable source for rutin and kaempferol rutinoside. Cultivars BJG, SJG and QS could be explored as a prominent source for theanine. The metabolites identified in the current study could potentially be used as chemical markers for tea plant fingerprinting, cultivar identification and tea authentication. It also provides valuable information to tea breeders in selecting breeding materials with desirable traits. On the other hand, chemical constituents of processed tea are influenced by both cultivars and processing techniques. Studies by other research groups have shown that both volatile and non-volatile compounds undergo substantial changes during the manufacture of oolong tea [1,4,52].Therefore, whether and how potential markers uncovered in the current study could be applied for analyzing processed Wuyi Rock tea samples warrants further investigations.

Supplementary Materials: The Supplementary Materials are available online, Figure S1: Distribution of metabolites in the methanol extracts of fresh leaves of 14 Wuyi Rock tea cultivars, Table S1: Filtered and normalized PCA data matrix generated from UPLC-QTOF MS in ESI$^-$.

Acknowledgments: This study was financially supported by the Natural Science Foundation of Fujian (2016J01108), the Distinguished Young Scholar Program of Fujian Agriculture and Forestry University (xjq201610) and the startup fund from Fujian Agriculture and Forestry University.

Author Contributions: X.Y. and Z.Y. conceived and directed the research; S.C. and M.L. performed the research; J.L., S.W. and X.W. operated MS instruments; S.C. and M.L. analyzed the data; G.Z., T.W., Q.C. and S.C. collected the samples; X.Y. and S.C. wrote the paper. All authors read and approved the final manuscript.

Conflicts of Interest: The authors declare no conflict of interests.

References

1. Chen, Y.L.; Duan, J.; Jiang, Y.M.; Shi, J.; Peng, L.; Xue, S.; Kakuda, Y. Production, quality, and biological effects of Oolong tea (*Camellia sinensis*). *Food Rev. Int.* **2011**, *27*, 1–15. [CrossRef]
2. Dou, J.; Lee, V.S.; Tzen, J.T.; Lee, M.R. Identification and comparison of phenolic compounds in the preparation of oolong tea manufactured by semifermentation and drying processes. *J. Agric. Food Chem.* **2007**, *55*, 7462–7468. [CrossRef] [PubMed]
3. Xiao, K.B. The taste of tea: Material, embodied knowledge and environmental history in northern Fujian, China. *J. Mater. Cult.* **2017**, *22*, 3–18. [CrossRef]
4. Ma, C.H.; Tan, C.; Li, W.L.; Chen, L.B.; Wang, Y.R.; Chen, X. Identification of the different aroma compounds between conventional and freeze dried Wuyi rock Tea (Dangui) using headspace solid phase microextraction. *Food Sci. Technol. Res.* **2013**, *19*, 805–811. [CrossRef]
5. Chaturvedula, V.S.P.; Prakash, I. The aroma, taste, color and bioactive constituents of tea. *J. Med. Plants Res.* **2011**, *5*, 2110–2124.
6. Ji, H.G.; Lee, Y.R.; Lee, M.S.; Hwang, K.H.; Kim, E.H.; Park, J.S.; Hong, Y.S. Metabolic phenotyping of various tea (*Camellia sinensis* L.) cultivars and understanding of their intrinsic metabolism. *Food Chem.* **2017**, *233*, 321–330. [CrossRef] [PubMed]
7. Zeng, L.; Luo, L.; Li, H.; Liu, R. Phytochemical profiles and antioxidant activity of 27 cultivars of tea. *Int. J. Food Sci. Nutr.* **2017**, *68*, 525–537. [CrossRef] [PubMed]
8. Cai, H.H.; Cheng, R.H.; Jin, Y.L.; Ding, S.W.; Chen, Z. Evaluation of oolong teas using ^{1}H and ^{13}C solid-state NMR, sensory analysis, and multivariate statistics. *J. Chin. Chem. Soc.* **2016**, *63*, 792–799. [CrossRef]
9. Zhao, F.; Lin, H.T.; Zhang, S.; Lin, Y.F.; Yang, J.F.; Ye, N.X. Simultaneous determination of caffeine and some selected polyphenols in Wuyi Rock tea by high-performance liquid chromatography. *J. Agric. Food Chem.* **2014**, *62*, 2772–2781. [CrossRef] [PubMed]
10. Daglia, M.; Antiochia, R.; Sobolev, A.P.; Mannina, L. Untargeted and targeted methodologies in the study of tea (*Camellia sinensis* L.). *Food Res. Int.* **2014**, *63*, 275–289. [CrossRef]

11. Dai, W.D.; Qi, D.D.; Yang, T.; Lv, H.P.; Guo, L.; Zhang, Y.; Zhu, Y.; Peng, Q.H.; Xie, D.C.; Tan, J.F.; et al. Nontargeted analysis using ultraperformance liquid chromatography-quadrupole time-of-flight mass spectrometry uncovers the effects of harvest season on the metabolites and taste quality of tea (*Camellia sinensis* L.). *J. Agric. Food Chem.* **2015**, *63*, 9869–9878. [CrossRef] [PubMed]

12. Liu, J.; Zhang, Q.; Liu, M.; Ma, L.; Shi, Y.; Ruan, J. Metabolomic analyses reveal distinct change of metabolites and quality of green tea during the short duration of a single spring season. *J. Agric. Food Chem.* **2016**, *64*, 3302–3309. [CrossRef] [PubMed]

13. Zhang, Q.; Shi, Y.; Ma, L.; Yi, X.; Ruan, J. Metabolomic analysis using ultra-performance liquid chromatography-quadrupole-time of flight mass spectrometry (UPLC-Q-TOF MS) uncovers the effects of light intensity and temperature under shading treatments on the metabolites in tea. *PLoS ONE* **2014**, *9*, e112572. [CrossRef] [PubMed]

14. Fraser, K.; Lane, G.A.; Otter, D.E.; Harrison, S.J.; Quek, S.Y.; Hemar, Y.; Rasmussen, S. Non-targeted analysis by LC-MS of major metabolite changes during the oolong tea manufacturing in New Zealand. *Food Chem.* **2014**, *151*, 394–403. [CrossRef] [PubMed]

15. Tan, J.; Dai, W.; Lu, M.; Lv, H.; Guo, L.; Zhang, Y.; Zhu, Y.; Peng, Q.; Lin, Z. Study of the dynamic changes in the non-volatile chemical constituents of black tea during fermentation processing by a non-targeted metabolomics approach. *Food Res. Int.* **2016**, *79*, 106–113. [CrossRef]

16. Lee, J.-E.; Lee, B.-J.; Chung, J.-O.; Kim, H.-N.; Kim, E.-H.; Jung, S.; Lee, H.; Lee, S.-J.; Hong, Y.-S. Metabolomic unveiling of a diverse range of green tea (*Camellia sinensis*) metabolites dependent on geography. *Food Chem.* **2015**, *174*, 452–459. [CrossRef] [PubMed]

17. Lin, J.; Zhang, P.; Pan, Z.; Xu, H.; Luo, Y.; Wang, X. Discrimination of oolong tea (*Camellia sinensis*) varieties based on feature extraction and selection from aromatic profiles analysed by HS-SPME/GC-MS. *Food Chem.* **2013**, *141*, 259–265. [CrossRef] [PubMed]

18. Alcazar, A.; Ballesteros, O.; Jurado, J.M.; Pablos, F.; Martin, M.J.; Vilches, J.L.; Navalon, A. Differentiation of green, white, black, Oolong, and Pu-erh teas according to their free amino acids content. *J. Agric. Food Chem.* **2007**, *55*, 5960–5965. [CrossRef] [PubMed]

19. Feng, L.; Gao, M.J.; Hou, R.Y.; Hu, X.Y.; Zhang, L.; Wan, X.C.; Wei, S. Determination of quality constituents in the young leaves of albino tea cultivars. *Food Chem.* **2014**, *155*, 98–104. [CrossRef] [PubMed]

20. Yang, X.R.; Ye, C.X.; Xu, J.K.; Jiang, Y.M. Simultaneous analysis of purine alkaloids and catechins in *Camellia sinensis*, *Camellia ptilophylla* and *Camellia assamica* var. *kucha* by HPLC. *Food Chem.* **2007**, *100*, 1132–1136. [CrossRef]

21. Tautenhahn, R.; Cho, K.; Uritboonthai, W.; Zhu, Z.; Patti, G.J.; Siuzdak, G. An accelerated workflow for untargeted metabolomics using the METLIN database. *Nat. Biotechnol.* **2012**, *30*, 826–828. [CrossRef] [PubMed]

22. Sawada, Y.; Nakabayashi, R.; Yamada, Y.; Suzuki, M.; Sato, M.; Sakata, A.; Akiyama, K.; Sakurai, T.; Matsuda, F.; Aoki, T.; et al. RIKEN tandem mass spectral database (ReSpect) for phytochemicals: a plant-specific MS/MS-based data resource and database. *Phytochemistry* **2012**, *82*, 38–45. [CrossRef] [PubMed]

23. Jiang, X.; Liu, Y.; Wu, Y.; Tan, H.; Meng, F.; Wang, Y.S.; Li, M.; Zhao, L.; Liu, L.; Qian, Y.; et al. Analysis of accumulation patterns and preliminary study on the condensation mechanism of proanthocyanidins in the tea plant [*Camellia sinensis*]. *Sci. Rep.* **2015**, *5*. [CrossRef] [PubMed]

24. Ferreres, F.; Silva, B.M.; Andrade, P.B.; Seabra, R.M.; Ferreira, M.A. Approach to the study of C-glycosyl flavones by ion trap HPLC-PAD-ESI/MS/MS: application to seeds of quince (*Cydonia oblonga*). *Phytochem. Anal.* **2003**, *14*, 352–359. [CrossRef] [PubMed]

25. Scoparo, C.T.; de Souza, L.M.; Dartora, N.; Sassaki, G.L.; Gorin, P.A.; Iacomini, M. Analysis of *Camellia sinensis* green and black teas via ultra high performance liquid chromatography assisted by liquid-liquid partition and two-dimensional liquid chromatography (size exclusion x reversed phase). *J. Chromatogr. A* **2012**, *27*, 29–37. [CrossRef] [PubMed]

26. Zheng, Y.; Hu, X.; Zhai, Y.; Liu, J.; Wu, G.; Wu, L.; ShenTu, J. Pharmacokinetics and tissue distribution study of camellianin A and its major metabolite in rats by liquid chromatography with tandem mass spectrometry. *J. Chromatogr. B* **2015**, *997*, 200–209. [CrossRef] [PubMed]

27. Xie, C.; Xu, L.Z.; Luo, X.Z.; Zhong, Z.; Yang, S.L. Flavonol glycosides from *Lysimachia capillipes*. *J. Asian Nat. Prod. Res.* **2002**, *4*, 17–23. [CrossRef] [PubMed]

28. Hanhineva, K.; Rogachev, I.; Kokko, H.; Mintz-Oron, S.; Venger, I.; Karenlampi, S.; Aharoni, A. Non-targeted analysis of spatial metabolite composition in strawberry (*Fragariaxananassa*) flowers. *Phytochemistry* **2008**, *69*, 2463–2481. [CrossRef] [PubMed]

29. Iijima, T.; Mohri, Y.; Hattori, Y.; Kashima, A.; Kamo, T.; Hirota, M.; Kiyota, H.; Makabe, H. Synthesis of (−)-epicatechin 3-(3-*O*-methylgallate) and (+)-catechin 3-(3-*O*-methylgallate), and their anti-inflammatory activity. *Chem. Biodivers.* **2009**, *6*, 520–526. [CrossRef] [PubMed]

30. Sano, M.; Suzuki, M.; Miyase, T.; Yoshino, K.; Maeda-Yamamoto, M. Novel antiallergic catechin derivatives isolated from oolong tea. *J. Agric. Food Chem.* **1999**, *47*, 1906–1910. [CrossRef] [PubMed]

31. Lv, H.-P.; Yang, T.; Ma, C.-Y.; Wang, C.-P.; Shi, J.; Zhang, Y.; Peng, Q.-H.; Tan, J.-F.; Guo, L.; Lin, Z. Analysis of naturally occurring 3"-Methyl-epigallocatechin gallate in 71 major tea cultivars grown in China and its processing characteristics. *J. Funct. Foods* **2014**, *7*, 727–736. [CrossRef]

32. Hashimoto, F.; Nonaka, G.; Nishioka, I. Tannins and related-compounds. XC. 8-C-ascorbyl(-)-epigallocatechin 3-*O*-gallate and novel dimeric flavan-3-ols, oolonghomobisflavans A and B, from oolong tea. *Chem. Pharm. Bull.* **1989**, *37*, 3255–3263. [CrossRef]

33. Hashimoto, F.; Kashiwada, Y.; Nonaka, G.; Nishioka, I.; Nohara, T.; Cosentino, L.M.; Lee, K.H. Evaluation of tea polyphenols as anti-HIV agents. *Bioorg. Med. Chem. Lett.* **1996**, *6*, 695–700. [CrossRef]

34. Nakai, M.; Fukui, Y.; Asami, S.; Toyoda-Ono, Y.; Iwashita, T.; Shibata, H.; Mitsunaga, T.; Hashimoto, F.; Kiso, Y. Inhibitory effects of oolong tea polyphenols on pancreatic lipase in vitro. *J. Agric. Food Chem.* **2005**, *53*, 4593–4598. [CrossRef] [PubMed]

35. Scharbert, S.; Hofmann, T. Molecular definition of black tea taste by means of quantitative studies, taste reconstitution, and omission experiments. *J. Agric. Food Chem.* **2005**, *53*, 5377–5384. [CrossRef] [PubMed]

36. Engelhardt, U.H. Chemistry of Tea. In *Reference Module in Chemistry, Molecular Sciences and Chemical Engineering*; Elsevier: Braunschweig, Germany, 2013.

37. Gullon, B.; Lu-Chau, T.A.; Moreira, M.T.; Lema, J.M.; Eibes, G. Rutin: A review on extraction, identification and purification methods, biological activities and approaches to enhance its bioavailability. *Trends Food Sci. Technol.* **2017**, *67*, 220–235. [CrossRef]

38. Ekborg-Ott, K.H.; Taylor, A.; Armstrong, D.W. Varietal differences in the total and enantiomeric composition of theanine in tea. *J. Agric. Food Chem.* **1997**, *45*, 353–363. [CrossRef]

39. Deng, W.-W.; Ogita, S.; Ashihara, H. Biosynthesis of theanine (γ-ethylamino-L-glutamic acid) in seedlings of Camellia sinensis. *Phytochem. Lett.* **2008**, *1*, 115–119. [CrossRef]

40. Deng, W.W.; Ogita, S.; Ashihara, H. Distribution and biosynthesis of theanine in Theaceae plants. *Plant Physiol. Biochem.* **2010**, *48*, 70–72. [CrossRef] [PubMed]

41. Deng, W.-W.; Ashihara, H. Occurrence and de novo biosynthesis of caffeine and theanine in seedlings of tea (*Camellia sinensis*). *Nat. Prod. Commun.* **2015**, *10*, 703–706. [PubMed]

42. Wu, Q.; Chen, Z.; Sun, W.; Deng, T.; Chen, M. De novo sequencing of the leaf transcriptome reveals complex light-responsive regulatory networks in *Camellia sinensis* cv. Baijiguan. *Front. Plant Sci.* **2016**, *7*, 322. [CrossRef] [PubMed]

43. Li, C.F.; Xu, Y.X.; Ma, J.Q.; Jin, J.Q.; Huang, D.J.; Yao, M.Z.; Ma, C.L.; Chen, L. Biochemical and transcriptomic analyses reveal different metabolite biosynthesis profiles among three color and developmental stages in 'Anji Baicha' (*Camellia sinensis*). *BMC Plant Biol.* **2016**, *16*, 195. [CrossRef] [PubMed]

44. Kato, M.; Kitao, N.; Ishida, M.; Morimoto, H.; Irino, F.; Mizuno, K. Expression for caffeine biosynthesis and related enzymes in *Camellia sinensis*. *Z. Naturforsch. C* **2010**, *65*, 245–256. [CrossRef] [PubMed]

45. Xia, E.-H.; Zhang, H.-B.; Sheng, J.; Li, K.; Zhang, Q.-J.; Kim, C.; Zhang, Y.; Liu, Y.; Zhu, T.; Li, W.; et al. The tea tree genome provides insights into tea flavor and independent evolution of caffeine biosynthesis. *Mol. Plant* **2017**, *10*, 866–877. [CrossRef] [PubMed]

46. Wei, K.; Wang, L.-Y.; Zhou, J.; He, W.; Zeng, J.-M.; Jiang, Y.-W.; Cheng, H. Comparison of catechins and purine alkaloids in albino and normal green tea cultivars (*Camellia sinensis* L.) by HPLC. *Food Chem.* **2012**, *130*, 720–724. [CrossRef]

47. Li, C.F.; Yao, M.Z.; Ma, C.L.; Ma, J.Q.; Jin, J.Q.; Chen, L. Differential metabolic profiles during the albescent stages of 'Anji Baicha' (*Camellia sinensis*). *PLoS ONE* **2015**, *10*. [CrossRef] [PubMed]

48. Wishart, D.S.; Jewison, T.; Guo, A.C.; Wilson, M.; Knox, C.; Liu, Y.; Djoumbou, Y.; Mandal, R.; Aziat, F.; Dong, E.; et al. HMDB 3.0–The Human Metabolome Database in 2013. *Nucleic Acids Res.* **2013**, *41*, D801–D807. [CrossRef] [PubMed]

49. Horai, H.; Arita, M.; Kanaya, S.; Nihei, Y.; Ikeda, T.; Suwa, K.; Ojima, Y.; Tanaka, K.; Tanaka, S.; Aoshima, K.; et al. MassBank: A public repository for sharing mass spectral data for life sciences. *J. Mass Spectrom.* **2010**, *45*, 703–714. [CrossRef] [PubMed]

50. Afendi, F.M.; Okada, T.; Yamazaki, M.; Hirai-Morita, A.; Nakamura, Y.; Nakamura, K.; Ikeda, S.; Takahashi, H.; Altaf-Ul-Amin, M.; Darusman, L.K.; et al. KNApSAcK family databases: integrated metabolite-plant species databases for multifaceted plant research. *Plant Cell Physiol.* **2012**, *53*, e1. [CrossRef] [PubMed]

51. Jiang, H.Y.; Engelhardt, U.H.; Thrane, C.; Maiwald, B.; Stark, J. Determination of flavonol glycosides in green tea, oolong tea and black tea by UHPLC compared to HPLC. *Food Chem.* **2015**, *183*, 30–35. [CrossRef] [PubMed]

52. Chen, Y.-J.; Kuo, P.-C.; Yang, M.-L.; Li, F.-Y.; Tzen, J.T.C. Effects of baking and aging on the changes of phenolic and volatile compounds in the preparation of old Tieguanyin oolong teas. *Food Res. Int.* **2013**, *53*, 732–743. [CrossRef]

Sample Availability: Samples of the compounds are not available.

molecules

MDPI

Article

Characterization of Odors of Wood by Gas Chromatography-Olfactometry with Removal of Extractives as Attempt to Control Indoor Air Quality

Ru Liu [†], Chen Wang [†], Anmin Huang * and Bin Lv

Research Institute of Wood Industry, Chinese Academy of Forestry, Beijing 100091, China;
liuru@criwi.org.cn (R.L.); wonderfulmorning@163.com (C.W.); zj3@caf.ac.cn (B.L.)
* Correspondence: ham2003@caf.ac.cn; Tel.: +86-10-6288-9437
† These authors contributed equally to this work.

Received: 28 December 2017; Accepted: 15 January 2018; Published: 18 January 2018

Abstract: Indoor air quality problems are usually revealed by occupants' complaints. In this study, the odors of two types of hardwood species, namely, Cathy poplar (*Populus cathayana* Rehd.) and rubberwood (*Hevea brasiliensis*) were selected and extracted with ethanol-toluene for removal of extractives in an attempt to eliminate the odors. The odorous components of neat and extracted woods were identified by gas chromatography-mass spectrometry/olfactometry (GC-MS/O). The results showed that about 33 kinds of key volatile compounds (peak area above 0.2%) were detected from the GC-MS, and about 40 kinds of odorants were identified from GC-O. The components were concentrated between 15 and 33 min in GC-O, which was different from the concentration time in GC-MS. Lots of the odors identified from GC-O were unpleasant to humans, and variously described as stinky, burnt, leather, bug, herb, etc. These odors may originate from the thermos-oxidation of wood components. After extraction, the amounts and intensities of some odorants decreased, while some remained. However, the extraction process resulted in a benzene residue and led to increased benzene odor.

Keywords: wood; odor; volatile organic compounds; gas chromatography-olfactometry

---

1. Introduction

Odor has a direct effect on human behaviors and can significantly affect the quality of life [1]. It certainly plays an important role in human attractions, memories, and emotions, and can be described as pleasant, unpleasant, or indifferent [2]. Therefore, odor is one of many important factors for indoor air quality. Generally, the indoor air quality is strongly affected by volatile compounds emitted from materials such as furniture, carpet, textile, plant and humans [3–5]. Among these, wood has been widely used indoors for a long time. Although the odor of wood is often described as pleasant with positive associations, some people are hypersensitive to certain species like padauk (*Pterocarpus indicus* Willd.). Besides, people are concerned with several issues regarding odors, such as the relationship between unpleasant odors and health problems [6]. The odor of wood originates from the volatile organic compounds (VOCs) which are highly relevant to the extractives [7].

Wood extractives are rich in volatiles, typically with several hundred different constituent VOCs in individual species [8–10]. Although many of the VOCs are not odorous in nature, a broad overview of the volatiles hitherto detected in diverse woods offers insights into prospective main odor-active candidates that contribute to wood odor [11]. Therefore, numerous studies have been carried out for identification of the specific class of extractives [12–16]. However, not all VOCs contribute to odor because it depends on whether the concentration of the VOC is higher than the human threshold odor concentration [2]. Hitherto, there are few reports on the odor of natural woods. One particular study on the odor-active substances of wood was carried out by Culleré et al. [17] who investigated the odor of acacia, chestnut, cherry, ash and oak chips by gas chromatography-olfactometry (GC-O). Different odorants were detected, including phenolic compounds, as well as compounds arising from the degradation of wood carbohydrates and lipids. Similar studies were performed based on GC-O on toasted and non-toasted oak woods by Díaz-Maroto et al. [18], Poplar, Pine, and Basswood by Wang et al. [19], and cedar by Schreiner et al. [20].

GC-O consists of gas chromatography-mass spectrometry (GC-MS) coupling with an olfactometric detection. It is essential to identify compounds with odor because they are usually a minor set of eluting compounds, which depends on the human nose and shows high sensitivity and reproducibility [21]. It has been widely used in many areas ranging from daily necessities to chemical products [22–27]. Therefore, it can be an optimal instrument for odor testing. With the developing of costumed furniture, the wood products are heavily used indoors, and the odor annoyance arises due to the woods. Cathy poplar (*Populus cathayana* Rehd.) is a kind of fast growing tree and widely used for costumed furniture in China. Rubberwood (*Hevea brasiliensis*) is also been used for solid wood flooring and furniture while it contains lots of gum with strong smells. Both of the woods can affect the indoor air quality. Therefore, in this study, the two types of woods were selected for odor identification. Considering the high connection between extractives and odor, the woods were extracted with ethanol-toluene as an attempt to eliminate odor. The aim of this study was to investigate the odors of the two kinds of wood and whether the extraction can be a good method to eliminate the odor, which is important for the controlling of indoor air quality. The odorous components of neat and extracted woods were identified by GC-MS/O. Besides, the chemical components of the woods are analyzed.

2. Results and Discussion

2.1. Chemical Components

The results of the chemical components of the wood samples are listed in Table 1. According to the related standards, the deviations in the parentheses should not be larger than 0.04%. The analysis can be helpful for better understanding the contents of chemical components as well as extractives in the woods. It can be seen that the extractives of RW were higher than that of CP. The α-cellulose contents of the two woods were almost the same, while the content of holocelluloses of RW was higher than that of CP. By calculating the content differences between holocelluloses and α-cellulose, a rough estimate content value of hemicelluloses can be obtained [28]. The results indicated the RW owned the higher hemicelluloses content than CP. As for the lignin contents, RW was higher.

Table 1. Chemical components of woods.

Labels	Extractives (%) *	Holocelluloses (%)	α-Cellulose (%)	Lignin (%)
CP	2.87 (0.02)	61.33 (0.01)	44.24 (0.01)	23.26 (0.03)
RW	4.67 (0.03)	67.11 (0.02)	43.78 (0.02)	26.07 (0.01)

* The extractives refer to benzene-ethanol soluble extractives. The values in the parentheses are the deviations of four replicates.

2.2. Identification of VOCs

As regards the origin of the compounds detected by GC-O, it must be taken into account that only some of the VOCs extracted from the woods are originally present in this material in significant amounts. Therefore, it was very important to know the key VOCs extracted from the woods. Figure 1 shows the GC-MS chromatograms of neat and extracted wood samples. The key compounds with peak area above 0.2% are detected from the chromatograms and the results are listed in Table 2. In general, these compounds can be classified into alkanes, aldehydes, alcohols, aromatics, ketones, carboxylic acids, esters, and miscellaneous (silane derivatives). About 33 kinds of key VOCs were identified in both CP and RW. However, the types and contents of the VOCs were varied like the content of 2,4-di-*tert*-butylphenol (peak 26) was much higher in RW than that in CP. CP and RW had some common VOCs, such as ethanol (peak 1), benzene (peak 2), acetic acid (peak 3), and so on. Some are unique for certain wood. For example, hexyltrimethoxysilane (peak 15), 4-oxononanal (peak 16), and (*E*)-2-decenal (peak 17) were not identified in RW, while tetradecanal (peak 28), 2-methyl-hetadecane (peak 30), and dibutyl phthalate (peak 34) were not identified in CP. Among these compounds, the tetradecane (peak 14) showed the largest intensity, but the percentages were not higher than acetic acid (peak 3). Hexanoic acid (peak 8), tridecane (peak 11), pentadecane (peak 19), 1-dodecanol (peak 23) also had relative high intensities compared with other compounds.

The ethanol-benzene extraction process resulted in significant reduction in the contents of VOCs. For example, the percentages of acetic acid (peak 3) were 14.8% and 16.3% in neat CP and RW, respectively. After extraction, the values both decreased to 1%. Besides, the contents of some compounds in CP like hexanal (peak 4), furfural (peak 5), 1,3-dichlorobenzene (peak 7), pentanoic acid (peak 6), heptanoic acid (peak 10), octanoic acid (peak 12), hexyltrimethoxysilane (peak 15), 4-oxononanal (peak 16), (*E*)-2-decenal (peak 17), *n*-octyltriethoxysilane (peak 21), and 1,2-benzene-dicarboxylic acid bis(2-methylpropyl) ester (peak 33) significantly decreased to lowest detecting limits. Therefore, there were only 20 kinds of key compounds identified after extraction. Similar results were found in RW, suggesting the removal of these compounds. Considering the classification, the contents of aldehydes, carboxylic acids, and miscellaneous considerably decreased. This should be explained by their non-polar characters and solubility in ethanol-benzene.

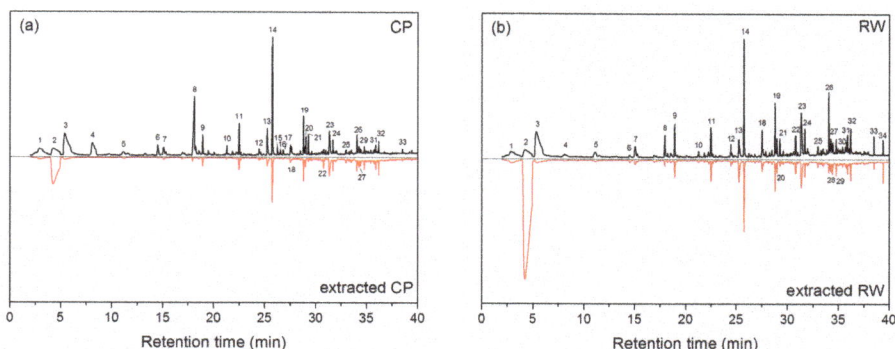

Figure 1. GC-MS chromatograms of neat and extracted wood samples. Peak numbers are listed in Table 2.

Pelaez-Samaniego et al. [29] had extracted wood fiber with hot water, and also mentioned the content of acetic acid and aromatics rapidly decreased. The amounts of alkanes, alcohols, ketones, and esters almost remained, which indicated the extraction method has little effect on the removal of these compounds. Aromatics of 1,3-dichloro-benzene (peak 7) also showed considerably reduced contents after extraction. It should be noted that the contents of pentadecane (peak 19), (*E*)-6,10-dimethyl-5,9-undecadien-2-one (peak 22), hexadecane (peak 24), 2,6,10-trimethylpentadecane (peak 25), 2,6,10,14-tetramethyl-pentadecane (peak 27), cedrol (peak 29), (4-octyldodecyl)-cyclopentane (peak 31), benzoic acid 2-ethylhexyl ester (peak 32) increased in CP after extraction, which might account for the removal of other compounds while the contents of these components were remained. Another remarkably increased intensity after extraction is found for benzene (peak 2), which should be associated with the solvent residue.

Table 2. Identified key compounds from GC-MS of neat and extracted wood samples.

Peak Number	Retention Time (min)	Compounds	Classification	Percentage (%) *			
				CP	Extracted CP	RW	Extracted RW
1	2.92	Ethanol	Alcohol	2.3 (0.2)	1.4	1.9 (0.1)	1.0
2	4.23	Benzene	Aromatic	5.6 (0.6)	32.8 (0.5)	5.9 (0.2)	68.5 (0.9)
3	5.41	Acetic acid	Carboxylic acid	14.8 (0.2)	1.0	16.3 (0.2)	1.0
4	8.10	Hexanal	Aldehyde	6.6 (0.2)	-	1.2	0.3
5	11.13	Furfural	Aldehyde	1.0	-	1.5 (0.1)	0.3
6	14.50	Pentanoic acid	Carboxylic acid	1.8	-	0.4	0.3
7	15.06	1,3-Dichlorobenzene	Aromatic	1.6	-	2.1 (0.2)	0.5
8	18.12	Hexanoic acid	Carboxylic acid	10.8 (0.4)	0.8	1.9 (0.1)	0.3
9	18.94	Nonanal	Aldehyde	2.2 (0.2)	1.3	2.8 (0.2)	1.0
10	21.32	Heptanoic acid	Carboxylic acid	0.9	-	0.4	0.2
11	22.50	Tridecane	Alkane	3.5 (0.2)	3.0 (0.2)	3.2 (0.1)	1.5
12	24.46	Octanoic acid	Carboxylic acid	0.5	-	0.9	0.2
13	25.26	(Z)-2-Decenal	Aldehyde	2.8 (0.2)	2.1 (0.1)	2.2 (0.1)	1.6 (0.1)
14	25.73	Tetradecane	Alkane	9.2 (0.5)	9.0 (0.5)	8.3 (0.1)	3.2 (0.1)
15	26.21	Hexyltrimethoxysilane	Miscellaneous	0.2	-	-	-
16	26.53	4-Oxononanal	Aldehyde	0.2	-	-	-
17	26.78	(E)-2-Decenal	Aldehyde	0.7	-	-	-
18	27.51	Nonanoic acid	Carboxylic acid	0.9	0.4	2.3 (0.2)	0.6
19	28.78	Pentadecane	Alkane	3.0 (0.2)	4.0 (0.3)	3.5 (0.2)	1.3
20	29.02	Dodecanal	Aldehyde	2.2 (0.1)	1.6 (0.1)	1.4	0.6
21	29.27	*n*-Octyltriethoxysilane	Miscellaneous	1.5	-	1.2	0.2
22	30.79	(E)-6,10-Dimethyl-5,9-undecadien-2-one	Ketone	0.7	1.2	1.4	0.7
23	31.35	1-Dodecanol	Alcohol	3.6 (0.1)	3.1 (0.1)	6.8 (0.3)	3.0 (0.2)
24	31.69	Hexadecane	Alkane	1.3	2.0	1.9	0.7
25	32.98	2,6,10-Trimethyl-pentadecane	Alkane	0.5	1.0	1.0	0.4
26	34.05	2,4-di-*tert*-Butylphenol	Alcohol	1.0	0.7	3.4 (0.1)	0.4
27	34.25	2,6,10,14-Tetramethyl-pentadecane	Alkane	1.1	2.3	1.3	0.8
28	34.45	Tetradecanal	Aldehyde	-	-	0.7	0.4
29	34.76	Cedrol	Alcohol	0.6	1.2	1.0	0.6
30	35.40	2-Methyl-hetadecane	Alkane	-	-	0.5	-
31	35.90	(4-Octyldodecyl)-cyclopentane	Alkane	0.5	0.9	1.1	0.4
32	36.20	Benzoic acid 2-ethylhexyl ester	Ester	0.7	1.2	1.3	0.6
33	38.51	1,2-Benzenedicarboxylic acid bis(2-methylpropyl) ester	Ester	0.2	-	0.7	0.2
34	39.40	Dibutyl phthalate	Ester	-	-	0.6	0.5

* The percentage was calculated based on the peak area. The values in the parentheses are the deviations of four replicates. Deviations lower than 0.5% are not listed in the Table.

2.3. Characterization of Odors

The odor images of neat and extracted wood samples tested by GC-O are shown in Figure 2. About 40 kinds of odors were identified. The time was concentrated from 15 to 36 min, which was different with the concentration time of VOCs in GC-MS, indicating that some VOCs did not contribute to the odors. Table 3 lists the odorants and odor descriptors of neat and extracted wood samples. These odors are described according to judgments of the panelists and then amended on the basis of the literatures [22–27]. From Table 2, it is clear that most odors are unpleasant. These odorants can be classified into alkanes, aldehydes, alcohols, aromatics, carboxylic acids, esters, and miscellaneous (silane derivatives). However, these compounds were not corresponding to the key VOCs. For example, some high concentration of alkanes like tridecane (peak 11), tetradecane (peak 14), pentadecane (peak 19), and hexadecane (peak 24) had small smells. Among the alkanes only 2,6,10-trimethylpentadecane (peak 27) offered an odor, which was described as gasoline and bronze-like. Félix et al. [30] investigated the odor of wood-plastic composites and also found little influence of alkanes on the odor of the composites. Some compounds lower than 0.2% caused significantly strong smells for both CP and RW, such as 2-nonenal (peak 15'), 1-methoxy-4-(2-propenyl)-benzene (peak 16'), 5,5,8-trimethyl-3,6,7-nonatrien-2-one (peak 17'), (E,E)-2,4-decadienal (peak 25') and 8-methyl-1-undecene (peak 32'). Between the retention times of 20 to 36 min, lots of odorants are emitted. However, these compounds are unpleasant to humans, except for hexadecanoic acid (21') at 26.3 min in CP and nonanoic acid (peak 24') at 27.4 min. These unpleasant compounds are usually described as stinky, burnt, leather, bug, herb, etc. Ezquerro et al. [31] mentioned that the unacceptable odor in packaging materials, such as the hexanoic acid is related to the thermo-oxidative degradation of cellulose. Therefore, the extraction method cannot remove it. At the later stage of testing from 37 to 40 min, some perfumed smells existed and the odors are described as milk, fruit, cake, and grass.

For neat CP, 34 kinds of odors were identified, where seven kinds were unique. Hexanal (bitter, green) was identified in both CP and RW. However, the percentage was only 1.2% in RW, which was 1/5 to that in CP. Therefore, the odor intensity was 0 in RW. Another six unique odors were not identified in GC-MS, while they all had strong smells. They were 2-furancarboxaldehyde (special, grain), octanal (fruit, soap), hexadecanoic acid (fruit, sour), 2-methylpropanoic acid, 3-hydroxy-2,4,4-trimethylpentyl ester (sour, bitter, herb), 1-tetradecanol (dirt, bronze), and 1-pentadecanol (bronze). The very strong odors are 12 kinds: octanal (fruit, soap), hexanoic acid (cheese, fishy), heptanoic acid (stinky, leather, bug), 2-nonenal (stinky, fat, iris), 1-methoxy-4-(2-propenyl)-benzene (bitter, herb, burnt), octanoic acid (stinky, fat, bug), (Z)-2-decenal (stinky, fat, herb), 4-oxononanal (herb), (E,E)-2,4-decadienal (animal, burnt), *n*-octyltriethoxysilane (burnt, bug), 8-methyl-1-undecene (paint, dirt), and 1-tetradecanol (dirt, bronze). These odors are all unpleasant. For neat RW, 32 kinds of odors were identified, where five kinds were unique. These compounds were 5-methyl-2-furancarboxaldehyde (chocolate, coconut), 2-octenal (bitter), α-methyl-α-2,5,7-octatrienylbenzenemethanol (bitter, nut), 3,7-dimethyl-2,6-octadien-1-ol, acetate (stinky, bug), and 2-undecenal (citrus, fishy, herb), which were trace in the amount of VOCs. The very strong odors of RW were heptanoic acid (stinky, leather, bug), octanoic acid (stinky, fat, bug), (Z)-2-decenal (stinky, fat, herb), (E,E)-2,4-decadienal (animal, burnt), 3,7-dimethyl-2,6-octadien-1-ol, acetate (stinky, bug), 2-undecenal (citrus, fishy, herb), *n*-octyl-triethoxysilane (burnt, bug), 8-methyl-1-undecene (paint, dirt), and 1,2-benzenedicarboxylic acid, bis(2-methylpropyl) ester (cake, grass), where 8/9 were unpleasant. By comparing the odors between CP and RW, the dominant odorants (intensity above 4) were 16 kinds: pentanoic acid (cheese, paint), hexanoic acid (stinky, leather, bug), nonanal (flower, fat), heptanoic acid (stinky, leather, bug), 2-nonenal (stinky, fat, iris), 1-methoxy-4-(2-propenyl)-benzene (bitter, herb, burnt), 5,5,8-trimethyl-3,6,7-nonatrien-2-one (stinky, chemical, bug), octanoic acid (stinky, fat, bug), (Z)-2-decenal (stinky, fat, herb), (E)-2-decenal (kerosene, herb), nonanoic acid (milk), (E,E)-2,4-decadienal (animal, burnt), dodecanal (mint, pungent), *n*-octyltriethoxysilane (burnt, bug), 8-methyl-1-undecene (paint, dirt), octadecanal (bronze, grease), and benzoic acid, 2-ethylhexyl ester (milk, fruit).

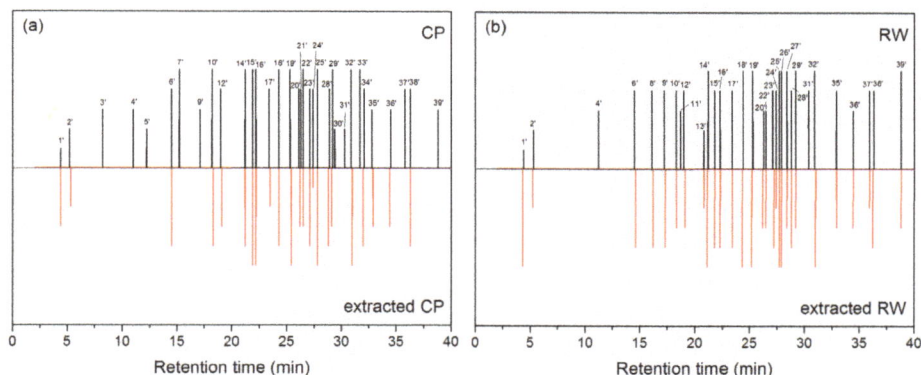

Figure 2. Odor images of neat and extracted wood samples tested by GC-O. Peak numbers are listed in Table 3.

Table 3. Odorants and odor descriptors of neat and extracted wood samples.

Peak Number in GC-O/MS	Retention Time (min)	Odorant	Odor Descriptor	Intensity			
				CP	Extracted CP	RW	Extracted RW
1'/2	4.4	Benzene	Gasoline, solvent	1	3	1	5
2'/3	5.2	Acetic acid	Irritant, sour	2	2	2	2
3'/4	8.2	Hexanal	Bitter, green	3	0	0	0
4'/5	11.1	Furfural	Nut	3	0	3	0
5'/-	12.2	2-Furan-carboxaldehyde	Special, grain	2	0	0	0
6'/6	14.5	Pentanoic acid	Cheese, paint	4	4	4	4
7'/-	15.2	Octanal	Fruit, soap	5	0	0	0
8'/-	16.1	5-Methyl-2-furan-carboxaldehyde	Chocolate, coconut	0	0	4	4
9'/-	17.1	Decamethyl-cyclopentasiloxane	Bread, nut	3	0	4	4
10'/8	18.2	Hexanoic acid	Cheese, fishy	5	4	4	3
11'/-	18.7	2-Octenal	Bitter	0	3	3	0
12'/9	19.0	Nonanal	Flower, fat	4	3	4	3
13'/-	20.8	α-Methyl-α-2,5,7-octatrienyl-benzenemethanol	Bitter, nut	0	0	2	2
14'/10	21.2	Heptanoic acid	Stink, leather, bug	5	4	5	5
15'/-	21.9	2-Nonenal	Stink, fat, iris	5	5	4	4
16'/-	22.2	1-Methoxy-4-(2-propenyl)-benzene	Bitter, herb, burnt	5	5	4	4
17'/-	23.4	5,5,8-Trimethyl-3,6,7-nonatrien-2-one	Stink, chemical, bug	4	2	4	4
18'/12	24.4	Octanoic acid	Stink, fat, bug	5	5	5	5
19'/13	25.3	(Z)-2-Decenal	Stink, fat, herb	5	5	5	5
20'/15	26.1	Hexyltrimethoxysilane	Irritant, stink, herb	4	3	3	3
21'/-	26.3	Hexadecanoic acid	Fruit, sour	4	0	0	0
22'/16	26.5	4-Oxononanal	Herb	4	3	3	3
23'/17	27.0	(E)-2-Decenal	Kerosene, herb	4	3	4	4
24'/18	27.4	Nonanoic acid	Milk	4	1	4	2
25'/-	27.8	(E,E)-2,4-Decadienal	Animal, burnt	5	5	5	5
26'/-	27.9	3,7-Dimethyl-2,6-octadien-1-ol, acetate	Stink, bug	0	0	5	5
27'/-	28.4	2-Undecenal	Citrus, fishy, herb	0	4	5	3
28'/20	28.8	Dodecanal	Mint, pungent	4	4	4	4
29'/21	29.2	n-Octyl-triethoxysilane	Burnt, bug	5	3	5	3
30'/-	29.4	2-Methylpropanoic acid, 3-hydroxy-2,4,4-trimethylpentyl ester	Sour, bitter, herb	2	0	0	0
31'/-	30.3	Decanoic acid	Herb	2	0	4	0
32'/-	31.0	8-Methyl-1-undecene	Paint, dirt	5	5	5	5
33'/-	31.7	1-Tetradecanol	Dirt, bronze	5	0	0	0
34'/-	32.1	1-Pentadecanol	Bronze	4	4	0	0
35'/25	32.8	2,6,10-Trimethyl-pentadecane	Gasoline, bronze	3	3	4	3
36'/29	34.6	Cedrol	Fat, wood	3	3	3	3
37'/-	35.9	Octadecanal	Bronze, grease	4	0	4	2
38'/32	36.3	Benzoic acid, 2-ethylhexyl ester	Milk, fruit	4	3	4	4
39'/33	38.8	1,2-Benzene dicarboxylic acid, bis(2-methylpropyl) ester	Cake, grass	3	0	5	3

The unpleasant odors are highlighted in character shading. The intensities were determined on agreements of four judges.

After ethanol-benzene extraction, both the amounts and intensities of odors decreased in both CP and RW, suggesting the extraction process had removed some kinds of odorants, such as hexanoic acid (cheese, fishy), nonanal (flower, fat), *n*-octyltriethoxysilane (burnt, bug), decanoic acid (herb), octadecanal (bronze, grease), and 1,2-benzenedicarboxylic acid, bis(2-methylpropyl) ester (cake, grass) and so on. However, some kinds odors remained their intensities, such as acetic acid (irritant, sour), pentanoic acid (cheese, paint), 2-nonenal (stinky, fat, iris), 1-methoxy-4-(2-propenyl)-benzene (bitter, herb, burnt), (Z)-2-decenal (stinky, fat, herb), (E)-2-decenal (kerosene, herb), (E,E)-2,4-decadienal (animal, burnt), dodecanal (mint, pungent), 8-methyl-1-undecene (paint, dirt), cedrol (fat, wood), benzoic acid, 2-ethylhexyl ester (milk, fruit) and so on though the amount percentages was reduced in GC-MS. In addition, consistent with the results of GC-MS, the residue of benzene caused increase of odor, which was described as gasoline and solvent.

As mentioned above, the uncomfortable odors originated from the thermos-oxidation. Maybe the temperature had a tremendous influence on these odorants. Further study can focus on the effects of temperature on the odor release or try to find another effective odor elimination process.

3. Materials and Methods

3.1. Materials

The two kinds of wood were harvested from forests and then bucked into lumbers. The lumbers were air dried for at least 3 months to eliminate excessive water. Both the kinds of wood were provided by at least three providers from a single origin and analyzed to ensure their uniformity. The average growth ring widths, densities and origins of place are presented in Table 4. The sapwood was chosen without bark and visible defects such as knots, decay, and so on. Figure 3 shows tangential sections of the two kinds of wood. The wood samples were ground into fibers of 40–60 mesh consistent with the particle sizes about 50 μm.

Table 4. Average growth ring widths, densities, and origins of place of the woods.

Label	Wood	Scientific Name	Average Growth Ring Width (cm)	Density (g/cm^3) *	Origin of Place
CP	Cathy poplar	*Populus cathayana* Rehd.	0.6	0.49–0.52	Hebei, China
RW	Rubberwood	*Hevea brasiliensis*	0.2	0.64–0.67	Hainan, China

* The density values of six replicates were tested at moisture contents about 12%.

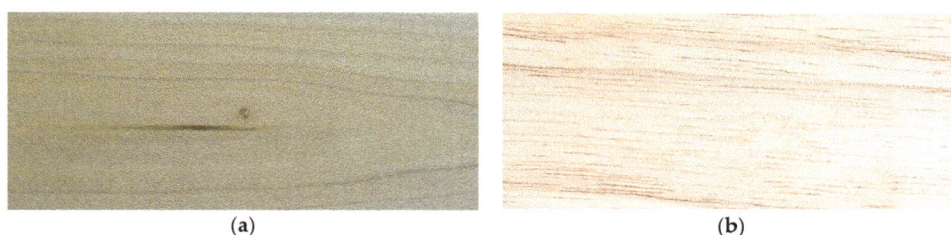

Figure 3. Images of tangential sections of the two kinds of wood. (**a**) CP; (**b**) RW.

3.2. Analyisis of Chemical Components

The wood fiber was extracted in a Soxhlet extractor with a 1:2 mixture of ethanol and toluene (*v*/*v*) for 6 h, followed by a second extraction with ethanol for 4 h to remove extractives. The extracted wood fiber was dried in an oven at $103 \pm 2\,°C$ to reach a constant weight. The content of extractives was calculated. The chemical components of the natural fibers for holocelluloses, α-cellulose, and lignin contents were performed according to chlorite method, TAPPI 203 cm-09, and TAPPI T 222 om-11, respectively [32].

3.3. Odor Characterization

The odor characterization experiment was tested by a GC-MS (QP-2010, Shimadzu, Shimane, Japan) combined with an olfactory port (OP 275, GL Sciences, Shimane, Japan) connected by a flow splitter to the column exit. Separation was achieved using a DB-WAX column (30 m × 0.25 mm × 0.25 μm, J&W Scientific Inc., Folsom, CA, USA) with a temperature program at 40 °C (3 min) to 230 °C (5 min) at 6 °C/min with helium as carrier gas (1.8 mL/min). The mass spectrometer was operated in electron impact mode (70 eV) and the masses were scanned over a range of 35–350 m/z. The transmission line temperature was 250 °C and the ion source temperature was 200 °C.

About 3 g wood fiber sample was added into a 15 mL headspace bottle and conditioned at 60 °C in a water bath for 40 min and the extraction occurred at the same temperature for 40 min. Desorption was carried out in the GC injection port.

Sensory assessments were carried out by a panel of four judges (two females and two males, 26 years old on average) from the Laboratory of Brewing Microbiology and Applied Enzymology at Jiangnan University. The panelists were trained for 3 months in GC-O using at least 30 odor-active reference compounds in a concentration 10 times above their odor thresholds in air. Sniffing time was approximately 40 min. During a GC run described above, the nose of a panelist was placed close to the sniffing port, responded to the aroma intensity of the stimulus, and recorded the aroma descriptor and intensity value as well as retention time. A six-point scale ranging from 0 to 5 was used for intensity judgment: 0 = none, 1 = very weak, 2 = weak, 3 = moderate, 4 = strong, and 5 = very strong.

The odorants were identified by comparing the MS spectra to the National Institute of Standards and Technology (NIST) library (https://www.nist.gov). The primary odor compounds were identified by mass spectrometry, retention time, and odor characterization.

4. Conclusions

About 40 kinds of odors and their relevant components of woods were identified by GC-O. These odorants were classified into alkanes, aldehydes, alcohols, aromatics, carboxylic acids, esters, and miscellaneous (silane derivatives). By comparing the time and intensity of odorants occurring in GC-O and that of VOCs occurring in GC-MS, no good relationship was found, which means that some high contents of VOCs did not contribute to the odors. The removal of extractives showed mostly reduced quantities and intensities of odors. However, some odors were unaffected. An increase of benzene residue was also observed. Many of the odors identified in woods were unpleasant to humans. The retention time was concentrated between 25 and 36 min described as stinky, burnt, leather, bug, herb, etc., which mainly originated from the thermos-oxidation of wood components. This study was helpful to understand the indoor air quality caused by wood. To mitigate the odor problem, a further study can focus on the effects of temperature on the odor release or try to find another effective odor elimination process.

Acknowledgments: This study was financially supported by the National Key Research and Development Program of China (No. 2016YFD0600706).

Author Contributions: Ru Liu and Chen Wang performed the experiments, analyzed the data, and wrote the manuscript; Anmin Huang conceived and designed the experiments; Bin Lv contributed to the materials and analysis tools.

Conflicts of Interest: The authors declare no conflict of interest.

References

1.	Kasper, P.L.; Mannebeck, D.; Oxbøl, A.; Nygaard, J.V.; Hansen, M.J.; Feilberg, A. Effects of dilution systems in olfactometry on the recovery of typical livestock odorants determined by PTR-MS. *Sensors* **2017**, *17*, 1859. [CrossRef] [PubMed]

2. Brattoli, M.; Cisternino, E.; Dambruoso, P.R.; de Gennaro, G.; Giungato, P.; Mazzone, A.; Palmisani, J.; Tutino, M. Gas chromatography analysis with olfactometric detection (GC-O) as a useful methodology for chemical characterization of odorous compounds. *Sensors* **2013**, *13*, 16759–16800. [CrossRef] [PubMed]

3. Peng, C.; Lan, C.; Wu, T. Investigation of indoor chemical pollutants and perceived odor in an area with complaints of unpleasant odors. *Build. Environ.* **2009**, *44*, 2106–2113. [CrossRef]

4. Bitter, F.; Müller, B.; Müller, D. Estimation of odour intensity of indoor air pollutants from building materials with a multi-gas sensor system. *Build. Environ.* **2010**, *45*, 197–204. [CrossRef]

5. Liu, W.; Zhang, Y.; Yao, Y.; Li, J. Indoor decorating and refurbishing materials and furniture volatile organic compounds emission labeling systems: A review. *Chin. Sci. Bull.* **2012**, *57*, 2533–2543. [CrossRef]

6. Rosenkranz, H.S.; Cunningham, A.R. Environmental odors and health hazards. *Sci. Total Environ.* **2003**, *313*, 15–24. [CrossRef]

7. Ka, M.H.; Choi, E.H.; Chun, H.S.; Lee, K.G. Antioxidative activity of volatile extracts isolated from *Angelica tenuissimae* roots, peppermint leaves, pine needles, and sweet flag leaves. *J. Agric. Food Chem.* **2005**, *53*, 4124–4129. [CrossRef] [PubMed]

8. Uçar, G.; Balaban, M. Volatile wood extractives of black pine (*Pinus nigra Arnold*) grown in Eastern Thrace. *Eur. J. Wood Prod.* **2001**, *59*, 301–305. [CrossRef]

9. Xu, K.; Feng, J.; Zhong, T.; Zheng, Z.; Chen, T. Effects of volatile chemical components of wood species on mould growth susceptibility and termite attack resistance of wood plastic composites. *Int. Biodeterior. Biodegrad.* **2015**, *100*, 106–115. [CrossRef]

10. Vainio-Kaila, T.; Hanninen, T.; Kyyhkynen, A.; Ohlmeyer, M.; Siitonen, A.; Rautkari, L. Effect of volatile organic compounds from *Pinus sylvestris* and *Picea abies* on *Staphylococcus aureus*, *Escherichia coli*, *Streptococcus pneumoniae* and *Salmonella enterica* serovar Typhimurium. *Holzforschung* **2017**, *71*, 905–912. [CrossRef]

11. Schreiner, L.; Beauchamp, J.; Buettner, A. Characterisation of odorants in wood and related products: Strategies, methodologies, and achievements. In Proceedings of the International Conference "Wood Science and Engineering in the Third Millennium"—ICWSE 2017, Brasov, Romania, 2–4 November 2017.

12. Risholm-Sundman, M.; Lundgren, M.; Vestin, E.; Herder, P. Emissions of acetic acid and other volatile organic compounds from different species of solid wood. *Holz Roh Werkst.* **1998**, *56*, 125–129. [CrossRef]

13. Manninen, A.M.; Pasanen, P.; Holopainen, J.K. Comparing the VOC emissions between air-dried and heat-treated Scots pine wood. *Atmos. Environ.* **2002**, *36*, 1763–1768. [CrossRef]

14. Ekeberg, D.; Flæte, P.O.; Eikenes, M.; Fongen, M.; Naess-Andresen, C.F. Qualitative and quantitative determination of extractives in heartwood of Scots pine (*Pinussylvestris* L.) by gas chromatography. *J. Chromatogr. A* **2006**, *1109*, 267–272. [CrossRef] [PubMed]

15. Widhalm, B.; Ters, T.; Srebotnik, E.; Rieder-Gradinger, C. Reduction of aldehydes and terpenes within pine wood by microbial activity. *Holzforschung* **2016**, *70*, 895–900. [CrossRef]

16. Sassoli, M.; Taiti, C.; Nissim, W.G.; Costa, C.; Mancuso, S.; Menesatti, P.; Fioravanti, M. Characterization of VOC emission profile of different wood species during moisture cycles. *iForest* **2017**, *10*, 576–584. [CrossRef]

17. Culleré, L.; de Simón, B.F.; Cadahía, E.; Ferreira, V.; Hernández-Orte, P.; Cacho, J. Characterization by gas chromatography–olfactometry of the most odor-active compounds in extracts prepared from acacia, chestnut, cherry, ash and oak woods. *LWT-Food Sci. Technol.* **2013**, *53*, 240–248. [CrossRef]

18. Díaz-Maroto, M.C.; Guchu, E.; Castro-Vázquez, L.; de Torres, C.; Pérez-Coello, M.S. Aroma-active compounds of American, French, Hungarian and Russian oak woods, studied by GC-MS and GC-O. *Flavour Fragr. J.* **2008**, *23*, 93–98. [CrossRef]

19. Wang, Q.; Shao, Y.; Cao, T.; Shen, J. Identification of key odor compounds from three kinds of wood species. In Proceedings of the 2017 International Conference on Environmental Science and Sustainable Energy (ESSE2017), Suzhou, China, 23–25 June 2017.

20. Schreiner, L.; Loos, H.M.; Buettner, A. Identification of odorants in wood of *Calocedrus decurrens* (Torr.) Florin by aroma extract dilution analysis and two-dimensional gas chromatography–mass spectrometry/olfactometry. *Anal. Bioanal. Chem.* **2017**, *409*, 3719–3729. [CrossRef] [PubMed]

21. Ai, N.; Liu, H.; Wang, J.; Zhang, X.; Zhang, H.; Chen, H.; Huang, M.; Liu, Y.; Zheng, F.; Sun, B. Triple-channel comparative analysis of volatile flavour composition in raw whole and skim milk via electronic nose, GC-MS and GC-O. *Anal. Methods* **2015**, *7*, 4278–4284. [CrossRef]

22. Pripdeevech, P.; Khummueng, W.; Park, S.K. Identification of odor-active components of agarwood essential oils from Thailand by solid phase microextraction-GC/MS and GC-O. *J. Essent. Oil Res.* **2011**, *23*, 46–53. [CrossRef]

23. Gao, W.; Fan, W.; Xu, Y. Characterization of the key odorants in light aroma type Chinese liquor by gas chromatography-olfactometry, quantitative measurements, aroma recombination, and omission studies. *J. Agric. Food Chem.* **2014**, *62*, 5796–5804. [CrossRef] [PubMed]

24. Soso, S.B.; Koziel, J.A. Analysis of odorants in marking fluid of siberian tiger (*Panthera tigris altaica*) using simultaneous sensory and chemical analysis with headspace solid-phase microextraction and multidimensional gas chromatography-mass spectrometry-olfactometry. *Molecules* **2016**, *21*, 834. [CrossRef] [PubMed]

25. Xiao, Z.; Fan, B.; Niu, Y.; Wu, M.; Liu, J.; Ma, S. Characterization of odor-active compounds of various *Chrysanthemum* essential oils by gas chromatography–olfactometry, gas chromatography–mass spectrometry and their correlation with sensory attributes. *J. Chromatogr. B* **2016**, *1009–1010*, 152–162. [CrossRef] [PubMed]

26. Denk, P.; Buettner, A. Sensory characterization and identification of odorous constituents in acrylic adhesives. *Int. J. Adhes. Adhes.* **2017**, *78*, 182–188. [CrossRef]

27. Elsharif, S.A.; Buettner, A. Influence of the chemical structure on the odor characters of β-citronellol and its oxygenated derivatives. *Food Chem.* **2017**, *232*, 704–711. [CrossRef] [PubMed]

28. Wang, W.; Cao, J.; Cui, F.; Wang, X. Effect of pH on chemical components and mechanical properties of thermally modified wood. *Wood Fiber Sci.* **2012**, *44*, 46–53.

29. Pelaez-Samaniego, M.R.; Yadama, V.; Lowell, E.; Amidon, T.E.; Chaffee, T.L. Hot water extracted wood fiber for production of wood plastic composites (WPCs). *Holzforschung* **2012**, *67*, 193–200. [CrossRef]

30. Félix, J.S.; Domeño, C.; Nerín, C. Characterization of wood plastic composites made from landfill-derived plastic and sawdust: Volatile compounds and olfactometric analysis. *Waste Manag.* **2013**, *33*, 645–655. [CrossRef] [PubMed]

31. Ezquerro, O.; Pons, B.; Tena, M.T. Development of a headspace solid-phase microextraction–gas chromatography–mass spectrometry method for the identification of odour-causing volatile compounds in packaging materials. *J. Chromatogr. A* **2002**, *963*, 381–392. [CrossRef]

32. Ou, R.; Xie, Y.; Wolcott, M.P.; Sui, S.; Wang, Q. Morphology, mechanical properties, and dimensional stability of wood particle/high density polyethylene composites: Effect of removal of wood cell wall composition. *Mater. Des.* **2014**, *58*, 339–345. [CrossRef]

Sample Availability: All the Samples of the compounds are available from the authors.

molecules

MDPI

Article

Determination of Branched-Chain Keto Acids in Serum and Muscles Using High Performance Liquid Chromatography-Quadrupole Time-of-Flight Mass Spectrometry

You Zhang, Bingjie Yin, Runxian Li and Pingli He *

State Key Laboratory of Animal Nutrition, College of Animal Science and Technology,
China Agricultural University, Beijing 100193, China; you_93@cau.edu.cn (Y.Z.);
youda_93@163.com (B.Y.); 18883870725@163.com (R.L.)
* Correspondence: hepingli@cau.edu.cn; Tel.: +86-10-6273-3688

Received: 4 December 2017; Accepted: 8 January 2018; Published: 11 January 2018

Abstract: Branched-chain keto acids (BCKAs) are derivatives from the first step in the metabolism of branched-chain amino acids (BCAAs) and can provide important information on animal health and disease. Here, a simple, reliable and effective method was developed for the determination of three BCKAs (α-ketoisocaproate, α-keto-β-methylvalerate and α-ketoisovalerate) in serum and muscle samples using high performance liquid chromatography-quadrupole time-of-flight mass spectrometry (HPLC-Q-TOF/MS). The samples were extracted using methanol and separated on a 1.8 μm Eclipse Plus C18 column within 10 min. The mobile phase was 10 mmol L^{-1} ammonium acetate aqueous solution and acetonitrile. The results showed that recoveries for the three BCKAs ranged from 78.4% to 114.3% with relative standard deviation (RSD) less than 9.7%. The limit of quantitation (LOQ) were 0.06~0.23 μmol L^{-1} and 0.09~0.27 nmol g^{-1} for serum and muscle samples, respectively. The proposed method can be applied to the determination of three BCKAs in animal serum and muscle samples.

Keywords: branched-chain keto acids; serum; muscle; HPLC-Q-TOF/MS

1. Introduction

The branched-chain amino acids (BCAAs) leucine, isoleucine, and valine are three essential amino acids that cannot be synthesized by animals and must be obtained from foods [1–3]. As a result of the stimulatory effect of BCAAs on protein synthesis and direct effects on glucose transport and insulin secretion, investigations into the metabolism of BCAAs have become a popular research area in recent years, with particular emphasis on the branched-chain keto acids (BCKAs) [4–10]. The BCKAs are derived from the initial step in the catabolism of BCAAs via BCAA amino transferase, which transfers the amino group of BCAAs to α-ketoglutarate to form the BCKAs (including α-ketoisocaproate (KIC, ketoleucine), α-keto-β-methylvalerate (KMV, ketoisoleucine) and α-ketoisovalerate (KIV, ketovaline)) in the muscle [11–13]. After BCAAs are catalyzed by amino transferase, BCKAs can be irreversibly oxidized to isovaleryl-CoA, isobutyryl-CoA and 2-methylbutyryl-CoA via the enzyme branched-chainketo acid dehydrogenase (BCKDH) [14–16]. The effect of BCKAs on stimulating protein synthesis has been reported [17,18]. And when the BCKDH complex is mutated and defective, the BCKAs will accumulate in tissues resulting in development of the maple syrup urine disease (MSUD) in animals [19]. Since BCKAs have such an important effect on the metabolic and nutritional state of animals, accurate determination of BCKAs in biological samples is highly valuable.

Several analytical methods have been applied to measure BCKAs in biological fluids such as plasma and urine. These include gas-liquid chromatography (GC) [20], gas chromatography-mass spectrometry (GC-MS) [21] and high performance liquid chromatography (HPLC) with fluoresce detection [22–24]. However, these techniques are either time consuming or lack sensitivity such that certain BCKAs are not detectable in tissue. Ultra-fast liquid chromatography-mass spectrometry (UFLC-MS) methods are quite sensitive with shorter analysis times and have become popular for the detection of keto acids. Nevertheless, this method relies on the similar derivatization of keto acids with *o*-phenylenediamine (OPD), which results in a tedious preparation procedure and detrimental exposure to toxic reagents [25].

In this study, we developed a simple and accurate method to determine the three BCKAs in serum and muscle samples using high performance liquid chromatography-quadrupole time-of-flight mass spectrometry (HPLC-Q-TOF/MS). Samples are extracted by methanol which can remove protein efficiently and the extracted solution was concentrated to improve the detection sensitivity. The high resolution mass spectrometry (HRMS) with high selectivity and resolution assured the accuracy of the results. The wide range of linearity with reproducibility over a broad range of concentrations makes it a convenient way to determinate BCKAs in serum and muscle samples. Compared with previous research [25], it could been seen that when multiple samples need to be analyzed, the method we described can considerably simplify pretreatment procedures.

2. Results and Discussion

2.1. Optimization of HPLC Q-TOF/MS Conditions

Chromatographic conditions such as analytical column, mobile phase composition and injection volume were optimized to obtain better performance on separating target compounds in standard solution and sample matrix. For example, the Waters BEH C_{18} (2.1 × 100 mm, 1.7 μm) column and the Agilent Eclipse Plus C_{18} (2.1 × 100 mm, 1.8 μm) column were compared to separate the three BCKAs. From Figure 1 we can see that, the Agilent Eclipse Plus (2.1 × 100 mm, 1.8 μm) C_{18} column was more suitable for separating the KMV and KIC in a 2.5 μmol/L standard solution for each compound, with better overall peak shapes and a better separation displayed by the resolution (>1.5 for KMV-KIC). Therefore, the Eclipse Plus C_{18} was chosen as the analytical column. In order to obtain an excellent separation and ionization effect, different concentrations (5, 10 and 20 mmol/L) of ammonium acetate were tested. A previous study had showed that the 10 mmol/L aqueous ammonium acetate improved the separation of the three BCKAs better than the lower concentration [26]. The 10 and 20 mmol/L ammonium acetate aqueous had similar performance in the separation and peak response of the three BCKAs. Considering the high concentration of ammonium acetate damaged the chromatographic column, the 10 mmol/L concentration of ammonium acetate was selected as the final aqueous phase. At the same time, in order to sufficiently elute the samples, we chose a long elution procedure with gentle change of gradient for this technique.

The optimization of mass spectrometry parameters included capillary and fragmentor voltages, nebulizer pressure and collision energy. Among these, fragmentor voltages had a dominant influence on the sensitivity of the compounds detection. By comparing the intensity of three BCKAs mixed standard solutions in different fragmentor voltages from 70 to 150 V, we determined that the three BCKAs had the maximum peak response when the fragmentor voltages were set at 100 V (Figure 2).

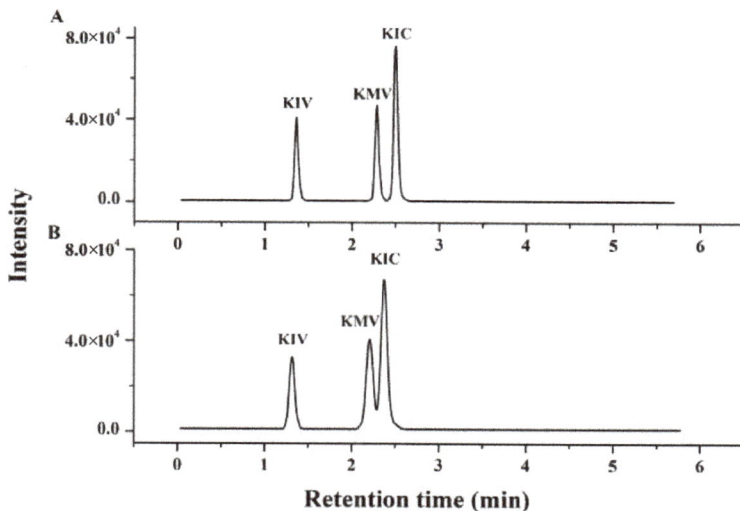

Figure 1. Influence of different columns on the HPLC performance of the branched-chain keto acids (KIV: α-ketoisovalerate; KMV: α-keto-β-methylvalerate; KIC: α-ketoisocaproate) mixed standard solution (2.5 μmol/L for each compound): (**A**) Agilent Eclipse Plus C_{18} column; (**B**) Waters BEH C_{18} column.

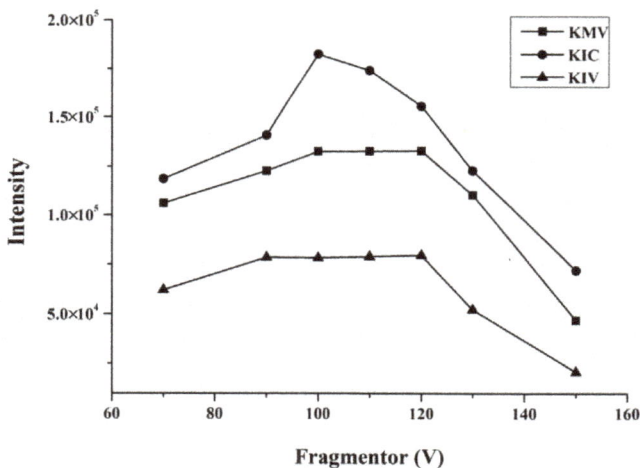

Figure 2. Intensity of 3 branched-chain keto acids (KIV: α-ketoisovalerate; KMV: α-keto-β-methylvalerate; KIC: α-ketoisocaproate) mixed standard solution (5 μmol/L for each compound) in different fragmentor (V).

The collision energy was tested at 10, 15, 20, and 25 V, and it showed that 20 V was most appropriate for the three BCKAs, so the collision energy was set as 20 V. Under these conditions, the mass spectra of three BCKAs standards in HPLC-Q-TOF/MS were presented in Figure 3, including full scan mass spectra (Figure 3A–C) and MS/MS spectrums (Figure 3D–F). The retention time of three BCKAs and the fragment ions *m/z* were presented in Table 1, which were used for characteristic determination.

Table 1. Chemical structure and tandem mass spectrometry parameters of the branched-chain keto acids.

Compounds	Molecular Formula	Structure	Retention Time (min)	Precursor Ion (*m/z*)	Fragment Ion (*m/z*)
α-ketoisocaproate	$C_6H_{10}O_3$		2.48	129.0542 (3.17) [a]	59.0158
α-keto-β-methylvalerate	$C_6H_{10}O_3$		2.25	129.0542 (3.17)	68.9959
α-ketoisovalerate	$C_5H_8O_3$		1.41	115.0391 (2.34)	59.0126

[a] Figures in brackets represent the error of the precursor ion theoretical *m/z* and measured value (ppm).

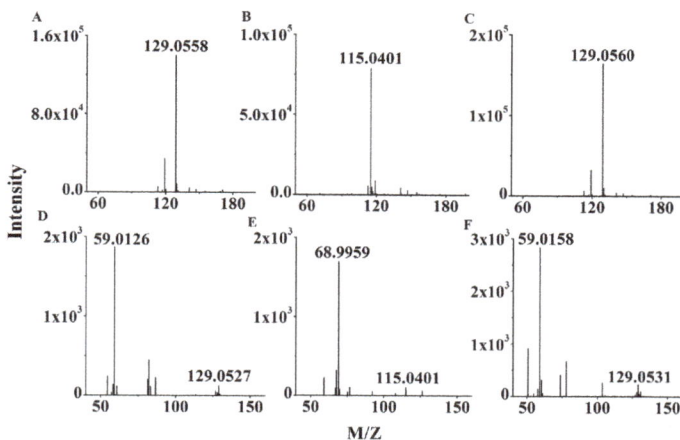

Figure 3. The mass spectra of the branched-chain keto acids standards using HPLC-Q-TOF/MS. (**A**) Full scan mass spectra of α-keto-β-methylvalerate (KMV); (**B**) Full scan mass spectrum of α-ketoisovalerate (KIV); (**C**) Full scan mass spectrum of α-ketoisovalerate (KIC); (**D**) MS/MS spectrum of KMV; (**E**) MS/MS spectrum of KIV; (**F**) MS/MS spectrum of KIC.

2.2. Optimization of Pretreatment Conditions

To accurately determine the three BCKAs in serum and muscle samples, an efficient extraction procedure is necessary. In this study, three extraction solvents including methanol, acetonitrile and methanol-acetonitrile (50:50, v/v) were compared. The results showed that most of the interfering proteins present in the two matrices were precipitated in the presence of methanol and can be efficiently removed by the subsequent refrigerated centrifugation procedure [26]. Thus, methanol was chosen as the extraction solvent. Figure 4 suggests that the recoveries were relatively lower when the samples were extracted without ultrasonic treatment. As a result, at least a 10-min ultrasonic treatment was required. In addition, drying of the extracted supernatant fluid and re-dissolution of the residues in a small amount of ultra-pure water removed organic solvent and concentrate samples, which improved the detected sensitivity to a large extent.

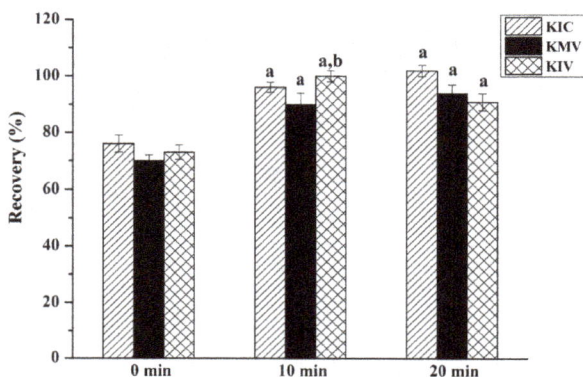

Figure 4. The influence of different ultrasonic time on the recoveries of three BCKAs (KIV: α-ketoisovalerate; KMV: α-keto-β-methylvalerate; KIC: α-ketoisocaproate). [a], [b] Means the same compounds different significantly ($p < 0.05$).

2.3. Method Validation

The sensitivity, accuracy and precision of an analysis method are usually validated using linearity, LOQ and recoveries of spiked samples. In this study, the HPLC-Q-TOF/MS-based method for the detection of three BCKAs showed satisfactory linearity within the concentration range of 0.1 to 100 μmol L^{-1} (R^2 > 0.998), which covered the range commonly observed in serum and muscle samples. The LOQ is defined as the concentration at which the signal-to-noise ratio is more than ten. Because the BCKAs are endogenous compounds, it is difficult to obtain serum and muscle samples free of them. Hence, the blank samples were surrogated by other artificial matrices. Often, phosphate buffered saline (PBS) is used for serum analyses because of its pH (7.4) and ionic strength. Bovine serum albumin (BSA) is frequently added to PBS to take the protein content of biological matrix into account [27]. In this study, the serum and muscle samples were surrogated by 2% BSA in PBS and 1 g/L of BSA, respectively. Table 2 showed that the LOQ of the three BCKAs ranged from 0.06 to 0.23 μmol L^{-1} and 0.09 to 0.27 nmol g^{-1} for serum and muscle samples, respectively.

Six replicates were tested for evaluation of spike recovery and relative standard deviation (RSD). The actual spiked concentration in the different matrices showed good consistency, and the calculated recoveries for three BCKAs when different concentrations of spiked solutions were used ranged from 78.4% to 114.3% with the RSD less than 9.7% (Table 2). The representative chromatograms of three BCKAs in no-spike, low-spike and high-spike serum and muscle samples are presented in Figure 5. The results showed that the BCKAs concentrations in the samples were all above the LOQ.

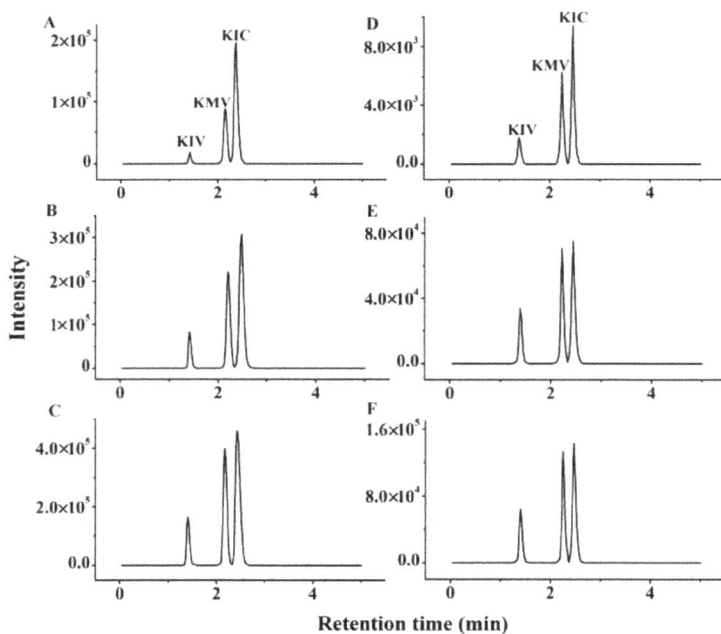

Figure 5. Representative chromatograms of BCKAs in different samples. (**A**) No-spike serum sample; (**B**) Low-spike serum sample; (**C**) High-spike serum sample; (**D**) No-spike muscle sample; (**E**) Low-spike muscle sample; (**F**) High-spike muscle sample.

Molecules **2018**, *23*, 147

Table 2. Recoveries (%) and limit of quantification (LOQ) for detection of the branched-chain keto acids in spiked samples using UPLC-Q-TOF/MS (*n* = 6).

Compounds	Recoveries (%)								LOQ	
	Serum (μmol/L)			Muscle (nmol/g)				Serum (μmol/L)	Muscle (nmol/g)
	10	20	50	1	2	5			
α-ketoisocaproate	114.3 (2.0)[a]	84.3 (4.0)	86.8 (1.9)	85.9 (6.9)	93.6 (7.9)	94.7 (1.6)		0.06	0.09
α-keto-β-methylvalerate	98.7 (7.6)	100.2 (4.8)	99.0 (2.9)	95.3 (5.4)	103.2 (9.7)	100.8 (0.6)		0.09	0.12
α-ketoisovalerate	89.2 (3.2)	85.1 (3.8)	83.1 (1.2)	82.7 (1.8)	78.4 (1.0)	80.0 (7.1)		0.23	0.27

[a] Figures in bracket represented relative standard deviation (%).

2.4. The Analysis of Authentic Sample

The proposed method was adopted for the determination of three BCKAs in authentic serum and muscle samples from pigs. Table 3 shows the concentrations ranges of three BCKAs in different samples, the reported variability is likely due to inherent variation among individual pigs and different levels of BCKAs in the diet. Nevertheless, the HPLC-Q-TOF/MS results were in good accordance with literature values, in which the concentration is about 20 nmol/L of KIV and KMV, 30 nmol/L of KIC [28–30]. Furthermore, we chose samples from pigs fed the 17% crude protein and 17% crude protein + BCAA diets which had significant differences in free AA concentrations of threonine, methionine, lysine and BCAA [9] to test the concentrations of three BCKAs in serum ($n = 12$). The results show that the addition of BCAA significantly increased the concentration of three BCKAs in precaval venous blood samples (Figure 6). The assay was able to detect differences in three BCKAs levels that reflected the differences in dietary BCAA supply, which demonstrates the availability and validity of the developed method.

Table 3. Concentration of three BCKAs in animal samples ($n = 6$).

Compounds	Serum (µmol/L)	Muscle (nmol/g)
α-ketoisocaproate	27.63 ± 5.39	2.61 ± 0.89
α-keto-β-methylvalerate	18.80 ± 4.92	1.45 ± 0.48
α-ketoisovalerate	10.20 ± 3.94	1.24 ± 0.37

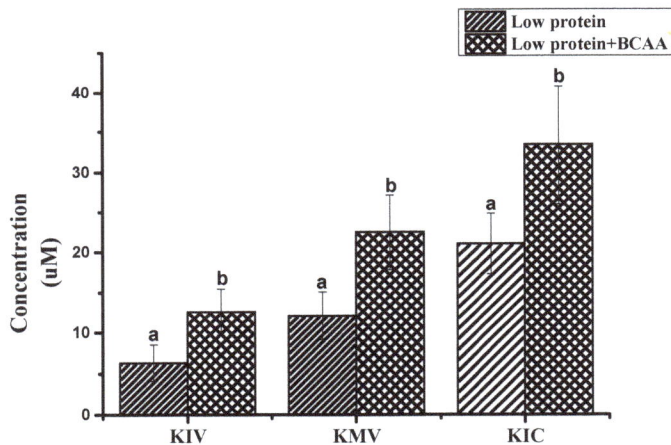

Figure 6. The influence of the addition of BCAA on the concentration of three BCKAs in serum of pigs fed different levels of dietary BCAA (KIV: α-ketoisovalerate; KMV: ketoisoleucine; KIC: α-ketoisocaproate). [a], [b] Means the same compounds different significantly ($p < 0.05$).

3. Materials and Methods

3.1. Materials and Reagents

The HPLC grade three BCKAs and the corresponding internal standard [^{13}C]-ketoisocaproate ([^{13}C]-KIC) were purchased from Sigma (St Louis, MO, USA). Ultra-pure water purified using a Milli-Q system (Millipore, Bedford, MA, USA) was used to prepare all aqueous solutions. HPLC grade methanol and acetonitrile were obtained from Fisher Scientific International (Hampton, NH, USA). HPLC grade ammonium acetate was purchased from Dikma Technology (Richmond Hill, ON, Canada). Nylon membrane filters (0.1 µm) were obtained from Whatman (Maidstone, UK).

3.2. Apparatus and Procedures

The chromatographic separation was developed using an Agilent 1290 Infinity HPLC system coupled to a 6520 quadruple-time of flight mass spectrometer (Q-TOF/MS) from Agilent Technologies (Santa Clara, CA, USA). An ultrasonic cleaner (Kunshan Ultrasonic Instrument Co., Kunshan, China) was used for promoting the sample dissolution and a vacuum concentrator (Eppendorf, Hamburg, Germany) was used to dry the supernatants.

3.3. Preparation of Standard Solutions

All standard stock solutions (10 mmol L^{-1}) were prepared in ultra-pure water and stored at -20 °C until use. A mixture of three BCKAs standard stock solution was prepared by mixing and diluting the single stock solutions at 1 mmol L^{-1} and stored at -20 °C until use. Working standard solutions were freshly prepared daily by diluting the standard stock solution with ultra-pure water to five different concentrations (from 0.1 to 100 μmol L^{-1}).

3.4. Sample Preparation and Extraction

Serum sample extraction was achieved by adding 800 μL methanol into 200 μL sample, then 4 μL of internal standard (0.5 mmol L^{-1} [^{13}C]-KIC) was added. Samples were put under vortex movement for 1 min, deproteined for 2 h at -20 °C, and centrifuged at 13,000 rpm for 10 min at 4 °C. Collection of 500 μL supernatant fluid was evaporated to dryness in a vacuum concentrator. The residues were resuspended in 100 μL of ultra-pure water, vortexed, and centrifuged again at 13,000 rpm for 10 min at 4 °C. The supernatant passed through 0.1 μm syringe filter was transferred to a conical insert in the sampler vial and then used for HPLC Q-TOF/MS analysis.

Muscle samples were crushed in liquid nitrogen. Sample extraction was achieved by adding 900 μL extraction solution (methanol:water, 8:2, *v:v*) into 100 mg muscle samples and 4 μL of internal standard (0.1 mmol L^{-1}[^{13}C]-KIC) was also added. Samples were put under vortex movement for 1 min, then placed in an ultrasonic bath for 10 min to achieve full extraction. Then deproteination, supernatant purification and HPLC Q-TOF/MS analysis followed the same procedure as for the serum samples.

3.5. HPLC Q-TOF/MS Conditions

Chromatographic separation of three BCKAs was achieved on an Agilent Eclipse Plus C_{18} column (2.1 × 100 mm, 1.8 μm). The column temperature was set at 30 °C and the flow rate was 0.3 mL/min. The mobile phase was 10 mmol/L ammonium acetate aqueous (solvent A) and acetonitrile (solvent B). The compounds were eluted with a linear gradient, consisting of 5–30% B over 0–3 min, to 90% B at 3.5 min, the constant gradient at 90% B for 3 min, and 5% B at 7 min for 3 min (re-equilibration of the column) before the loading of the next sample, so the total run time was 10 min. The injection volume was 5 μL.

The separated components from HPLC were subsequently analyzed by Q-TOF/MS, which was in line with the HPLC and operated in an electrospray ionization negative mode. The settings were as follows: drying gas temperature at 350 °C, a desolvation flow rate of 12 L/min, and a nebulizer pressure of 60 psig. The capillary and fragmentor voltages were set at 3500 and 100 V, respectively. Data were acquired in a full scan mode (m/z 70–300) at the rate of 2 spectra/s. To ensure the mass accuracy of detected ions, reference molecules (m/z 121.050873 and 922.009798) were continuously introduced into the electrospray ionization source to perform internal calibration. The MS/MS spectra of the compounds were recorded at a rate of 800 ms/spectra with the collision energy in 20 V. The characteristic fragment ions were used for confirmation.

3.6. Validation Procedure

Method validation was performed with serum and muscle samples, including linearity, sensitivity, as well as recovery and accuracy of the three BCKAs in six replicates. The recoveries were determined by spiking samples with a mixture of BCKA standards at 10, 20, 50 µmol L^{-1} in serum or 1, 2, 5 nmol g^{-1} in muscles. The internal standard solution was added in both spiked samples and standard solution. The concentration of mixed standard solution was gradually diluted and analyzed in order to measure the limit of quantitation (LOQ, S/N = 10). The m/z of the 3 precursor ions were presented in Table 1 which were used for quantitative determination.

3.7. Analysis of Actual Animal Samples

Animals breeding experiments were carried out at Huazhong Agricultural University (Wuhan, China) as previously described [9]. All piglets used in this study were housed and handled according to the established guidelines of Huazhong Agricultural University. All procedures performed on the animals were approved by Huazhong Agricultural University Animal Care and Use Committee (approval permit number 30700571). After the experiments, blood samples from precaval veins were collected after fasting 12 h and centrifuged (3000× g) for 10 min. Then the supernatant was transferred into new tubes and stored at −20 °C. After blood was collected, pigs were slaughtered and longissimus dorsi was collected and stored at −20 °C until use.

3.8. Statistical Analysis

Data were analyzed using the analysis of variance (ANOVA) procedure of SAS system (version 9.2; SAS Institute Inc., Cary, NC, USA). A p value less than 0.05 was considered statistically significant.

4. Conclusions

The objective of this study was to develop a simple method to measure BCKAs in serum and muscle samples without derivatization. The HRMS offered excellent selectivity under complex matrices and ensured the collection of high quality data and the accuracy of quantitative data. The wide range of linearity of the assay exceeded the levels that were commonly observed in serum and muscle samples. Meanwhile, the LOQ of 0.06~0.23 µmol/L and 0.09~0.27 nmol/g for serum and muscle samples meets the generally desired quantifiable levels in physiological samples. The recoveries for the three BCKAs range from 78.4% to 114.3% with RSD less than 9.7%, which demonstrates the accuracy and precision of this procedure. Furthermore, the developed method has been validated using authentic serum and muscle samples. Thus the study provides a simple, reliable and effective method for the determination of three BCKAs in serum and muscle samples using HPLC Q-TOF/MS and the proposed method can be applied to the determination of three BCKAs in actual serum and muscle samples.

Acknowledgments: The financial support from the National Natural Science Foundation of China (31472126) is gratefully acknowledged.

Author Contributions: P.H. conceived and designed the experiments; B.Y. performed the experiments; Y.Z. and B.Y. analyzed the data; P.H. contributed analysis tools; Y.Z., B.Y. and R.L. wrote the paper.

Conflicts of Interest: The authors declare no conflict of interest.

References

1. Tom, A.; Nair, K.S. Assessment of branched-chain amino acid status and potential for biomarkers. *J. Nutr.* **2006**, *136*, 324S–330S. [PubMed]
2. Lynch, C.J.; Adams, S.H. Branched-chain amino acids in metabolic signalling and insulin resistance. *Nat. Rev. Endocrinol.* **2014**, *10*, 723–736. [CrossRef] [PubMed]
3. Lu, J.; Xie, G.; Jia, W.; Jia, W. Insulin resistance and the metabolism of branched-chain amino acids. *Front. Med.* **2013**, *7*, 53–59. [CrossRef] [PubMed]

4. Holecek, M.; Siman, P.; Vodenicarovova, M.; Kandar, R. Alterations in protein and amino acid metabolism in rats fed a branched-chain amino acid- or leucine-enriched diet during postprandial and postabsorptive states. *Nutr. Metab.* **2016**, *13*. [CrossRef] [PubMed]
5. Hattori, A.; Tsunoda, M.; Konuma, T.; Kobayashi, M.; Nagy, T.; Glushka, J.; Tayyari, F.; Cskimming, D.M.; Kannan, N.; Tojo, A.; et al. Cancer progression by reprogrammed BCAA metabolism in myeloid leukaemia. *Nature* **2017**, *545*, 500. [CrossRef] [PubMed]
6. Wilkinson, D.J.; Hossain, T.; Hill, D.S.; Phillips, B.E.; Crossland, H.; Williams, J.; Loughna, P.; Churchward-Venne, T.A.; Breen, L.; Phillips, S.M.; et al. Effects of leucine and its metabolite β-hydroxy-β-methylbutyrate on human skeletal muscle protein metabolism. *J. Physiol.* **2013**, *591*, 2911–2923. [CrossRef] [PubMed]
7. Wisniewski, M.S.W.; Carvalho-Silva, M.; Gomes, L.M.; Zapelini, H.G.; Schuck, P.F.; Ferreira, G.C.; Scaini, G.; Streck, E.L. Intracerebroventricular administration of alpha-ketoisocaproic acid decreases brain-derived neurotrophic factor and nerve growth factor levels in brain of young rats. *Metab. Brain Dis.* **2016**, *31*, 377–383. [CrossRef] [PubMed]
8. Ananieva, E.A.; Van Horn, C.G.; Jones, M.R.; Hutson, S.M. Liver BCATm transgenic mouse model reveals the important role of the liver in maintaining BCAA homeostasis. *J. Nutr. Biochem.* **2017**, *40*, 132–140. [CrossRef] [PubMed]
9. Wang, X.; Wei, H.; Cao, J.; Li, Z.; He, P. Metabolomics analysis of muscle from piglets fed low protein diets supplemented with branched chain amino acids using HPLC-high-resolution MS. *Electrophoresis* **2015**, *36*, 2250–2258. [CrossRef] [PubMed]
10. Zheng, L.; Wei, H.; Cheng, C.; Xiang, Q.; Pang, J.; Peng, J. Supplementation of branched-chain amino acids to a reduced-protein diet improves growth performance in piglets: Involvement of increased feed intake and direct muscle growth-promoting effect. *Br. J. Nutr.* **2016**, *115*, 2236–2245. [CrossRef] [PubMed]
11. Rietman, A.; Stanley, T.L.; Clish, C.; Mootha, V.; Mensink, M.; Grinspoon, S.K.; Makimura, H. Associations between plasma branched-chain amino acids, beta-aminoisobutyric acid and body composition. *J. Nutr. Sci.* **2016**, *5*, e6. [CrossRef] [PubMed]
12. Islam, M.M.; Nautiyal, M.; Wynn, R.M.; Mobley, J.A.; Chuang, D.T.; Hutson, S.M. Branched-chain amino acid metabolon: Interaction of glutamate dehydrogenase with the mitochondrial branched-chain aminotransferase (BCATm). *J. Biol. Chem.* **2010**, *285*, 265–276. [CrossRef] [PubMed]
13. Cole, J.T.; Sweatt, A.J.; Hutson, S.M. Expression of mitochondrial branched-chain aminotransferase and alpha-keto-acid dehydrogenase in rat brain: Implications for neurotransmitter metabolism. *Front. Neuroanat.* **2012**, *6*, 18. [CrossRef] [PubMed]
14. Lee, A.J.; Beno, D.W.A.; Zhang, X.; Shapiro, R.; Mason, M.; Mason-Bright, T.; Surber, B.; Edens, N.K. A ^{14}C-leucine absorption, distribution, metabolism and excretion (ADME) study in adult Sprague-Dawley rat reveals β-hydroxy-β-methylbutyrate as a metabolite. *Amino Acids* **2015**, *47*, 917–924. [CrossRef] [PubMed]
15. Pimentel, G.D.; Rosa, J.C.; Lira, F.S.; Zanchi, N.E.; Ropelle, E.R.; Oyama, L.M.; Oller Do Nascimento, C.M.; de Mello, M.T.; Tufik, S.; Santos, R.V.T. β-Hydroxy-β-methylbutyrate (HMβ) supplementation stimulates skeletal muscle hypertrophy in rats via the mTOR pathway. *Nutr. Metab.* **2011**, *8*, 11. [CrossRef] [PubMed]
16. Green, C.R.; Wallace, M.; Divakaruni, A.S.; Phillips, S.A.; Murphy, A.N.; Ciaraldi, T.P.; Metallo, C.M. Branched-chain amino acid catabolism fuels adipocyte differentiation and lipogenesis. *Nat. Chem. Biol.* **2016**, *12*, 15. [CrossRef] [PubMed]
17. Escobar, J.; Frank, J.W.; Suryawan, A.; Nguyen, H.V.; Van Horn, C.G.; Hutson, S.M.; Davis, T.A. Leucine and alpha-Ketoisocaproic acid, but not norleucine, stimulate skeletal muscle protein synthesis in neonatal pigs. *J. Nutr.* **2010**, *140*, 1418–1424. [CrossRef] [PubMed]
18. Columbus, D.A.; Fiorotto, M.L.; Davis, T.A. Leucine is a major regulator of muscle protein synthesis in neonates. *Amino Acids* **2015**, *47*, 259–270. [CrossRef] [PubMed]
19. Knerr, I.; Weinhold, N.; Vockley, J.; Gibson, K.M. Advances and challenges in the treatment of branched-chain amino/keto acid metabolic defects. *J. Inherit. Metab. Dis.* **2012**, *35*, 29–40. [CrossRef] [PubMed]
20. Crowell, P.L.; Miller, R.H.; Harper, A.E. Measurement of plasma and tissue-levels of branched-chain α-keto acids by gas-liquid chromatography. *Method Enzymol.* **1988**, *166*, 39–46.
21. Fernandes, A.A.; Kalhan, S.C.; Njoroge, F.G.; Matousek, G.S. Quantitation of branched-chain α-keto acids as their *N*-methylquinoxalone derivatives: Comparison of *O*- and *N*-alkylation versus-silylation. *Biomed. Environ. Mass Spectrom.* **1986**, *13*, 569–581. [CrossRef] [PubMed]

22. Pailla, K.; Blonde-Cynober, F.; Aussel, C.; De Bandt, J.P.; Cynober, L. Branched-chain keto-acids and pyruvate in blood: Measurement by HPLC with fluorimetric detection and changes in older subjects. *Clin. Chem.* **2000**, *46*, 848–853. [PubMed]

23. Hara, S.; Takemori, Y.; Yamaguchi, M.; Nakamura, M.; Ohkura, Y. Determination of alpha-keto acids in serum and urine by high-performance liquid-chromatography with fluorescence detection. *J. Chromatogr. B* **1985**, *344*, 33–39. [CrossRef]

24. Anumula, K.R. Rapid quantitative-determination of sialic acids in glycoproteins by high-performance liquid-chromatography with a sensitive fluorescence detection. *Anal. Biochem.* **1995**, *230*, 24–30. [CrossRef] [PubMed]

25. Olson, K.C.; Chen, G.; Lynch, C.J. Quantification of branched-chain keto acids in tissue by ultra fast liquid chromatography-mass spectrometry. *Anal. Biochem.* **2013**, *439*, 116–122. [CrossRef] [PubMed]

26. Yin, B.; Li, T.; Li, Z.; Dang, T.; He, P. Sensitive analysis of 33 free amino acids in serum, milk and muscle by ultra-high Performance liquid chromatography-quadrupole-orbitrap high resolution mass spectrometry. *Food Anal. Method* **2016**, *9*, 2814–2823. [CrossRef]

27. Thakare, R.; Chhonker, Y.S.; Gautam, N.; Alamoudi, J.A.; Alnouti, Y. Quantitative analysis of endogenous compounds. *J Pharm. Biomed. Anal.* **2016**, *128*, 426–437. [CrossRef] [PubMed]

28. Kand'Ar, R.; Zakova, P.; Jirosova, J.; Sladka, M. Determination of branched chain amino acids, methionine, phenylalanine, tyrosine and alpha-keto acids in plasma and dried blood samples using HPLC with fluorescence detection. *Clin. Chem. Lab. Med.* **2009**, *47*, 565–572. [CrossRef] [PubMed]

29. Cree, T.C.; Hutson, S.M.; Harper, A.E. Gas-liquid chromatography of alpha-keto acids: Quantification of the branched-chain-alpha-keto acids from physiological sources. *Anal. Biochem.* **1979**, *92*, 159–163. [CrossRef]

30. Hutson, S.M.; Harper, A.E. Blood and tissue branched-chain amino and alpha-keto acid concentrations-effect of diet, starvation, and disease. *Am. J. Clin. Nutr.* **1981**, *34*, 173–183. [PubMed]

Sample Availability: Samples of the BCKAs are available from the authors.

![molecules logo] *molecules*

MDPI

Review

Recent Advances in Applications of Ionic Liquids in Miniaturized Microextraction Techniques

Maria Kissoudi and Victoria Samanidou *

Laboratory of Analytical Chemistry, Department of Chemistry, Aristotle University of Thessaloniki,
541 24 Thessaloniki, Greece; marikiss@chem.auth.gr
* Correspondence: samanidu@chem.auth.gr; Tel.: +30-231-099-7698

Received: 14 May 2018; Accepted: 12 June 2018; Published: 13 June 2018

Abstract: Green sample preparation is one of the most challenging aspects in green analytical chemistry. In this framework, miniaturized microextraction techniques have been developed and are widely performed due to their numerous positive features such as simplicity, limited need for organic solvents, instrumentation of low cost and short time of extraction. Also, ionic liquids (ILs) have unequivocally a "green" character, which they owe to their unique properties including the re-usage, the high reaction efficiency and selectivity in room temperature, the ability to dissolve both organic and inorganic compounds, and thermal stability. In the present review, the recent advances in the application of ionic liquids in miniaturized liquid and solid phase extraction techniques as extractants, intermediate solvents, mediators and desorption solvents are discussed, quoting the advantages and drawbacks of each individual technique. Some of the most important sample preparation techniques covered include solid-phase microextraction (SPME), dispersive liquid-liquid microextraction (DLLME), single-drop microextraction (SDME), stir bar sorptive extraction (SBSE), and stir cake sorptive extraction (SCSE).

Keywords: ionic liquids; sample preparation; microextraction; solid-phase microextraction; dispersive liquid-liquid microextraction; single-drop microextraction; stir bar sorptive extraction; stir cake sorptive extraction

---

1. Introduction

Nowadays, sample preparation procedures are more than ever linked with the protection of the environment following the philosophy of Green Analytical Chemistry (GAC). A typical analytical procedure consists of three main parts: sampling, sample preparation and final analysis. It is generally known that almost the 75% of its time is spent in the stage of preparation, so it is a critical part of the analytical procedure. According to GAC, new environmentally friendly instrumentation and methodologies with the minimum emissions of pollutants and environmental sustainability in terms of cost and energy of chemical laboratories are the major aims in the field of sample preparation [1–3].

Among the sample preparation methods used to clean up and concentrate analytes, liquid-liquid extraction (LLE) and solid-phase extraction (SPE) are the most famous and widely used. Despite their universal application, these methods are accompanied with some drawbacks, such as time-consumption, high cost, inability to extract polar compounds (mainly for LLE), use and disposal of great amount of toxic solvents, complication to automate and potential evaporation and dissolution in a proper solvent prior the analysis, which adds an extra step in the whole procedure. Miniaturization in sample preparation techniques is the key to overcome these above drawbacks. Furthermore, the combination of ionic liquids with the miniaturized sample preparation techniques could be the panacea for the pre-stated limitations [4,5].

Ionic liquids (ILs) are organic molten salts with melting point lower than 100 °C in contrast with the common molten salts (fused salts), such as sodium chloride which becomes liquid at

801 °C. "Room temperature ionic liquids" (RTILs) is a term which describes a group of ionic liquids with melting points at or below room temperature (25 °C). ILs are composed of positive and negative ions, usually a bulky organic nitrogen-containing cation, such as imidazolium, pyridinium, pyrrolidinium, phosphonium or ammonium, and a halogen-based organic or inorganic anion, including trifluoromethylsulfonate $[CF_3SO_3]^-$, bis[(trifluoromethyl)sulfonyl]imide $[(CF_3SO_2)_2N]^-$ (i.e., NTf_2), trifluoroethanoate $[CF_3CO_2]^-$ or Cl^-, PF_6^-, BF_4^-, respectively, whereas water and other organic solvents are composed of molecules [6,7]. The first discovered RTIL is the ethylammonium nitrate $[EtNH_3][NO_3]$ with melting point at 12 °C by Walden in 1914 [8]. However, remarkable progress has been achieved in the field of ionic liquids since 1992, when Wilkes and Zaworotko introduced a promiscuous generation of ionic liquids, the air and water stable 1-ethyl-3-methylimidazolium-based ILs, $[EtMeim]BF_4$ and $[EtMeim]MeCO_2$ [9]. The future of ILs belongs to the deep eutectic solvents (DESs), an interesting subclass of ILs which are less toxic, they have a stronger ecofriendly profile and they can be produced easier and cheaper than the ILs. One of their main drawbacks is their solid state in room temperature and their high viscosity [10].

The structures of the most commonly used ILs are represented in Figure 1. ILs owe the fact that they remain liquid in temperatures below 100 °C to the large size of the ions that they consist of, avoiding the packing of ionic lattice, which happens in inorganic salts. Compared with conventional solvents, ILs have some special physicochemical properties, such as low volatility, high electrical conductivity, long-term thermal and chemical stability, low flammability, and low vapor pressure. The negligible vapor pressure is one of the properties that makes ILs "green solvents", as they contribute to the reduction of atmospheric pollution and the health risks that are entailed. The chemical and physical properties of ILs, such as polarity, hydrophobicity, viscosity, depend on the anionic and cationic constituents giving them the characterization, "designer solvents" [11–15].

Among the approximate 10^{18} anion-cation combinations that can synthesize ionic liquids, there is a great majority of combinations that have toxic effects on water, environment, bacteria, plants, fish and human [16]. In a recent quantitative structure-property/activity relationship study of Zhao et al., a database for the toxicity of ILs has been established [17]. The authors conclude that the toxicity of ILs is related to the number of oxygen atoms that are present in the molecule. Also, the anion $[NTf_2]^-$ influences the level of toxicity of ILs. However, the relationship of the toxicity and the length of the alkyl side chains is not always proportional. Therefore, it is crucial researchers get advice from relevant databases before synthesis and use of ILs either for industrial or laboratory use. Furthermore, there are guides not only for the selection of ionic liquids but for the solvents generally. One of the most useful tools for solvent selection has been proposed by scientists of Pfizer and it is presented in Table 1 [18]. ILs are not always "green" solvents. Some of them are made from toxic ions and they are not biodegradable, despite their promiscuous physicochemical properties. Of course, not all the organic solvents that are used in the laboratories are toxic and harmful for the environment, but some of them can be replaced by other types of solvents with lower environmental impact.

This review summarizes the recent advances in application of ionic liquids in miniaturized liquid and solid phase extraction techniques as extractants, intermediate solvents, mediators, and desorption solvents. As it is shown in Figure 2, the amount of published studies associated with ionic liquids has grown rapidly within the last decade. Some of the most important sample preparation techniques covered include solid-phase microextraction (SPME), dispersive liquid-liquid microextraction (DLLME), single-drop microextraction (SDME), stir bar sorptive extraction (SBSE), stir cake sorptive extraction (SCSE).

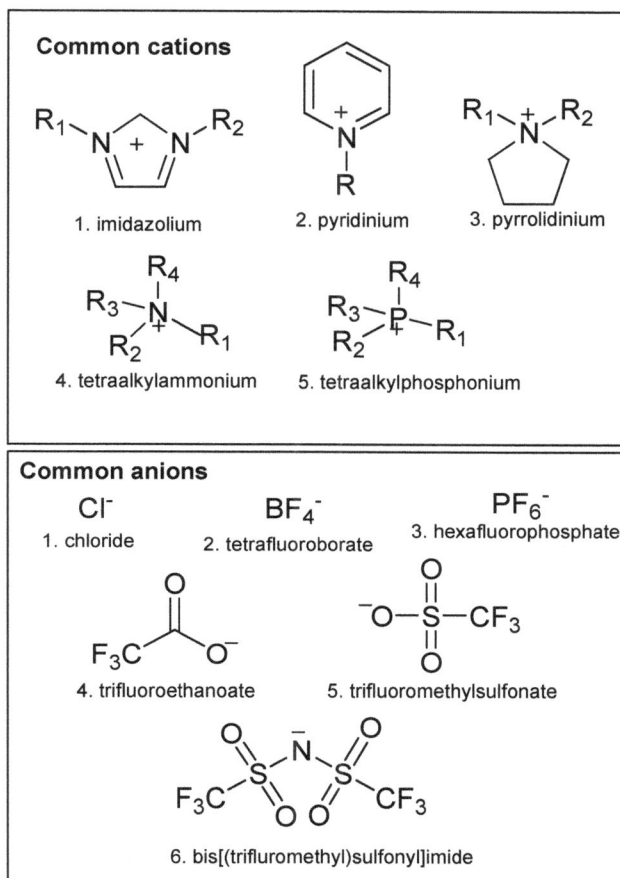

Figure 1. Structures of common cations and anions of ionic liquids.

Table 1. Solvent selection guide.

Preferred	Usable	Undesirable
Water	Cyclohexane	Pentane
Acetone Ethanol	Heptane	Hexane(s)
2-Propanol	Toluene	Di-isopropyl ether
1-Propanol	Methylcyclohexane	Diethyl ether
Ethyl acetate	Methyl t-butyl ether	Dichloromethane
Isopropyl acetate	Isooctane	Dichloroethane
Methanol	Acetonitrile	Chloroform
Methyl ethyl ketone	2-MethylTHF	Dimethyl formamide
1-Butanol	Tetrahydrofuran	N-Methylpyrrolidinone
t-Butanol	Xylenes	Pyridine
Ils (nontoxic combinations of ions)	Dimethyl sulfoxide	Dimethyl acetate
	Acetic acid	Dioxane
	Ethylene glycol	Dimethoxyethane
		Benzene
		Carbon tetrachloride

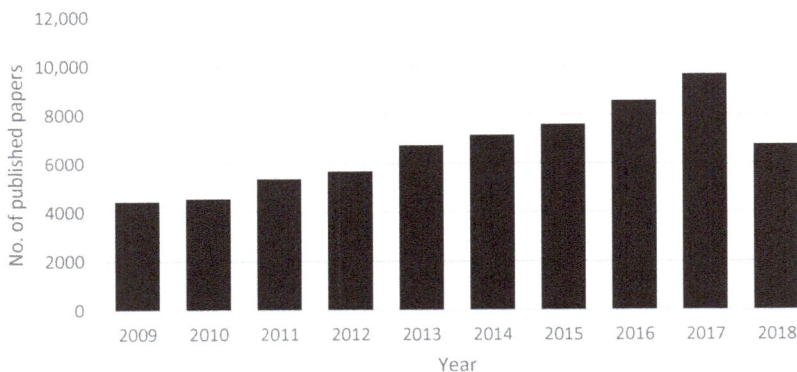

Figure 2. Number of published scientific papers worldwide per year (from 2009 to 2018) on the applications of ionic liquids in sample preparation techniques (based on Scopus and ScienceDirect).

2. Ionic Liquids in Miniaturized Microextraction Techniques

2.1. Solid-Phase Microextraction (SPME)

Solid-phase microextraction (SPME) is a widely known sample preparation technique, combining sampling and pre-concentration, due to its quickness and cost-effectiveness with absence of organic solvents. Since its establishment by Pawliszyn and co-workers it has been performed combined with several types of commercial sorbent coatings, such as polydimethylsiloxane (PDMS), polyacrylate, carboxen, PDMS-carboxen or divinylbenzene (DVB) depending on the analyte [19,20].

In a recent study of Tang and Duan, a porous polymeric ionic liquid, poly(1-vinyl-3-(4-vinyl-benzyl)imidazolium chloride), was synthesized and used as a sorbent coating for SPME for the analysis of polar organic acids [21]. The results showed that the fiber was more sensitive and practical for the extraction of polar compounds in comparison with the commercial fibers.

Ionic liquids compromise another effective type of sorbent coating that can be coupled with SPME technique in two different modes, headspace (HS-SPME) and direct immersion (DI-SPME). In the first mode, the sorbent coating is exposed to the headspace of the sample where the target analyte is present, while in the second the SPME-fiber is pushed out of the hollow needle and immersed into the sample directly. Following the sorption of the analyte in both cases, the fiber is drawn into the needle, the needle is withdrawn from the sample vial and transferred to the injection port of an analytical instrument, where desorption of the analyte takes place and the analysis is carried out [4,22].

A benzyl-functionalized crosslinked polymeric ionic liquid (PIL) was developed by Merdivan et al. and successfully used as a sorbent coating in headspace solid-phase microextraction (HS-SPME) coupled to gas chromatography with flame-ionization detection for the determination of seven volatile polycyclic aromatic hydrocarbons (PAHs) in environmental water samples [23]. The VBHDIM-NTf$_2$ IL monomer and (DVBIM)$_2$C$_{12}$-2NTf$_2$ IL cross-linker were used for the synthesis of the crosslinked PIL-based sorbent coating. Figure 3 represents the schematic illustration of HS-SPME procedure using the benzyl-PIL fiber for the extraction of PAHs. Compared to the commercial PDMS fiber coating for the PAHs, the crosslinked PIL fiber showed higher log K$_{fs}$ due to the presence of benzyl moieties within the IL monomer and IL crosslinker of the PIL sorbent coating indicating its superior affinity, higher sensitivity with low LODs, ranged from 0.01 to 0.04 µg L^{-1} for the PIL–benzyl fiber and from 0.01 to 0.07 µg L^{-1} for the PDMS fiber. The linearity, the RSD values and the relative recovery values were more satisfactory for the PIL–benzyl fiber than the PDMS fiber. The presence of benzyl moieties in the PIL coating material enhances π-π interactions between the target compounds (PAHs), which are composed of aromatic rings, and the sorbent coating.

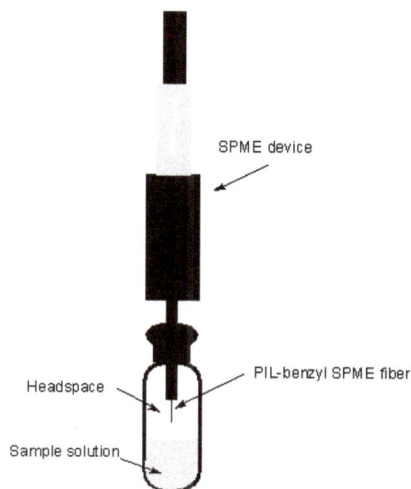

Figure 3. Headspace-solid phase microextraction procedure for the extraction of PAHs in water sample using a polymeric ionic liquid-based fiber.

In-tube SPME is an improved mode of SPME sample preparation technique introduced by Pawliszyn in 1997, where the coupling of SPME with HPLC is achieved more conveniently. Sun et al. developed a fiber-in-tube solid phase microextraction device with copper support modified via chemical bonding with ionic liquids. Especially, one copper tube was filled with eleven copper wires functionalized with ionic liquids and it was combined online with a high-performance liquid chromatography system to strengthen extraction capacity and eliminate dead volume, building an online in-tube SPME-HPLC system. In this framework, an in-tube SPME-HPLC method for the determination of estrogens in water samples was developed with high enrichment factors and satisfactory sensitivity (LODs: 0.02–0.05 g L^{-1}). The presence of ionic liquids in the metal support was crucial for the effectiveness of the developed method increasing the stability of in-tube SPME device and improving the sensitivity by increasing sample volume with a high sampling rate [24].

2.2. Dispersive Liquid-Liquid Microextraction (DLLME)

Dispersive liquid-liquid microextraction (DLLME) was first introduced by Rezaee et al. in 2006 as a novel alternative technique for the extraction and preconcentration of organic compounds in water samples [25]. The fundamentals of DLLME technique are based on the mixing of an aqueous sample containing the analytes with a non-miscible with water organic solvent, used as an extractant, together with a small amount of a dispersive solvent, which is miscible in both water and the extractant solvent. The mixing of the two solvents should be preceded their injection into the sample by a syringe or a micropipette. Gentle manual shaking of the mixture disperses the organic extractant as fine droplets to form a homogenous cloudy solution in which partition of the analytes takes place. Then, the sample is centrifuged, and the sediment phase is collected and determined with the appropriate analytical method [26].

Since its introduction, DLLME has been performed as an extraction technique in various fields of chemistry, such as analytical, environmental, biochemistry, medicinal, pharmaceutical, toxicology and many others. Its increasing popularity is owed to some significant properties, such as the simplicity, the low cost, and the environmentally friendly profile. Furthermore, the dispersive mode improves the extraction kinetics by increasing the contact surface between the extractant and the sample. The most critical part of this technique is the choice of the extraction solvent, which should have a density greater than water's density and to form a cloudy solution with the dispersive solvent [27,28].

Taking into consideration the properties of ionic liquids as extraction solvents described before, the combination of ILs in DLLME process could be characterized promiscuous. Zhou et al. and Baghdadi and Shemirani were the first researchers who performed the first successful applications of ILs in DLLME process, namely IL-DLLME, for the determination of organophosphorus pesticides in environmental samples and for the extraction of mercury from water samples, respectively [29]. While, Liu and coworkers introduced the conventional IL-DLLME combined with HPLC-DAD for the determination of heterocyclic insecticides in water using the IL 1-hexyl-3-methylimidazolium hexafluorophosphate, ($[C_6MIM][PF_6]$), as the extractant and methanol as the dispersive solvent [30].

2.3. Stir Bar Sorptive Extraction (SBSE)

Stir bar sorptive extraction (SBSE) was initially introduced in 1999 by Baltussen and co-workers [31]. It uses a stir bar for the extraction consisting of a magnet covered with glass, which in turn is coated by a layer (typically 0.5–1 mm) of sorptive material, usually polydimethylsiloxane (PDMS). The bar is subsequently inserted into a vial, which contains the aqueous sample and it is stirred until equilibrium of analytes concentration between sorbent and sample matrix is reached. After the extraction, the bar is removed and transferred to a clean vial, where the target compounds are analyzed by liquid or gas chromatography by liquid or thermal desorption [32].

Fan and co-workers introduced a novel approach for the extraction and determination of nonsteroidal anti-inflammatory drugs (NSAIDs) by high-performance liquid chromatography-ultraviolet detection (HPLC-UV) [33]. The researchers synthesized an ionic liquid, 1-allylimidazolium tetrafluoroborate ($[AIM][BF_4]$) chemically bonded sol-gel coating for stir bar sorptive extraction using γ-(methacryloxypropyl)trimethoxysilane (KH-570) as a bridging agent, which showed excellent mechanical strength and chemical/thermal stability compared to the conventional PDMS-based or C18 coating materials. After the optimization of the critical parameters of the technique, such as stirring rate, extraction time, desorption solvent, pH, salt effect, the developed method showed satisfactory reproducibility with RSDs lower than 7.6% and high sensitivity (LODs: 0.23–0.31 $\mu g\ L^{-1}$) for the determination of three NSAIDs. The proposed method is applicable in environmental, food and biological samples

In the framework of the improvement and development of microextraction techniques, the combination of two or more techniques could induce significant enhancements to the extraction procedure reducing the extraction time, the cost, or the sensitivity. Two microextraction techniques that can be easily conflated is the stir bar sorptive extraction (SBSE) and dispersive liquid-liquid microextraction (DLLME) introducing the stir bar dispersive liquid microextraction (SBDME). In this approach, Chisvert et al. used a magnetic ionic liquid (MIL) and a neodymium-core magnetic stir bar as the extraction phase [34]. At low stirring rates it acts similar to SBSE while the MIL remains in the stir bar surface. At higher stirring speed the MIL disperses into the solution acting similar to DLLME. When the extraction is over, the stirring stops and the MIL retrieves onto the stir bar surface due to its magnetic properties. The desorption of the MIL-coated stir bar contained the preconcentrated analytes performed thermally in a gas chromatography-mass spectrometry (TD-GC-MS) system. The above approach was applied for the extraction of lipophilic organic UV filters from aqueous environmental water samples.

In a recent work by Benedé et al. another successful combination of stir bar sorptive extraction (SBSE) and dispersive liquid-liquid microextraction (DLLME) has been achieved and the stir bar dispersive liquid microextraction (SBDLME) was created for the determination of ten polycyclic aromatic hydrocarbons (PAHs) in natural water samples [35]. A neodymium stir bar magnetically coated with a magnetic ionic liquid (MIL), the $[P_{6,6,6,14}{}^+][Ni(II)(hfacac)_3{}^-]$, was used as extraction device. After the stirring, the MIL is magnetically retrieved onto the stir bar and then subjected to thermal desorption-gas chromatography-mass spectrometry system (TD-GC-MS). In contrast to other methods for the determination of PAHs, this one requires less time and manipulation for the sample preparation, solvent evaporation is not a mandatory step and the sensitivity level is more than satisfactory allowing the determination of PAHs in aqueous sample at the low ng L^{-1} levels.

2.4. Single-Drop Microextraction (SDME)

Single-drop microextraction technique (SDME) was first introduced by Liu and Dasgupta in 1995 as an alternative extraction process eliminating the problems of solvent evaporation, which exists in LLE and SPE, and the degradation of SPME fiber [36]. The basis of this technique is the distribution of the target compounds and a microdrop of solvent that is suspended in the tip of a microsyringe needle. It uses small volumes of organic solvents and simple equipment reducing the total cost of each individual application. SDME can be performed with direct immersion (DI-SDME) in the aqueous sample or in the headspace of the sample (HS-SDME). The main limitation of this technique is the fact that the stability of droplet is highly depended on the used solvent. Thus, solvents with low viscosity, high vapor pressure and low surface tension decrease the effectiveness of the extraction process. ILs could be used as an alternative to the common organic solvents used in SDME.

Jiang et al. used the SDME technique for the extraction and concentration of flavor and fragrance substances in fruit juices [37]. A hydrophilic IL, 1-hexyl-3-methylimidazolium tetrafluroborate, used successfully as an extraction solvent. The method was validated after the optimization of the main parameters, such as the volume of solvent microdrop, the extraction time, the enrichment time, the temperature, the pH of the sample solution, the height of the microdrop above the solution surface. The method demonstrates satisfactory sensitivity and accuracy with relative standard deviations lower than 9.1%.

A headspace single-drop microextraction (HS-SDME) method for the determination of aromatic analytes by HPLC was developed by An et al. using as solvents tetrachloromanganate ($[MnCl_4^{2-}]$)-based magnetic ionic liquids [38]. A rod magnet was used to sustain the microdroplet of MIL during HS-SDME. High stability of the microdroplet under high temperature and long extraction times was achieved due to the magnetic susceptibility of the MILs. The method showed high sensitivity and precision for the target compounds, and satisfactory relative recoveries from an application in real sample.

He and co-workers developed an ionic liquid-based headspace single drop microextraction combined with high-performance liquid chromatography (HS-SDME-HPLC) method for the determination of camphor and trans-anethole in licorice tablets [39]. The ionic liquid used as the extracting medium was 1-bityl-3-methylimidazole hexafluorophosphate. The method showed satisfactory stability and sensitivity after setting the volume of the IL microdrop to 12 µL. The method is simple, rapid, selective, precise, accurate and linear for the target compounds. SDME technique has the advantage that allows the single step separation, purification and enrichment improving the signal-to-noise ratio and ensuring the accuracy of the method as there was no loss of volatile components. The method is expected to be widely performed for the preparation of volatile compounds of drugs with high boiling points.

2.5. Stir-Cake Sorptive Extraction (SCSE)

Stir-cake sorptive extraction technique is an improved version of SBSE introduced in 2011 [40]. The SCSE device consists of a holder made of iron where the stationary phase is placed. The most common used extractive mediums in SCSE are monolithic cakes designed and prepared properly according to the extracted analytes. In the literature the most often used extraction phases are poly(4-vinylbenzoic acid-divinylbenzene) (VBADB) sorbents based on polymeric ionic liquids.

Wang et al. proposed a new SCSE approach using as sorbent a polymeric ionic liquid monolith obtained by the in situ copolymerization of an ionic liquid, 1-allyl-3-methylimidazolium bis[(trifluoro methyl)sulfonyl]imide (AMII) and divinylbenzene (DB) in the presence of *N,N*-dimethylformamide [41]. By coupling SCSE–AMIIDB with high performance liquid chromatography/diode array detection (SCSE–AMIIDB–HPLC/DAD) a simple and effective method for the determination of trace benzimidazoles (Bas) residues in water, milk and honey samples was established with low LODs and high levels of recovery. The SCSE-AMIIDB extractive medium can effectively extract polar BAs through multi-interaction such as hydrophobic, π-π, hydrogen-bonding and dipole-dipole interactions because of the multiple functional groups in the sorbent.

Another preparation of a polymeric ionic liquid-based sorbent for SCSE was developed by Zhang et al. for the extraction of metal ions for the first time [42]. The SCSE sorbent was prepared in situ polymerization of 3-(1-ethyl imidazolium-3-yl) propyl-methacrylamido bromide and ethylene dimethacrylate and was used for the extraction of trace antimony in environmental water samples and combined with hydride generation atomic fluorescence spectrometry (HG-AFS) for the determination of trace antimony. The developed method has some advantages such as convenience, sensitivity, good reproducibility and cost-effectiveness, satisfactory linearity, and high recoveries.

In another study, Chen and Huang prepared a polymeric ionic liquid-based adsorbent as the extraction medium of stir cake sorptive extraction (SCSE) of three organic acid preservatives, namely, *p*-hydroxybenzoic acid, sorbic acid and cinnamic acid [43]. The synthesis of the adsorbent was carried out the copolymerization of 1-ally-3-vinylimidazolium chloride (AV) and divinylbenzene (DVB) in the presence of a porogen solvent containing 1-propanol and 1,4-butanediol. After a long study to obtain the optimized conditions, the SCSE/AVDVB could extract the preservatives effectively through multiply interactions. In this framework, a simple and sensitive method by combining SCSE/AVDVB and HPLC/DAD was developed for the simultaneous analysis of the target preservatives in orange juices and tea drinks. The proposed method showed high sensitivity, good reproducibility, high cost-effectiveness and environmental friendliness.

Apart from the application of SPME for the determination of estrogens, SCSE has been used an alternative for the determination of estrogens in water samples was developed by Chen and coworkers [44]. In the framework of their study, a new PIL-based, a poly (1-ally-3-vinylimidazolium chloride-co-ethylene dimethacrylate)-AVED, monolith cake was prepared and used as sorbent of SCSE to extract trace estrogens with multi-interactions such as hydrophobic, π-π, hydrogen-bonding and dipole-dipole interactions effectively before their injection in HPLC-DAD system. The developed AVED/SCSE-LD-HPLC/DAD method showed a wide linear range, low LODs, satisfactory reproducibility and good recoveries for real water samples.

3. Conclusions

Over the last decade, many researchers have been attracted from the versatile character of ILs and PILs and their potentials. As discussed above, one of the most challenging applications of ILs is their usage in microextraction techniques as extractants, intermediate solvents, mediators, and desorption solvents. Table 2 summarizes the superior properties and characteristics of ILs in comparison with the common sorbent materials that are used in the miniaturized sample preparation techniques described in the present review [4,17–44]. As the IL-based microextraction techniques are budding gradually, some limitations should be noted, such as the high cost of their synthesis, their incompatibility with GC due to their low volatility and their potential toxicity. In general terms, the research in the field of ionic liquids will not stop evolving as the need for green analytical procedures is the priority of sample preparation. Taking into consideration their promiscuous properties and advantages, not only will the microextraction processed be improved, but also separation techniques, such as liquid and gas chromatography or electrophoresis, will expand their potentials.

Table 2. Limitations of the conventional sorbents and superior properties of ILs when they are used in the most commonly used miniaturized microextraction techniques.

SPME		DLLME		SBSE		SDME		SCSE	
Conventional Sorbents (PDMS, DVB, CAR, PEG, CW)	ILs	Conventional Organic Sorbents	ILs	Conventional Sorbents (PDMS, EG-silicone, PA)	ILs	Conventional Organic Sorbents	ILs	Conventional Sorbents (MIPs)	ILs
- low thermal stability of the fiber - short expiry date - small selectivity - fragility - limited operating time - decomposition during heating	- high thermal stability - high boiling points - no decomposition with heating	- low density - low viscosity and evaporation results to high instability of the drop	- high density allows phase separation - hydrophobic or hydrophilic nature miscible or immiscible with the disperser solvent - high viscosity and surface tension allows to form larger, more stable droplets	- limited extraction efficiency towards polar and less polar compounds - limited number of available extraction solvents - physical damage to the extraction phase when stirring at high speed	- thermal stability - chemical stability	- low viscosity and evaporation results to high instability of the drop and poor precision levels	- high viscosity (formation of a larger-volume drop) - low vapor pressure - good thermal stability without evaporation - immiscibility with water	- low thermal stability - limited number of available extraction solvents	- mechanical stability - improved processability - durability - spatial controllability

PDMS: Polydimethylsiloxane, DVB: divinylbenzene, CAR: carboxen, PEG: polyethylene glycol, CW: Carbowax, EG: ethylene glycol, PA: polyacrylate.

Conflicts of Interest: The authors declare no conflict of interest.

References

1. Tobiszewski, M.; Mechlińska, A.; Namieśnik, J. Green analytical chemistry—Theory and practice. *Chem. Soc. Rev.* **2010**, *39*, 2869–2878. [CrossRef] [PubMed]
2. Spietelun, A.; Marcinkowski, Ł.; De La Guardia, M.; Namieśnik, J. Green aspects, developments and perspectives of liquid phase microextraction techniques. *Talanta* **2014**, *119*, 34–45. [CrossRef] [PubMed]
3. Tobiszewski, M.; Mechlińska, A.; Zygmunt, B.; Namieśnik, J. Green analytical chemistry in sample preparation for determination of trace organic pollutants. *TrAC-Trends Anal. Chem.* **2009**, *28*, 943–951. [CrossRef]
4. Płotka-Wasylka, J.; Szczepańska, N.; de la Guardia, M.; Namieśnik, J. Miniaturized solid-phase extraction techniques. *TrAC-Trends Anal. Chem.* **2015**, *73*, 19–38. [CrossRef]
5. Poole, C.F.; Lenca, N. Green sample-preparation methods using room-temperature ionic liquids for the chromatographic analysis of organic compounds. *TrAC-Trends Anal. Chem.* **2015**, *71*, 144–156. [CrossRef]
6. Vičkačkaite, V.; Padarauskas, A. Ionic liquids in microextraction techniques. *Cent. Eur. J. Chem.* **2012**, *10*, 652–674. [CrossRef]
7. Welton, T. Room-Temperature Ionic Liquids. Solvents for Synthesis and Catalysis. *Chem. Rev.* **1999**, *99*, 2071–2084. [CrossRef] [PubMed]
8. Walden, P. Ueber die Molekulargrösse und elektrische Leitfähigkeit einiger geschmolzener Salze (Molecular weights and electrical conductivity of several fused salts). *Bull. Acad. Imp. Sci.* **1914**, *8*, 405–422.
9. Wilkes, J.S.; Zaworotko, M.J. Air and Water Stable I-Ethyl-3-methylimidazolium Based Ionic Liquids. *Chem. Commun.* **1992**, *13*, 965–967. [CrossRef]
10. Shishov, A.; Bulatov, A.; Locatelli, M.; Carradori, S.; Andruch, V. Application of deep eutectic solvents in analytical chemistry: A review. *Microchem. J.* **2017**, *135*, 33–38. [CrossRef]
11. Aguilera-Herrador, E.; Lucena, R.; Cárdenas, S.; Valcárcel, M. The roles of ionic liquids in sorptive microextraction techniques. *TrAC-Trends Anal. Chem.* **2010**, *29*, 602–616. [CrossRef]
12. Sun, P.; Armstrong, D.W. Ionic liquids in analytical chemistry. *Anal. Chim. Acta* **2010**, *661*, 1–16. [CrossRef] [PubMed]
13. Ghandi, K. A Review of Ionic Liquids, Their Limits and Applications. *Green Sustain. Chem.* **2014**, *4*, 44–53. [CrossRef]
14. Prado, R.; Weber, C.C. Applications of Ionic Liquids. In *Application, Purification and Recovery of Ionic Liquids*; Elsevier: New York, NY, USA, 2016. [CrossRef]
15. Berthod, A.; Ruiz-Ángel, M.J.; Carda-Broch, S. Recent advances on ionic liquid uses in separation techniques. *J. Chromatogr. A* **2018**, *1559*, 2–16. [CrossRef] [PubMed]
16. Plechkova, N.V.; Seddon, K.R. Applications of ionic liquids in the chemical industry. *Chem. Soc. Rev.* **2008**, *37*, 123–150. [CrossRef] [PubMed]
17. Zhao, Y.; Zhao, J.; Huang, Y.; Zhou, Q.; Zhang, X.; Zhang, S. Toxicity of ionic liquids: Database and prediction via quantitative structure-activity relationship method. *J. Hazard. Mater.* **2014**, *278*, 320–329. [CrossRef] [PubMed]
18. Alfonsi, K.; Colberg, J.; Dunn, P.J.; Fevig, T.; Jennings, S.; Johnson, T.A.; Kleine, H.P.; Knight, C.; Nagy, M.A.; Perry, D.A.; et al. Green chemistry tools to influence a medicinal chemistry and research chemistry based organisation. *Green Chem.* **2008**, *10*, 31–36. [CrossRef]
19. Pawliszyn, J. (Ed.) *Solid Phase Microextraction: Theory and Practice*; Wiley-VCH, Inc.: New York, NY, USA, 1997.
20. Silva, E.A.S.; Risticevic, S.; Pawliszyn, J. Recent trends in SPME concerning sorbent materials, configurations and in vivo applications. *Trends Anal. Chem.* **2013**, *43*, 24–36. [CrossRef]
21. Tang, Z.; Duan, Y. Fabrication of porous ionic liquid polymer as solid-phase microextraction coating for analysis of organic acids by gas chromatography—Mass spectrometry. *Talanta* **2017**, *172*, 45–52. [CrossRef] [PubMed]
22. Gionfriddo, E.; Souza-Silva, E.; Pawliszyn, J. Headspace versus Direct Immersion Solid Phase Microextraction in Complex Matrixes: Investigation of Analyte Behavior in Multicomponent Mixtures. *Anal. Chem.* **2015**, *87*, 8448–8456. [CrossRef] [PubMed]

23. Merdivan, M.; Pino, V.; Anderson, J.L. Determination of volatile polycyclic aromatic hydrocarbons in waters using headspace solid-phase microextraction with a benzyl-functionalized crosslinked polymeric ionic liquid coating. *Environ. Technol.* **2017**, *38*, 1897–1904. [CrossRef] [PubMed]

24. Sun, M.; Feng, J.; Bu, Y.; Luo, C. Ionic liquid coated copper wires and tubes for fiber-in-tube solid-phase microextraction. *J. Chromatogr. A* **2016**, *1458*, 1–8. [CrossRef] [PubMed]

25. Rezaee, M.; Assadi, Y.; Milani Hosseini, M.R.; Aghaee, E.; Ahmadi, F.; Berijani, S. Determination of organic compounds in water using dispersive liquid-liquid microextraction. *J. Chromatogr. A* **2006**, *1116*, 1–9. [CrossRef] [PubMed]

26. Trujillo-Rodríguez, M.J.; Rocío-Bautista, P.; Pino, V.; Afonso, A.M. Ionic liquids in dispersive liquid-liquid microextraction. *TrAC-Trends Anal. Chem.* **2013**, *51*, 87–106. [CrossRef]

27. Mansour, F.R.; Khairy, M.A. Pharmaceutical and biomedical applications of dispersive liquid-liquid microextraction. *J. Chromatogr. B Anal. Technol. Biomed. Life Sci.* **2017**, *1061–1062*, 382–391. [CrossRef] [PubMed]

28. Ma, J.; Lu, W.; Chen, L. Recent Advances in Dispersive Liquid-Liquid Microextraction for Organic Compounds Analysis in Environmental Water: A Review. *Curr. Anal. Chem.* **2012**, *8*, 78–90. [CrossRef]

29. Zhou, Q.; Bai, H.; Xie, G.; Xiao, J. Trace determination of organophosphorus pesticides in environmental samples by temperature-controlled ionic liquid dispersive liquid-phase microextraction. *J. Chromatogr. A* **2008**, *1188*, 148–153. [CrossRef] [PubMed]

30. Liu, Y.; Zhao, E.; Zhu, W.; Gao, H.; Zhou, Z. Determination of four heterocyclic insecticides by ionic liquid dispersive liquid-liquid microextraction in water samples. *J. Chromatogr. A* **2009**, *1216*, 885–891. [CrossRef] [PubMed]

31. Baltussen, E.; Sandra, P.; David, F.; Cramers, C. Stir bar sorptive extraction (SBSE), a novel extraction technique for aqueous samples: Theory and principles. *J. Microcolumn Sep.* **1999**, *11*, 737–747. [CrossRef]

32. Camino-Sánchez, F.J.; Rodríguez-Gómez, R.; Zafra-Gómez, A.; Santos-Fandila, A.; Vílchez, J.L. Stir bar sorptive extraction: Recent applications, limitations and future trends. *Talanta* **2014**, *130*, 388–399. [CrossRef] [PubMed]

33. Fan, W.; Mao, X.; He, M.; Chen, B.; Hu, B. Development of novel sol-gel coatings by chemically bonded ionic liquids for stir bar sorptive extraction—Application for the determination of NSAIDS in real samples. *Anal. Bioanal. Chem.* **2014**, *406*, 7261–7273. [CrossRef] [PubMed]

34. Chisvert, A.; Benedé, J.L.; Anderson, J.L.; Pierson, S.A.; Salvador, A. Introducing a new and rapid microextraction approach based on magnetic ionic liquids: Stir bar dispersive liquid microextraction. *Anal. Chim. Acta* **2017**, *983*, 130–140. [CrossRef] [PubMed]

35. Benedé, J.L.; Anderson, J.L.; Chisvert, A. Trace determination of volatile polycyclic aromatic hydrocarbons in natural waters by magnetic ionic liquid-based stir bar dispersive liquid microextraction. *Talanta* **2018**, *176*, 253–261. [CrossRef] [PubMed]

36. Liu, S.; Dasgupta, P.-K. Liquid Droplet. A Renewable Gas Sampling Interface. *Anal. Chem.* **1995**, *67*, 2042–2049. [CrossRef]

37. An, J.; Rahn, K.L.; Anderson, J.L. Headspace single drop microextraction versus dispersive liquid-liquid microextraction using magnetic ionic liquid extraction solvents. *Talanta* **2017**, *167*, 268–278. [CrossRef] [PubMed]

38. Jiang, C.; Wei, S.; Li, X.; Zhao, Y.; Shao, M.; Zhang, H.; Yu, A. Ultrasonic nebulization headspace ionic liquid-based single drop microextraction of flavour compounds in fruit juices. *Talanta* **2013**, *106*, 237–242. [CrossRef] [PubMed]

39. Huang, X.; Chen, L.; Lin, F.; Yuan, D. Novel extraction approach for liquid samples stir cake sorptive extraction using monolith. *J. Sep. Sci.* **2011**, *34*, 2145–2151. [CrossRef] [PubMed]

40. He, X.; Zhang, F.; Jiang, Y. An improved ionic liquid-based headspace single-drop microextraction-liquid chromatography method for the analysis of camphor and trans-anethole in compound liquorice tablets. *J. Chromatogr. Sci.* **2012**, *50*, 457–463. [CrossRef] [PubMed]

41. Wang, Y.; Zhang, J.; Huang, X.; Yuan, D. Preparation of stir cake sorptive extraction based on polymeric ionic liquid for the enrichment of benzimidazole anthelmintics in water, honey and milk samples. *Anal. Chim. Acta* **2014**, *840*, 33–41. [CrossRef] [PubMed]

42. Zhang, Y.; Mei, M.; Ouyang, T.; Huang, X. Talanta Preparation of a new polymeric ionic liquid-based sorbent for stir cake sorptive extraction of trace antimony in environmental water samples. *Talanta* **2016**, *161*, 377–383. [CrossRef] [PubMed]

43. Chen, L.; Huang, X. Analytica Chimica Acta Preparation of a polymeric ionic liquid-based adsorbent for stir cake sorptive extraction of preservatives in orange juices and tea drinks. *Anal. Chim. Acta* **2016**, *916*, 33–41. [CrossRef] [PubMed]

44. Chen, L.; Mei, M.; Huang, X.; Yuan, D. Talanta Sensitive determination of estrogens in environmental waters treated with polymeric ionic liquid-based stir cake sorptive extraction and liquid chromatographic analysis. *Talanta* **2016**, *152*, 98–104. [CrossRef] [PubMed]

MDPI

St. Alban-Anlage 66

4052 Basel

Switzerland

Tel. +41 61 683 77 34

Fax +41 61 302 89 18

www.mdpi.com

Molecules Editorial Office

E-mail: molecules@mdpi.com

www.mdpi.com/journal/molecules

www.ingramcontent.com/pod-product-compliance
Lightning Source LLC
Chambersburg PA
CBHW051716210326
41597CB00032B/5503